What Do We Know about War?

What Do We Know about War?

Revised Third Edition

Edited by Sara McLaughlin Mitchell
and John A. Vasquez

ROWMAN & LITTLEFIELD
Lanham • Boulder • New York • London

Executive Acquisitions Editor: Michael Kerns
Assistant Editor: Elizabeth Von Buhr
Sales and Marketing Inquiries: textbooks@rowman.com

Credits and acknowledgments for material borrowed from other sources, and reproduced with permission, appear on the appropriate pages within the text.

Published by Rowman & Littlefield
An imprint of The Rowman & Littlefield Publishing Group, Inc.
4501 Forbes Boulevard, Suite 200, Lanham, Maryland 20706
www.rowman.com

86-90 Paul Street, London EC2A 4NE

Second edition 2012. Third edition 2021.

British Library Cataloguing in Publication Information Available

Library of Congress Cataloging-in-Publication Data Available

ISBN 978-1-5381-9315-0 (cloth)
ISBN 978-1-5381-9316-7 (paperback)
ISBN 978-1-5381-9317-4 (ebook)

™ The paper used in this publication meets the minimum requirements of American National Standard for Information Sciences—Permanence of Paper for Printed Library Materials, ANSI/NISO Z39.48-1992.

To

My Undergraduate and Graduate Students—SM

My Mother, Helen J. Vasquez—JV

Contents

Preface to the Revised Third Edition

This book reviews the major scientific findings on interstate war for students and the interested lay reader in a non-technical fashion while still preserving the subtlety of the original analyses. This revised third edition differs from the original third edition in that it includes a new afterword on the Russia-Ukraine War; the rest of the volume remains the same. We included a chapter on the war to make the book timelier and illustrate how the many findings can be applied to explain and analyze an ongoing war. In this way, we hope that the abstract findings will not only become more comprehensible, but also demonstrate how they are policy relevant to what goes on in the "real world."

We also engaged in an analysis of the Russia-Ukraine War because so little of the commentary on the war has been informed by general empirical research on the causes of war. This is because discussion of the war has been primarily focused on moral and/or political condemnation of the Russian invasion rather than explanation. While we do not argue that moral discussion has no place in analyzing this war, our purpose here is to see what quantitative conflict research can tell us about war that normative evaluation cannot. Normative and empirical analyses are different, each with their own rules and criteria for justification. Our purpose in the afterword is empirical, and we attempt to see if general findings about the factors that bring about interstate wars are applicable to this war.

What we find is that the Russia-Ukraine War is not very unique. In fact, it is the most common type of war fought in the post- Napoleonic era; namely, a war between neighbors over bordering territory. Slightly over 50 percent of the interstate wars in this era are of this type (see Vasquez and Valeriano 2010, 300). The Russia-Ukraine War follows a long line of wars, and the territorial explanation of war can illuminate its causes. In addition, this war is similar to recent wars in the region, specifically Nagorno-Karabakh, the Russian-Georgia War of 2008, and of course the 2014 Crimean Seizure. We also show that several other factors increased the chances for war in this case including power preponderance by a nuclear power, differences in regime type, diversionary and historical legacy motives, and a history of rivalry and multiple contentious diplomatic issues. We also discuss the potential for the Russia-Ukraine War to escalate into a broader and even more costly conflict involving major states on opposing sides. We conclude that nuclear weapons and other Cold War norms make escalation less likely.

As before, it has been a complete pleasure working with Rowman & Littlefield. Our appreciation to Michael Kerns, our editor, who has always been supportive and aided us throughout the process. We also thank our production editor, Felicity Tucker, and our copy editor, Katy Whipple. We have learned so much from the contributors to our edited volume and used many of those insights in writing the new Russia-Ukraine War chapter. We encourage readers to consult the original preface to the third edition, which follows, for more detailed thanks to all other individuals and institutions that assisted us in the broader volume and for an overview of the other chapters in the book.

We hope the discussion of a current case can help readers see the relevance of statistical research and demonstrate what peace science scholarship can tell us about recent wars. Instructors might even consider having students examine other cases, especially well-studied historical cases, to see what statistical patterns are present in these cases as a way of exploring the utility of the scientific study of peace and war. Alternatively, they might examine ongoing wars. As we go to press, the Hamas-Israel War has broken out. This war has its roots going back to the 1948 Arab-Israeli War. Like the Russia-Ukraine War, it involves territory and border issues, but unlike that war it involves several parties and a non-state actor. The war has generated strong emotions on both sides justifiably, in particular on US college campuses. This will make it difficult to study empirically, but the key to establishing peace is to have an accurate understanding of what causes war.

Sara McLaughlin Mitchell
John A. Vasquez
October 28, 2023

Preface to the Third Edition

Many problems have plagued humanity, and war has surely been among them. Mitigating war and maintaining peace are two of the key tasks for those concerned about international relations. To handle these issues, it is necessary to know the causes of each. This book brings together the leading scholars within international relations to review the current theory and research on peace and war—what is known, the quality of the evidence, and future research strategies. This book reports on one approach to the study of peace and war: the use of the scientific method to identify those factors that bring about the outbreak of interstate war and those factors that promote peace.

One of the reasons knowledge about the causes of war and the conditions of peace has advanced in recent decades has been the creation of an international community of peace science scholars using common data to test hypotheses focused on a limited number of questions. It is not an exaggeration to say that in the past fifty years more social scientists have been working on this set of related research questions than at any other time in history. The scientific approach permits a division of labor and the creation of a body of research findings that encourages the cumulation of knowledge. Having many minds work on the same problem in a sustained manner may have payoffs that have eluded the previous efforts of isolated individuals. The struggle of humanity to end the scourge of war has always rested on the shoulders of those who could provide answers as to why war occurs and how it can be prevented. The scientific community of scholars is one of the best hopes for solving the puzzle of war.

The third edition of this book appraises what has been learned about the causes of war and the conditions of peace in the twenty years since the publication of the first edition. All of the chapters are new, and several new authors have been included along with a new section on current trends in research. This edition also sees the addition of Sara McLaughlin Mitchell as coeditor. She willingly accepted my invitation to come on as coeditor as I transition to retirement. She raised the funds for and hosted the conference upon which this book is based, and she greatly improved the book's intellectual content.

Readers should note that the focus of this volume is on interstate war. We have not treated civil war because that would require its own volume, and we are pleased to report that since the second edition of this book, Rowman and Littlefield released in 2016 the companion volume, *What Do We Know about Civil Wars?* edited by T. David Mason and Sara McLaughlin Mitchell.

Trying to summarize the findings on a particular subject is an ambitious goal, and to have any hope of achieving it scholars have to have sustained interaction and discussion. As with the previous edition, this one was prepared by having a conference with the contributors. This time the conference was hosted at the University of Iowa by Sara McLaughlin Mitchell in December 2019 with the aim to see what we had learned as a community of peace. Little did we know how lucky we were to get together just before COVID-19 shut everything down.

Except for some graduate students, all of the scholars involved in this project had known each other for a number of years and had regularly seen each other at annual meetings, especially that of the Peace Science Society (International), which has provided an important forum for discussion and exchange of research. A paper on each of the major topics related to the causes of war and the conditions of peace was presented, reviewed, and discussed. Each paper was then revised for inclusion into this book.

As before, it has been a complete pleasure working with Rowman and Littlefield. Our appreciation to Susan McEachern, our editor, who has always been supportive and aided us throughout the process. Our thanks also to others at the press: Katelyn Turner, Alden Perkins, our production editor, and early on Traci Crowell. A special thanks to our copy editor, Ami Naramor, and our project manager at Integra Software, Sudantradevi Mohan.

Many organizations and individuals helped make the conference and this book a reality. The conference was made possible with funding support from the University of Iowa Department of Political Science's Shambaugh Memorial Fund and the College of Liberal Arts and Sciences Perry A. and Helen Judy Bond Fund for Interdisciplinary Interaction. We are grateful to Liz Cecil, Solomon Fenton-Miller, and Wendy Sandersfeld for staff support for hosting the conference. We also thank Iowa faculty members and students who served as discussants and helped take notes during the discussions, including Darrell Carter, Addison Huygens, Willow Kruetzer, Brian Lai, Bomi Lee, Elizabeth Menninga, Cody Schmidt, Nathan Timbs, Joshua Tschantret, and Yufan Yang. Sara McLaughlin Mitchell wishes to thank John Vasquez for inspiring her to study war and for inviting her to serve as an editor for this edition.

Lastly, I (Sara) dedicate this book to all of my undergraduate and graduate students who have inspired me to learn more about what causes war. Seeing war through their perspectives has pushed me to think about conflict from different viewpoints and to study causes of wars in ways that I had not anticipated before stepping into the classroom.

John Vasquez wishes to thank Monica Bielawiec, who expertly compiled and formatted various chapters and references into a manuscript suitable for production. My thanks to her for this and for being my research assistant for the past two years during which she completed multiple projects, several related to this volume. My thanks to Yuxin Victoria Chen, who with great care tracked down a number of citations that were needed at the last minute. My thanks also to the University of Illinois for financial support provided through the Thomas B. Mackie research fund.

I would also like to express my appreciation to Marie T. Henehan, who has provided support during my professional career and has been someone with whom I could share and test out ideas, including many of those featured in this book. She has also always been willing to do some of the meticulous work, like check endless references here and going back to 1980 with my first book. Some twenty books later, it is good to still have the love of my life at my side as we enter our forty-first year of marriage and our eighth winter on Block Island.

Finally, I dedicate this book to my mother, Helen J. Vasquez. Born in the year the United States entered World War I and married shortly after Pearl Harbor, she, like millions of other Americans, was touched by war, even though she avoided its worst scourges. She was always supportive of me and instilled in me a sense of work that has been immensely useful in my academic life. More importantly, she provided a sense of love without which no life would be worth living. We were saddened to lose her in 2010, but we were glad that at ninety-two she still had all her wits about her.

Introduction

Sara McLaughlin Mitchell and John A. Vasquez

In the West, at least since the time of Thucydides, scholars have pondered the causes of war. Despite this history, it was only in the twentieth century with the efforts of Lewis F. Richardson, Pitirim Sorokin, and Quincy Wright that scientific techniques began to be rigorously applied to the problem. By mid-century these early efforts inspired group projects to collect and analyze data. In 1963 J. David Singer founded the Correlates of War project with the intention of collecting replicable data that would serve as a foundation for a body of scientific knowledge on war. Many of the contributors to this book use the data originating with that project. Around the same time, scholars developed several other projects through which they began to study war scientifically. Among them were the 1914 studies of Robert North (Holsti, North, and Brody 1968; Choucri and North 1975), the Inter-nation Simulation project of Harold Guetzkow (1968), the Dimensionality of Nations project of Rudolph Rummel (1979), and the International Crisis Behavior Project (Brecher and Wilkenfeld 1989, 1997). It has been a little over fifty years since these projects and the numerous efforts of the individual researchers who have analyzed the resultant data about war. What have we learned from these efforts?

This book has been written to answer this question. Rather than producing a secondary study of that research, this book brings together those scholars actively engaged in research to answer the question themselves. The emphasis here is on international relations scholars, but they bring different perspectives and theories to bear on the question. This book, however, is not just another compilation of reports of various projects; it is an attempt to provide a systematic discussion of what we know about the onset of war and what we need to do to extend this knowledge. Now in its third edition, this book itself provides a record of the progress made in the past twenty years.

Of course the knowledge we have about war is not as definitive as knowledge produced in biology or physics. Nonetheless, that does not mean that we know nothing or that everything is a matter of perspective. The word *knowledge* is used in this book to indicate that certain hypotheses have passed at least a modicum of rigorous tests following scientific procedures and using replicable data—in other words, that the knowledge claims are based on systematic examination of evidence and not on mere speculation or intellectual argument. This does not mean that further inspection may not lead to different conclusions. Science is a process, not necessarily an end product. This is especially the case when one is working at the frontiers of knowledge in a still-young science, as are the contributors to this book. In a sense this is a book about what we know *now*.

This book is divided into four parts. Part I examines what has been learned about the factors that bring about interstate war. A chapter is devoted to each of the major factors thought to promote war. In chapter 1, Hensel and Goemans examine contentious issues—with a focus on territorial disputes—in order to see which are related to armed conflict and war, and which escalate to war and which do not. In chapter 2, Morey and Kadera look at the research done on power and capability with a focus on findings regarding parity and preponderance and their association with war onset. In chapter 3, Kenwick and McManus investigate alliances and their role in deterrence. In chapter 4, Sample explores the role of arms races in making militarized interstate disputes (MIDs) escalate to war. In chapter 5, Akcinaroglu and Radziszewski review the findings on rivalry, a well-known factor in producing war among states. In chapter 6, Fuhrmann assesses the role of nuclear weapons in conflict and peace and how they have changed international politics since 1945. Each of these factors—territory, power, alliances, arms races, rivalry, and nuclear weapons—have been seen as the most important factors related to the onset of war and the ones about which we know the most. Part I ends with chapter 7 by Quackenbush, who provides an assessment of what we know about the outcome of war (who wins and why) and its long-term impact.

Part II examines the factors that promote peace. This part begins with Mousseau's evaluation in chapter 8 of the liberal peace, at the core of which is the democratic peace. In chapter 9, Gibler and Miller review the main challenger to this theory—the territorial peace. In chapter 10, Owsiak, Diehl, and Goertz argue that peace as an explicit dependent variable has generally been understudied in quantitative research and outline a research program for enhancing our knowledge of how peace comes about. In chapter 11, Powell and Wiegand evaluate the research on how nations can peacefully resolve territorial and maritime disputes. Although the research on peace is less extensive than that on war, some things have been learned, and this is a research agenda from which much can be expected in the next decade.

Part III is a new section in this book. It covers areas that have recently been a focus of work, and while the knowledge in these areas is not as firm as the knowledge addressed in the first two parts of the book, these areas represent emerging trends in research that deserve attention. In chapter 12, Valeriano, Maness, and Jensen look at the newest area in peace science—cyber warfare, how it can be studied, and what is known about it. In chapter 13, McLaughlin Mitchell and Yang discuss the environment and how it might produce conflict, with a focus on renewable resources like water. In chapter 14, Wolford analyzes what we know about the role of leaders and their decisions for war. Recently, war financing and debt and their role in war have received renewed attention; Cappella Zielinski and Poast review this work in chapter 15. In chapter 16, Braumoeller concludes Part III with an assessment of trends in warfare since 1816 (for example, is interstate war declining?) and some of the key debates over our understanding of these trends.

Part IV presents three concluding chapters that reflect on what we have learned. In chapter 17, Thompson ponders the big picture and what research has told us and can tell us, giving us several insights on the project along the way. In chapter 18, Owsiak and Atkinson provide some guidance on how we should approach our subject in order to get better answers to what we do not know about war. This volume concludes in chapter 19 with an assessment of what we know about war by McLaughlin Mitchell and Vasquez.

Part I

FACTORS THAT BRING ABOUT WAR

Chapter One

Territory and Contentious Issues

Paul R. Hensel and Hein Goemans

TERRITORY: WHAT DO WE KNOW ABOUT CONTENTIOUS ISSUES?

In the past few decades the study of contentious issues—"disputed point[s] or question[s], the subject of a conflict or controversy" (Randle 1987, 1)—has grown into a prominent empirical research program. While the study of contentious issues has a relatively long and fruitful history dating from the 1960s literature on foreign policy analysis, it did not begin to have a major impact in conflict research until the 1990s. Thereafter, with the the collection of new data (see Hensel 2001) the study of contentious issues has moved to the forefront of empirical research on conflict.

In this chapter, we first trace the early development of the study of contentious issues from its origins to its current prominence in research on conflict and war. We then explore the theoretical and empirical background for the recent surge in interest and proceed to review the rapidly developing literature on territorial issues. We next expand our scope to consider research on non-territorial issues and how these other issues compare with the issue of territory. We conclude the chapter by discussing avenues for future research.

EARLY DEVELOPMENT OF THE ISSUES APPROACH

The first theoretical work on contentious issues can be found in the literature on foreign policy analysis. It focused on how foreign policy and domestic issues were dealt with differently in domestic policymaking (e.g., Lowi 1964; Rosenau 1967; Zimmerman 1973). While this earliest work tended to focus on different policymaking processes, it inspired other scholars to reconsider issues from a more substantive perspective.

Researchers who examined foreign policy behavior from this more substantive perspective often challenged the typical realist characterization of world politics as a quest for power or security, or at best as a distinction between "high politics" issues involving security and "low politics" issues involving everything else. O'Leary (1976, 320–22), for example, suggested that state behavior is motivated by the desire to resolve specific contentious issues, and that state behavior will vary across types of issues (in other words, "issues make a difference"). Similarly, scholars such as Mansbach and Vasquez (1981) have contended that patterns of conflict and cooperation between states should vary substantially based on the nature of the specific issues at stake.

This early work revealed preliminary evidence that states' interactions do indeed vary based on the issues at stake. In an important early study that relies on event data involving the United States or West Germany, Mansbach and Vasquez (1981) identified seventy-eight distinct stakes (issues) and revealed a great deal of variation in their conflict-cooperation patterns.[1] Extending this research to an early proposition from Rosenau's (1966) pre-theory, Vasquez (1983) distinguished between issues based on the tangibility or intangibility of the policy goals sought and the means used to attain these goals. Exploring a data set that focused primarily on diplomatic conflict and cooperation and that included few events related to armed conflict, he found that more tangible issues appeared to yield more cooperation while more intangible issues appeared more conflict prone.

DATA ON ISSUES IN ARMED CONFLICT

The first work to focus specifically on contentious issues and interstate conflict involved the identification of issues at various levels of armed conflict, up to and including war. Luard (1986) examined the issues surrounding wars since 1400 and Holsti (1991) did the same for wars since 1648, based primarily on the statements of leaders at the time. Both Luard and Holsti found territory to be a very prominent issue in modern wars. Table 1.1 examines Holsti's data, limiting analysis to primarily interstate conflicts since 1816.[2] Combining Holsti's classifications of "territory," "strategic territory," "territory (boundary)," "national unification," or "irredentism" into one overarching "territory" category, for at least one of the participants, fifty-one of the sixty-eight interstate wars (75.0 percent) involved at least one territorial issue. The prominence of territory remained consistent across historical periods, as it was an issue in twenty-nine of thirty-eight wars (76.3 percent) between 1816 and 1945 and twenty-two of thirty wars (73.3 percent) between 1946 and 1989.

This early work prompted other scholars to develop research projects to consider contentious issues, particularly territorial issues, in more depth and produced major advances in the availability of data on issues in armed conflict. For example, while the initial release of the Correlates of War (COW) project's

Table 1.1. Territorial Issues in Militarized Conflict

	Proportion of Conflicts Involving Territorial Issues:			
Type of Conflict	*1816–1945*	*1946–1989*	*1990+*	*Total*
Holsti Interstate Wars (1816–1989)	29/38 (76.3%)	22/30 (73.3%)	–	51/68 (75.0%)
COW Militarized Interstate Disputes (1816–2010)	222/709 (31.3%)	286/1,016 (28.2%)	137/590 (23.2%)	645/2,315 (27.9%)
COW Fatal MIDs (1816–2010)	51/104 (49.0%)	101/211 (47.9%)	34/102 (33.3%)	186/417 (44.6%)
COW War-Level MIDs (1816–2010)	30/56 (53.6%)	12/22 (54.6%)	2/4 (50.0%)	44/82 (53.7%)
ACD Interstate Conflicts (1946–2018)	–	35/39 (89.7%)	6/9 (66.7%)	41/48 (85.4%)
ICOW Territorial Claims Under Way (1816–2001)	550	300	159	844 (in 1+ era)
ICOW Claims with 1+ MID (1816–2001)	209/550 (38.0%)	117/300 (39.0%)	43/159 (27.0%)	345/844 (40.9%)

Militarized Interstate Dispute (MID) data set (Gochman and Maoz 1984) had not coded any aspect of the issues at stake in each militarized dispute, version 2 of the data set coded which state(s) in each dispute had sought to revise the status quo through militarized action (Jones, Bremer, and Singer 1996). The new data identified four types of status quo revision, more than one of which may be at stake in any given dispute: territorial sovereignty, government policies, the makeup of a state's governing regime, and a residual "other" category. This allowed scholars to begin to test hypotheses about the role of territorial issues in the escalation of armed conflict and led to a rapid expansion of issue-related research in the mid-1990s.[3] Another widely used data set, the UCDP/PRIO (Uppsala Conflict Data Program/Peace Research International Olso) Armed Conflict Data, identifies the incompatibilities—as stated by the parties—that are the subject of each armed conflict.The data set distinguishes between conflicts over territory and conflicts over governments, with some conflicts involving both.

Using the most recent publicly available versions of the MID and Armed Conflict Data (MID 4.30 and Armed Conflict 19.1), Table 1.1 reports the prominence of territorial issues by historical era. More than one-fourth of all militarized disputes (27.9 percent) involved challenges to the territorial status quo. This ranges from 23.2 percent of MIDs since the end of the Cold War up to 31.3 percent in the pre–World War II era. Territorial issues are more frequent in each era when looking at higher levels of conflict, suggesting that while other issues may lead to the threat or display of force, those are not as likely as territory to lead to serious escalation. Of all militarized disputes that led to at least one battle-related fatality in each era, 44.6 percent involved territorial issues. Similarly, in each era at least half of all full-scale interstate wars (53.7 percent) involved territorial issues. The results are similar for interstate conflicts in the Armed Conflict Data, which require a minimum of twenty-five fatalities in a year to qualify to enter the data. During the Cold War territorial incompatibilities were at stake in 89.7 percent of such armed conflicts, although this dropped to 66.7 percent in the post–Cold War era.

DATA ON CONTENTIOUS ISSUES

A second major advance in the study of contentious issues data was the collection of data on contentious issues themselves, independent of the occurrence of militarized disputes or wars. Huth (1996; Huth and Allee 2002) and the Issue Correlates of War (ICOW) project (Hensel 2001; Hensel et al. 2008) both released data on territorial issues identified based on diplomatic disagreements between countries, independent of how the claimant states try to manage their claims. The importance of these data is demonstrated by the fact that fewer than half of all territorial claims in the ICOW data set experience even a single militarized dispute. These new data sets can therefore be used to test propositions about the initiation and termination of disputes and disagreements over territory, the choice to militarize the territorial claim, the peaceful management of territorial claims, and many other research questions.

Categorizing by the same historical eras that we used to compare the various militarized conflict data sets, the bottom portion of Table 1.1 examines the frequency and militarization of territorial claims in the latest version of the ICOW data set (1.30). Between 1816 and 1945, 550 territorial claims were active at some point, 38.0 percent of which experienced at least one militarized dispute. Between 1946 and 1989, 300 were active, 39.0 percent of which experienced militarized conflict. In the twelve-year period between 1990 and the end of the currently available data in 2001, 159 were active, 27.0 percent of which were militarized. In total, between 1816 and 2001, 40.9 percent of the 844 territorial claims in the data set experienced militarized conflict at least once. With this broad overview of the development of the empirical study of contentious issues in the past sixty years, we now turn to theoretical developments in the study of war that are culminating in a recent surge in interest in the theoretical status of contentious issues.

CONTENTIOUS ISSUES AND THEORY

Until Fearon (1995) made the bargaining model of war the dominant paradigm through which to study international conflict, contentious issues were just one of a variety of theoretical and empirical approaches. For better or for worse, the theoretical work that established the so-called bargaining model of war (Fearon 1995; Powell 1996, 2006) had a devastating effect on the empirical study of conflict.

As the scholarly community overwhelmingly accepted the claim that if war is costly, two and only two causal mechanisms offer rational explanations for war for unitary rational actors, it quickly became obvious that it would be difficult to put these theoretical mechanisms to work empirically. Thus to date there is little to no consensus on indicators to measure variation in private information, in incentives to misrepresent, and in anticipation of commitment problems. As a result less than a decade after Fearon's seminal contribution the empirical literature on conflict had fractured and lost focus.

A focus on contentious issues, we argue, has the potential to unify several of these disparate strands of empirical conflict research and offer a coherent research agenda. Nothing in this approach contradicts the logic of the bargaining model of war. First and foremost, the current bargaining model of war literature leaves as exogenous the issue put on the bargaining table. For exposition's stake, scholars (Powell 1999, 86–87) represent the issue on the bargaining table as the total pie that could possibly be at stake, for example, the combined territory of two states. This would be innocuous if states always wanted as much as they possibly could get. Indeed, if this were true, the only issue that is put on the bargaining table always is the total pie. Some scholars have articulated such an extreme stance. Holsti (1991, 14) thus proposes that issues might not be relevant because every state wants as much as it can get. This argument is also embodied in debates between "offensive realists" who argue that states have unlimited aims, "defensive realists" who portray states as primarily concerned with maximizing security, and liberals who argue that states have limited and specific aims (Jervis 1976; Kydd 1997; Moravcsik 1997; Schweller 1998; Mearsheimer 2001; Taliaferro 2001; Snyder 2002; Glaser 2010; Narizny 2017, 162). In an effort to finally address this question in empirical terms, Schultz and Goemans (2019) show that in the context of territorial disputes, states generally have genuinely limited aims. This finding suggests that it would be a mistake to assume all issues boil down to the same underlying unlimited demands.

A research agenda that focuses on contentious issues holds much promise. For one, it could address the question of when and why states and leaders have genuinely limited aims. For another, it offers some attractive ways out of the constraints of the current consensus on the bargaining model of war. Fearon (1995, 379n1) explicitly excluded considerations of "private benefits" from his rationalist explanations for war. A focus on contentious issues is well situated to identify which (types of) issues can be exploited for a leader's private gains. For example, in their study of territorial disputes in Africa Goemans and Schultz (2017) find that leaders make limited territorial claims for areas with ethnic brethren, but *only* if those would likely bolster the leaders' hold on power. Otherwise, borders and areas that partition ethnic brethren are just as unlikely to be contested as areas and borders that do not. In other words, sometimes maritime claims, sometimes river claims, and sometimes territorial claims can be exploited for leaders' (or their supporters') private gains. Drilling down into specific issues makes it possible to explore when, how, and why specific issues become important to different domestic coalitions and how these domestic coalitions affect the likelihood of international conflict.

More generally, different issues are likely to result in different bargaining dynamics. In the literature on territorial disputes it has long been acknowledged that prior historical formations are one of the few legitimate bases for territorial claims (Murphy 1990; Abramson and Carter 2016). These claims typically frame the issue in the "domain of losses" and are therefore likely to induce risk tolerance (Kahneman and Tversky 1979), a departure from Fearon's basic model that assumes risk neutrality or risk aversion. Others have argued that some territorial claims revolve around discontinuities in the players' utility function and the resulting issue indivisibilities (Hassner 2003; Toft 2003; Goddard 2009; see also Carter 2010; Leonard 2020). In other words, the future of the contentious issues research agenda looks very bright and offers new ways forward.

To illustrate the breadth and promise of the study of contentious issues, we now focus in depth on the literature on *territorial* issues and unpack how territorial issues are not unidimensional but cover a wide variety of mechanisms that could link territory to conflict. In the final section we briefly discuss research that examines other issues or compares the management of multiple issue types.

TERRITORIAL ISSUES

We can say, along with Schultz (2015, 129, following Vasquez 2009), that *some* "territory is the kind of good for which states and individuals are (or can be) highly motivated to fight." To disentangle why some but not other issues—in this case territory—are worth fighting over and the specific mechanisms that link different contentious issues to conflict, it will prove helpful to establish a baseline of the importance of territory in the modern international system.

Territory is central to modern international relations. Territory is the sine qua non of states: in Weber's (2015) famous definition, states exercise authority "within a given area—*and the area is integral to its definition*" (emphasis added).[4] No territory, no state. The rediscovery of cartography and significant advances in mapping in the eighteenth century fundamentally shaped the role and importance of territory and territoriality—the principle of exclusion by area (Sack 1986)—in modern domestic and international politics (Branch 2014). This marked a break with the early non-territorial systems of political authority. "The key difference between territorial and non-territorial strategies of rule," Goettlich (2019, 208) argues, "is whether authority is applied to people and things based on *where* they are"—classification by area—"rather than by *who* or *what* they are"—classification by type. "Territoriality is essentially 'rule by geography' rather than universal rule, or rule by networks or personal relationships." In this world political spaces are overwhelmingly compact and fit into each other as a completed jigsaw puzzle; empty spaces—*terra nullius*—are abhorred. Clear territorial limits—depicted and disseminated on maps—fulfilled "the goal of nations to be exclusionary communities: to distinguish 'us' from 'them.'"[5] This bordered order did more than demarcate spaces of political authority; it created new forms of collective action and collective identity.[6] Between the seventeenth and eighteenth centuries the territorial nation-state became a *system-defining* characteristic with lasting consequences for state and social behavior. Territory now became the container of peoples, of resources, of identity, and of meaning in new and powerful ways.

In this territorial order borders can affect conflict in several ways (see Goemans and Schultz 2017), and conflict over borders is not necessarily over the territory they enclose. States sometimes become involved in territorial disputes because they dispute the border; it is not the specific territory that is so much contested here as it is the specific location of the border. This can be for a variety of reasons. Sometimes the border is poorly demarcated and needs better specification; sometimes the text of an established border treaty is ambiguous or unclear in the face of new cartographic discoveries.[7] Sometimes states contest the specific principle used to delimit the border, as, when early in the twentieth century Argentina and Chile disputed whether the watershed or the highest peaks in the Andes should form their border. As the border moves, so of course does the territory under control of the contesting states, but in these cases the dispute is more over the location of the border than the territory it encloses. Since the conflict is over the location of the border, treaties and agreements that pin down and clearly specify those borders are likely to end the conflict. (This contrasts with prior historical claims to territory.) As such, clear borders have been credited with reducing the costs of conflict among a wide range of social groups, from insects (Eason, Cobbs, and Trinca 1999) to urban gangs (Brantingham et al. 2012) to nation-states (Goemans 2006; Carter and Goemans 2011; Goemans and Schultz 2017).[8]

Second, the physical design of borders, rather than the territory they enclose, also affects conflict.[9] Recent research (Atzili 2012; Carter and Poast 2015; Hassner and Wittenberg 2015; Getmansky, Grossman, and Wright 2019) has examined the role of border walls and border fortifications and concluded that these walls have often a stronger economic than security dimension and are designed to counter the forces unleashed by globalization. Such borders are thus expressly designed to minimize conflict among states.

Third, as noted earlier in this chapter, borders shape states and identities because borders coordinate expectations about group membership (Goemans 2006). Borders not only coordinate behavior among states (Acharya and Lee 2019); they are posited to bolster the project of collective action *from within*. Elinor Ostrom (2000, 149) argues that a clear boundary is the first essential principle for collective action because it "enables participants to know who is in and who is out of a defined set of relationships and thus with whom to cooperate." The project of defining the group itself is fundamental to creating and reinforcing group identities, themselves essential to group formation.[10] As Abbott (1995, 860) famously contends, "Social identities come into existence when social actors tie social boundaries together in certain ways. *Boundaries come first, then entities*."[11] In the modern nation-state territorial identities often can form the basis of territorial claims. Claims on the basis of "prior historical formation" (Murphy 1990) and claims to territory on the basis of its status as part of the "homeland" (Shelef 2019) are almost always based on identity claims. This is how the intangible value of territory, often identified as a major factor in territorial disputes, can be linked with identity. Borders and territoriality help create and establish identity; to maintain and uphold that identity then requires states and leaders to maintain the borders that uphold that identity.

Fourth, just as different pieces of territory produce and are demanded by different identities, they also often have quite different material value. For example, Goertz and Diehl (1992b) find that the risk of armed conflict during and after a territorial exchange increases with the intrinsic value of an exchanged territory (measured by area and population) as well as its relational value (whether it is considered homeland or dependent territory). Huth (1996) finds that three different indicators—strategic location, ethnic similarity between the states, and ties between the challenger state and an ethnic group along the border—increase the likelihood of armed conflict over a territorial claim, although the presence of economic resources decreases conflict; Huth and Allee (2002) explore domestic politics further and note that ethnic ties to the challenger primarily increase conflict for democratic challengers. On the other hand, Tir (2006a) finds that when two states engage in a territorial claim after exchanging territory, they are more likely to begin armed conflict when the territory has economic value; ethnic ties to the territory have a weaker effect on conflict and the strategic value of the territory has little systematic impact on future conflict.

Fifth, scholars have used aggregated measures of salience to measure the overall value of territory rather than treating each individual salience indicator separately in empirical models. Hensel (2001; Hensel et al. 2008) measured the salience of territorial claims with a twelve-point index that includes a number of characteristics of the claimed territory that should increase its value to the challenger and/or target state; higher values on this salience index significantly increase the probability of militarized conflict over a territorial claim. Hensel and Mitchell (2005) break this twelve-point salience index into separate indices of tangible salience (indicating the presence of resources, strategic location, or permanent population) and intangible salience (indicating for each state whether the territory is considered part of the national homeland rather than a colony or dependency, whether the state has identity ties to the territory and its population, and whether the state has had sovereignty over the territory in the relatively recent past). The overall salience of the claimed territory significantly reduces the likelihood of reaching agreement in peaceful settlement attempts, while increasing the likelihood of three different forms of militarized conflict over the claim. Tangible salience has the same effects, reducing the likelihood of agreement while increasing all three forms of militarized conflict. Intangible salience has an even stronger impact on the two most severe forms of militarized conflict, with roughly double the substantive impact on escalation, although to the authors' surprise, agreement seemed to be even more likely when intangible salience was higher.

This brief overview offers but some ideas to show how "territory" can be linked to conflict through various causal mechanisms. The promise of the contentious issues approach is not just that it can establish new empirical foundations on which to build. The contentious issues research agenda goes further and suggests fertile ground for theoretical innovation. Why do some issues arise but not others? Who benefits from promoting this contentious issue? Why do some issues seem prone to war while other issues apparently can be resolved peacefully? We are on the cusp of a breakthrough in our ability to collect fine-grained data, which will allow us to probe deeper than ever before into the mechanisms that link contentious issues to international conflict.

The rapid advancement of Geographical Information Systems (GIS) technology is beginning to transform research on territorial conflict by adding much greater precision to these earlier studies, which depended on rough indicators of whether any economic resource or shared ethnic group was present. Schultz (2017) developed a new data set, built on the territorial disputes data collected by Huth (1996), that precisely maps post–World War II territorial disputes. This is a valuable resource that opens up a broad range of new research possibilities. Goemans and Schultz (2017) rely on these data to examine the basis for territorial claims in Africa. They slice all borders into one-kilometer segments and use other geospatial data—from Geo-EPR on ethnic groups (Wucherpfennig et al. 2011), to MRDS (Lujala, Rød, and Thieme 2007) for natural resources—to assign attributes to each segment. This approach makes it possible to analyze substantial variation both across and within dyads; for example, why some states make claims on other states and why in the same dyad a state may claim some areas but not others. This geospatial approach makes it possible to identify local territorial effects. Goemans and Schultz (2017) find that it is not the mere partition of ethnic groups but the political status of partitioned groups that conditions territorial claims in Africa. Natural resources, if anything, decreased the likelihood of a territorial claim, except perhaps for claims articulated early in the existence of the dyad. Precise GIS data allow us to link specific territorial claims to specific domestic constituencies who could mobilize in favor of conflict or resolution.

To round out this brief overview of how the value of territory can vary along several dimensions, we note that the settlement of a territorial issue does seem to produce more peaceful relations between former enemies (e.g., Vasquez 1993). Simmons (2005) and Schultz (2014) note that bilateral trade between the former claimants appears to increase substantially after the settlement of territorial claims. Owsiak (2012) and Schultz (2014) find that settling the border between two states significantly reduces the risk of future militarized conflict between them. Gibler (2012) also reports a "territorial peace," noting that settling borders and ending territorial threats greatly increases the prospects for democratization; he suggests that settled borders may be a virtual precondition for the well-known democratic peace (see chapter 8 on the liberal peace). These findings are important in their own right, but also because they strongly suggest that specific issues—rather than the endless demands for "more"—lay at the very root of contentious relations. Contentious issues appear to lay at the root of questions of war and peace.

MULTIPLE ISSUES

The first wave of empirical research on contentious issues and armed conflict focused specifically on territorial issues. This made a great deal of sense, as work by scholars such as Luard and Holsti had shown that territorial issues were the most prominent issue types in interstate wars throughout the modern era. More recently scholars have begun to broaden their data collection efforts to also study the management of *non*-territorial issues. This broadening of scope has the potential to discover new patterns and to significantly advance our understanding of contentious issues more generally.

Data on non-territorial issues can be compared directly to the data on territorial claims collected by the ICOW project. The ICOW project began collecting data on territorial claims but subsequently expanded to river and maritime claims. Drawing from work by Hensel et al. (2008) and Hensel and Mitchell (2017), Table 1.2 summarizes contentious issues based on the relative prominence of their tangible and intangible salience dimensions. Territorial issues, the subject of most early work, are often described as unique because they typically have high tangible salience (due to the value of their physical contents) and high intangible salience (due to the connection to an individual's or a country's identity).

River claims involve contention over the use of a shared international river and typically have high tangible salience because of the potential for human consumption, irrigation, hydroelectric power generation, and commercial navigation, but they typically have low intangible salience because most rivers do not have a strong connection to national identity. Maritime claims involve contention over the ownership or usage of a maritime zone and, like river claims, typically have high tangible salience because of the resources such as fisheries or oil that might be found in the zone, as well as the potential for navigation, but typically have low intangible salience because the waters do not have as strong a connection to identity as mainland territory. Chapter 13 in this volume offers more detail on these water-related issues.

As data collection on these first three issues approached completion, the ICOW project began collecting data on identity claims, which fills a third cell in the

Table 1.2. Typology of Contentious Issues by Claim Salience

	Low Intangible Salience	*High Intangible Salience*
High Tangible Salience	• **River** (Turkish dam projects on Euphrates River) • **Maritime** (Cod Wars, Eastern Mediterranean)	• **Territory** (Golan Heights, Alsace-Lorraine)
Low Tangible Salience	• **Firms or industries** (Airbus subsidies, shrimp imports) • **Treatment of individuals** (caning of Michael Fay, arrest of Maria Butina)	• **Identity** (Germans in South Tyrol, Russians in Ukraine or Estonia)

table. These claims involve contention over the status of an ethnic group that is shared by both states. They involve relatively high intangible salience because of the importance of defending one's ethnic kin from a foreign threat. They typically have low tangible salience, though, as most identity claims involve demands for better treatment of the group in its current state rather than incorporation into the state making the demands, so there is little prospect of tangible gains to be had by making the claim.

There are many similarities in the ways in which different issues are managed. Hensel et al. (2008) find that with higher issue salience all three types of issues are more likely to become militarized. At the high end of issue salience we find territory with strategic or economic value or intangible value because the territory is considered part of a state's homeland or identity, the usage of a disputed river for such purposes as irrigation or hydroelectric power generation, and the usage of a disputed maritime zone for such purposes as fishing or oil/gas extraction. At the same time, when the issue in question is more salient all three issue types are also more likely to see peaceful conflict management efforts.

While there are many similarities in the ways that territorial, river, and maritime claims are managed, there are also important differences.[12] Hensel et al. (2008) note that territorial claims—widely regarded as the most salient issue type overall since the early days of the contentious issues literature—are the most likely to become militarized. The data sets used for that initial study have been expanded, though, offering an opportunity to update the analysis.

Version 1.3 of the ICOW data covers the whole world from 1816 to 2001 for territorial claims; the Americas, Europe, and the Middle East from 1900 to 2001 for river claims; and the Americas and Europe from 1900 to 2001 for maritime claims. Table 1.3 relies on these data to examine the frequency of militarized conflict over contentious issues. Analyses of variance (ANOVAs) indicate a statistically significant difference between issue types in the number of total militarized disputes per issue ($F = 6.83$, $p < 0.01$), and in the number of fatal militarized disputes per issue ($F = 12.66$, $p < 0.001$). There should be little doubt that territorial claims are more conflict prone than river or maritime claims, although much more can be learned from examining the specific situations in which conflict does or does not occur.

Table 1.4 examines the likelihood of armed conflict when two or more states dispute a contentious issue. This model is largely consistent with previous work

Table 1.3. Militarization of Territorial, River, and Maritime Claims

Issue Type	*Any MID* *Mean (S.D.)*	*Fatal MID* *Mean (S.D.)*	*N*
Territorial claim	1.01 (2.42)	0.49 (1.32)	844
River claim	0.20 (0.63)	0.08 (0.46)	103
Maritime claim	0.66 (1.72)	0.05 (0.25)	143
Total	0.89 (1.72) F = 6.83 (p < 0.01)	0.39 (1.19) F = 12.66 (p < 0.001)	1,090

Table 1.4. Accounting for Issue Militarization

Variable	*Any MID over Issue Est.(S.E.)*	*Fatal MID over Issue Est.(S.E.)*
Intercept	−4.46 (0.24)**	−5.60 (0.36)***
River issue	−0.52 (0.39)	−0.72 (0.68)
Maritime issue	−0.41 (0.18)**	−2.21 (0.46)***
Within-issue salience	0.14 (0.02)***	0.19 (0.03)***
Recent MIDs	0.59 (0.06)***	0.54 (0.06)***
Recent failed talks	0.12 (0.04)***	0.05 (0.04)
Joint democracy	−0.56 (0.24)**	−1.13 (0.37)***
Challenger capabilities	0.50 (0.24)**	0.63 (0.289)**
N:	16,842	16,842
Likelihood Ratio X^2:	246.54	253.46
	$p < 0.001$ (7 d.f.)	$p < 0.001$ (7 d.f.)

* $p < 0.10$; ** $p < 0.05$; *** $p < 0.01$ (standard errors clustered by dyad)

such as Hensel et al. (2008) and Hensel (2012) and examines some prominent control variables while emphasizing details of the disputed contentious issue. Table 1.4 omits territorial issues because they are the referent category, allowing us to observe any differential impact of river or maritime issues. Drawing from past research, we expect within-issue salience—variation in the salience or importance of a given issue relative to other issues of the same type (Hensel et al. 2008)—to increase the risk of armed conflict, as should the amount of recent armed conflict and the number of recent failed negotiations over the issue. The analysis controls for joint democracy between the two claimants (where joint democracy is coded where the two states both have Polity scores of 6 or greater) and for the relative capabilities of the challenger state (measured as the percentage of total COW CINC capabilities in the dyad held by the challenger, or the state that is making demands to revise the status quo).

The results in Table 1.4 suggest that even after controlling for other relevant variables, the differences between territorial, river, and maritime disputes, as presented in Table 1.3, are indeed significant. River issues are not systematically different from the referent category of territorial issues, which may be influenced by the relatively small number of river claims (only 848 of the 16,842 observations involve river claims), but maritime issues are significantly less likely than territorial ones to become involved in both armed conflict generally ($p < 0.02$) and in fatal conflict ($p < 0.001$). Within-claim salience, indicating the relative salience of the claim compared to other territorial, river, or maritime claims, significantly increases the likelihood of claim escalation ($p < 0.001$). The weighted amount of recent armed conflict (MIDs) over the same claim significantly increases the risk of future conflict over the claim ($p < 0.001$), and the weighted number of unsuccessful recent negotiations over the claim—which may have failed due to an inability to reach agreement, to ratify the agreement, or to carry out the agreement—reduces the likelihood of armed conflict in general ($p < 0.001$) but has no impact on fatal conflict ($p < 0.27$).[13] Among control variables, joint democracy significantly reduces armed conflict, as does a greater advantage in relative capabilities for the challenger state making demands in the claim.

With these similarities in mind, other research reveals that systematic differences also emerge in claim management across issues. For river claims, for instance, the level of water scarcity in a river basin greatly increases the risk of armed conflict over any issue between riparian states (Tir and Stinnett 2012), as well as the risk of armed conflict over river claims specifically (Hensel, Mitchell, and Sowers 2006; Brochmann and Hensel 2009). On the other hand, formal treaties governing the usage of shared rivers tend to reduce conflict over the rivers and increase cooperation between the riparian states (Brochmann and Hensel 2009, 2011; Tir and Stinnett 2012).

For maritime claims, Nyman (2015) finds that armed conflict has become more likely with advances in offshore drilling technology, which has opened up the possibility of exploiting previously unreachable resources and increased incentives to ensure sovereignty over offshore oil fields. Neither ratification of the United Nations Convention on the Law of the Sea (UNCLOS) nor declaring exclusive economic zones (EEZs) appears to reduce the risk of armed conflict over maritime issues. Membership in UNCLOS, however, reduces the risk of new maritime claims and increases third-party management of such claims. Exclusive economic zones increase the chances of success in bilateral negotiations over the maritime issue (Nemeth et al. 2014).

Finally, the peaceful management and settlement of contentious issues will not be emphasized in the present chapter because that topic is covered by Powell and Wiegand in chapter 11. One important element that affects peaceful management (and perhaps armed conflict), though, is the role of institutionalization in managing each issue. Owsiak and Mitchell (2019), for example, find that territorial claims—which are highly salient but are not governed by strong international rules or institutions—are managed the most informally, typically through bilateral negotiation or arbitration. Reflecting the existence of the UN Convention on Non-navigational Uses of International Watercourses—which offers rules and principles for managing river claims, river claims are managed somewhat more formally. Note, however, that this convention has not been adopted by many relevant states. As a result such claims are managed more often with nonbinding third-party activities such as mediation. Maritime claims are governed by UNCLOS, which institutionalizes rules and creates a formal institution to resolve maritime questions that could not be resolved otherwise, and which most relevant states have adopted. These claims are managed the most formally, often through multilateral negotiations and legal processes.

While there are no global institutions to deal with territorial issues that are comparable to UNCLOS or the International Watercourses Convention, Zacher (2001) has suggested that a norm of territorial integrity has emerged and strengthened over the past century under which borders are to be respected and states should be unlikely to seek to (and even less likely to succeed) seize territory from other states. Zacher reports preliminary evidence to support this suggestion, with successful territorial conquests becoming very rare by the late twentieth century. The power and prevalence of this norm has now become a lively debate in the literature. On one hand, in line with Zacher, Fazal (2004) notes that cases of state death have become far less common since 1945. Frederick, Hensel, and Macaulay's (2017) overview of territorial claims between 1816 and 2001 notes several patterns that are consistent with the alleged territorial integrity norm. The number of territorial claims per state in the international system is now the lowest that it has been since the early nineteenth century as many claims have ended and few new claims have arisen to replace them, and fewer territorial claims have been militarized in the post–Cold War era (1990–2001) than in any earlier era since 1816. On a related note, Hensel and Macaulay (2016) find that as the observable measures of the territorial integrity norm have strengthened, states sharing ethnic groups have been less likely to make irredentist demands for the cession of the territory where the shared group lives, and more likely to make identity claims demanding better treatment of the group in its current home state, suggesting that the territorial integrity norm is modifying the types of demands that states make against potential adversaries. On the other hand, Altman (2017, 2020) has recently thrown doubt on the existence and relevance of the norm of territorial integrity by pointing out that after 1945 states are increasingly likely to grab small pieces of territory with fait accompli strategies, thus avoiding full-fledged war. Hensel, Allison, and Khanani (2009) attempted to measure the strength of the norm quantitatively by coding shared memberships in international treaties and institutions that require respect for the territorial integrity of fellow members, and found only weak support for the territorial integrity norm with respect to the outbreak of militarized disputes over territorial issues.

LOOKING FORWARD

In the past half century the scholarly literature on contentious issues and armed conflict has come a long way. Early propositions about general patterns of conflict and cooperation were tested with event data, and then advances in data collection allowed the study of issues in armed conflict as well as issue management and settlement. Research also advanced from an atheoretical identification of issues in event data or wars to comprehensive data sets of territorial claims, and then to other issues such as river and maritime claims and now identity claims.

One important direction for future research involves the ever-improving availability of new data on territory and borders. Geographical Information Systems and related technologies have advanced and are starting to be used to good effect. Many of the benefits of GIS are limited to recent decades when scholars have access to a multitude of data layers that can be used, with far less data available for historical analyses. Even if much of the data is limited to a recent time frame, though, much can be done to use GIS analysis to understand the salience of claimed territories, the exact locations along a border that are likely to be claimed, and the exact locations in a claim where armed conflict is most likely to break out.

For each issue type, more research should be done on the origins of contention over the issue in question. Up to this point most work has focused on the management of issues once they have emerged, whether militarized conflict or peaceful management. There are a few exceptions for each type of issue, but more systematic work in this area would be very useful.

Future data collection could also extend this research agenda in new directions. Returning to the typology of issues discussed earlier in this chapter, which classifies issues by tangible and intangible salience levels, nobody has collected systematic data on any type of issue that has generally low levels of both tangible and intangible salience. This omission makes some sense, as these cases are the least likely to produce national security concerns of the type that feature in Vasquez's steps to war model, and issues like this rarely if ever showed up in studies of war issues by Luard and Holsti. Data on such issues could nevertheless be useful for studies of negotiation and conflict management in order to gain a more complete understanding of how issue salience affects the onset and effectiveness of negotiations.

Finally, a different direction for future data collection might involve the collection of data on contentious issues within nation-states. For example, US states regularly disagree over the use of shared rivers, and numerous disagreements have arisen between Indian states over borders—some of which have actually led to the threat or use of violence between local residents. Disagreements between units within a single state would appear likely to be managed differently from disagreements between sovereign states, where international anarchy leaves no sovereign actor above the disagreeing units that might be able to impose a settlement or otherwise direct the management of the issue.

NOTES

1. Note that they were not yet testing a detailed categorization of issue types that should be expected to have specific events on conflict or cooperation.

2. We present data only from 1816 onward for the sake of consistency with the other data sets in the table. Similar results hold if internal wars or earlier time periods are added (Vasquez 1993).

3. Research was still limited to issues that had become militarized, with no collection of data on issues outside of armed conflict. This categorization of issues in militarized conflict was broadened by Gibler (2017), who disaggregated the COW "territorial" issue type into numerous more specific types of territorial issues in armed conflicts.

4. "Staat ist diejenige menschliche Gemeinschaft, welche innerhalb eines b[es]timmten Gebietes—dies: das 'Gebiet,' gehört zum Merkmal—das Monopol legitimer physischer Gewaltsamkeit für sich (mit Erfolg) beansprucht." Max Weber, 1926. *Politik als Beruf*, second edition, Muenchen and Leipzig, Von Dunker and Humblot, p. 8.

5. Herb (1997, 7).

6. Cerny 1995; Nexon 2009; Scott 2009; Wilson and Donnan 2016; Atzili and Kadercan 2017. The process whereby a border creates a new social identity is traced in detail in Sahlins 1989.

7. For a fascinating example, see Kyle Gardner's "The Elusive Watershed of Ladakh: Explaining India and China's Missing Border." https://www.orfonline.org/expert-speak/elusive-watersheds-ladakh-explaining-india-china-missing-border/.

8. An important caveat is in order: when nation-states agree on a border, a different problem sometimes arises: they do not agree on their citizens. If the state cannot legitimately challenge the border, then perhaps it could move unwanted people in order to achieve a more homogenous nation.

9. For a discussion of the new concept of *border orientation*, see Beth A. Simmons and Michael Kenwick, "Border Orientation in a Globalizing World: Concept and Measurement," unpublished manuscript, 2019.

10. Wilson and Donnan 2012.

11. Abbott 1995, 860; emphasis added. See also Nail 2016.

12. Some of the control variables have different effects across the issue types, such as only reaching statistical significance for the management of one or two issues but not the other(s). Future research could benefit from closer investigation of the reasons for these sometimes inconsistent results.

13. This weighting scheme is meant to emphasize the impact of more recent events over more distant events. Each event's impact decays by 10 percent per year, with full 100 percent weight in the first year after the event occurred and only 10 percent ten years afterward.

Chapter Two

Dyadic Power Distributions and War

Daniel S. Morey and Kelly M. Kadera

When do states turn to interstate wars to establish a winner and loser in the ubiquitous competition over international affairs? This chapter explores answers that rely on the relative distribution of capabilities between states. These answers, almost by definition, have realist roots, beginning with arguments proposing that preponderance leads to war and advocating for balance of power (BoP) as a path to peace. They also include BoP's chief competitor, power transition (PT) theory, and associated arguments about hegemonic stability and the potentially stabilizing and peaceful benefits of preponderance.

The importance of power distributions goes well beyond the PT-BoP debate or questions of polarity and war. Current research uses power (dis)parity as a control variable in almost all dyadic studies of war and epoch or era controls to represent polarity conditions (e.g., post–Cold War dummy after 1991). Further, power and power distributions underlie many leading arguments about the causes and consequences of war: capability aggregation in alliances, Boulding's (1962) loss of strength gradient, arms races, deterrence, and war outcomes.[1] Understanding the relationship between power distributions and conflict is fundamental to understanding war and peace.

In this chapter we advocate for a return to theorizing about power and about its causal connections to war and other forms of political violence. Despite more than fifty years of research, basic questions regarding power distributions and war remain unanswered. Frustrated by a lack of progress brought on by conflicting theoretical camps that often appear to ignore each other and the inability of empirical results to crown a champion, most researchers have moved on from these questions. However, the stakes are too high to throw up our hands. The role of power is too central to the field of international relations to leave it undertheorized and unresolved.

In this chapter we discuss the conceptualization and measurement of power, compare the theoretical arguments for BoP and PT as explanations of the relationship between power distributions and conflict at both the dyadic and systemic levels of analysis, examine the state of the empirical evidence, and make recommendations for future scholarship on power distributions and conflict.

DEFINING AND MEASURING POWER

Dahl (1957) proposed one of the most intuitively appealing definitions of power: "A gets B to do what B otherwise would not do." Dahl's definition underscores the inherently relational feature of power. It makes little sense to say A is powerful because power does not exist in a vacuum. Instead, A is powerful vis-à-vis B. Dahl's definition also emphasizes another dimension of power: its presence should be noted in comparison to what would happen when it is not exerted. In a sense, Dahl asks us to observe what would happen in the absence of A's efforts, which might prove to be an elusive task in global politics.

Dahl also developed a measure of power that he used to assess a legislator's ability to pass a bill. It assessed the bill's likelihood of passage with or without that legislator's sponsorship, leveraging the empirical frequencies of sponsorship and bill passage. Scholars of international politics could use a measure of power that similarly compares the "otherwise" condition to what happens when A exerts effort. We don't yet have a way to measure what state B would do in the absence of A's efforts. What would B do, for example, in the absence of a material capabilities advantage of a rival, A? Effectively, we just use controls by putting A and B in an analysis set with lots of other dyads and include many other explanatory variables. But a more conceptually linked measure would be able to directly compare cases where A exerts some effort to control B with those in which it does not.[2]

Morgenthau (1978) offers one of the deepest treatments of the elements of national power. He identifies nine primary components of power: geography, natural resources, industrial capacity, military preparedness, population, national character, national morale, quality of diplomacy, and quality of government. Many categories include subcategories for consideration. Morgenthau's expansive views on power cover traditional ideas on power such as the size of the military, but also includes nonmaterial ideas such as diplomacy and quality of government. In fact, Morgenthau stresses the idea that qualitative factors such as morality and character play an important role in determining state power. Morgenthau also sought to understand the interrelationships between different factors of power. For instance, he examines the importance of population as a latent source of power. Large populations allow states to field large fighting forces and build the economic infrastructure to support a modern army. A large population could also drain state power if a state lacks the resources to provide for the military and the general population. In sum, Morgenthau provides a more complex picture of power than we would expect of most classic realists.

Approaches to measuring state power have primarily leveraged gross domestic product (GDP) and the Correlates of War (COW) project's aptly named Composite Indicator of National Capability (CINC). While Organski and Kugler (1980) favored GDP, subsequent scholars have leaned toward the CINC score. Gibler (2017a, 22), citing Singer, Bremer, and Stuckey (1972), claims that CINC is "a measure we have used for more than forty years," and we find more than 6,000 GoogleScholar results for "CINC score" in research published since 2000. De Soysa, Oneal, and Park (1997) use both GDP and CINC scores to test the comparative performance in predicting.

The CINC score combines six components of state capability to create a single state-year measure of capability. The measure allows scholars to compare both across states and across time for the same state. The six components include two military measures, two economic measures, and two demographic measures. To capture military strength, CINC uses the total number of military personnel and the amount of military spending. Economic strength is measured using iron and steel production and energy consumption. Finally, demographic strength is a combination of a state's total population and its urban population. The total for each of the six categories is summed for all states for each year. An individual state's total for each category is then divided by the world total—this determines the share of each category a state possesses. The final CINC score for each state-year is the average across the six categories. In the end CINC represents the average system share of the material capabilities of a state. With its heavy military spending and large economy, the United States' CINC score was 0.139 in 2012 (the last data currently available). By comparison China's score was 0.218, fueled by its large population and economic growth.

While CINC has gained wide popularity, it faces some technical and conceptual drawbacks. First, if any component has missing data, that component normally gets excluded from a state's overall average score for the year in which the data are missing. Thus, the score can represent the average across the six categories or across a subset of only two or three categories. Second, and relatedly, CINC scores are sensitive to which states are included in the group of referent states, which can artificially change a state's score or a dyad's balance (Kadera and Sorokin 2004). Third, we need to ask whether the same dimensions should be included for every year, given that CINC aims to be comparable across time. Should we measure a state's power level the same way today as we do for the year 1816? For example, iron and steel production may be an outdated indicator of productivity in the service and information age, leading to a feeling that CINC cannot adapt to contemporary understandings of power. This is especially true in the area of technology, which is excluded from the CINC score. Advances in military technology have allowed smaller military forces to exert strength on the battlefield that far exceeds the number of troops.

While the CINC score has gained dominance within conflict studies, some have used other indicators of national power. Various measurement schemes have employed as few as two and up to twenty different variables. Researchers have also employed linear and nonlinear formulas to translate power inputs into measures of state power (see Merritt and Zinnes 1988). Despite differences in inputs and functional forms, the measurement schemes all correlate highly, especially when measuring the strength of the most powerful states (Merritt and Zinnes 1988, 26).

One critique of the CINC score focuses on its exclusively material view of capabilities and how and whether these translate into state power. Arendt (1958) proposed a view of power built upon the idea of cooperation. In order to have power, people must learn to work together. Power thus belongs to the group, not to a specific leader, and the group maintains power only as long as its members continue to work together. Here material capabilities are only a starting point to power; the ability of the group members to work together determines their final level of power. This view of power fits nicely with emerging views on the power of states fighting as coalitions that can equal more or less than the sum of the parts based on the ability of the members to coordinate and function together (Morey 2016, Forthcoming).

Another weakness of the CINC score is that it only measures power held by states. Goddard and Nexon (2016) argue that non-state actors, especially collectives of individuals, can mobilize themselves and nonmilitary elements of power (e.g., cultural, symbolic, and diplomatic) to alter outcomes.

Power can also come in the form of ideas, which remain unmeasured by CINC. Nye (2005) highlights the importance of soft power (e.g., cultural attraction, ideology). By using soft power, a state can get others to want the same things and achieve its desired outcome without resorting to military force (hard power). Ideas also create limits on state action by making certain actions illegitimate, such as the taboo against the first use of nuclear weapons (Tannenwald 1999).

Finally, power measures emphasizing material capabilities do not account for the power of relational networks (Hafner-Burton, Kahler, and Montgomery 2009; Cranmer, Desmarais, and Menninga 2012; Akcinaroglu and Radziszewski 2017) or the perceptions of status that can better establish systemic polarity properties (Zala 2017).

PREPONDERANCE VERSUS PARITY

Research at the dyadic level of analysis has focused on the BoP versus PT debate. These two theories make opposite assumptions and contradictory predictions (Kadera 1999, 2001), yet fifty-plus years of research has failed to adjudicate between them. One reason for this is that as research in this area has evolved, it has become clearer that these two opposing arguments may in fact be presenting the extreme ends of the same story and they both may be correct within a limited range.

Arguments that parity creates stability come in various forms and at different levels of analysis. However, they all share at their core the idea that an equilibrium, in terms of power, is necessary to avoid one state gaining ascendancy and eroding the weaker states' rights and eventually destroying them (Morgenthau

1978, 175). The goal is to create an environment where states generally survive (Morgenthau 1978) and where violence is limited (Waltz 1964). Drawing on the central tenets of realism, most theories of parity take a dim view of anarchy and assume states must defend themselves if they are to survive. In the end the best way to achieve this is through creating a BoP. This perspective on parity and peace has a long lineage starting with Thucydides (1972) and continuing in various forms in works by Waltz (1964), Morgenthau (1978), Walt (1987), and Mearsheimer (2001), to name a few.

DYADIC PARITY AND PEACE

Morgenthau (1978) articulates the classic version of balancing at the dyadic level. He argues that two patterns of balancing among states provide protection for all states. First, when two states engage in a rivalry to control each other they will directly resist each other. Each side matches power gains by the other in a continual process over time. Through this process the two sides in the rivalry sustain a dynamic equilibrium of power between them. Each state is getting stronger; however, the relative power of the two states remains balanced. This process ends when the states change their goals, the weaker state yields, or war determines a winner. The balance is precarious and in constant need of maintenance; states can never have final security and they must be ready to match the moves of their opponent. However, as long as the balance is maintained each state will remain free. While direct competition keeps strong states free from domination, the need to balance also protects the freedom of weaker states. If one state seeks to dominate a third state, the other state in the competition must aid the third state in order to maintain the power balance between the original competitors. As with the pattern of direct competition, the process repeats over time with an increasing investment in resources from both states. From this process both of the primary states remain balanced and the third state remains free from domination. In the end BoP rests on the idea that uncertainty maintains peace. Under a BoP neither state can gain a measureable power advantage. Under this condition neither state can be certain of victory and equality in power reduces the odds of victory to a coin flip. Given this uncertainty over who will win each state is cautious and peace is preserved (Filson and Werner 2002, 2004; Reiter 2003).

How states maintain the BoP is complicated and often misunderstood. The traditional focus is on states balancing the scales through adding to their own power. The study of how states do this has created a niche literature on the arms versus allies debate. A state can respond to a rising competitor by either expanding its own military forces (internal balancing) or seeking allies against its challenger (external balancing). The decision is not easy and depends on different circumstances. In the end arming takes time and resources, but the state maintains control over the tools of its own security. Allies provide a fast method of augmenting power, but one has to question if an ally will honor its agreements (Mearsheimer 2001; Leeds and Savun 2007).

The other approach to balancing is to reduce an opponent's power; balance is created by lightening the other side of the scale. The primary tool here is war. States will engage in conflict to reduce their opponents' level of power and ensure they cannot gain dominance (Kadera 2001; Mearsheimer 2001; Levy 2004). This is an important point as BoP predicts conflict as a potential outcome if the balance starts to shift. The results of the conflict should be to restore the balance and maintain stability. Allen and Martinez Machain (2018) highlight force-specific conflictual consequences that emerge when a balance breaks down. They find that in international crises, power-preponderant states are "more likely to use air power instead of ground troops in a conflict" (157), and when facing high-salience threats, strong states are in fact "willing to utilize ground troops alongside air power" (159), suggesting that preponderant states liberally use impressive levels of force to resolve disputes. Taking a different approach, Kadera and Morey's (2008) dynamic model incorporates three different types of conflicts: counterforce, industrial competition, or peaceful competition (cold war). Their model thus suggests power relationships can determine the type of conflict, provides an alternative logic for why weaker states fight, and demonstrates how the type of conflict can affect power distribution dynamics.

Classic versions of BoP theory assumed that states gaining power would be seen as threatening and create pressure for counterbalancing. However, cases of powerful states not facing a counterbalance led to refinements of the basic theory. Walt (1987) argued that states do not blindly balance power; instead, they balance against threats. Power is certainly a part of threat, but so are intentions. States that were powerful but lacked hostile intent would not be seen as threatening and therefore would not require balancing. In a similar vein, Levy (2004) contends that states balance against land power and not against sea power. This is a refinement of the idea of threat as land powers have the ability to occupy territory and threaten the existence of states. Sea powers lack the ability to take and hold land, so they are less threatening.[3]

The lack of balancing against the United States has forced an expansion in balancing theorizing. Paul (2005b, 2018) claims that states are using "soft balancing" against the United States. Soft balancing is when states "develop diplomatic coalitions or ententes with one another to balance a powerful state" (Paul 2005b, 58). Working through diplomacy and international institutions, states constrain the powerful, potentially threatening state without directly resorting to formal alliances of military conflict. The powerful state cannot achieve its goals and is denied the fruits of its dominant position. The soft balancing argument keeps the motivation of states to avoid domination, but changes the tools of balancing.

SYSTEMIC PARITY AND PEACE

Taking a different approach, several systemic theories also argue that parity creates peace. Early theories of conflict, heavily influenced by Waltz (1964), focused on systemic conditions of war and peace. The primary focus of these studies was the polarity of the system. Scholars primarily emphasize the number of major powers in the system when conceptualizing polarity (Deutsch and Singer 1964; Waltz 1964; Wohlforth 1999). However, others envision polarity as the number of blocs in the system (Wayman 1984). Logics propose that the level of global conflict rises or falls as the number of global power centers shifts.

Deutsch and Singer (1964, 390–91) argue that the international system will be more stable under multipolarity. They see stability as a world with no large-scale wars, with a low probability of limited wars, with most states continuing to survive, and with no single state rising to dominance.

According to Deutsch and Singer, two primary factors produce peace under multipolarity. First, as the number of actors increases, so do the opportunities to interact. In these interactions states will not firmly align with others: sometimes they will side with another state and other times they will disagree. The pattern of agreement and disagreement creates cross-cutting pressures between states. States will avoid being too conflictual with others on issues on which they disagree in order to maintain a working relationship for issues upon which they do agree. Second, Deutsch and Singer argue that as the number of actors increases, the amount of attention each state can focus on another actor decreases. As a result, states cannot devote time and energy to escalating hostilities. While these two processes form Deutsch and Singer's core argument, the underlying driver of both is the systemic BoP. Furthermore, they state that in a multipolar system, states realign to balance out any rising state. With a large number of poles, states can readily realign to achieve a new balance. Peace is secured through the power balance and the uncertainty it creates for states.

Interestingly, despite arguing that multipolarity is stable, Deutsch and Singer (1964, 405) do not believe it is durable. It is prone to breaking down and unable to last for the long term. Drawing on the case of Prussia's unchecked rise in 1864, Goddard (2009) suggests a particular mechanism by which a multipolar system shifts to a more polarized one: a rising state might use legitimation strategies to justify its expansion and stave off balancing coalitions from forming against it, much like a rising challenger co-opts the hegemon in PT scenarios.

Waltz (1964) begins by arguing that stability is best defined as peaceful adjustments (limited violence) and durability (the system's ability to remain in the same state). He contends that bipolarity, or a system with two large powers that are roughly balanced with a clear preponderance over other states, is the most stable. Like Deutsch and Singer (1964) he focuses on the role of uncertainty, but forms a much different theoretical argument.[4] For Waltz, two superpowers reinforce an existing bipolarity as they contest each other globally (no peripheries) and across issues (e.g., economics and education), so that neither state can gain an advantage. Unlike under multipolarity, one state gaining in any area becomes obvious to the other that has fallen behind. This zero-sum situation motivates each side to keep up and maintain the balance. Further, the superpowers engage in recurring crises, demonstrating their willingness to resist each other. Knowing the other side will resist tempers each superpower's level of threats, helping to avoid full-scale war. Crisis involvement makes each side credible, and that credibility makes both poles cautious. Each of the bipolar states is so powerful it can absorb changes in the BoP without resorting to war. Small gains for one side or the other do not mean the other has lost its security. Because of sheer size, each state can absorb small shifts without turning to military conflict. In the end the certainty of meeting resistance and the uncertainty of the outcome of a war create a precarious peace where crisis is constant, but war is avoided. The system is durable because each of the leading states is so strong there is little hope of other states catching up and shifting the system to multipolarity.

DYADIC PREPONDERANCE AND PEACE

As with theories connecting parity with peace and stability, arguments that peace is best preserved through one state holding a preponderance of power come in many forms. The theories propose a series of assumptions that go beyond the basic idea that an imbalance of power maintains peace.

Organski (1958), breaking from classical realism, assumed that order exists within the international system. States are arranged in a hierarchy based upon their power. At the top is the system leader, the unquestionably most powerful state in the system. In the next level are the challengers. Contenders are strong states that cannot directly rival the system leader but are close enough in power that with time, they could challenge the hegemon. Below the contenders come the major powers, followed by all of the other states. The hegemon determines the rules of the international system, which it structures to benefit itself. States' level of satisfaction with the hegemon's rules further divides states. Some contenders are satisfied with the system and seek to maintain it, but the dissatisfied contenders wish to alter the system to their advantage. War becomes likely in the presence of states that are both strong and dissatisfied. Organski envisioned a world with states all growing at different rates. States start by growing very slowly, then move into a transitional S-shaped growth phase where economic development accelerates, and finally the rate of growth declines as the state reaches maturity. Since different states grow at different rates and they start at different levels of initial development, the outcome is a mix of states growing at various rates. While a power hierarchy characterizes international politics, that hierarchy is unstable. Over time states rise or fall within the hierarchy as their rate of growth changes. Thus power transitions occur when a state from the challengers closes the gap with the system leader.

The second core part of PT, satisfaction, has proven harder to define and the lack of clarity has been criticized as a weakness within PT (see DiCicco and Levy 1999).[5] Organski (1958) conceptualized that states will be dissatisfied when they do not have an adequate share of the system's resources, but this defines satisfaction in a way that overlaps with capabilities, rather than creating a distinct conceptualization. Lemke and Werner (1996) use military buildups to measure dissatisfaction: quick rises in military capabilities signal a state is unhappy and wishes to change the current distribution of resources. Another approach leverages alliance portfolios: the more a state shares similar alliances with the system leader, the greater its satisfaction (Kim 1991; Lemke and Reed 1996). Both approaches have been criticized because neither discusses the sources of dissatisfaction or measures it directly (Danilovic and Clare 2007). Lemke and Reed (2001) argue that regime type indicates satisfaction and states with similar regime types share interests with the system leader. Danilovic and Clare (2007, 292–93) claim that states define their interests in terms of influence within certain regions and that "dissatisfaction arises if two or more major powers clash over their influence in the same regional area." This argument provides a direct source and measure for dissatisfaction. Sample (2018b) develops a multilevel argument for satisfaction that incorporates the level of satisfaction with the system overall and the distribution of benefits within the dyad.

Lemke and Werner (1996) expand the scope of PT by describing a series of regional hierarchies as well as one global hierarchy. The power transition process that operates globally also operates within each regional hierarchy. When a dissatisfied regional challenger reaches parity with a regional hegemon, it will initiate a conflict to establish regional dominance. Eventually the logic was applied to any dyadic BoP. Thus the logic of PT could be expanded to explain a larger number of wars between a diverse number of states, and measures of parity are now a standard control variable in most models of war (Geller 2000; Gibler 2017a).

SATISFACTION AND PHOENIXES IN PREPONDERANCE AND PEACE

Several scholars examine nuances of PT theory and offer important extensions. Organski (1958) originally postulated that challengers initiate wars against the hegemon because they are dissatisfied with the hegemon's regime and want to establish their own and are finally strong enough to do it. If the joint occurrence of dissatisfaction and power transition predicts wars, then removal of either condition should help states avoid wars. Some analysts (e.g., Papayoanou 1997) have argued that China's contemporary rise in military and economic power should not worry the United States as long as it co-opts China by engaging it in the rewards of economic interdependence and free trade. Bussmann and Oneal (2007) examine whether hegemons do take a satisfaction approach by providing private goods and find that while the dominant state's allies have improved chances of winning defensive wars, they do not appear to be protected from aggression and do not enjoy economic growth benefits. These findings question the universality of PT's core assumption about what motivates states to fight.

For similar reasons, some scholars have attempted to measure dissatisfaction and use it with power parity to predict dyadic war initiation. De Soysa et al. (1997) argue that power is a proxy for satisfaction, meaning no separate measure is needed. But Lemke and Reed (1998) contend that Organski (1958) clearly intended for the two concepts to be distinct, pointing out that some states prosper despite the hegemon's regime while others prosper because of it, leading to a mix of dissatisfied and satisfied potential challengers. Sample (2018b) successfully tests her theory positing that dyadic conflicts over territory generate dissatisfied potential challengers whereas systemic wealth generation produces satisfied potential challengers. She contends that growth in wealth distinguishes satisfaction from power. At the system level satisfaction is about a state's position in and preferences for the dominant state's regime. At the dyadic level, Sample (2018b) says that satisfaction is about how each state is positioned globally and about the uncertainty of the dyadic relationship and whether the two states have a history of conflict.

A second extension questions the impact of war on power levels. Both BoP and PT see direct conflict as a means of altering the power balance between states. Research found that the power levels of war winners (and states that did participate) are only marginally different than what we would expect had the war not occurred (Organski and Kugler 1977). Further, Organski and Kugler found that defeated states tend to have accelerated growth rates after the war, like the mythical phoenix rising from the ashes. After about fifteen to twenty years power distributions return to what they were prior to the war. In essence, war causes only short-term alterations in power growth and overall power levels.

If true, this finding calls into question the motivation driving both PT and BoP. In both theories states use war to alter the power balance. But if war has no appreciable impact on long-term power balances, what motivates states to fight? Many answers are possible; leaders may misunderstand the true impact of war or short-term benefits may be what they seek (this would be partially consistent with PT's explanation that the rising state uses war to speed a transition). It may also be the case that altering the power balance is not the goal and that the shifting power between states creates or intensifies bargaining problems that eventually lead to war (Fearon 1995).

SYSTEMIC PREPONDERANCE AND PEACE

Hegemonic stability theory has many similarities with PT but proposes a different theoretical process. Hegemonic stability theory (Modelski 1987; Gilpin 1988; Modelski and Thompson 1988) sees a dominant state, the hegemon, as the key to global peace. The hegemon emerges from a global war that determines the new system leader (hegemon) and a new stable hierarchy is formed. Over time this hierarchy erodes either through the rise of new global issues that call into question the hegemon's ability to lead or through the rise of new challengers via a process similar to what Organski (1958) described. Eventually the system breaks down into competing sides, resulting in another global war and the determination of the new hegemon. In between the cycles of hegemonic war, the hegemon rules over a peaceful system and its leadership role ensures general levels of peace.

Taking a polarity approach, Wohlforth (1999, 7–8) argues that the most stable system is where there is one leading state, or unipolarity. He defines stability similar to Waltz (1964) as prone to peace and durability. The presence of one clearly dominant state generates peace through certainty of outcome similar to PT. No other state is strong enough to defeat or even to compete with the leading state. Absent any hope of victory, no state risks war with the unipolar power and instead bandwagons. The presence of a unipolar power also reduces pressure on the other system members as it eases the pressures of the security dilemma. If another state begins to rise in power other states do not have to fear or take countermeasures. If war comes the unipole will essentially determine the winner of any war by choosing a side. Once the leading state takes a side the other state has no hope of victory. Shifts in the BoP are thus no longer threatening as states enjoy the safety the unipolar power provides. In the end, certainty of outcome maintains the peace systemwide.

Kadera, Crescenzi, and Shannon (2003, 240) break with past research and argue that who holds power internationally matters. Building a dynamic model of democratic peace at the system level, they contend that as the strength of the democratic community increases, democratic states are more likely to survive and the negative effects of conflict on democratic survival decreases. Here the focus is not on poles or power concentration, but on who holds power. Thus the system is reconceptualized away from realist notions to focus on international communities of states. Further, Kadera et al. (241–42) redefine power as the strength of a democratic community measured by interacting how much military power democratic states possess, the number of democracies in the system, and the intensity of the democratic regimes. In the end power is more than raw military power; it is at least partially determined by ideas and by how strongly a state holds an idea.

Goh (2013) also advances the systemic preponderance research agenda, arguing that hegemony maintains peace not through material capabilities alone, but by leveraging social relationships. She focuses on the United States' continuing hegemony in East Asia, maintained by co-opting a rising China and other regional states by renegotiating the international order along the way.

SUMMARIZING PARITY AND PREPONDERANCE

From these different perspectives theories of preponderance and parity come to very different conclusions (Table 2.1). First, theories of parity are based on the idea that balance creates peace while theories of preponderance side with the view of power imbalance as peaceful. Further, parity theories vary in hypothesizing whether the weaker or stronger state will attack, with BoP being the clearest in predicting that the stronger state poses the threat. Power transition theory sees the weaker state initiating a conflict before the transition in order to speed the transition and gain dominance faster. From these viewpoints policymakers interested in maintaining peace are given clashing recommendations. Balance of power theory emphasizes states investing national resources to ensure the balance is kept. Power transition theory does not believe states have much control over growth rates, at least in the long run, so the focus is on satisfying a rising challenger.

Table 2.1. A Summary Comparison of BoP and PT Logics

	BoP Theory	*PT Theory*
When is war most likely?	Inequality	Equality
Which state is most conflictual?	Stronger	Rising (Weaker)
Who wins?	Stronger	Challenger (Weaker)
How to win	Have strongest military	Prevent challenger from gaining too much strength OR Satisfy challengers through policy concessions
How to avoid conflict	Maintain power balance	Maintain power preponderance, satisfaction

EMPIRICAL EVALUATIONS OF PARITY, PREPONDERANCE, AND PEACE

Despite the different assumptions and conclusions, it has been difficult to adjudicate empirically between PT and BoP. Organski and Kugler bemoan the stubbornness of the data to provide a conclusive answer. Looking at all major power and contender dyads, Organski and Kugler (1980) find support for both BoP and PT. Overall, they find that a system that includes equal and overtaking states is in the most likely condition to spawn war. Using a wider sample, Houweling and Siccama (1988) find essentially the same thing. Power transitions among great powers are dangerous, but preponderance or stable balances are peaceful. Geller (2000) reviews the empirical evidence regarding parity and war and finds two works that support or partially support BoP compared to nine studies that find support for PT. Geller goes further and reviews the literature on the rate of change and the probability of conflict. Here the evidence was again mixed, with studies finding support for and against the proposition that rate of change may drive war onset. Geller finds that the weight of the evidence supports PT: "The analysis indicates a growing and cumulative body of evidence pointing to the salience of both static and dynamic capability balances for the occurrence and initiation of militarized disputes and warfare" (268).

Empirical work since 2000 has continued in the same inconclusive pattern, but may have shifted the argument somewhat back toward BoP. Sweeney (2003) studies dispute severity instead of onset and finds that when a dyad has dissimilar interests, parity is associated with less severe conflicts. Alsharabati and Kugler (2008) introduce a formal model of PT in order to provide a coherent theory bringing together many disparate parts of the PT research agenda. Empirically the novel contribution is the finding that the further you are from a transition in time, the more the probability of war declines. So transitions are dangerous, and the closer you are to one, the greater the probability of war. Gibler challenges many past studies finding that most of the parity and war relationship can be accounted for by a few outliers. By eliminating three rivalry outliers (United States-Russia,

China-Russia, and North Korea-South Korea) or removing small island nations with little opportunity for conflict, the effect of parity on conflict disappears statistically. Looking exclusively at interstate war Gibler (2017a, 35) finds twelve cases of war between roughly equal powers compared to twenty-nine wars where one side had a power preponderance. Finally, while not directly addressing the PT-BoP debate, work by Bell and Johnson (2015) recasts the debate by focusing on future power shifts. Using a rationalist frame, Bell and Johnson argue that states worry about the state of future power relations instead of maintaining an exclusive focus on current power levels. Dyads that expect a large future shift in power balances suffer from commitment problems and cannot reach peaceful settlements. Facing a continuing deterioration in the power balances, a dominant state is more likely to launch a preventive strike knowing it will face future demands from a challenger once the challenger is stronger.

Overall, empirical tests of preponderance versus parity have proven inconclusive, either supporting none of these arguments (see Bueno de Mesquita 1978) or providing support for one perspective without necessarily ruling the other out.

One reason for the confusing state of findings regarding power parity and preponderance is that past research has largely not taken into account the role that research design plays in reported findings. Specifically, the decisions of how to measure the concept of power ratio and how to structure the sample have important implications. First, the existing literature presents a number of different ways the power ratio has been operationalized. The most basic form is to divide one state's power by another; however, even this simple form shows disagreement with some studies dividing the weaker by the stronger (to create a variable ranging between 0 and 1), while other works divide the strong state's power level by the weaker state's level (creating a measure with a minimum of 1). A second group divides one state's power level by the total power level within the dyad. Again, studies are split on if the stronger or weaker state goes in the numerator. A third approach is to take the natural log of one of the measures already mentioned and use that as the measure of the power ratio. When using a directed-dyad design the same patterns repeat but with the referent state always in the numerator and the denominator being either the power ratio of the other state in the direct dyad or the sum of both states' power levels. The same pattern appears in some works employing the natural log of these values. This plethora of measurement schemes makes comparison, and thus cumulative knowledge, very difficult. The problem is made worse by the fact that few scholars explicitly discuss why they adopt one measure over another or consider how it relates back to the theoretical proposition under examination.

A second issue is the choice of dependent variable. Reed (2000) finds that power parity is positively associated with the onset of militarized interstate disputes (MIDs), but reduces the odds of escalation to war when modeled jointly. Thus, the relationship between parity and conflict depends on how one conceptualizes conflict (initiation or escalation).

The third issue is whether to use a dyad or directed-dyad research design. The majority of studies use dyads, as has been the norm in conflict studies for a number of years. However, a number of studies test PT/BoP using directed dyads. This is an important consideration that has not received enough attention, especially since PT and BoP not only predict when a dyad should experience conflict but also predict who should initiate it. The dyadic design has clear limitations when it comes to answering questions of initiation. The issue of sample construction also adds to the confusion of the current state of the evidence. Using a comparable measure of power ratio, we draw opposite conclusions when looking at dyads or directed dyads.[6] Table 2.2 demonstrates this using a standard model of conflict.

In Table 2.2 we report results for two different probit models. In column 1 the sample is relevant direct dyads while in column 2 the sample is relevant dyads. The dependent variable is MID initiation. For the dyadic analysis, the power ratio equals the stronger state's CINC score divided by the sum of the two states' CINC scores. To aid in comparison, the same procedure is used to measure the power ratio for directed dyads only. State 1's (the referent state) CINC score is always in the numerator. In both of these calculations a score of 0.5 represents a dyad where states are equal in power. The further the score is from 0.5, the more asymmetric the power relationship.[7] The control variables are standard to conflict models and represent dyadic traits that do not vary between dyads or directed dyads, keeping the models as similar as possible.

Table 2.2. Power Ratio and MIDs: Dyads versus Directed Dyads

	(1) *Directed Dyads*	*(2)* *Dyads*
Capability Ratio	0.2300***	−0.9223***
	(0.0001)	(0.0001)
Major Power Dyad	0.5763***	0.4412***
	(0.0001)	(0.0001)
Joint Democracy	−0.2931***	−0.3527***
	(0.0001)	(0.0001)
Allies	−0.1656***	−0.2076***
	(0.0002)	(0.0002)
Shared Border	0.6471***	0.5826***
	(0.0001)	(0.0001)
Year	0.0019***	0.0019***
	(0.0001)	(0.0001)
Constant	−6.2823***	−5.0490***
	(0.0001)	(0.0001)
Observations	163,110	81,555
log likelihood	−10,396	−8,677

The dependent variable in Model 1 is Initiate; in Model 2 it is Occurrence.
The sample is politically relevant dyads.
*** $p < 0.01$ ** $p < 0.05$ * $p < 0.1$ (two-tailed), p-values reported
Errors are clustered by dyad or direct dyad.

In both models the capability ratio is significant at the 0.0001 level; however, the signs on the coefficients reverse. When looking at directed dyads the probability of a state initiating a MID increases in relation to its power advantage. This would appear to support BoP since conflict is more likely when states are unequal and grows in probability as the difference increases. However, in the sample of dyads we would draw the opposite conclusion and find support for PT. As the stronger state increases its advantage, the probability of a MID occurring decreases. The simple design choice between dyads and directed dyads changes the inferences we draw and shifts support from one theory to the other. The fact that most studies in the PT-BoP debate use a dyadic design partially explains why past reviews of the literature find a preponderance of the evidence supporting PT. However, this appears to be based less on true scientific evidence and more on the early popularity of dyadic research designs.

Examining empirical patterns at the systemic level, Mansfield (1992) argues that the level of concentration of power within the system determines the levels of systemic conflict. Connecting pieces from Waltz (1964) and Deutsch and Singer (1964), Mansfield argues that power concentration and war exhibit a non-monotonic relationship with an inverted-U shape: high and low levels of concentration experience low levels of war, and mid ranges experience high levels of war. When power concentration is low many potential counter coalitions can form to oppose an aggressive state, and the benefit of fighting would be low as capturing a small state would not greatly alter the power level. When power concentration is high powerful states must check each other, and the gain from defeating a smaller state would not greatly alter the power balance. However, in the middle range fewer counterbalancing coalitions can form and no strong states are available to oppose rising powers. Mansfield finds empirical support for the inverted-U hypothesis and that large changes in power concentration are more war prone.

Resolving these issues is beyond the scope of this chapter; however, a few conclusions are clear. First, careful thought needs to be directed toward the best way to measure the capability ratio between states. The variety of different measures and methods makes it difficult to compare across studies and develop a true sense regarding the state of the literature. Simply adopting the "standard" approach can no longer be considered sufficient justification for employing one scheme over the others. Researchers must carefully consider the options and defend their choices. Second, the same careful examination process must occur when selecting the sample for a research design. Relying on dyads because that is what has been done in the past is insufficient. The justification of sample in relation to theory needs to take on greater importance when testing between BoP and PT.

Motivated by the contradictory evidence favoring both BoP and PT, Kadera (1999, 2001) created a model incorporating both theories into a general model of power and conflict. Using a system of differential equations, she drew elements from each theory to model the relationship between power and conflict. Results from the model indicate three different types of power transitions, each with its own conflict dynamics. In the first type the challenger surpasses the dominant state as envisioned by Organski (1958). In the second type the challenger rises close, but is unable to complete a transition. Finally, in some cases a transition occurs, but the rising state cannot maintain its position and is in turn passed by the dominant state (a double transition). The model also finds that the conflict levels between the two states will become coupled; in other words, each state will mirror the conflict level of the other. Instead of a clear initiator as the BoP and PT theories predict, Kadera's model expects that each state builds in their level of conflict directed at the other. Finally, in many cases the winner of a transition is not the most conflictual state. In the end the lack of clear empirical results may be an issue of needing better theory that can explain the seeming contradiction in the data. As Kadera shows, this can also lead to novel insights found in neither BoP nor PT.

WHAT WE NEED TO KNOW

Several avenues for future research on the relationship between power distributions and war emerge from this review.

First, we lack a clear theory for the functional form of the relationship between (dis)parity and war propensity. Most statistical tests of power and war assume that war becomes monotonically more likely as power shifts away or toward a certain point (balance or preponderance). This does not fit with a close reading of the major theories. Power transition theory says war becomes more likely as the dissatisfied state approaches parity, while balance of power theory predicts conflict as one state gains an advantage. This is different than saying the probability of war increases the closer a challenger gets to the zone of parity or dominance. A weak state may gain power relative to the leader but remain well short of parity; in this case the power shift should have little or no impact on the probability of war. However, once a state reaches rough parity with the leader, a small increase in relative power could have a major impact on the probability of war (see also Hegre 2008). The same logic applies to balanced dyads. This mode of testing also assumes that the change in the probability of conflict continues to change in the same direction and at the same rate as power balances continue to shift. In the case of a balanced dyad, as one state gains dominance war becomes more likely and the probability of conflict continues to increase as the dominant state gains an ever-larger power balance. This is unlikely as at some point the now dominant state would possess such a large power advantage that the probability of war would start to decline. Future work needs to take these dynamics into account and either build linear theories of power and war or adopt a nonlinear approach.

Second, an integrative or noncompetitive view of BoP and PT theories suggests two peaceful conditions, not one. Preponderance is peaceful, and so is a stable balance. In Organski and Kugler's book (1977) the "equal, no overtaking" condition rarely led to war. Similar results are obtained in de Soysa et al. (1997). However, most empirical studies of war only measure and include preponderance as a predictor of war.

Third, not only do we need to consider the differences between stable parity (peaceful balances) and unstable parity (war-prone power transitions), we need to consider another way in which equality conditions can vary. Specifically, strong states that are equal may behave differently from mid-tier states that are equal and from weak states that are equal. Kadera, Kim, and Ring (2010) suggest that strong and equal dyads can escalate to higher levels of violence and endure longer militarized disputes.

Fourth, we currently have little theoretical or empirical guidance on who wins power transitions and how. Challengers, such as Germany, sometimes lose. What predicts who wins a power transition, and is conflict initiation or escalation by that actor a reliable predictor? And if initiators tend to lose, why do they initiate in the first place? It seems irrational for them to do so.

Fifth, scholars should theorize more about power as a dependent variable. Recent advancements in the literature suggest this path would be fruitful. For example, Gibler (2017a) demonstrates that the state-building process produces state entry into the system, and that this is associated with regional convergence to parity.

Sixth, and relatedly, scholars should theorize more about war as an independent variable. War produces winners and losers, creating dyadic power relationships and systemic power distributions. As empirical work on the phoenix phenomenon demonstrates, the complex relationship between war and power shifts produces outcomes not predicted in leading formal models. How then do power and war at the dyadic and global levels constitute a feedback system?

Finally, how should we limit or expand the scope conditions? And what general framework can we adopt to guide the answers across different types of states, different types of dyads, different regions, different types of actors, and different types of conflict? Some work has been done extending BoP and PT arguments to state dyads in a Latin American hierarchy (Lemke and Werner 1996), to Asia (Kim 2002), and to political actors in Africa (Lemke 2002). Some scholars also argue that we can even apply similar logics to civil wars (Toft 2007). Should we adopt unifying theories that leverage core assumptions around which all such extensions can be made? Or should we let each application evolve in unique ways that produce a family of loosely affiliated power-based theories?

NOTES

1. We thank Paul Diehl for this observation.

2. This is similar to methodological problems associated with the study of deterrence. How do we know when a state actually attempts deterrence? One standard approach has been to limit the sample to rivals and assume they would both attempt to deter one another. While this approach is widely accepted, it does limit our ability to draw general inferences.

3. From this logic Levy (2004) argues that BoP should be thought of as a theory of behavior for the European subsystem and may not be applicable to other regions.

4. See Wayman (1984) for an attempt to reconcile these two perspectives.

5. One possible way forward would be to link the idea of satisfaction with understandings of revisionists and status quo states within the rationalist literature. Essentially dissatisfied states in PT are revisionist states in rationalist models; drawing clear connections would bring rationalist insights into the PT-BoP debate. We thank Paul Poast for this suggestion.

6. We thank Sara Mitchell for bringing this to our attention.

7. For directed dyads, the range of the capability ratio is 0 to 1; for dyads (using the stronger state in the numerator), the range is between 0.5 and 1.

Chapter Three

Deterrence Theory and Alliance Politics

Michael R. Kenwick and Roseanne W. McManus

Do alliances cause war or promote peace? Variants of this question have persisted since antiquity (Thucydides 1972), and some of the earliest contemporary efforts to scientifically analyze military conflict were dedicated to examining the link between alliances and war (Singer and Small 1966b). Nonetheless, few questions in the study of war have been so frequently asked and yet remain so clearly unsettled.

Much evidence appears to suggest alliances contribute to peace. The bulk of recent history has been characterized by peace, and a wealth of empirical findings indicate peaceful times coincide with the presence of strong, defensive alliance structures (Levy 1981; Siverson and Tennefoss 1984; Leeds 2003). The apparent strength of this relationship has led some to present alliances as a potential "prescription for peace" (Johnson and Leeds 2011). And yet we have also seen cataclysmic deterrence failures and instances where alliances seem to exacerbate tensions and hasten conflict onset (Senese and Vasquez 2008). Great power alliances did not prevent the onset of the First and Second World Wars and actually catalyzed their spread.[1] More broadly, evidence suggests that forming new alliances may instigate a period of hostility before any deterrent effect takes hold (Morgan and Palmer 2003; Kenwick, Vasquez, and Powers 2015; Kenwick and Vasquez 2017).

The decades-long debate and competing evidence suggest that the question of whether alliances promote war or peace is overly simplistic. Even those most optimistic about extended deterrence would permit that defensive alliances do not always deter and that their failure can have catastrophic effects.[2] On the other hand, skeptics must contend with the question of why states would invest significant resources in forming alliances if they were clearly inefficacious (Morrow 2000; Poast 2019a). In our review of the literature on alliances and war we therefore reframe the question in what we think is a more fruitful way: *under which conditions* are alliances most likely to either incite or deter the initiation of conflict?[3]

Our review highlights several conditions that might be relevant. Existing theory suggests that alliances are more likely to deter when they are associated with stronger sunk cost signals that enhance their credibility and when the political and reputational costs for abrogating the alliance are higher. They are less likely to deter when the distribution of power or the interests of leaders begin to shift. Alliances become more likely to actually incite violence when they embolden their signatories to act provocatively through a moral hazard problem or exacerbate mutual tensions by hastening a security dilemma, although it is difficult to predict precisely when these conditions will exist. Although this literature review enables us to begin to answer the question of when alliances provoke or deter, further research is necessary to more precisely predict when alliances will provoke, deter, or have little effect on the probability of war.

Analyzing the effect of alliances on war is particularly important to contemporary international relations. Recent findings that the deterrent capacity of alliances weakens amidst shifts in power sit uncomfortably with the fact that the international alliance system has remained surprisingly fixed amidst substantial changes to the distribution of power in the past several decades.[4] The advent of nuclear weaponry has seemingly buttressed deterrence, but also raises the costs of failure to an unfathomable level. Against this backdrop states are increasingly turning to alternative means of signaling resolve to ensure joint security, but little is known about how these dynamics will shape the prospects for peace. Our review engages with these tensions as it seeks to identify avenues for future research that can illuminate the problem of extended deterrence in the modern international environment.

ALLIANCE FUNCTIONS AND EXTENDED DETERRENCE

Few states can guarantee their own security without coordinating with others. Alliances provide focal points for this coordination and are especially critical to states lacking the conventional or nuclear capabilities to deter adversaries. In keeping with previous research, we define alliances as written agreements between official representatives of two or more countries committing one another to some form of military action, in the case of some contingency.[5] Alliances are more than an informal, unwritten understanding. Their terms are explicitly written down and typically made public in an attempt to deter security threats from third parties. The contingent nature of alliances also differentiates them from defense cooperation agreements, whose terms are typically targeted at enhancing cooperative behavior without committing to any action in the event of armed conflict (Kinne 2020, 735). By contrast, the cooperative nature of military alliances is inseparable from their relation to conflict. Alliances can be compellent or deterrent, defensive or offensive, but all contain contingencies related to conflict.

Alliances serve many purposes. Canonical treatments of alliance formation typically interpret alliances as mechanisms of capability aggregation—states form alliances in an attempt to pool their resources against the power or threat posed by a common foe (Waltz 1979; Walt 1987; Mearsheimer 2001; Fordham and Poast 2016).[6] Beyond pooling military resources, alliances also improve states' ability to coordinate with each other and plan mutual operations in the event of war (Poast 2019). Yet alliances do more than aggregate and coordinate. They also formalize hierarchies in the international system (Lake 2009). Minor powers may give up some autonomy to a major power, exchanging favorable policies for security guarantees (Morrow 1991). Even more broadly, certain types of alliances can reduce territorial threat and facilitate transitions to democracy (Gibler and Vasquez 1998; Gibler and Wolford 2006).

The alliance function that is most relevant to this volume is extended deterrence. Successful deterrent alliances should make potential adversaries less likely to initiate and escalate hostilities. How do alliances induce deterrence? For classic models of deterrence, the reason lies in how an alliance affects the adversary's expected war outcome (Morrow 1994; Smith 1995, 1998; Fearon 1997). Assuming for the moment that an alliance is known to be reliable—that is, there is no doubt about whether allies will fulfill their obligations—its presence should make the cost of war higher and the expected outcome less favorable for any attacker. Instead of fighting one state, the potential attacker faces the prospect of also fighting its allies. While the most resolved adversaries might still be undeterred, less resolved adversaries will back away from conflict.

The Bargaining Model Approach

Considering alliances within the context of the bargaining model of war complicates this story. This perspective assumes that attacks result from the failure of peaceful bargaining. At the bargaining table, an alliance may reduce the likelihood of a challenger making demands of a target state, but the added capabilities on the target's side increase the target's propensity to resist demands. This makes theoretical predictions about alliances' effect on the outbreak of war ambiguous (Smith 1995).

In order to better understand this ambiguity, a stylized expansion of Fearon's (1995) bargaining model is insightful. Fearon contends that war is costly for both parties and therefore inefficient. Potential disputants can, and often do, avoid the costs of war by reaching a peaceful bargain. This framework is stylized in the top left panel of Figure 3.1, where State A and State B are bargaining over some divisible good, represented by the line. Each side prefers a maximal amount of the good. The good could be divided through war, but war imposes costs on each side, which are equivalent to losing some portion of the good. These costs create the bargaining range. Both sides would prefer deals in the bargaining range to fighting because these deals give them a settlement roughly approximate to the likely war outcome without paying the costs of fighting.

To extend this logic to the domain of deterrence, consider three states with slightly modified roles: a potential challenger (State A) that desires to change the status quo, a potential target (State B) that seeks to defend the status quo, and a third party (State C) that has an alliance with State B. Even if all parties view

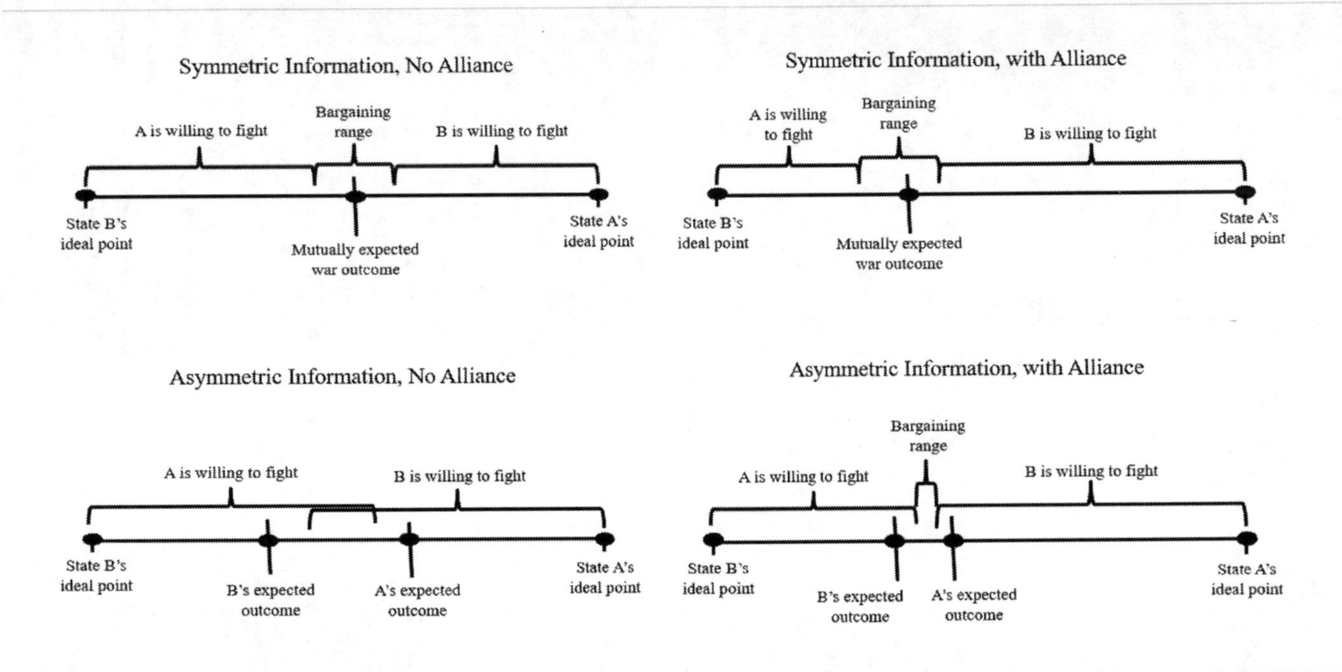

Figure 3.1. Alliances and the Risk of War.

the alliance as reliable and agree on how the addition of State C changes the expected war outcome, it is not clear that the addition of the alliance will either increase or decrease the size of the bargaining range. Rather, it may simply shift the bargaining range in a way that is more favorable to State B, as illustrated in the top row of Figure 3.1.

In keeping with the bargaining model of war, Figure 3.1 assumes State A initiates bargaining prior to deciding to attack State B. If State B has an alliance, then the expected outcome of fighting is less favorable to State A, which therefore becomes willing to accept less at the bargaining table. The reverse is true of State B, which now expects more and whose refusal to offer an acceptable bargain can prompt State A to attack. The shift in the expected war outcome caused by the alliance therefore does not necessarily imply any change in the probability of war, although it does increase the likelihood that any negotiated bargain will be more favorable to State B.

While the expected outcome determines the location of the bargaining range, the width is determined by each side's cost of war (Fearon 1995). Having an alliance is likely to lower the cost of war for State B and raise it for State A. This is represented in the upper-right panel of Figure 3.1 by the fact that the bargaining range is no longer exactly centered around the expected war outcome. While the increase in A's cost need not be exactly symmetric with the reduction in B's cost, the costs will move in opposite directions, with the changes partially or even entirely cancelling each other out. The number of possible bargains that both sides prefer to war is therefore not expected to reliably increase or decrease.

The bargaining model is useful because it demonstrates the tenuousness of a common line of reasoning in the alliance literature: that by increasing the costs of conflict, alliances induce deterrence and prevent war. This model highlights the importance of mutual expectations. Provided all parties agree on how a war would end, the addition of an ally does not necessarily increase or reduce the risk of an attack. Therefore, if we define deterrence success as reduced likelihood of conflict initiation, it is not yet clear that alliances improve deterrence success.

Adding Incomplete Information

Continuing within the bargaining model of war framework, we argue that if alliances reduce the likelihood of war, the reason must be related to one of the rationalist causes of war that Fearon (1995) contends prevents states from reaching a bargain. Out of Fearon's rationalist causes of war—the indivisibility problem, the commitment problem, and the information problem—we claim that alliances are most likely to ameliorate the information problem. The information problem refers to a situation in which two parties disagree about the expected outcome of war and therefore cannot agree on what would constitute an equitable bargain ex ante. This occurs because both sides have private information about their capabilities and resolve and have an incentive to misrepresent this information to their adversary. Therefore, instead of the universally agreed-upon expected outcome portrayed in the top panels of Figure 3.1, a more common situation is probably for each side to have different expectations about the outcome, as portrayed in the bottom left panel.

The question becomes under which conditions alliances can effectively reduce information asymmetries. In most bargaining situations, regardless of whether there is an alliance, one of the factors that potential disputants will consider is whether a third party will assist one side or another. When there is high uncertainty about third-party involvement, the disputants' expectations about the result of fighting are likely to be farther apart, and the risk of conflict is heightened. For example, the bottom left panel of Figure 3.1 depicts a situation in which State B is confident of State C's intervention, but State A is not. This results in States A and B having very different expected outcomes and no bargaining range. However, if State B has an alliance with State C that is viewed as reliable, then instead of having to guess about whether State C will intervene, both State A and State B will be confident that it will. As shown in the bottom right panel of Figure 3.1, this shifts State A's expected outcome closer to State B's and enables a bargaining range to exist.

This suggests that alliances can induce deterrence by lowering information asymmetries and helping both sides agree on the expected conflict outcome. Considering the informational benefits of alliances enables us to square the traditional wisdom that alliances deter with the bargaining model framework. It also demonstrates how deterrence success may depend more upon an alliance's ability to credibly convey information than upon its ability to shift the balance of capabilities.

ALLIANCE RELIABILITY AND CREDIBILITY

"Don't we have our flank covered by Germany?"
—Franz Conrad, Chief of the General Staff
"Are you sure about Germany?"
—Austrian Emperor Franz Joseph I, July 5 1914[7]

When will alliances reduce uncertainty about who will fight together in a possible war? Two concepts are central to this question: reliability and credibility. Reliability refers to the *actual* likelihood that a third party will fulfill its alliance obligation to intervene if its ally is attacked, whereas credibility is the *perceived* likelihood of such intervention. Credibility, unlike reliability, is in the eye of the beholder; it can never be known with certainty. For deterrence to succeed, potential attackers must believe an alliance is credible.[8]

Many scholars have investigated the general reliability of alliances. Early research found that although allies are more likely to fight together than non-allies (Singer and Small 1966a), allies aided each other in war only about one quarter of the time (Sabrosky 1980; Siverson and King 1980). After taking into account that alliances often have specific provisions that obligate allies to intervene in some situations but not others, Leeds, Long, and Mitchell (2000) found allies fulfilled their obligations around 75 percent of the time between 1816 and 1944. In more recent years, however, alliance reliability appears to have dropped remarkably, with commitments only honored about 22 percent of the time between 1945 and 2003, according to a study by Berkemeier and Fuhrmann (2018b) that uses similar methodology to Leeds, Long, and Mitchell. Yet these numbers must be interpreted with caution since adversaries are disproportionately likely to challenge alliances they do not believe are credible (Smith 1995, 1996b).

Of course alliance reliability can be observed only after war breaks out. Prior to that point, the actual intention of an ally to intervene is private information known only to the ally itself. Since allies abandon each other at least some of the time, there will always be some uncertainty among observers about whether an ally will keep its commitment. Therefore, prior to the outbreak of war, what matters more than alliance reliability is alliance *credibility*. Because reliability is private information, having a reliable alliance is neither necessary nor sufficient for the alliance to be viewed as credible. In some cases a third party might be fully committed to assisting its ally, but it does not successfully signal this and observers doubt the alliance's credibility. In other cases a third party might *not* be committed to its ally, but it successfully bluffs observers into thinking it is, thereby averting an attack. If credibility does not automatically flow from reliability, this begs the question of how alliances can establish credibility. High capabilities alone are insufficient, as potential challengers also have to believe that a third party will actually use those capabilities to defend its ally. How can allies signal their commitment to do so?[9]

Morrow (2000) argues that by writing down an alliance in treaty format and broadcasting its terms, states can impose the possibility of reputational damage upon themselves if they abrogate the alliance obligations. Studies have found that countries with better reputations for upholding alliances find it easier to make new alliances and obtain more favorable alliance terms (Mattes 2012; LeVeck and Narang 2017; Crescenzi 2018).[10] Alliances therefore function as a hand-tying mechanism by increasing the costs states pay for abandonment. All alliances invoke some hand-tying costs, but there is variation in how large these costs are, which in turn can create variation in alliance credibility. States with many international commitments might be more concerned about their international reputation than states with fewer commitments, giving their alliances more credibility. Domestic politics also affects the level of hand-tying costs invoked by alliances. Tomz and Weeks (2019) find that an alliance increases willingness to intervene in support of a foreign country among the US public for both reputational and moral reasons. This implies that leaders who are accountable to the public might find it politically untenable to abandon allies. Therefore, more accountable leaders may also be able to invoke greater hand-tying costs and make more credible alliance commitments.

Alliance credibility can also be buttressed by the sunk costs associated with alliance formation and maintenance. The first such cost is paid during the treaty negotiation process, in which states expend resources in an attempt to form mutually compatible war plans (Poast 2019a). Alliances also require coordination in foreign and military policy (Morrow 2000; Poast 2019a) and are often buttressed by coordinating socio-legal expectations and understandings (Powell 2010). Such coordination is costly because it requires states to give up some autonomy, but the very fact that it is costly allows it to function as a sunk cost signal. Because allies that cared little about each other would be unwilling to undertake such coordination costs, it can be inferred that allies who coordinate extensively are committed to each other. At the same time, peacetime military coordination makes fighting together less costly, which should decrease the attraction of abandonment. Alliances that are associated with greater policy and military coordination should therefore be viewed as particularly credible.

In short, the ability of alliances to address the information problem is contingent upon credible signaling. This section has discussed various factors that can serve to enhance credibility and make alliances more than mere scraps of paper. This begins to shed light on our overarching question of *under which conditions* alliances deter. We can conclude that alliances that invoke strong hand-tying costs and entail heavy sunk costs should be more credible than those that do not. Since alliance partners would be unlikely to make such signals without a true underlying commitment to each other, these signals should have informational value to outside observers and should therefore contribute to deterrence. At the same time, one must also consider other causes and contingencies that lead to failed deterrence.

CAUSES OF DETERRENCE FAILURE

As mentioned at the outset, alliances have the capacity to fail in their objectives and perhaps even to incite the very conflicts they seek to deter. While weak credibility is one reason this might occur, we identify three additional mechanisms driving deterrence failure and provocation: (1) changes in the distribution of power and interests, (2) the moral hazard problem, and (3) the security dilemma. All three mechanisms relate to the problem of misperception and disparate estimates of credibility.

Changes in the Distribution of Power and Interests

The passage of time can fundamentally alter the relationship between alliances and war and harm the prospects for deterrence success. It is known that the passage of time can broadly weaken institutional effectiveness. For example, Anderson, Mitchell, and Schilling (2016) find that the pacifying effect of joint intergovernmental organization (IGO) membership has dissipated in recent decades as hegemonic influence wanes. Applying this to alliances, one can reasonably assume that when alliances are formed their terms reflect the interests of the signatory parties and are relevant to their contemporary security environment. As time passes, however, this may cease to be the case. Perhaps structural changes take place that alter the distribution of power among allies and adversaries. Perhaps old threats recede and new threats emerge. It may also be the case that the political elites who signed an alliance leave office and their successors have different interests. For these reasons, leaders often struggle to project hand-tying mechanisms (such as alliance commitments) over distant time horizons and against unknown adversaries (Fearon 1997, 82).

These concerns are borne out in the empirical record. Siverson and King (1980) find that allies are less likely to aid each other when the alliance has persisted longer, while Ostrom and Hoole (1978) find that the relationship between alliances and war is initially positive, then becomes negative, and finally becomes null. The reduced reliability of alliances over time may be driven by changes in governing coalitions, which significantly raise the risk of abrogation (Leeds and Savun 2007; Leeds, Mattes, and Vogel 2009). Since democracies experience the most frequent leadership changes, it might be tempting to view them as less reliable allies. On the other hand, the effect of leadership and governing coalition changes may be moderated in democracies because legislatures check the leader's ability to make unilateral policy changes (Martin 2000). Sensing these tensions, potential challengers and targets may disagree about the strength of an alliance, producing dangerous informational asymmetries that may lead to failed deterrence. These problems will be further exacerbated when the military capability of alliance partners is in flux, calling into question their willingness or ability to abide by their commitments (Johnson and Joiner 2019). In extreme cases alliances whose terms no longer reflect political realities may not only fail to deter but may actually increase the risk of war if disagreements about their credibility lead to widely disparate estimates of the war outcome.

Of course, there is likely to be variation in the extent to which interests change over time. If potential challengers can observe that interests remain aligned, deterrence is more likely to succeed. This may be why Russett (1963) and Huth and Russett (1984) find that economic and political-military ties between a defender and client state bolster extended deterrence success. Allies may also signal their continued interest in each other's security through public gestures such as leadership visits (McManus 2018).

Moral Hazard

The *moral hazard problem* is a term for the expectation that people will engage in more risky behavior when some or all of the potential consequences of their behavior are borne by a third party. For example, people with car insurance may drive more recklessly, knowing that the costs of damaging the car will be incurred by the insurance company rather than themselves. A similar logic governs alliance politics. As noted in the bargaining model discussion, forming an alliance improves a state's expected war outcome, enabling it to bargain harder. The moral hazard problem represents an extreme version of this dynamic in which a third party seeks to avoid conflict by making an alliance with a protégé, but achieves the opposite result. By shifting some of the risks of war from the protégé to its ally, the alliance emboldens the protégé to launch an attack or become so diplomatically aggressive or militarily provocative that it invites an attack upon itself (G. Snyder 1984; Benson 2012; Kang 2012).

Alarmingly, such dynamics can drag a major power ally into a war it does not want. Morgenthau highlights this peril in arguing that oblivious major powers may "lose their freedom of action by identifying their own interests completely with those of a weak ally," who may then drag them into wars (Morgenthau and Thompson 1985, 589). Indeed, he argues that the Ottoman leadership deliberately steered its allies toward the Crimean War, knowing that Britain and France would have preferred a peaceful resolution, but would nevertheless offer military support if war occurred. Why would a major power fight such conflicts against its interests? While these states may prefer negotiated compromise, they can become entrapped by a desire to preserve an alliance or to maintain their reputation for reliability, or because—conditional on war occurring—the costs of a protégé's military defeat are intolerably high (G. Snyder 1984, 467).

The moral hazard problem is, however, far from a forgone conclusion. Fang, Johnson, and Leeds (2014) argue that third-party defenders are not only able to avoid the moral hazard problem but can often pressure a protégé into making bargaining concessions *even after* a challenger has initiated a militarized dispute. The logic underpinning this dynamic is that protégés will be more accommodating at the negotiating table if they believe doing otherwise would risk an ally's willingness to defend them against larger future threats. Yuen (2009) predicts a similar dynamic, but under different circumstances. Her analysis incorporates the possibility that alliances can either incite or restrain protégés, and restraint can occur *even when intervention is certain* if the amount of third-party assistance is large enough for the adversary to minimize its demands, yet too small for the protégé to prefer war. The implication of these arguments is that restraint can occur either because a protégé depends on its ally for future security or because the extent of the ally's support does not offset the costs of fighting.

Finally, Benson, Meirowitz, and Ramsay (2014) put forward a compelling, if counterintuitive argument that moral hazard can sometimes directly *induce* deterrence if the aggressive posture of an alliance protégé effectively communicates resolve to adversaries who are then deterred from instigating crises. Third parties offer unconditional alliances knowing that this may increase their protégés' propensity to initiate conflicts, but believing that this provocative behavior is necessary to deter challengers from further escalation (Benson, Bentley, and Ray 2013). This highlights how the moral hazard problem results from more than a defender's strategic miscalculation—it may be part of a deliberate, albeit high-risk strategy for inducing deterrence.

Taken together, this discussion reveals that the question of whether an alliance will produce a moral hazard problem is no less conditional than the question of whether an alliance will deter. Nevertheless, two general strategies for alliance management emerge, each of which has risks. The first pertains to third-party defenders seeking to avoid the moral hazard problem. Here they must be credible enough that challengers moderate their demands, but not so credible that they cannot subsequently pressure protégés to concede with the threat of future abandonment. Second, states may lean into the moral hazard problem by making ironclad, unconditional commitments in the hope that their protégés' bellicosity will deter aggression, but doing so creates a risk that aggression will beget escalation instead of deterrence.

The Security Dilemma

The third, related, mechanism linking alliance formation to war onset is the security dilemma, wherein nominally defensive actions taken by one state undermine the perceived security of another, resulting in a hostile spiral of escalatory interactions and heightened tensions (Herz 1950). Jervis (1976) explains how this situation, which he calls the spiral model, results from misperception.[11] Each state builds arms or forms alliances for defensive purposes and assumes that its own good intentions are obvious to others, yet each perceives the other's activities as aggressive. Several scholars have posited that alliances may have such an effect, even if their stated terms are expressly "defensive" in nature (G. Snyder 1984; Vasquez 1993). When a state forms a defensive alliance its rivals are likely to believe that the alliance reduces their own latitude in securing their national interests. Naturally the rivals may try to offset the effect of the alliance through building armaments or forming a counter-alliance. The end result arguably leaves both parties worse off: the balance of power remains the same, but the aggregate destructive impact of any potential dispute increases (G. Snyder 1984; Vasquez 1993, 166; Senese and Vasquez 2008, 49).

But why do these dynamics increase the risk of war? One answer lies in the rising level of perceived threat; even defensive alliances may reinforce perceptions of a potential opponent as hostile, which increases bargaining difficulties. Another answer might be the magnification of information problems as the number of potential parties in a dispute grows, particularly when alliances are not fully credible. Vasquez (1993, 167) argues, "Counter-alliances … produce greater uncertainty, because of the possibility that one or more allies will not honor their commitments. Such uncertainty complicates last-minute calculations about who might win in a war, and therefore whether it is 'rational' to initiate a war." In either scenario, alliances are perhaps better seen as a catalyst for rather than an underlying cause of conflicts.

These issues may have been at play in the lead-up to the First World War. J. Snyder (1984) points to an "integrative spiral" in which alliance blocs, each fearing abandonment, inched toward increasingly rigid commitments and heightened hostilities. He notes that the Triple Entente originated primarily as a means of resolving colonial issues, but that Germany feared a more generalized threat and so attempted to break it up during the 1905 Morocco crisis.[12] This only served to strengthen the alliance and create the very monster that Kaiser Wilhelm II had feared (477). As time passed, the alliance networks in Europe spread, overlapping with existing rivalries and heightening the risk that crises that might otherwise have remained local would spread throughout the continent (Vasquez et al. 2011; Vasquez 2018). Uncertainty again seemed to be at play, as hopes persisted among

German elites throughout the July Crisis that Great Britain would stay out of the coming war despite the Entente. So acute was this hope that, as late as August 1, Kaiser Wilhelm attempted briefly to halt mobilization in the west after receiving a telegram suggesting Britain would stay out of the war under this contingency (Williamson and Van Wyk 2003, 107). These hopes were soon dashed, but the incident highlights the cloud of uncertainty under which political elites were operating. Alliances did little to lend clarity to the situation and may have actively obscured the constellation of combatants who would fight in the war to come.

A challenge in understanding the relationship between the security dilemma and deterrence is defining the precise conditions under which hostility and uncertainty will take root. The historical evidence supporting spiral models often comes from great power rivalries in the pre-nuclear era, but it is less clear whether such spirals of fear will be stoked among states with less acrimonious relations or those operating under a nuclear umbrella. Further, it is not clear that uncertainty will always increase as hostilities persist. It may be that there are junctures in the stream of interactions where an ironclad commitment reduces rather than exacerbates uncertainty. Along these lines, Morrow (2017, 343) suggests that the early rather than later stages of a hostile spiral may be most conflict-prone since states' uncertainty will be highest at this time.

DETERRENCE OR PROVOCATION: THE EMPIRICAL RECORD

The empirical study of alliance effectiveness faces inherent obstacles. First, both the argument that alliances deter and the argument that alliances provoke are potentially compatible with the empirical finding that alliance formation is correlated with war onset (Levy 1981). The provocation story suggests that alliance formation exacerbates mistrust, uncertainty, and moral hazard, sparking war. However, if we observe a war occurring *after* alliance formation, we often cannot be certain if it is due to the alliance itself or due to the preexisting tensions that created the demand for an alliance. If a state fears that it will be attacked by an adversary, it will likely seek out an ally in the hope of inducing deterrence. This results in treatment selection bias, which makes it difficult to disentangle the direct impact of some intervention or treatment from the factors that led an entity to seek that treatment in the first place. In this case it may be that alliance formation is disproportionately observed in instances where conflict is already likely to occur. This might make alliances look provocative even if they actually reduce the probability of war compared to what it would otherwise be.

A second, less recognized form of treatment selection bias may also be at work.[13] If a state anticipates that one of its adversaries is seeking an outside alliance, it may preemptively instigate hostilities either in an attempt to deter the outside protector from forming an alliance, or simply because fighting the conflict today is more appealing than it would be in the future once the alliance is formed. An example of this might be Russia's provocations toward Ukraine and Georgia as they have pursued membership in the North Atlantic Treaty Organization

(NATO). This has two possible implications. First, it suggests that alliances may be more provocative than they appear since the prospect of an alliance may incite conflict even before any alliance is actually observed. Second, it suggests that the alliances we do observe empirically are those that adversaries were not able to prevent from being formed, perhaps indicating that alliances formed in crisis environments are actually more reliable. This might explain why alliances have appeared less reliable during the relatively peaceful post–World War II era (Berkemeier and Fuhrmann 2018).

Both of these issues are examples of the inherent endogeneities that plague observational studies of war (For discussions, see King, Keohane, and Verba 1994, 197–98; Senese and Vasquez 2008, 28–31). While these issues complicate our efforts to interpret empirical findings, this does not imply that we should simply dismiss existing findings out of hand, as this might result in rejecting some valid conclusions. Instead, as randomized control trials remain out of reach, scholars must infer causality by attempting to triangulate across different empirical designs and bases of evidence. While existing studies have relied heavily on regression analyses and dyad-year designs, the field must continue to develop a greater diversity of creative techniques to untangle the web of causality.

The next obstacle to studying whether alliances succeed at deterrence is that although we can easily observe deterrence failures, we have more difficulty observing deterrence success, and there is some risk we may incorrectly attribute peace to deterrence rather than equally plausible alternative mechanisms. Deterrence requires more than an absence of militarized hostilities. It refers to a situation in which a nontrivial possibility of conflict exists, but one side successfully convinces the other that the costs of military force will outweigh its benefits (Huth 1988, 17–18). Assuming every peaceful situation is a deterrence success is clearly not appropriate. For example, Canada is not deterring the United States from attacking it. Rather, the United States simply does not want to attack Canada. Often deterrence may be confused with homophily, particularly in parts of the world where many states sign on to regional accords with mutual defense provisions.

It might be more reasonable to assume deterrence success when war does not break out between known rivals or during a crisis (i.e., immediate deterrence). Yet even this assumption can be controversial. The debate over whether the US alliance commitment to NATO (coupled with US nuclear and troop deployments in Europe) deterred a Soviet invasion illustrates this. The predominant view among US policymakers during the Cold War was that this commitment was essential to deterring the Soviets from realizing their expansionist ambitions. However, Mueller (1988) argues that the US nuclear alliance commitment was simply irrelevant because the Soviet Union was essentially satisfied with the status quo and had no appetite for large-scale military engagement after its experience with World War II. Still others argue that attempts at deterrence actually exacerbated and prolonged the Cold War (Lebow and Stein 1995). Thus, even in this thoroughly analyzed case, it is not entirely clear whether deterrence succeeded.

Given these difficulties it is perhaps not surprising that the quantitative literature on deterrence appears at first glance to have produced a kaleidoscope of seemingly contradictory findings. Yet, beneath the surface, we find some informative patterns that illuminate when alliances incite and when they deter. First, the relationship between alliances and conflict has varied across different historical eras (Levy 1981; Senese and Vasquez 2008; Kang 2012). While few studies have evaluated the impact of alliances prior to the Congress of Vienna in 1815, Levy (1981) finds that alliance formation was frequently followed by war during this time. Nevertheless, the endogeneity problem may have been particularly acute in this era, as alliance formation was tied particularly tightly to war preparation.[14] After the Congress of Vienna alliances appeared to exert a more stabilizing role, perhaps facilitating cooperation during the Concert of Europe (Levy 1981, 605). Nevertheless, the stabilizing effects of alliances waned as the European system continued to suffer repeated crises throughout the late nineteenth century, culminating in the massive episodes of deterrence failure underlying World Wars I and II. Overall, the spiral model seems to better characterize the link between alliances and war between 1816 and 1945, with several studies finding a positive relationship when examining this time period (Senese and Vasquez 2008; Kenwick, Vasquez, and Powers 2015). After World War II, however, there is more agreement that alliances are either weakly or negatively associated with war—at least during the Cold War era (Vasquez and Kang 2013). Scholars have attributed this relationship to various causal mechanisms, including nuclear weaponry (Fuhrmann and Sechser 2014) and the territorial integrity norm (see Morrow 2017, 345). While there is uncertainty about which causal mechanisms drive this temporal variation, these findings show that the deterrent impact of an alliance is a function of the international system in which it operates. This fact should give scholars pause when making firm declarations about the likely impact alliances will have in contemporary politics.

Second, one of the primary innovations of the past two decades has been the further disaggregation of alliances based on the types of commitments they contain. While it may be true that even "defensive" treaties can appear threatening, several recent studies have found that alliances with firm defensive arrangements are more likely to deter than those with offensive provisions or nonaggression pacts (Leeds 2003; Benson 2011, 2012; Johnson and Leeds 2011; Fang, Johnson, and Leeds 2014).

A third important factor is the time since alliance formation. Studying the effects of *forming* rather than *having* an alliance is important both because alliances are often formed in the shadow of a military threat one hopes to deter, and because spiral-based arguments suggest it is the act of formation specifically that may spark escalating hostilities. Studies centered upon formation have more often found that alliances incite rather than deter. Morgan and Palmer (2003), for example, find that states initiate more conflicts and increase defense spending following alliance formation. Kenwick, Vasquez, and Powers (2015) look at instances where states form a defensive alliance when none was previously in place; they find that alliance formation is associated with both militarized disputes and war between 1816 and 1945, but negatively associated with militarized disputes after 1945, which they attribute to nuclear weaponry. As Leeds and Johnson (2017) note, these relationships must be interpreted with caution since new alliances are more prone to treatment selection bias than established ones. To get around this problem, instead of looking at time since formation Kenwick and Vasquez (2017) divide alliances into those formed at independence and those formed after, finding that the former reduce the risk of conflict, while the latter have the opposite effect, largely irrespective of historical era. Overall, these studies highlight that new alliances seem to be more associated with instability than established ones.[15]

Fourth, the strategic environment affects the prospects for deterrence success. Factors such as geography and capability matter—allies fare better when defended from nearby and by highly capable partners, though the effect of geographic proximity appears to be attenuated by recent technological innovations (Bak 2018). Subtler features of the strategic environment are more difficult to unpack. Much of the evidence linking alliances to heightened risk of conflict comes from great power rivalries and/or states contending over disputed territory (e.g., G. Snyder 1984; Vasquez 1993), but these results may not generalize. In some instances alliances appear particularly effective at reducing the risk of territorial conflict (Wright and Rider 2014), perhaps owing to their ability to facilitate dispute resolution (Gibler and Vasquez 1998). Morrow (2017) contends that it is not hostility per se but the level of uncertainty that conditions an alliance's proclivity to incite or deter; the greater the uncertainty, the more likely an alliance is to incite. Huth's (1988) broader analysis of deterrence further underscores the importance of the strategic environment, finding that a defender's behavior must also be taken into account—deterrence is buttressed when a defender engages in tit-for-tat or firm-but-flexible diplomatic strategies that balance credibility and stability (203).

Finally, the deterrent impact of alliances appears to vary depending upon the level of hostility being analyzed. The distinction begins at the theoretical and conceptual levels. Many formal models of deterrence characterize challenger-defender relations in terms of the former's decision whether or not to "attack" the latter (e.g., Smith 1995). By contrast, spiral models of war more often focus on long sequences of interactions that potentially culminate in war. As a result, deterrence theorists more often operationalize an "attack" by analyzing the onset of militarized interstate disputes (Palmer et al. 2020), which are meant to approximate an initial armed probe of the status quo (e.g., Leeds 2003; Benson 2011; Johnson and Leeds 2011; Fuhrmann and Sechser 2014). By contrast, those testing spiral models have more often focused on the onset of interstate war (Senese and Vasquez 2008; Kenwick, Vasquez, and Powers 2015; Vasquez and Rundlett 2016). Overall, studies analyzing dispute initiation have tended to find a greater deterrent effect of alliances than studies analyzing war initiation. Studies that have directly compared the deterrent effect of alliances at different levels of hostility have also found variation (e.g., Benson 2011; Johnson and Leeds 2011; Kenwick, Vasquez, and Powers 2015; McManus 2018), but the pattern of variation is not always consistent. Therefore, there is tentative evidence that alliances may deter some levels of hostility better than others, but more research is needed on this topic. Probably the most promising direction for future research is to bridge the gap between disputes and war by analyzing the process of escalation.

It would be misleading to suggest that there are no inherent tensions or fundamental disagreements in this decades-long debate, but many aspects of this summary of the empirical literature point us back to a central claim of this chapter: the success or failure of a defensive alliance hinges critically on credibility and uncertainty. Alliances can better deter when they are well established and clearly supported by capable and resolved defenders; they are more likely to incite when they are newly formed and foster uncertainty among potential adversaries skeptical of each other's intent.

CHANGES IN ALLIANCE PATTERNS AND IMPLICATIONS

The richness of data on alliances has allowed researchers to frequently analyze deterrence in statistical evaluations covering hundreds of years of international history. Such a capacious domain helps establish robust correlations and mitigates the risk that inferences will be driven by a small set of historic events. This approach is also well suited to answering questions about general deterrence because it avoids the inferential pitfalls of studying international crises, where the ex ante risk of deterrence failure is already high (Achen and Snidal 1989). And yet, if scholars are to draw policy implications for the contemporary international system from these studies, it is important to evaluate whether and how the alliance system has evolved over the past two centuries. In this section we highlight two trends that have important and potentially countervailing influences on the prospects for future deterrence success.

The first trend relates to the frequency of different types of alliances. Figure 3.2 shows the prevalence of different types of alliances over time, as recorded by both the Correlates of War Formal Alliances data set (Gibler 2009) and the Alliance Treaty Obligations and Provisions (ATOP) data set (Leeds et al. 2002). Defense pacts are of primary interest for studying deterrence because they explicitly commit allies to defend each other. Ententes and consultation pacts are also relevant to deterrence because these agreements require the members to consult with one another in the event of conflict, implying the likelihood of some form of assistance. However, these pacts offer a weaker commitment that may undermine credibility and deterrence. Offensive alliances and neutrality pacts are also included, though neither is intended for deterrence. Offensive agreements clearly make it easier for a state to engage in aggression, and neutrality pacts can also facilitate aggression by guaranteeing that a potential rival will not intervene.

In the earliest years in which data are recorded, defense and consultation pacts are the dominant type of agreement by far. However, in the period between the Franco-Prussian War and World War I (1872–1914), both data sets document a major shift in which nearly a quarter of alliances feature offensive provisions and up to 41 percent feature neutrality provisions. Meanwhile, only slightly more than half of alliances contain a defensive commitment. Thus the period leading up to the First World War was characterized by a shift away from alliances centered around deterrence. The interwar period (1919–1939) continues to have

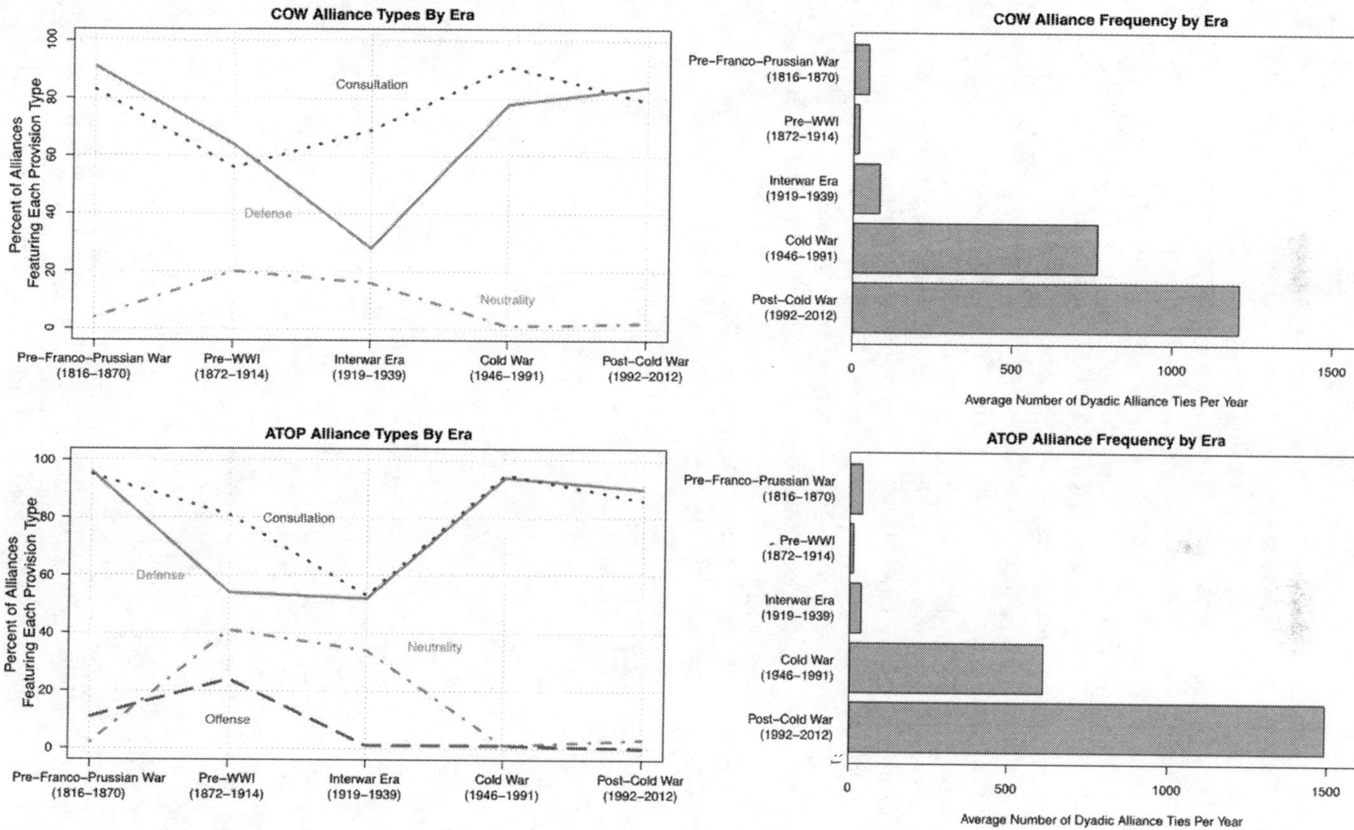

Figure 3.2. Change in Alliance Types and Frequency Over Time.

few defensive commitments compared to other eras, but the number of offensive pacts notably declines. During the Cold War (1946–1991) we see a major surge in defensive agreements. At this point 98 percent of pacts are purely defensive or consultative. In the post–Cold War era offensive agreements become completely extinct. The fact that alliances are becoming nearly exclusively focused on deterrence should arguably make the international system more stable.

The second notable trend in the international alliance system relates to the rate at which new alliances are formed and terminated. Figure 3.2 shows seemingly exponential growth in dyadic alliance ties. However, looking at the number of dyadic ties is deceptive for considering change in the system because of the outsized role played by the expansion of large alliances such as NATO and the Organization of American States (Gibler and Wolford 2006). As of 2016, NATO had twenty-eight members, so when Montenegro joined in 2017 this event alone resulted in twenty-eight new dyadic alliance ties. Yet this represented only a trivial change to the international alliance system.

An arguably better measure of change in the international alliance system is the number of entirely new alliances formed. Figure 3.3 displays the number of new defense pacts formed in each year, 1946–2016, using the ATOP data. This figure shows a steep drop-off in the number of new pacts formed since the 1970s, with the exception of a short-lived spike after the end of the Cold War. The reduction in pact formation has been particularly dramatic recently, with only one new defensive alliance formed between 2007 and 2016. There are also increasingly few instances of alliance termination. According to the ATOP data, only two defensive alliances ended in 2007–2016, and only six in the decade prior. The lower rate of alliance births and deaths has led to the existence of a

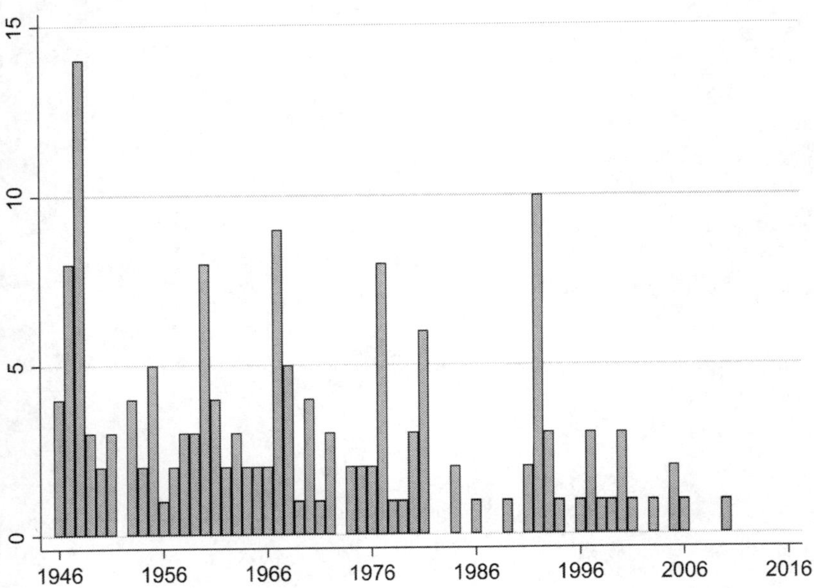

Figure 3.3. Number of New Defense Pacts in ATOP by Year.

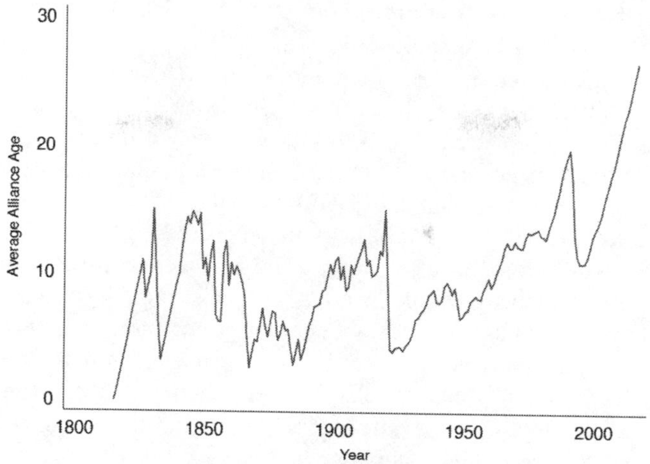

Figure 3.4. Average Alliance Age Over Time, According to ATOP.

global network of alliances that are older on average than at any point in recent history. This is illustrated in Figure 3.4.

This discussion reveals a trend in the international system that scholars have yet to fully grapple with. It appears that the international alliance system is stagnating. Many of the world's alliances are remnants of the Cold War, and the interests and priorities of members may have drifted apart since the alliances' formation. It is not clear why alliance formation has declined. McManus and Yarhi-Milo (2017) suggest that it could be because publicly establishing alliance ties is domestically controversial when countries with shared strategic interests do not also have shared values. It might also be due simply to changing international norms and trends. Regardless of the reason, the stagnation of the alliance system could have worrisome implications for alliance credibility given how shifting power and interests can undermine alliances.

THE RISE OF NEW SIGNALING MECHANISMS

If forming new alliances is no longer a common practice, how can states whose interests have recently come into alignment establish extended deterrence? Scholars are increasingly considering alliances as just one tool in a country's signaling toolkit. McManus and Nieman (2019) treat alliances as one variable in a larger set that can be used to estimate the level of support that a major power signals for a client state. The other variables include nuclear deployments, troop deployments, arms transfers, joint exercises, leader visits, and statements of support. McManus and Nieman find that alliances are a meaningful signal of support when given by Russia, China, and the United Kingdom. However, US alliances appear to be less meaningful because the United States has alliance commitments to many countries, mostly in Latin America, that it does not regularly send other signals of support for.

Kinne identifies defense cooperation agreements (DCAs) as another way to both signal support and establish security ties. Kinne (2018, 801) claims that DCAs are essentially replacing alliances as the main way countries institutionalize their security relationships, pointing to the stagnation in new alliance formations and the meteoric rise in the number of DCAs since the 1980s. He notes that DCAs provide greater flexibility than alliances in an increasingly complex security environment. Wolford (2015) also points to military coalitions as an alternative to alliances. Like alliances, coalitions are a mechanism for aggregating capabilities, but in contrast to the permanency of alliances, coalitions are "crisis specific phenomena" (7) and are typically used for compellence rather than deterrence.

Scholars have also begun to explore the deterrent effect of other types of signals. Alliances appear to be more effective for deterrence than sunk-cost military signals (Krause 2004; Fuhrmann and Sechser 2014), while the effect of military coalitions on crisis bargaining success is highly conditional (Wolford 2015). On the other hand, McManus (2018) finds that leader visits do a better job than alliances of deterring small-scale challenges. She also finds that visits boost the deterrent effect of alliances, and it is plausible that other signals might also enhance alliance credibility by communicating that the alliance relationship is still meaningful.

CONCLUSION: WHAT DO WE KNOW AND NOT KNOW ABOUT ALLIANCES AND WAR?

The research reviewed in this chapter has undoubtedly enhanced our understanding of alliances and war. Theoretical work has explained why alliances might enhance deterrence and also why they might lead to moral hazard, distrust, and a spiral toward war. While these arguments might seem like polar opposites, more nuanced theorizing and careful empirical work has demonstrated that the ultimate effect of alliances is conditional upon a variety of actions. For example, when the risk of deterrence failure stems primarily from uncertainty about third-party intervention and an alliance can credibly signal that a third party will intervene, the presence of an alliance is likely to contribute to deterrence success. On the other hand, when there is a shift in the distribution of power, suspicion of hostile intentions, or failure to restrain an ally from demanding too much, an alliance may actually make conflict more likely. Empirical work suggests that the deterrent effect of alliances may depend upon the geopolitical situation during the time period when they exist, the terms of the alliance itself, how long the alliance has existed, and the level of hostility analyzed.

We also know that the alliance system has changed in recent years. Since World War II a shift has taken place toward primarily defensive alliances. Furthermore, since the 1970s the international alliance system has stagnated significantly, with fewer alliances being born or dying. This stagnation of the alliance system has coincided with the increased use of new types of signals such as defense cooperation agreements and leadership visits.

Despite all this, there is still much that we do not know about alliances and war. First, we do not know exactly how the stagnation of the alliance system will alter the benefits and dangers of alliances. Stagnation seems likely to ameliorate the potential security dilemma caused by alliances because the security dilemma argument focuses on the formation of new alliances. On the other hand, the stagnation may undermine deterrence as the contemporary alliance system no longer reflects the current balance of power or interests.

Second, the existing literature would benefit from more granular analyses of the role alliances play in repeated engagements among rivals and in conflict escalation more broadly. Both deterrence theory and spiral models of war seek to explain a series of complex interactions, often unfolding across repeated crises over the course of several years or decades. While existing data sources provide a sense of how hostile conflicts ultimately become and whether they are reciprocated, we know less about the precise sequence of actions linking conflict initiation to escalation.[16] Thus better data are required to understand alliances' impact on conflict processes, and how the actions and behaviors of alliance partners shape crisis bargaining. Such analyses hold the promise of resolving some of the apparent contradictions observed across analyses of dispute and war onset.

Third, we know little about how well alliances will be able to address contemporary challenges such as terrorism and intrastate conflict. Since the collapse of the Soviet Union NATO has been involved in humanitarian operations in Bosnia, Kosovo, and Libya and has fought against the Taliban and al-Qaeda in Afghanistan. This goes well beyond NATO's founding purpose of deterring a Soviet attack on Western Europe. Scholarly work has yet to devote much attention to how NATO's attributes as an alliance aid or impede the success of such missions and how the undertaking of such missions affects external perceptions of NATO and NATO's internal cohesion. It is also unclear if the alliance attributes that contribute to the deterrence of state adversaries also help deter non-state actors.

NOTES

1. For specific treatments of the First World War, see J. Snyder (1984), Vasquez et al. (2011), and Vasquez (2018). For broader analyses linking alliances with the contagion of conflict, see Sabrosky (1980), Leeds (2005), and Vasquez and Rundlett (2016).

2. We focus our discussion on the particular contributions of alliances to extended deterrence. For broader discussions of deterrence theory, see Zagare and Kilgour (2000) and Quackenbush (2011).

3. Morrow (2017) argues for a similar reframing of the debate.

4. See Johnson and Joiner (2019).

5. This is broadly consistent with Gibler (2009), who defines alliances as "a formal contingent commitment by two or more states to some future action," which is consistent with the criteria of Leeds et al. (2002), who in addition look at the actual written obligations taken on by the signatories.

6. Alliances can also function as a substitute for arms building (Most and Starr 1989; Morgan and Palmer 2000; Kang 2020).

7. Quoted in Williamson and Van Wyk (2003, 59).

8. The problem of credibility persists even in the domain of extended nuclear deterrence. Here the grave consequences of war are more apparent, but potential challengers will still question whether or how a defender will bring its nuclear arsenal to bear in defense of an ally (See Fuhrman, chapter 6 in this volume, on credibility in nuclear deterrence). Therefore, Henry Kissinger argued that few Cold War crises had their outcome determined by the strategic balance of nuclear capabilities alone (Gaddis 2005, 275).

9. We assume that allies generally want to signal the reliability of their commitments in order to avoid challenges, although Quackenbush (2011) points to a downside of this: when two alliance partners are unequally reliable, the more reliable ally will be a more attractive target for military challenges because this minimizes the possibility of a multinational conflict.

10. Other scholars, for example, Mercer (1996), are skeptical of the importance of reputation.

11. Jervis contrasts this with the deterrence model in which there are no misperceptions, and tough policies to deter an adversary are actually warranted.

12. The informal nature of the Entente means it falls short of most definitions of defensive alliances, perhaps contributing to information problems. The arrangement was, however, public and seen as reasonably credible, which we believe warrants its discussion within this context.

13. This point is based on an insightful comment from Hein Goemans.

14. Levy (1981) also notes that while alliances were often followed by war, the majority of wars were not preceded by alliance formation, suggesting alliances were not a central cause of war during this time (604–5).

15. The findings also point toward an interesting tension—alliances are most likely to incite when they are young, but their terms are most likely to be questioned when they age—future work is necessary to unpack how these countervailing forces interact.

16. See Braithwaite and Lemke (2011) and Terechshenko (2020) for measurement-based analyses of hostility and escalation.

Chapter Four

Arms Races

Susan G. Sample

Few specific questions about war causation and conflict processes have attracted so much public attention and academic inquiry as that of the impact of arms races on the likelihood of war. Military spending is highly visible compared to other policy choices, and citizens are likely to have a better feel (or think they do) for the spending of money than they might for the more abstract or indirect consequences of policies like alliance making. Beyond that, the weapons we associate with increased military spending create the most visible damage of war, and are the most direct source of fear for most people. Citizens might feel protected or threatened by military increases, but one way or another, it is likely they will notice and care about arming in a way that makes it the subject of popular political debate.

Beyond the understandable public interest, the contemplation and analysis of the general role of arms buildups in war by scholars of history and politics dates back as far as people have been leaving written records of political commentary. Thucydides cites the growth of the Athenian navy and the power of its empire as the primary contributing factor to the Peloponnesian War more than twenty-four hundred years ago, there is evidence of competition between the Romans and the Carthaginians leading to the First Punic War, and England and Spain were in a naval competition at the time of the ill-fated Spanish Armada (Stevenson 2016b, 11). In modern times two devastating world wars were followed by a cold war that risked the sustainability of human civilization altogether, and these dangers provided plenty of incentive for intense study of arms races and their potential consequences.

Theorizing around arms races during the twentieth century largely paralleled the political events of the time: different historical lessons were drawn from the experience of the two world wars, leading to very different theoretical arguments about the likely consequences of rapid arms buildups. World War I had been proceeded by conscious and deliberate competitive arms races. Britain and Germany, for example, actually had explicit and specific policies of response to each other's naval buildups. Other countries too had built up their militaries before going to war with what seemed in retrospect a cavalier attitude, wholly ignorant of the devastation before them. From these events, many drew the lesson that those arms races had encouraged the behaviors that led to war and that they should be avoided at all costs to avert similar devastation in the future. The lesson of World War II was just the opposite: if only countries had shown a willingness to meet Hitler's military buildups with their own, the argument went, he might have better understood the likely cost of a second world war and been deterred from aggression. Arming rapidly and seriously could have spared the tragedy of more than seventy million deaths.

These two very different counterfactual lessons about the consequences of military buildups then played out on both political and analytical fronts during the Cold War. People latched on to the different risks: on one hand, arming leading to increased hostility and reckless conflictual behavior; on the other, a lack of a clear, strong arms buildup failing to signal resolve in the face of opportunistic aggression. At stake were the lives of countless millions of innocent people. If the wrong policy were pursued, the outcome would be disastrous in a nuclear world.

In the attempt to resolve this theoretical conflict, the larger arming literature provides a clear entrée into exploring some of the key questions in political science. Studies of arming have examined the way internal political institutions and processes determine policy, and they have investigated the link between domestic politics and foreign policy choices and outcomes. As a result of the emergence of robust and consistent findings around key initial questions, the literature has evolved from specific examination of the independent impact of arms races to more complex exploration of the way arms policies are embedded in larger interactive processes of state interaction. The arms race literature also intersects directly with the question of whether political behavior can best be understood through theories presupposing rational behavior, or whether incorporation of nonrational processes into our systematic analyses gains us valuable additional understanding. As a consequence the literature is a microcosm for appreciating the theoretical and empirical complexity that can lie at the heart of studying what seems at first blush to be a fairly simple question, as well as for appreciating some of the real and substantive gains we have made in the past decades in what we know about war. It demonstrates well how science advances in the context of a web of complex theoretical and empirical relationships.

WHAT IS AN ARMS RACE?

The models for explaining arms buildups tend to divide into three basic frames: buildups driven by technological change, by domestic political dynamics, and by action-reaction cycles (Stevenson 2016b, 11). While domestic political dynamics are part of the broader arming literature (Allison 1971, Allison and Morris 1975; Cockburn 1983; Stoll 1982; Bitzinger 2010; Chapman, McDonald, and Moser 2015), arms races have primarily been conceptualized within the international relations literature as consciously interactive and escalating arms buildups between states. Given the dynamics of the US-Soviet dyad and the incredible stakes involved, that case in particular garnered a lot of attention, as did the cases associated with the two world wars (Huntington 1958; Kennedy 1983; Miller 1984; Zagare 1987; Maurer 1992; Stoll 1992; Podvig 2008). Conceptually, as an interactive contest between states, arms races were assumed to be embedded in a dyadic, conflictual relationship with the buildup itself reflecting high levels of hostility. For military increases to be considered a race, the pattern had to be unusual, and that unusually high military growth hinted at a potential limit on the ability to continue the buildup indefinitely. That potential limit, economic or political, formed a kind of deadline that was seen as an additional risk factor in the relationship. Faced with an inability to continue spending at that level indefinitely, and absent a major qualitative technological change to "reset the clock," states might feel compelled to act immediately to achieve their objectives, to avoid the risk of falling behind in the race and losing their opportunity altogether for the foreseeable future (Huntington 1958; Stevenson 2016b, 12).

Many of these ideas were formalized in the models of Lewis F. Richardson (1960a). He assumed that the driving dynamics of a country's buildup could be explained by the grievances the country had against the opposing state, the military holdings of that state, and the current level of armaments within the country itself. The grievances would seem to reflect a variable level of issues at stake, their salience, and general historically defined hostility. The military holdings of the other state in comparison to one's own could reflect a comparative power balance, and the domestic military holdings could also tap into the range of domestic economic and political limits on spending. The traditional conceptualization and Richardson's models have face validity—they make a certain intuitive sense in the way we generally think about arms races and how they work, and they proved fertile ground for the development of testable empirical hypotheses related to both the causes and consequences of arms buildups.

That said, this conceptualization also highlights a lot of the complication of uncovering the real causes and effects of arming. It assumes an already hostile relationship or rivalry between the states, and it embeds or elides questions of the impact of a buildup on the distribution of capabilities and on the defense burden of a particular state. The problem comes from the fact that these things that seem inherent to the concept of an arms race and its consequences may be theoretically linked to the chance of war between states *independent* of the occurrence of rapid arms buildups themselves: rivalry, the distribution of capabilities, and state defense burdens may well increase the chance of war in the absence of rapid arms buildups. As a result much work has been done trying to untangle the question of whether the arms race, compared to these other possibilities, was even the relevant variable to be examined if the real goal was to make good policy and avoid war.

DO ARMS RACES LEAD TO WAR?

Over time the high-stakes policy dilemma at the heart of the arms race literature has provided fruitful ground for systematic and progressive work on arms. The empirical question at the center of this analysis was whether arms races increased or decreased the likelihood of war. Systematic empirical testing of the relevant hypotheses was dependent on the large-scale data gathering and construction that began in earnest in the latter part of the twentieth century. As a consequence of the scientific shift, we now have relatively solid answers to some of the foundational empirical questions about arms races, and that knowledge has led to clear theoretical advances in the very questions we ask about arms races at this point. Substantively, we know that arms races are associated with a higher probability of states getting into militarized disputes and escalating those conflicts to war rather than deterring war. Settling this dispute has enabled a shift toward more complex questions that involve untangling the links between arms races and other variables in theories of war causation, discerning how arms buildups fit into larger processes, and exploring the way the study of arming can contribute to larger debates in the discipline.

Answering that initial critical question of whether arms buildups were associated with an increased risk of conflict demonstrated the complexity with what seemed like a fairly simple question. One issue was how to measure arms buildups in the first place: as with most variables, there is no perfect measure, only ones that prove more generally useful across a range of studies. Attempts to measure military buildups through weapons or capital stocks seemed valid since a weapon has the ability to do a specific amount of damage and thus would seem to reflect a measurable concrete threat. However, the variation in weapons systems across countries and years made military capital impractical as a generalizable measure for arms buildups; military spending, on the other hand, while a less perfect measure of actual military capability, proved useful at signaling national intent around military capability, and was far more generalizable as a measure across time and space (Ward 1984; Diehl and Crescenzi 1998; Sample 1998, 2002; Bolks and Stoll 2000).

Employing a measure based on changes in military expenditures, a first dramatic empirical finding that linked arms races with an increased chance of war was a study by Wallace (1979). Taking militarized interstate disputes (MIDs) between major states as his unit of analysis, he found a very strong correlation between ongoing arms races at the time of the dispute and subsequent escalation to war. Empirically, however, there were doubts about both the way arms races were being measured and the sample of disputes used to test the hypothesis since many of the disputes were related to the two world wars. Subsequent tests addressing those issues (Altfeld 1983; Diehl 1983, 1985) questioned the findings and conclusions altogether when different measures of arming and test samples were used; however, since these studies changed both variable measures and test samples, uncertainty remained about the answer to the basic question.

Sample's research (1997, 1998) addressed some of the key methodological questions raised. To evaluate the potential impact of the methodological choices on the findings, she systematically employed different empirical measures of arms buildups to see if that could explain the earlier findings, then evaluated what happened to the findings when testing using different samples of militarized disputes. Taking each of these changes in turn allowed her to demonstrate that while Wallace's measure of arms races (itself controversial for various reasons) might account for the absolute strength of the relationship he found, his conclusion that arms races were positively and significantly associated with war was not a simple artifact of his methodology. Moreover, she looked at the patterns of disputes that did not escalate. Some were Cold War disputes where one or both states had nuclear weapons. She found a clear pattern among the other non-escalating disputes: the overwhelming majority of them were between states who *did* go to war within five years as the result of another dispute between them escalating. These findings made clear that a real association exists between states engaging in mutual military buildups and dispute escalation, though perhaps not as strong as Wallace had found.

Resolving that methodological problem simply highlighted the importance of a number of theoretical critiques of the studies, including several specific issues related to deterrence theory. Studying the impact of arms races has always been complicated by the fact that arms buildups are inherently tied to other critical variables of interest in conflict studies. The initial theoretical challenges to Wallace's study centered on the issue of whether it really provided a legitimate test of the key questions and assumptions of deterrence theorists who argued that arms buildups should not promote conflict but rather deter aggressive behavior. Weede (1980) and Wallace (1980, 1982) debated whether an examination of the simple correlation between arming and escalation adequately addressed what was for deterrence theorists the more significant question of whether a given arms buildup represented a status quo power being passed by a revisionist power.

The difference between the arguments can be illustrated easily by thinking about different (familiar) interpretations of World War I. On one hand, World War I is often thought of as the quintessential lesson in the dangers of arms races. While Britain and Germany engaged in a naval race, the competition on the Continent between the Central Powers (Germany and Austro-Hungary) and the Dual Alliance (France and Russia) was at full bore. By 1910 the Russians were rapidly increasing their military, and the speed of that change caused explicit reactions in Berlin and Vienna, which led to more aggressive buildups by Russia. These actions all around produced a series of crises, including in the Balkans, which precipitated more reactive behavior on the part of all four countries, linking the arms-racing behavior to the ultimate outbreak of World War I (Stevenson 2016a, 45). It is easy to imagine those arms races as increasing the likelihood that militarized conflict would escalate compared to a situation where they were absent.

On the other hand, it is also easy to argue that the real problem the international system faced was that Germany was dissatisfied with its international status, and the arms buildups simply reflected its changing ability to force the world (including, significantly, the dominant United Kingdom) to recognize its power. The arms race then is not causing the conflict, but mirroring it. In this case it might well be that the focus on the arms race itself simply masked the shift in the power distribution that then allowed the revisionist power to feel impunity to act aggressively. The fact that arms buildups can shift the distribution of power leaves open the question of whether any link between arming and war is the result of the arming itself or simply changes in the distribution of capabilities, which then provides an alternative theoretical explanation both for the war and for the statistical relationship between arming and war. Bivariate tests of the relationship simply did not address these distinctions.

It is critical to remember that these are not just esoteric, academic distinctions: untangling the web of relationships and understanding the real cause-and-effect relationships is necessary to create reasonable policy. If arms races independently increase the chance of conflict, then arms control agreements are a rational strategy for decreasing the likelihood of war. If, on the other hand, arms races are simply a mechanism for shifting the distribution of capabilities, or simply reflect the underlying relationship and do not have an independent impact on the chance of war, then arms control agreements can be likened to treating the symptom of a disease and not dealing with the real ailment (Kydd 2000). For Weede's (1980) critique in particular, the *political* problem would not be that there was an arms race between the two states, but rather that the status quo country had not spent enough to prevent the revisionist one from being in this position—a very different implication for policy.

A second theoretical problem is that by looking at the impact of arming on disputes, deterrence had *already* failed before the test began. Fearon (1994b) had previously argued that the problem with addressing the question of whether arms races encourage militarized disputes to escalate was that the test missed the possibility that arms buildups deterred disputes from occurring in the first place. The disputes that then did occur only happened *because* one or both states had already chosen to escalate to the dispute *in spite of* the known threat presented by the arms buildup of the other state. This selection created a biased sample for the test and thus would hardly make for a proper test of deterrence theory since, by selection, the states had not been deterred. Addressing this possibility required a test of whether prior arms buildups were associated with dispute occurrence in the first instance, not escalation once they had occurred. Sample (2002) tested this possibility by looking at the relationship between mutual arming in dyad years and dispute occurrence, and found a positive relation here as well: when both countries in a pair of states engaged in an unusual arms buildup, that dyad was nearly twice as likely to experience a militarized dispute in a given year. The relationship was not evidently simply an artifact of states overcoming a real deterrent effect to select into the dispute.

The question of whether arms races were masking shifts in the distribution of capabilities, or a revisionist country having new power to advance its interests through war, required a shift from bivariate to multivariate testing. It also necessitated a fundamental expansion of the data beyond the original focus only on major states. By only examining major states, earlier studies risked a real selection effect that any conclusion they drew about arms races or any other variable was really a reflection of the way a small group of major states at a particular point in time interacted among themselves, and not really a theoretical conclusion about conflict processes more generally.

To address this problem, Sample (2000, 2002) gathered data on all states, both major and minor, for all years from 1816 forward. Beyond that, she divided the data by era (1816–1944 and 1945–1993) and ran the tests separately on each. One of the assumptions of using the whole Correlates of War period was that the structure of the system did not change enough to impact conflict processes; by dividing the data, that assumption became empirically testable. In this case the new structure of the test allowed her to determine if the post–World War II era (because of bipolarity, nuclear weapons, or something else) was significantly different from the previous period. And, finally, she designated disputes by those including only major states, only minor states, and mixed dyads.

She then incorporated variables intended to address different power-based explanations for the correlation between arming and war, including whether or not the states were at power parity (from the balance of power/power preponderance debate), power transitions (which should get to Weede's critique as well as reflecting power transition theory's argument generally), and whether there had been a rapid approach toward parity (getting at the question of whether the arms race might be destabilizing the relationship because of the speed of the change).

The results of the tests indicated meaningful distinctions in conflict processes around arms races and changes in the distribution of capabilities when accounting for the historical era and the status makeup of the dyads. Arms buildups were strongly and significantly related to war for disputes between major states and between minor states through World War II, but not after. The capabilities variables, in contrast, showed much weaker relationships when they were significant: power parity was significantly related to dispute escalation for major states before World War II, and power transitions were related to escalation for minor states after World War II, but in other cases the capabilities variables were not significant. Critically, accounting for changes in the balance of capabilities due to arming did not undermine the robust relationship between arming and war. Arms races were not simply reflecting or masking power shifts.

EMBEDDING ARMS RACES IN HISTORICAL PROCESSES

The resolution to the basic empirical question of whether arms races really were independently related to the occurrence and escalation of disputes immediately shifted the ground of the research. Arms races in these tests, and in much of the debate of the twentieth century, had been treated as theoretically central to the conflict dynamic, certainly by those concerned that arms races could lead to war. That centrality was reflected in case studies and was reinforced by the (necessary) empirical testing used to isolate the impact of arms races (and other variables).

Methodological choices that treated dyadic disputes as independent cases, a choice that has legitimately advanced what we understand about arms races and other variables in conflict processes, can have the curious impact of bracketing out history for methodological simplicity. Bringing that history back in while maintaining scientific rigor has been critical to theoretical advance in how we think about the role of arms races in conflict processes.

As soon as the statistical relationship between arming and war was convincingly isolated, the larger processes in which arming is embedded immediately retook the foreground in our studies. Both the steps-to-war and rivalry theses offered theoretical reintegration of historical process with the scientific orientation of recent decades through arguing that individual events such as militarized disputes are incidents occurring within larger ongoing relationships between states. The theories then offered different explanations for the role of arms races within that larger process.

Steps-to-war theory (Vasquez 1993, 2009) argues that the association between arms races and escalation to war is a significant causal link in a chain of interactions between states. For a variety of reasons states tend to pursue realpolitik solutions to territorial conflicts, and are thus likely to seek allies and build up arms in an effort to advance their chances of winning recognized and uncontested sovereignty over the contested territory. While perhaps intended to signal resolve and deter the opponent from escalation of the conflict, the real impact of these policy choices is to make further militarization likely, including more disputes, the domestic empowerment of hard-line political actors, the possible generation of real rivalry, and war. For states in conflict, choosing to rapidly build arms directly increases the chance of war between the states. This dynamic and the role arms races play in it is evident for disputes over any issue, but the probability of war is greater with territorial disputes because those disputes are more conflict-prone in the first instance (Senese and Vasquez 2008). Arms races are embedded in a causal chain that depends on increasing levels of hostility as states interact with one another over time, with each step independently and in combination increasing the probability of war. If some diplomatic intervention or other event halts the iterative move toward increased conflict, the chance of war decreases.

The territorial dispute between Pakistan and India over Kashmir and Jammu has precipitated several wars, including one in 1965, as well as multiple near misses such as the crisis over Pakistan's capture of an Indian pilot in 2019. In the early 1960s multiple diplomatic attempts to resolve the ongoing conflict over Kashmir had failed, and repeated militarized conflicts led Pakistan's leaders to think that Pakistan had a chance at victory in what they estimated would be a localized war, assuming as they did that the Kashmiris themselves would rise up in rebellion against Indian rule. Both states engaged in rapid military increases in this period, though they were not solely reflective of this specific dyadic conflict: Pakistan, seen as a strategic link to the central Asian USSR, was the recipient of considerable American aid, and India, reacting to its 1962 war with China, initiated a large buildup in the wake of that conflict (Chaudhuri 2016, 239). Despite their multicausal nature, these military buildups increased the chance of escalation, almost certainly contributing both to Pakistan's belief that it had a real chance of victory, and to the likelihood that India would not keep the conflict localized to the Kashmir region (the very thing Pakistan was banking on) (Paul 2005a).

In contrast to this formulation, the rivalry research argued that the relationship between arming and war was not primarily causal (Diehl and Crescenzi 1998). The research on rivalry held that much of the literature empirically testing the relationship between different variables and war had turned into a classic case of missing the forest for the trees. Between breathing the air of realism arguing that all states face the same struggles in anarchy, and making methodological choices like treating disputes as independent cases, scholars had ended up leaving unrecognized, or at least undertheorized, a strong and clear pattern in international conflict: the fact that violent conflict was not randomly distributed through the international system, but rather was taking place among the same states repeatedly. And it was that pattern that was theoretically critical to understanding war (Diehl 1998; Diehl and Goertz 2000; Colaresi, Rasler, and Thompson 2007).

In this context arms races might be part of the larger historical relationship of a pair of states in rivalry, but the rivalry itself should be considered central to the investigation of conflict processes. Rivals see each other as the enemy by definition, and have both a history of conflict and an expectation of future conflict. In rivalry states will exhibit a range of behaviors reflecting that enmity, including a range of competitive policies, and those intended to prepare for possible war, like arms races. The rivalries between Israel and its Arab neighbors are useful examples here. The rivalries were the result in the first instance of an original determination by Jews in Palestine to create a state, and they can be dated from Israeli independence. All subsequent behavior in the rivalries, including arms races (both sides were generously supplied by the superpowers through years of shifting alliances), and the multiple wars that ensued, reflected that initial enmity (Kober 2016, 209–12).

The core argument of theories of rivalry has critical implications for how we understand the relationship between arms races and war, including the possibility that the empirical relationship between arming and war might well be theoretically trivial if not entirely spurious (Diehl and Crescenzi 1998). While this possibility was addressed directly in the rivalry literature, it also indirectly reflected much of the debate around arms races during the Cold War. Many of the arguments in favor of deterrence as policy were really arguments about the nature and policy implications of rivalry. Arguments in opposition to focusing political attention on arms reduction efforts often argued that such policies were pointless because the arms race was spurious (Kydd 2000)—as long as the underlying rivalry between the United States and the Soviet Union continued, reducing arms of a particular type or in a particular moment would not substantially affect the chance of war between them. And, in the case of the Arab-Israeli conflict, the probability of war has likely declined, not because the arms races have subsided (they have not), but because the United States is the sole superpower patronizing all the parties and discourages escalation between them, and the conflict between

Iran and its Sunni rivals is diverting attention from and overshadowing that original rivalry between Israel and its neighbors (Kober 2016, 221).

The intersection of these issues highlights the way that the arms race literature provides an excellent reflection of conflict studies as a whole. The stakes are high in political terms, and our understanding of the key variable is sufficiently wrapped up with other critical concepts that empirically untangling the threads to get to the different theoretical arguments is challenging, but is also absolutely critical to understanding how conflict processes work and making appropriate policy. In reality, arms races have typically been *conceptualized* as happening among rivals, though the language was not necessarily used; the unusually intense, competitive, and interactive nature of the conceptualized relationship is difficult to imagine outside of an ongoing hostile relationship between the states. Despite the definition, however, it is also true that most of the empirical testing evaluating the role of arming in conflict did so using test designs and variable measures that were absent any requirement for demonstrating that the countries were in fact intentionally competing and engaging in interactive behavior. As a result the studies were evaluating mutual military buildups that might or might not be actual arms races.

These methodological choices are important because they mean that while we knew that arming was related to war, it was still unclear why. It remained to be seen if the statistical relationship with war might simply be a statistically spurious artifact of the rivalry causing a range of behaviors, including both arms races and war; if arms races provide an additional venue for competition between already hostile states and thus interact with rivalry to increase the chance of war between rivals; or if awareness of mutual arms buildups can cause threat and hostility between states and thus increase the chance of war, even if those states are not rivals. Obviously the answer to this question has important implications for both scholarship and policy considerations.

These critical questions were approached from different directions. Gibler, Rider, and Hutchison (2005) specifically operationalized arms races on criteria based on their demonstratively being actual arms races as traditionally conceptualized, not simply mutual military buildups. To meet that conceptualization the countries had to be rivals, they had to meet an 8 percent per year military expenditure growth rate for three years (a common measure), and there had to be historical evidence of intended arms competition. This formulation then provided a means of controlling for the possibility that the relationship between arming and war found in prior research was spurious, as well as settling the nagging question of whether mutual military buildups (the indicator most research had used to measure arms races) might somehow have a different effect on war than historically verified arms races. The study concluded that the relationships between arming and war found in previous studies still held. Rivals experiencing arms races were more likely to both experience disputes and go to war than rivals who were not (even though most wars among rivals did not have attendant arms races). Rider, Findley, and Diehl (2011) also found that arms races were more likely to happen in rivalry, but a modest positive relationship remained between arms races and war once rivalry was accounted for.

Another methodological option for approaching these questions is treating rivalry as a control variable to test the relationship between arming and war. This achieves the purpose of testing whether the relationship is spurious while allowing for the possibility that arms buildups outside of rivalry are not trivial events. In their multilayered testing of the steps-to-war thesis Senese and Vasquez (2008) treat rivalry in this way, and find that the probability of war increases with the addition of each step, including arms buildups, even in the absence of the dyad ever meeting the criteria of a rivalry. Sample (2012) too finds that mutual military buildups are related to escalation among non-rivals as well as rivals. An examination of the specific cases where there was a mutual buildup but no rivalry reveals different sorts of cases. In some cases two parties were preparing to join a larger war (World War I or II), but they were not directly in conflict with each other (though sometimes they were in dispute with an ally of the other party). In other cases the specific source of the mutual buildup is not clear (and may well be related to domestic politics), but it is then part of the context of subsequent militarized disputes between the states. For example, Kwame Nkrumah was overthrown in Ghana in 1966 in a context in which several states, including Ghana and Guinea, were engaged in military buildups. His overthrow and Guinea's desire to reinstate him led to multiple disputes involving the two countries as well as other regional states.

Further, there were multiple cases where there was an active territorial claim and the dispute ended up either in war or takeover of the weaker state of the pair by the stronger. A complex territorial conflict over Albania between that country, Italy, and Greece involved mutual military buildups all around, but no declared rivalry within any pair. It ended with Albania's capitulation to Italian occupation in 1940 and to war between Italy and Greece in the same year (Sample 2018a) Germany and Lithuania were engaged in mutual military buildups and opposition before Lithuania was occupied by the Soviet Union in 1939, and Finland and the Soviet Union were not official rivals in the lead-up to the Winter War, though they were mutually arming and had decades-old territorial disputes (Sample 2018a). In all of these cases the relationship between the states follows the pattern proposed by the steps-to-war theory, including arms racing, though the relationship never reached the threshold of rivalry, and thus would be discounted in studies of rivals alone. The evidence indicates that arms races are more likely among rivals, but these findings suggest the likelihood that the arming process itself affects the relationship between states even in the absence of rivalry and is worth exploring in those cases, a conclusion that would make theoretical sense given the theorized step increase of the steps-to-war theory as well as the premise of the hostile spiral argument regarding arms races.

It is an unavoidable truth, however, that arming policies are expensive. States cannot undertake them lightly. Ongoing military buildups can have a negative impact on a country's economic growth, and as a highly visible policy, they can bring direct political costs as policymakers have to deal with publics who object to guns-versus-butter types of trade-offs (Barnett and Levy 1991; Kang 2020). In the case of rivalry leaders are likely to pay fewer political costs, and the policy will be more generally accepted because they can build a ready defense for the necessity of the policy (Rider 2013). Few other circumstances warrant choosing such a costly policy. Unstable domestic politics in a regime where a leader might direct the use of force toward his or her own population could do it, though it is an open question as to whether that would have any international ramifications at all. It is arguable, however, that in international relationships any conflict between states with high enough salience should satisfy the dual need for perceived necessity of the policy and political justification. Territorial disputes, given their demonstrated propensity to violence, clearly fit those criteria. The relationship between territorial disputes, arming, and war has been demonstrated through several lines of research related to the steps-to-war theory.

In an attempt to unravel the complex relations between arming, steps to war, rivalry, and territorial conflict, Sample (2014, 2018a) traced the history of forty-two cases in which territorial conflicts had led to war. She argued that previous statistical studies had incorporated the critical variables and demonstrated their correlation with war, but had failed to clearly examine the timing of the emergence of the different elements of the conflict. The methodology of those studies made it effectively impossible to determine if arming and other steps to war were causing the conflict, or if rivalry was shaping all of those choices. Tracing the historical patterns of each case would allow us to make those distinctions and to more directly examine each of the theoretical claims. If rivalry came before the other steps in a case, then clearly it is impossible to refute the idea that rivalry is the primary element of the relationship, causing the other "steps." Likewise, if rivalry followed the other steps, then it is logical to conclude that the conflict process could not be explained primarily by rivalry.

Sample's findings indicate that part of the long-standing controversy over the relationship between arming and war may be the result of arms races legitimately playing different roles in different types of conflicts. In the cases of territorial disputes it was clear that discernible patterns emerged in cases that represented multiple causal paths to war. There were both clear categories of cases where the timing of the emergence of the policies in the historical relationship between states directly reflected the theoretical argument of the steps-to-war thesis, and those where rivalry preceded all other variables. Among the former, Cambodia and Vietnam had a two-way territorial claim from 1954 that led to politically relevant alliances in addition to concrete military buildups, multiple territorial disputes, and war in the 1970s, even before they were officially rivals. In contrast, the relationship between Egypt and Israel was characterized by intense rivalry immediately, and that rivalry included relevant alliances as well as arms buildups and multiple wars. These different paths indicate that in some cases arms races can be understood as impacting the likelihood of states taking further steps toward war, and in some cases the arms race may be at least partially an effect of the underlying rivalry.

Collectively this research allows us to conclude that arms races increase the chance of war between states, though they also almost certainly must be understood in the context of larger conflict processes. The implication for policymaking would seem to be that policies directed toward arms reduction can be usefully employed to reduce the probability of war between states in conflict or manage the level of conflict over time, but that arms reduction should not be thought of as the primary tool for ending that conflict in most cases. Diplomatic efforts directed at resolving the underlying territorial or other issues at the heart of the conflict are critical for long-term conflict resolution.

ARMS RACES AND RATIONALITY

While the arms race literature has convincingly answered many of its originating questions, an unanswered theoretical debate at the heart of that puzzle is relevant to our studies of political behavior more generally—the extent to which the political outcomes we see are best explained while maintaining the assumption that political behavior is inherently rational. A key disagreement in the arms race debate has always been the question of whether arms races introduce a nonrational dynamic into the dyadic relationship between states that leads them to choose war neither explicitly intended before the final escalation. This notion reflects common understandings of World War I, among other events, associated as it is with arms races (like that between Germany and the United Kingdom) that seemed more about status conflicts than specific rational attempts to guarantee security or control the system. To some extent, shifting our focus toward the larger historical processes in which arms races are embedded may suggest an acceptance of psychosocial or nonrational processes in the historical relationship between states in conflict. Rivalry, for example, seems to be more than just the sum of its parts; states come to see each other as "the enemy" in ways that would seem to owe as much to social construction as to salient conflicts over specific issues. Steps-to-war theory assumes that the political context in which each decision is made is shifting toward increased hostility or an increased belief that further escalation is necessary and proper, which may or may not reflect rational calculations of any reasonable sort; hard-liners gain power for reasons related to the political and emotive appeal of hard-line rhetoric against the other state, not because that is the only choice possible given rational calculations.

Many of our traditional theories have assumed, however, that state behavior is inherently rational and that states have clear interests and pursue them systematically through cost-minimizing strategies given the information available to them in the context, with missing information then providing the explanation for decisions that were disastrous in hindsight. Deterrence theory reflects important elements of the rationalist framework for understanding arming policies: arms races are dramatic and visible, certainly, but fundamentally, they are attempts to manage or shift the balance of capabilities. That was the premise behind the initial debates between Wallace (1980, 1982) and Weede (1980) in the wake of the former's original finding linking arms races and war. The arms race literature has always represented a clear problem to the rationality assumption and this formulation of the interstate dynamic during arms races. If we assume that states intend to deter opponents with their arming policies and should rationally be deterred, the link between arming and war indicates that the outcome is quite different from the intention, which is a problem for the theory, and empirical testing demonstrates that this connection remains true even when accounting for shifts in the capability distribution.

It can be argued that the security dilemma can account for the seemingly nonrational outcome. The essence of the dilemma is that states face threats and engage in behavior that seems rational in the short term, but has unpredictable and even nonrational long-term consequences. The security dilemma alone, however, is unsatisfying as an explanation for arms races because it points to a general threat in the system that may explain why states typically arm, but not why their arming policies vary enormously from one time period to another. The rarity of unusual arms buildups requires a more precise pinpointing of when and why states make such a costly decision, and that is where rivalry and territory (as a demonstrably highly salient issue) come in. Still, narrowing the circumstances for arms policies explains why states engage in arms buildups at particular times, but not how something intended to manage conflict leads so often to escalation. The question of why remains.

One solution is that the behavior and outcomes can still be explained through rational behavioral choices understood in a more complex way than the basic maintenance and manipulation of the distribution of capabilities. Rider (2013) argues that consideration of *when* states arm is critical to understanding why they overcome the dual cost of expense and the known inherent risk of the arming policy. In his analysis rivalry provides the context of a specific threat that legitimates the cost, though arming is still expensive, and rivalries can continue for decades, leaving the question of why nations arm at some times and not others.

The missing piece is the other element of the rationalist argument regarding arming: signaling (Fearon 1994; Rider 2013). For deterrence to work, states must have an idea of both the capability of the enemy and their resolve. Uncertainty about either requires "testing" behavior to gain that information. The onset of an arms race in this formulation suggests that a country is missing information either about capability or resolve, and the arms buildup is intended to give a country that information. In the 1920s and 1930s, when there was both pressure toward naval racing and pressure to avoid war, British naval diplomacy emphasized transparency and information exchange about naval building programs as part of its balancing act between arms control and arms racing in the interwar years (Maiolo 2016, 114), suggesting it was deliberately intending to make the information about capability and intentions clear to stabilize its relationships.

It is arguable that capabilities are likely to be more generally known, but resolve can be more or less clear. Rider (2013) argues that uncertainty about resolve, and a lack of prior information that answers that question, are most likely when a new leader comes to power. A new leader may be an unknown quantity and therefore needs to signal resolve to the rival state. Alternatively, a rival state may want to test the resolve of the new leader. In either case a new leader introduces an element of uncertainty into a high-stakes situation, requiring a specific policy to resolve that uncertainty. An arms buildup is a clear signal on the part of either state, and may well be reciprocated in a high-stakes situation.

If the goal is signaling, then the resolution to the lack of information should decrease uncertainty and the chance of war, but we know that is not the case. Rider (2013) argues that this is because arms races by their nature alter countries' preferences for war and peace because they inevitably change state power and preparedness, thus affecting the chance of victory in case of war. Arming may increase their chance of victory, but not enough to deter or invoke capitulation by the opposing state, so war becomes more attractive, especially given the fact that resources are finite and arms races are costly. He thus contends that war is not the result of a hostile spiral, but of the shift in state preferences over time due to the arming.

While incorporating signaling is a valuable addition to understanding the process, it is not clear that it resolves the question of whether nonrational dynamics are at play beyond the rational calculations of states regarding their preferences for war and peace. It makes sense that arms races can alter the relative capabilities of a state and its defense burden and thus change the preference for war. Those arguments have been central to the debates within the arms race literature from the beginning. If the only escalations in the context of arms races are in cases where there is also a power transition or a shift toward parity—both of which offer a clear signal that the costs and benefits of war have changed because of the arms race—then this outcome would suggest that the assumptions of rationality offer the best explanation of how arms races are related to war outcomes. Empirical testing, however, has shown that while those shifts in the distribution of capabilities (measured by changes in relative power balance, power transitions, and rapid movement toward power) have some explanatory value in explaining which disputes will go to war, the link between arms races and war clearly exists when these variables and defense burdens are accounted for (Sample 2002, 2012).

The alternative explanation for why something intended to avoid conflict disproportionately leads to war is that the link between arms races and war is best explained by nonrational elements in the interaction between the states engaging in the arms race. Policymakers themselves have never been convinced that the process was an entirely rational one: Soviet premier Khrushchev, in 1961, in the wake of a missile test that resulted in a Soviet missile going far off course, imagined that such an accident could easily result in misinterpretation and escalation to war. He feared that the psychological stresses in circumstances of declared emergency situations would make it impossible for a leader to realistically choose between whether a hostile act was the result of intention or accident, and war could be the outcome (Radchenko 2016, 166). Other nonrational dynamics are captured in the theoretical framework of the hostile spiral argument of the security dilemma, and findings about cognitive biases in decision making from psychology (Jervis 1976, 66–68; 1985). In the hostile spiral argument one or both states in a dyad build up their arms, perhaps motivated by a specific security concern. The state(s) see(s) the increase and assumes, rightly or wrongly, that it is the center of the other state's planning, and thus feels increasingly threatened and commences or increases its own buildup, thus precipitating an arms race that independently increases the chance of a conflict, even when such a conflict was not rationally intended.

McDermott, Cowden, and Koopman (2002) tested several specific hypotheses related to rational and nonrational elements in decision making, including the role of uncertainty around capabilities and the impact of framing and perceived hostility on arming choices. Uncertainty about capabilities would plausibly offer a rational explanation for the emergence of a hostile spiral. In contrast, evidence from prospect theory suggests that decision makers respond very differently to objectively identical questions when they are framed differently, suggesting that people do not in fact process information in ways expected by rationality theories. The authors ran a laboratory simulation game with university students making decisions on military expenditures, manipulating four variables that would relate to signaling/rationality: the frame of the situation, uncertainty over the characteristics of the weapons, uncertainty over the overall capacity of the weapons, and the tone of the messages themselves (hostile, conciliatory, or neutral). They found that uncertainty over the nature of weapons systems and their capability was not in fact related to the choice to allocate more money to weaponry.

They further found, however, that the initial framing of the problem and the tone of the messages both were significantly related to the decision to increase military spending among the experimental subjects (McDermott et al. 2002). When the subjects were told that their country must attain superiority rather than parity, that frame drove decisions independent of other factors, and messages perceived as hostile led to increased spending. Other evidence suggests that the hostility of the messages itself may not be objectively understood: in their historical study of decision making around the commencement of World War I, Holsti, North, and Brody (1968) found that the countries involved in the initial conflicts misread behavior and messages from their opponents as more hostile and deliberately challenging than they were likely intended, and responded with more escalatory language, precipitating increasingly hostile responses until war became seen as inevitable.

This evidence indicates that nonrational or psychological dynamics are at play in the decision to build up arms in the first instance, and potentially also in the role they play in moving small conflicts into full-scale confrontation or war. The results certainly suggest that the rational formulations of the process are not capturing all of the political dynamics in the arming process. Teasing out the rational and nonrational components of the process may require more in-depth analysis of specific cases and their interactive trajectory.

QUESTIONS WORTH THINKING ABOUT

While many of the originating questions in the arming research have largely been settled, several logical avenues remain to explore for future research regarding arming processes and their impact on war. One is the aforementioned question of the interplay of rational and nonrational elements in political processes, and the way that manifests particularly in arming policies. History tells us that politicians themselves have been concerned about the risk of misperception and accident in conflict dynamics, and the empirical evidence demonstrates that while rational calculations of the changing chances of victory may play a role in moving from arms race to war, they do not solely account for the multifaceted impact of arming behavior. Understanding how these different dynamics work may require new hypotheses or careful, theoretically informed analyses of particular cases. Addressing specific cases may allow us, for instance, to tease out distinctions between arms buildups that are produced for general signaling objectives and those that arise specifically from hard-liners coming to power with the intention to escalate from the current state of the dispute. Tracing of particular cases might also allow for an historical accounting of the intentions of leaders and their calculations regarding the likelihood of victory that would give us a fuller understanding of the interplay between rational calculation and nonrational dynamics at play in crisis situations.

Another path worth considering is whether state arming policies are worth exploring outside of the context of rivalry and/or territorial conflicts. We know that mutual buildups are rare outside of these circumstances, though there is some evidence that they may be linked to conflict in these cases as well (Sample 2012). We also know that there are cases where there were unilateral buildups, but not mutual ones. Are these cases theoretically relevant in international relations? The literature has largely dismissed them because they do not meet the conceptual definition of an arms race. However, if arming has an independent effect on state relations, then prematurely dismissing it may result in lost information. Arming is certainly a high-cost policy, but outside of international conflict states may initiate military buildups for domestic reasons. Could buildups driven by domestic dynamics affect relationships with contiguous states, potential rivals, etc.? This question of course gets to the bigger issues of the links between domestic and international politics. The literature on civil war addresses numerous ways that civil wars can spill over into other states such as movements of arms, refugees, transnational learning, etc. (Forsberg 2016), but it may be worth thinking about the way a policy like arming can have spillover effects into a country's foreign relations as well.

Regional or neighborhood effects as well as network effects of arming might be worth investigating here as well. There are historical circumstances when nearly every country in a region is engaged in high levels of arming simultaneously. This is the case with Central America in the 1970s and 1980s (driven largely by internal factors) and with the former Soviet countries at the turn of the twenty-first century. Multiple countries in Southeast Asia have had high levels of arming from the 1960s until well into the twenty-first century. Questions regarding both the causes and effects of these regional dynamics might well be addressed, including asking whether the regional arming patterns have any discernible impact on dispute behavior, or are just reflective of the complex conflict dynamics of the region, such as the multifaceted territorial disputes over the South China Sea.

Finally, it may be worth circling back to Huntington's distinction between qualitative and quantitative buildups and the role of technological change in arms races. With intense state interest in technology changes in cybertechnology or militarization of space, would races in these venues stabilize or destabilize relationships between participating states? Presumably, Huntington's conclusion would suggest that arms races would constantly be starting anew as each qualitative change in technology occurs, and rivals would never feel at risk of permanent disadvantage, but in light of rapid technological advance, the ideas might be worth exploring in a new century.

As suggested in the beginning, the literature on arms races provides an excellent entrée into the scientific literature on conflict processes. The political stakes are high: the theoretical arguments regarding the causes and consequences of arms races lead to very different implications for appropriate policies to reduce the occurrence of international violence, and the costs of getting it wrong are potentially disastrous. A number of deeper questions require continued research, but this literature is also a case where we can absolutely say that we have significantly advanced the frontier of knowledge on this issue. We know that arms races do independently increase the probability of conflict; we know that this independent effect is also typically occurring within a larger rivalry or ongoing conflict over a highly salient dispute over territory. We can also assert that political interventions in favor of arms control are likely to be valuable moderating influences on the risk of violent conflict, even while we may best think of them as measures intended to buy time for concentrated diplomatic intervention to resolve the territorial conflict or underlying issues that define the rivalry. The more we know about war, the more confident we can be in crafting policy that makes the world safer—and that the scientific study of arming has taught us a great deal.

Chapter Five

Rivalries and Crisis Bargaining

Seden Akcinaroglu and Elizabeth Radziszewski

INTRODUCTION

How do international rivalries affect crisis bargaining, both the escalation to war and de-escalation? Crises are defined as "disputes where at least one of the participants threatens another with the use of armed force" (Leng and Wheeler 1979, 655) or, in other words, where at least one party to the conflict indicates willingness to go to war over its interests (Leng and Singer 1988, 159). Crisis bargaining or action states take to influence each other's behavior during these serious disputes (Leng 2000) leads to challenges in finding a mutually acceptable settlement that would make de-escalation possible. Individual, domestic, and international dynamics as well the saliency of the contentious issues affect the extent to which the dispute is likely to escalate, while the existence of shocks creates an opening for improvement in enemy relations with the potential to ease bargaining challenges. When a crisis escalates it may involve, for example, additional threats to use force, the use of force that results in casualties but falls short of a full-blown war, or the breakdown of bargaining that ends in war. War, as the worst outcome, refers to sustained combat involving organized armed forces that results in either one thousand battle-related deaths (Sarkees and Wyman 2010) or, as other scholars have adopted in the quantitative conflict literature, twenty-five battle-related deaths (Gleditsch et al. 2013). While all states experiencing a crisis with another state are likely to face constraints during bargaining, rivalries—defined as militarized competitions that could endure over a course of many years (Diehl and Goertz 2000), or as competitive relationships where rivals perceive each other as enemies and a source of threats that could become militarized (Colaresi, Rasler, and Thompson 2007)—in particular, have a unique impact on crisis bargaining.[1] Focusing on rivals as unique antagonists merits special attention. Disputes involving rivals likely involve multiple issues, more so than those between non-rival states, which amplifies bargaining failures (Colaresi and Thompson 2002; Mitchell and Thies 2011). Rivals also face a higher risk of military disputes than other pairs of states.[2] In the past two centuries rivals have been responsible for 80 percent of interstate disputes and wars, and their warfare is directed toward each other rather than randomly (Maoz and Mor 2002; Rasler, Thompson, and Ganguly 2013).

The rivalry approach to explaining crisis bargaining implies that pairs of states that see each other as rivals are likely to perceive their confrontation as deserving of special attention. They approach their interaction under the shadow of preexisting hostility. Whether the hostility involves prior militarized disputes (Diehl and Goertz 2000) or nonmilitary ones, many, though not all, rivals see each other as threats to national security (Hensel 1996b), with the potential to inflict future damage to the state's interests (Rasler, Thompson, and Ganguly 2013). Thompson (1995), for example, focuses on principal rivals that identify each other as primary threats to national security. Diehl and Goertz (2000) identify enduring rivalries or states that have experienced at least six militarized disputes over the course of at least twenty years as particularly dangerous. Finally, conflicts among rivals are linked and repeated over time (Klein, Goertz, and Diehl 2006). Crisis bargaining between rivals thus has a different situational background than bargaining between states that do not specifically see each other as competitors or rivals that have not fought any militarized disputes.

What are the conditions under which the context of rivalry results in crisis escalation? And how does international rivalry impact crisis de-escalation? This chapter relies on the frequently utilized levels of analysis analytical approach to explaining outcomes in international relations. Introduced by Kenneth Waltz and modified by J. David Singer, the analytical framework makes sense of phenomena by looking at three sources of explanations, individual, domestic (state), and international. Other levels, such as the dyadic level, have also been utilized (Mingst 2004). In this chapter we examine how individual, domestic, dyadic, and international dimensions of international rivalry reduce bargaining challenges between states. Additionally, we consider the effect of issues and shocks in international rivalry. Collectively these factors have played a role in increasing and sustaining rivalry and thus creating the milieu of suspicion and distrust under which crisis bargaining takes place. Furthermore, as a crisis may unfold over the course of months, individual, domestic, and international forces as well as domestic and systemic shocks and specific types of issues may create additional constraints or opportunities that explain conditions under which international rivalries affect the outcome of crisis bargaining. We present a review of the existing research on how these rivalry characteristics impact crisis bargaining (Table 5.1), both escalation and escalation, and then bring many, though not all, of these factors together to empirically examine their significance for crisis bargaining from 1918 to 2011. The chapter also highlights promising avenues for future research on understanding how international rivalries impact crisis escalation and de-escalation.

Table 5.1. A Summary of the Literature on Escalation/Deescalation in Crisis Bargaining among Rival States

Variables		*Expectation*	*Citation*
International and Dyadic Factors			
Capabilities	Symmetric power	Escalate	Leng 2000
	Asymmetric power	Escalate	Colaresi, Rasler, and Thompson 2007
Weak actor	Weak actor as the trigger	Escalate	Clare and Danilovic 2010
Multiple actors	Number of crisis actors	Escalate	Gochman and Maoz 1984;
	Number of all actors involved	Escalate	Brecher 1993; Colaresi and Thompson 2002
Superpower involvement	Superpower intervention	Escalate	Leng 2000
Resolve	Hostile threats	Escalate	Clare and Danilovic 2010
		De-escalate	Akcinaroglu, Radziszewski, and Diehl 2014; Akcinaroglu and Radziszewski 2017
	Many rivals	Escalate	Clare and Danilovic 2010
		De-escalate	Akcinaroglu, Radziszewski, and Diehl 2014; Akcinaroglu and Radziszewski 2017
Domestic Factors			
Domestic constraints	Public perceptions of the rivalry or of the rival	Escalate	Vasquez 1993; Thompson 2001; Bak, Chavez, and Rider 2019
	Audience costs	De-escalate	Russett 1993; Hensel, Goertz, and Diehl 2000; Colaresi and Thompson 2002
	Democratic dyad	De-escalate	
Leadership changes	Leader change	Escalate	Werner 1999; Maoz and Mor 2002; Clark and Nordstrom 2005; Colaresi 2005; Weiss 2013; Wu and Wolford 2018; Smith and Spaniel 2019
	Irregular leader change	Escalate	
	Regime change	Escalate	
Preferences of leaders	Hawkish	Escalate	Leng 1983, 2000; Vasquez 1993
Individual Factors			
Issues	Territory	Escalate	Hensel 1996b; Werner 1999; Hensel and Mitchell 2005; Senese and Vasquez 2005; Rasler and Thompson 2006; Hensel et al. 2008; Vasquez 2009; Carter 2010; Lektzian, Prins, and Souva 2010; Mitchell and Thies 2011; Dreyer 2012; Owsiak and Rider 2013; Moon and Souva 2016
	Strategic territory	Escalate	
	Contiguity and territory	Escalate	
	Frequently militarized issues	Escalate	
	Multiple issues	Escalate	
	Territory with intangible interests		

History	Previous crises	Escalate	Diehl and Goertz 2000; Leng 2000; Prins 2005; Chari 2012
		De-escalate	Wagner 2000
Shocks			
Systemic shocks	Post–Cold War	De-escalate	Bennett 1996; Levy and Ali 1998; Diehl and Goertz 2000; Rasler 2000; Colaresi 2001; Akcinaroglu, DiCicco, and Radziszewski 2011; Morey 2011; Rasler, Thompson, and Ganguly 2013
International shocks	Third-party intervention Global wars: World War I, World War II	De-escalate	
Domestic shocks	Regime change	De-escalate	
	Natural disasters	De-escalate	

FROM CRISIS TO ESCALATION

International and Dyadic Factors

Seeing relations among states as driven by their interests to survive in an anarchic world, the realist school of thought emphasizes the importance of realpolitik foreign policies focused on demonstrating power and the willingness to use it (Leng 2000). Rationalists assume that when faced with threats, the enemy will back down rather than risk fighting a costly war. Rivalry scholars note the other possibility is that such threats can elicit the opposite response and invite a defiant counterthreat (Vasquez 1993), especially when the enemy has equal capabilities (Leng 1983) and can credibly respond to coercive strategies. The negative, escalatory outcome of realpolitik strategies in crisis bargaining in the context of international rivals, however, follows a different pattern. Backing down in the face of threats is not common even when the risk of war is high. Even more surprising, states that are weaker in power reciprocate with their own coercive strategies that further increase the odds of the crisis escalating to war. In their steps-to-war model Senese and Vasquez (2005) find that such coercive strategies increase the security dilemma, which pushes states with territorial disputes to build up arms and alliances and puts them on the path of experiencing repeated militarized disputes involving the use and show of force. As these disputes increase so does the risk of war. Only when the disputes are low in frequency or have reached a very high number do states face a limited risk of experiencing war. Although the steps-to-war model does not specifically focus on international rivalries, it shows that this pattern applies to states with territorial disputes, and most international rivalries emerge because of territorial issues such as the recent Kashmir-related border skirmishes between Pakistan and India, for example.

In a world of incomplete information about true resolve, where resolve is the extent to which states are willing to go to war to keep promises and threats, the enemy must decide whether the rival's resolve is real or merely a bluff. Resolve is defined as the extent to which a state will risk war in order to keep its promises and uphold its threats (Mercer 1996, 15). As states strive for the best outcome, they exaggerate their resolve and capabilities (Fearon 1995). States often struggle to communicate and to decipher each other's resolve. This problem is more significant, however, when bargaining involves rival states and thus makes the risk of escalation particularly high. The history of rivalry clouds objective information and obfuscates the leaders' ability to understand their enemies' signals. What could be seen as objective when negotiating with non-rivals is processed through the lens of a rivalry (Colaresi, Rasler, and Thompson 2007). As such, resolve communicated by a militarily weak state might be correctly interpreted as a bluff in a non-rivalry setting but not in the context of a rivalry, leading to negotiation breakdown and to the type of repeated escalatory pattern culminating in war that Senese and Vasquez (2005) describe in the steps-to-war model.

Empirical findings show that rivals escalate their crises to war despite asymmetry in their capabilities (Colaresi and Thompson 2002). Weaker states in a rivalry context might, in fact, be more sincere when standing tough as they seek to communicate their resolve not only during a specific crisis but also in future conflicts (Clare and Danilovic 2010). Thus, it follows that states facing other hostile threats and rivalries are more likely to demonstrate resolve in a crisis. They do so because they have an expectation of a long-term hostile relationship with the rival(s) (Diehl and Goertz 2000). This expectation is greater with salient issues involving territory that has a broad public significance (Dreyer 2012), issues that have multiple dimensions to them, and issues that have been previously militarized (Mitchell and Thies 2011). In these contexts the strategy of resolve is embraced before the new crisis appears (Leng 2000) and increases the likelihood that the dispute escalates militarily rather than being resolved peacefully.

The strategy of demonstrating resolve becomes particularly important for weaker states that have failed in past disputes and that face multiple rivals. The challenger, however, might perceive the weaker state's escalation of the crisis as too risky and dismiss it as a bluff. Yet when crisis bargaining takes place in the context of international rivalry, constrained and weak states are on average 30 percent more likely to escalate the crisis with a challenger (Clare and Danilovic 2010). This suggests that seemingly objective factors that states might rely on to discern resolve, such as level of capabilities and resource constraint, might carry a different, possibly contradictory meaning in the shadow of rivalry, thereby increasing information breakdowns that limit the parties' ability to de-escalate the initial outbreak of a crisis. Overall, weak states in a crisis situation with a rival have a lot more at stake in backing down, and their aggressive posture can be understood only within the framework of sustained competition embedded in the nature of the rivalry. The pressure to escalate increases.

Broader regional and international dynamics have the potential to affect the distribution of power capabilities between rivals and to impact the nature of crisis bargaining. While a good number of crises between rivals have broader regional or systemic relevance, not all disputes have the potential to directly involve an intervention from outside states. Yet when a crisis situation expands beyond the dyad, the level of uncertainty increases about multiple actors' capabilities, domestic constraints, and strategies (Gochman and Maoz 1984; Brecher 1993). That uncertainty will be especially heightened in the rivalry context as a past history of threats and the unique competitiveness of interactions could make states more likely to misinterpret objective signals of resolve and assume worst-case scenarios. Not surprising, then, Colaresi and Thompson (2002) show that as the number of actors involved in a crisis with rivals increases, so do the odds of a crisis escalating to the point of war. As rivals pay more attention to each other than to other states, they are likely to attach more significance to any developments involving other actors' presence in the conflict. In such contexts of heightened uncertainty rivals might be especially motivated to resort to realpolitik tactics to signal their resolve. In doing so they are more likely to face a higher risk of crisis escalation.

At other times, and more relevant for regional powers, mere expectations that a superpower might intervene to help with settlement before the crisis reaches the point of war could embolden states to risk war, as was the case with the Indo-Pakistani and Egyptian-Israeli rivalries (Leng 2000). Overall, both dynamics involving either direct or expected involvement of other actors are linked to greater risk of crisis escalation.

Domestic Factors

While international competition drives realpolitik policies, leaders often face significant domestic pressure to stand tough or risk their removal from power. Nowhere is this truer than in the context of rivalry. Consequently even militarily weak rivals may escalate the crisis. This explains why, for example, despite its military weakness relative to India, Pakistan has frequently embraced coercive strategies that have intensified the crisis with its enemy (Diehl, Goertz, and Saeedi 2005).

Domestic pressure on rivals to stand tough depends on the magnitude of the rivalry. Although Long (2003) examines variation in the duration of war among enduring rivals depending on domestic audience costs, his insights and findings can be extended to prewar crisis bargaining. In contexts where rivals had either recent and frequent conflicts of low intensity or more distant but intense conflicts, the selectorate or the group capable of removing the leader from power (Bueno de Mesquita and Siverson 1997) exhibits less flexibility in an acceptable peaceful settlement. Domestic factions are then likely to gain political advantage and use the new crisis to outbid each other (Colaresi 2005). They may agree on the basic idea that a rival represents a security threat, but push for more aggressive responses in a crisis as a way to appeal to domestic audiences (Ganguly and Thompson 2011). The policymakers' hands are tied in such contexts if they are predisposed to accommodative behavior, or the domestic conditions favor the rise of hard-line leaders who themselves push for aggressive tactics.

Additionally, the extent to which past realpolitik policies have failed or succeeded determines shifts in societal beliefs that favor hard-liners and their coercive policies. Public perceptions that previous war has been worth the cost, as was the case in Germany after World War I, creates ripe conditions for hard-liners to dominate domestic politics (Vasquez 1993). Domestic audiences are more likely to identify threats from rivals as out-group threats, which are more likely to be considered high-stakes risks connected to security and welfare (Thompson 2001; Bak, Chavez, and Rider 2019). When hard-liners face a crisis situation in a high-stakes environment, their own beliefs about the value of realpolitik strategies coupled with domestic support result in a dangerous bargaining dynamic. Vasquez (1993) argues that even if the target is more accommodative when pressed with coercive tactics, it will have to respond in the same way or risk losing power. The domestic context in which hard-liners dominate will consistently push both sides to take more escalatory steps.

Hawks are also more likely to believe that the adversary will offer concessions and de-escalate the crisis when faced with a militarily superior enemy, regardless of its own domestic conditions (Leng 2000). Such leaders, for example, would be closed to any incoming information about the adversary's potential for responding with escalation due to that adversary's domestic pressure to stand tough or risk losing influence at home. This greatly limits the hawk's understanding of how the enemy's hands might be tied and, how in such contexts, their own coercive stand will inevitably lead to war by turning the domestic public of the rival state even more in favor of war.

The Crimean War crisis bargaining that pitted Russia against Turkey and Britain illustrates how quickly the situation can escalate when a leader blindly follows a strategy of coercive diplomacy without considering how such tactics can push the domestic public in the other state toward antagonism and eliminate an opening for compromise. To maintain the rights of the Greek Orthodox community in the Ottoman Empire, which appeared under siege after Turkey gave the Catholic Church access to disputed holy places, Russia made a series of provocative moves. In December 1852 Russia mobilized two army corps on the Turkish border and its two war vessels arrived in the Black Sea. In March 1853 Russia began an invasion of Wallachia and Moldavia (Peterson 1996). When the Turks initially asked the British for assistance, the latter were reluctant to provide any support. However, as Russia's coercive strategy intensified the public began to turn against the doves in the British government. Russia's occupation of two Turkish principalities shifted the balance of power in the direction of hard-liners to such an extent that Russia's later attempts at compromise were rejected as Britain now shifted to its own coercive strategy. Tsar Nicholas underestimated the extent to which aggressive tactics designed to demonstrate resolve could shift domestic debates in the rival state and tie policymakers' hands. In the case of the Crimean War the public empowered domestic hawks to such a degree that the policymakers had to choose between escalating the crisis with Russia or being voted out of office for making a compromise. In the end they chose the former (Peterson 1996). The presence of more hawkish leadership during a crisis, which increases the likelihood that coercive bargaining dominates as a strategy, thus increases the possibility of escalation. When accommodative leaders' previous attempts to de-escalate the tension have been met with resistance, such leaders must now demonstrate toughness by increasing the stakes or risk dire consequences. As general findings on rival cooperation from 1950 to 1990 show, unreciprocated cooperation culminates in leaders' removal from power (Colaresi 2004). Thus it is too costly for leaders who attempted to lower the tensions to repeat the same strategy when it initially backfired.

Other domestic conditions have an impact on rivals' strategies in crises and the crises' potential to escalate. Notable here are leadership turnover and regime change. If both occur during an interstate crisis, they are likely to exacerbate information gaps about states' resolve and to contribute to escalation of tensions immediately after the occurrence (Wu and Wolford 2018). A major domestic change injects private information into the bargaining dynamic that leads to uncertainty about new leaders' resolve and the extent to which transformation in political institutions will constrain or limit policymakers' actions toward the rival. Changes in domestic institutions can affect rules by which leaders can secure benefits or be punished as well as alter the size of relevant domestic audiences, factors that impact resolve in crisis bargaining (Weiss 2013; Wu and Wolford 2018). Furthermore, many unobservable factors accompany the arrival of new leadership, such as leaders' perceptions of institutional constraints (Clark and Nordstrom 2005), that increase uncertainty about whether the new leaders would have an easier path to war. The extent of the new leaders' hawkish tendencies might also be unknown. While hawkish leaders on average embrace realpolitik strategies, it is usually extreme hawks who are willing to bear the cost of war (Vasquez 1993). Furthermore, changes in public beliefs might create an opening for new leaders to embrace more accommodative policies. These different predispositions and the domestic context that affects leaders' strategies are not immediately clear to the enemy. Consequently, a rival state might try to test the new leaders' resolve by standing tough and resisting compromise.

Empirical evidence supports these insights and shows that regardless of whether they involve rival or non-rival states, crises with new leaders last longer and result in a higher level of fatalities than crises with leaders who have been in power for over four years (Smith and Spaniel 2019). Werner (1999), however, finds that only specific types of leadership changes, those that are unexpected and unconstitutional, dramatically increase the risk that the belligerents will seek to renegotiate previous settlements by resorting to violence. The risk dissipates when change is expected and marginal.

When crisis bargaining takes place between international rivals, empirical results show a high probability that disputes will escalate to the use of force immediately following leadership and regime change (Maoz and Mor 2002; Wu and Wolford 2018), but the risk declines over time when information about resolve is revealed (Wu and Wolford 2018). This happens because the passage of time reveals to the opponent that some of the new leaders are more dovish or accommodative political entrepreneurs than their predecessors or that some of them challenge existing strategies and have more control over institutions. As such, with time the possibility for conflict de-escalation and even the termination of the rivalry may emerge (Rasler 2000). Initial uncertainty regarding national leaders' beliefs may be replaced with greater understanding of what to expect from the enemy as their strategies become clearer. The change in leadership from Nasser to Sadat in Egypt thus favored de-escalation of tensions between Egypt and Israel as time revealed that Sadat was open to new strategies (Leng 2000). Overall, these findings suggest that time can offset the initial negative impact of domestic transitions on crisis bargaining between rivals. While initial leadership and regime change increases information asymmetries about resolve and elevates the risk of crisis escalation, this risk goes down as the passage of time reduces uncertainty and creates opportunities for peaceful resolution.

Individual

At the individual level, it is worth considering the role of learning and misperception on crisis bargaining between rivals. These factors have been explored in the general literature on the causes of war (e.g., Jervis 1994, 2002; Duelfer and Dyson 2011), but studies show they might play an even greater role in the context of rivalry, often complicating objective assessment of rivals' resolve and intentions that intensifies uncertainties and escalates crises to the point of war. Past crises weigh heavily on current disputes among rivals in influencing the lessons that leaders draw from interactions (Leng 2000). Rationalist literature on war argues that wars and crises reveal information about enemies that reduces uncertainty and thus could provide more credible information regarding capabilities and resolve for future interactions (Wagner 2000). Yet this form of rational information updating is somewhat muted among rival states. In his focus on twelve militarized crises in the context of Soviet-American, Egyptian-Israeli, and Indo-Pakistani rivalries, Leng (2000) concludes that instead of reducing uncertainty that could create an opening for negotiated settlements, experiences from past crises led to more coercive bargaining and violent escalation. After fourteen or more dyadic crises, leaders react to new crises with militarized force almost 90 percent of the time instead of seeking peaceful accommodation (Prins 2005).[3] Put simply, more so than in any other state interactions, rivals are predisposed to draw only selective information and lessons from their previous crises, from events during a crisis, or from events occurring between the crises. In the case of the Indo-Pakistani rivalry, this meant that leaders were unable to derive diplomatic lessons from their interactions, often repeating the same coercive patterns that have resulted in yet more escalation (Chari 2012). Since the early 1980s these escalations have stopped short of war due to fears of nuclear exchange. Possession of nuclear weapons by rivals ultimately prevents the crises from turning into wars involving the use of force and its resultant casualties (Ganguly and Hagerty 2005). However, nuclear weapons have not dissuaded India and Pakistan from threating to use force in their subsequent crises before ultimately de-escalating (Ganguly and Hagerty 2005). Overall, the failure to update beliefs and the determination to pursue coercive strategies in order to demonstrate resolve explain why crisis escalation is more likely to occur in the context of international rivalries.

The escalatory pattern usually begins when a rival state misperceives its enemy's intentions and posture as aggressive and responds with coercive tactics. This in turn pushes the target to respond in like manner and without considering that the initiator might have wrongly misperceived the dynamic that led to its initial aggressive tactic (Maoz and Mor 2002). While that hostility need not always end in war (Leng 2000), the crisis nevertheless escalates to a dangerous point. The repetitive pattern of coercive behavior is also consistent with Diehl and Goertz's (2000) concept of enduring or long-term rivalry; when a pattern of hostility sets in, the rivalry itself is then "locked in" and becomes resistant to changes. It is worth noting that while this dynamic shapes the nature of crisis bargaining between international rivals in general, the escalatory pattern is likely to be greater depending on whether the leaders of rival states have a hawkish or dovish view of the world. Hawkish leaders believe in standing tough no matter the cost (Leng 2000). Such leaders display greater risk acceptance for war than leaders with more accommodative or dovish beliefs (Leng 1983; Vasquez 1993). While not all rivalries involve hawkish leaders, those that are enduring are most likely to involve leaders with lower aversion to war (Leng 2000). The escalatory pattern in such rivalries is thus likely to be most dangerous.

Issues

In his discussion of the connection between commitment problems and war, Fearon argues that "mutually preferable bargains are unattainable because one or more states would have an incentive to renege on the term" (1995, 3). This happens because of actual or expected changes in the future balance of power and in the opportunities of states. In an anarchic world states have no guarantees that current deals will not empower one party more in the future to then back away from the agreement and have more power to leave its enemy worse off than before the deal was signed. Enforcing agreements is difficult. When concerns of this nature emerge states might be willing to risk a war now to avoid fighting a more costly one later. Even if the parties have a sincere interest in settling peacefully, the possibility that the agreement could empower one of them enough to later abandon the deal could be sufficient to deter effective crisis management (Wagner 2000; Powell 2006).

Commitment problems are rampant when states' disputes involve territories, such as the competition between India and Pakistan over Kashmir (Owsiak and Rider 2013; Moon and Souva 2016). If the territory in question has economic or military value, a bargain that involves even a small transfer might leave one of the states better off in the future with greater bargaining leverage. About half of all rivals share unsettled borders, and these territorial disputes sustain the rivalry (Owsiak and Rider 2013). Rivals with disputed borders have an expectation of facing threats from each other in the future (Diehl and Goertz 2000), which means that they must pay attention to how the current deal will affect their enemy's power in future interactions. Past interactions between rivals that are marked by mistrust and fear are more likely to make such states assume the worst from an enemy and thus exacerbate commitment problems. Overall, while territory is the most war-prone issue among states in general, it is even more salient in the context of contiguous rivals as it generates severe concerns about future exploitation. The interplay between rivalry, contiguity, and contested territory is supported by empirical research, which shows that crises involving territory and contiguous rivals are more likely to escalate to war than crises involving other types of issues or noncontiguous rivals (Senese and Vasquez 2005; Rasler and Thompson 2006). Last, strategic territory matters as it has offensive and defensive value which can change the balance of power between disputants. Scholars have found that disputes over salient territory are thus less likely to lead to peaceful settlement and more likely to turn into militarized interactions among the claimants (Huth 1996, Hensel 2001, Hensel et al. 2008, Carter 2010).

Once an issue becomes intractable due to commitment problems and its significance is further linked to national identity and the evolving mistrust that sustain the rivalry (Dreyer 2012, Huth 2000), political leaders begin to face significant constraints from the public to compromise. In addition to material considerations that may drive territorial disputes, territories can have intangible salience or a deep psychological connection for the population in rival states (Rasler and Thompson 2006), which creates greater reputational losses for leaders who compromise in such disputes (Hensel and Mitchell 2005). Domestic attachment to territory linked to national identity and the ensuing public pressure to defend explain why leaders are more likely to resort to militarized strategies when handling a crisis involving contestation over territories (Moon and Souva 2016, Vasquez 2009) than crises related to maritime or river issues (Lektzian, Prins, and Souva 2010). Moreover, Hensel and Mitchell (2005) find that among states with territorial disputes, though not exclusively international rivals, it is contestations over intangible issues that are more likely to escalate to higher-level military conflict involving fatalities and interstate war than disputes over territories linked to tangible issues.

Commitment problems over territory, however, are not insurmountable and rivals are at times able to overcome them with the help from third parties that can offer security guarantees. For example, the presence of the United Nations troops in what would become a demilitarized zone in the Sinai Peninsula made it easier for Israel to give up occupation of the territory, overcome commitment problems with Egypt, and sign the 1979 peace treaty (Benziman 2019). Mediators can use leverage and support from international organizations to make agreements enforceable and alleviate future security concerns (Walter 2002). In the context of rivalry, however, commitment problems will be stronger not only because of underlying territorial issues but also because of the conflict-prone nature of international rivalry itself that makes mediation challenging, particularly when rivals develop high levels of hostility toward each other (Greig 2001). When rivals engage in mediation, as was the case with Egypt and Israel, they are more likely to do so in later stages of the rivalry after the issue has already been the source of prior violent disputes. More often, however, intense and frequent territorial disputes between rivals may put more domestic pressure on leaders to embrace hawkish policies in future interactions with the enemy. As such, domestic conditions characterized by lack of structural changes and a reanalysis of existing policies could push states to resist mediation efforts and choose instead to respond aggressively (Greig 2001). These dynamics impede de-escalation of tensions involving territory more so in the context of rivalry than among other dyads.

FROM CRISIS TO DE-ESCALATION

Given the level of hostility and the expectation of threats in rivalries, it is not surprising that crisis bargaining in the context of rivalry is particularly challenging. Yet not all disputes escalate to the point of war. Rare as it may be, rival states do engage in some form of accommodation in response to threats (Akcinaroglu and Radziszewski 2017), which might create more favorable conditions for crisis de-escalation. Leng (2000) also finds that when rivals want to avoid war, they can tolerate high levels of escalation but still de-escalate enough to avoid the crisis turning into a militarized dispute.

Domestic Factors

Several conditions favor movement away from militarized response toward de-escalation. In general, commitment problems and uncertainty about opponents' resolve are less severe and thus less likely to escalate the crisis when rival states are democracies. Due to greater institutional and domestic transparency in democracies, leaders' resolve in crisis bargaining is easier to decipher (Ramsay 2004). For example, a leader who tries to extract a high level of concessions from a rival state and threatens war might be considered more sincere if the domestic public and the opposition are also supportive of coercive tactics. In such a case the enemy might be more inclined to de-escalate the tension to avoid war and to believe that the rival's threats are indeed sincere.

Scholars have explored extensively the existence of domestic audience costs in democracies for not following through on commitments to stand tough (e.g., Fearon 1995; Partell and Palmer 1999; Tomz 2007a), with the most notable connection to leaders in democracies where high levels of accountability to domestic audiences make such leaders especially vulnerable.[4] Audience costs can also help explain why commitment problems might be less severe among democratic rivals. If the public exhibits strong support for the settlement, leaders who might consider reneging on such agreements in the future might be politically vulnerable. As such, rival states might have more confidence in the enforceability of an agreement when they are both democracies. Last, due to the values shared between democracies (Russett 1993), escalation of conflict with another democracy may make the leaders politically vulnerable to their domestic public. While the public might recognize the existence of the rivalry and see it as a threat, it may nevertheless be reluctant to pressure the leaders to escalate the crisis to the point of war. The leaders of rival states would thus have an interest in de-escalating the tensions. Overall, explanations based on transparency, audience costs, and democratic values bolster existing empirical findings regarding democratic rivals experiencing a greater likelihood of crisis de-escalation (Colaresi and Thompson 2002) and a lower propensity to engage in militarized disputes (Hensel, Goertz, and Diehl 2000). As rivalries often, though not always, exist when one or both states are not democratic, the positive effect is largely observed when one or both states experience a transition to democracy.

International

When managing international crises, rivals at times face concurrent tensions on multiple fronts. This vulnerable situation presents an opportunity for de-escalation as rivals embedded in multiple rivalries must avoid the risk of costly wars with more than one state at a time. Akcinaroglu, Radziszewski, and Diehl (2014) show that when a state faces not one but multiple severe threats in the form of militarized disputes, it is likely to pursue a series of positive gestures toward the state that is most militarily capable. This suggests that possibility for crisis de-escalation is greater for states with multiple and militarily superior rivals. Furthermore, a crisis on one front can push a state to engage in accommodative behavior toward another rival and offset the potential for future crises to escalate with another enemy. Thus even if a state engages in coercive behavior toward one enemy, it likely pursues some form of cooperation with another. As Akcinaroglu, Radziszewski, and Diehl (2014) and Akcinaroglu and Radziszewski (2017) show, the choice of which rival to accommodate is based on the enemy's military capability as well as the potential that the accommodation of the less threatening rival can de-escalate a more pressing crisis on another front. For example, they find that accommodating a rival state with extensive economic connections to other states in the world can be useful in managing crises with another rival due to the economically connected rival's material leverage.

While Akcinaroglu, Radziszewski, and Diehl's (2014) and Akcinaroglu and Radziszewski's (2017) analyses apply to a subset of the most dangerous rivalries, the enduring ones, and stops short of investigating whether rivals reciprocate strategic cooperative behavior, they show how the international environment affects decisions about when to refrain from realpolitik strategies. When the broader crisis environment is taken into consideration at the time of the dispute, states might adopt different strategies to handle multiple crises, which in turn could have significant implications for understanding why some crises have the potential to de-escalate.

Shocks

De-escalation of crises between rivals is also possible with the existence of shocks or some type of environmental crisis (Rasler 2000). Shocks may reveal previously unquestioned internal and strategic vulnerabilities or free leaders from domestic audience costs that limit compromise. They alter the parties' expectations by reducing the perceptions about the level of threat the enemy poses, which allows the rivals to break the continuity of coercive foreign policy (Rasler, Thompson, and Ganguly 2013). Failures in Afghanistan (an international shock) and the arrival of reform-oriented Gorbachev as a new political entrepreneur created an opening for questioning Soviet policies and increased awareness among some elites about the need to adopt a new approach toward the United States (Mandelbaum 1989). This fresh outlook ultimately led to the end of the US-Soviet rivalry. The literature mostly connects shocks to the termination of rivalry, which is not surprising given that shocks precede rivalry termination approximately 77 percent of the time (Diehl and Goertz 2000). Among the shocks empirically associated with rivalry termination are systemic power shifts, regime changes, global war, high-concentration conflicts or conflicts with a high level of casualties over a short period of time for both opponents, and domestic crises (e.g., Bennett 1996; Levy and Ali 1998; Diehl and Goertz 2000; Colaresi 2001; Morey 2011; Rasler, Thompson, and Ganguly 2013).

Shocks can affect individual crisis dynamics within the rivalry even if they do not lead to rivalry termination. While this line of research is considerably underdeveloped, there is some evidence that the positive, de-escalatory effect of shocks on crisis management depends on the timing and level of violence between rivals prior to the onset of the crisis or during the crisis itself. As mentioned earlier, a shock that involves a change in political institutions introduces more uncertainty into an existing crisis, making it difficult for the rival state to understand how the shock might affect policymakers' resolve. In the long run, however, as information regarding the shock's consequences becomes clearer, enemies can reduce information gaps regarding resolve, making a settlement possible. Over time, these shocks also reveal to the enemy whether an opening has been created for more dovish policymakers to challenge the status quo. For example, Rasler (2000) examined the impact of three international shocks—Israel's invasion of Lebanon, the Gulf War, and the intifada—on patterns of de-escalation between Palestine and Israel from 1978 to 1998. She found that shocks had a positive effect in the long run—five years—but not in the short or medium run. The intifada, a shock that resulted in more violence between the rivals, initially hardened public attitudes and encouraged more hard-line strategies. In the long term, however, the shock had a more moderating impact on the public, which created an opportunity for moderate policy entrepreneurs to question existing strategies toward the rival.

In cases where rivals have not experienced recent violence prior to the occurrence of a shock and where the shock itself does not create violence between enemies, public perceptions of the enemy are likely to change and create a more immediate opening for moderation. After the 1999 Izmit earthquake hit Turkey, post-disaster cooperation at the grassroots level and empathy toward the victims combined with government officials' disaster-related talks to create a window of opportunity for rapprochement between Greece and Turkey (Akcinaroglu, DiCicco, and Radziszewski 2011). When shocks shift public opinion toward cooperation, leaders are freed to pursue more dovish or accommodative policies. Such policies in turn are likely to be interpreted as sincere information because they mirror domestic attitudes and enable the rival state to reciprocate with positive gestures. Yet the extent to which shocks can shake the inertia of rivalry and lead to rapprochement depends on whether the enemies had experienced recent violence. In the case of Israel and Palestine the shock of intifada violence generated a positive effect in the long run only. In the India-Pakistan rivalry a powerful earthquake that hit Kashmir was not a sufficient domestic shock to elicit empathy and to inspire change in public attitudes due to the high occurrence of Hindu-Muslim violence at the communal level. However, in the Turkey-Greece rivalry the absence of recent violence made it possible for the shock to induce the rivals' cooperation. Although Turkey and Greece were not experiencing a crisis at the moment the earthquake struck, this case nevertheless shows the potential that shocks have to de-escalate rivalry.

EMPIRICAL ANALYSIS OF CRISIS ESCALATION: COMPREHENSIVE MODEL

To summarize the findings in the literature, among the international and dyadic factors, weak and crisis-initiating states (Clare and Danilovic 2010), the number of actors (Colaresi and Thompson 2002), the power imbalance/balance (Colaresi,

Rasler, and Thompson 2007), the expectation of or actual superpower involvement (Leng 2000), and the existence of multiple hostile threats and rivals (Clare and Danilovic 2010) are all linked to crisis escalation. Domestic factors such as leadership changes, shorter leader tenure, and various types of leader exits in the dyad (Clark and Nordstrom 2005; Wu and Wolford 2018) lead to informational problems and to the exacerbation of violence. However, rivals that are democratic and are subject to severe audience costs are less likely to escalate the crisis (Russett 1993; Colaresi and Thompson 2002). Among the individual factors, past crises can add information, lead to faulty learning and the perpetuation of beliefs and attitudes, and keep the actors locked in rivalry, adding to mistrust (Leng 2000; Wagner 2000). Additionally, issues matter. Territorial issues, specifically those between contiguous rivals, should exacerbate fears of agreement enforceability, hence leading to an escalation of the crisis (Moon and Souva 2016). Territorial issues that involve identity (intangible issues) or crises that involve multiple issues are harder to resolve (Hensel and Mitchell 2005; Vasquez 2009; Mitchell and Thies 2011). Last, scholars talk about shocks. Among them are regime change (Wu and Wolford 2018) and natural disasters as domestic shocks. The former is expected to exacerbate informational problems while the latter is likely to lead to rapprochement (Akcinaroglu, DiCiccio, and Radziszewski 2011). In addition, the ending of the Cold War was a systemic shock.

We bring all of these variables together in a comprehensive model of crisis escalation.[5] Using this model, we can establish the relationship between a binary outcome variable, in this case escalation to violence in the dyad during the crisis, and the group of predictor variables listed in Table 5.2. The model, based on a logistic regression, thus explains the probability of a dispute escalating to a military clash between rival states in the dyad in crisis.[6] The advantage of employing a statistical analysis is that we can establish results that are generalizable across countries.

Our results are summarized in Table 5.2 (see the appendix for the statistical model).[7] Escalation is a common phenomenon in rival dyads during a crisis, with almost 60 percent of our observations involving escalation to important degrees of violence as opposed to no violence and minor clashes. The results confirm that both informational and commitment issues are an obstacle to peaceful settlement, contributing instead to aggressive behavior. The analysis shows that most of the international and dyadic factors affect escalation. Except for superpower involvement and hostile threats in the dyad, all the variables were statistically significant. Multiple actors involved in the crisis exacerbate informational problems, causing bargaining failures and hence escalation (Colaresi and Thompson 2002). Though statistically significant, we did not find support for escalation when the crisis-triggering actor was the weaker party. This ran contrary to our expectations. Indeed, we found that war was more likely when the rivals had symmetric capabilities (Leng 2000). On the plus side, our results demonstrated that the existence of multiple rivalries may sometimes drive states to avoid escalation. Though the need to show resolve is important, these simultaneous constraints may drive actors toward accommodation to avoid multiple wars (Akcinaroglu, Radziszewski, and Diehl 2014; Akcinaroglu and Radziszewski 2017).

Table 5.2. A Summary of Findings from the Statistical Analysis

Variables	*Operationalization*	*Expectation*	*Statistical Significance*	*Expected Direction?*
International and Dyadic Factors				
Capabilities	Symmetric power (CINC ratio)	Escalate	Yes	Yes
	Asymmetric power (CINC ratio)	Escalate	Yes	No
Weak actor	Weak actor as the trigger	Escalate	Yes	No
Multiple actors	Number of crisis actors	Escalate	Yes	Yes
	Number of all actors involved	Escalate	Yes	Yes
Superpower involvement	Superpower intervention	Escalate	No	Yes, but not significant
Resolve	Sum of hostile threats	Escalate	No	Yes, but not significant
		De-escalate		No
	Sum of rivals	Escalate	Yes	No
		De-escalate		Yes
Domestic Factors				
Domestic constraints	Minimum audience costs	De-escalate	Yes	Yes
	Democratic dyad	De-escalate	Yes	Yes
Leadership changes	Leader change	Escalate	Yes	Yes
	Irregular leader change	Escalate	Yes	Yes
	Regime change	Escalate	No	No
Individual Factors				
Issues	Territory	Escalate	Yes	No
	Contiguity	Escalate	Yes	No
	Contiguity and territory	Escalate	Yes	Yes
	Multiple issues	Escalate	No	No
	Territory with intangible interests	Escalate	Yes	Yes
History	Previous crises	Escalate	Yes	Yes, but not robust
		De-escalate		No
Shocks				
Systemic shocks	Post—Cold War	De-escalate	Yes	No
Domestic shocks	Regime change	Escalate	No	No
	Natural disasters	De-escalate	Yes	Yes

Our expectations about the domestic factors and escalation were also satisfied. Mainly, the results show that democracies and those regimes with high audience costs are better able to reveal resolve and are taken seriously by their rivals. Knowing that the rival political leaders are likely to be punished for bluffing by their domestic audiences, states filter out credible actions from those that are not. Leadership changes lead to informational issues. Specifically, unexpected and unconstitutional changes increase the commitment issue by raising expectations that new leaders will renege on previous agreements. In both cases leaders are likely to resort to violence so that their rivals can form beliefs about their resolve.

Among the individual factors, territorial claims were robustly linked to escalation. Disputes that revolve around contiguous territory are more likely to endanger the balance of future power between rivals, making it less likely any will settle down peacefully during a crisis (Senese and Vasquez 2005; Rasler and Thompson 2006). Similarly, intangible territorial claims, those that involve an ethnic identity, make it less likely for rivals to give in. In that situation, leaders find it politically impossible to make concessions to the rival when the public is psychologically connected to the population in that territory (Hensel and Mitchell 2005; Rasler and Thompson 2006). Evidence then shows that the salience of the territorial issue matters. Results also confirm that rivalry has a self-perpetuating escalatory pattern. Previous crisis among rivals does nothing to alleviate informational problems, but instead it amplifies misperceptions that occur in locked-in rivalries (Leng 2000; Prins 2005).

Last, we found that some shocks provide an opening for rivals to find mutually acceptable bargains. Natural disasters provide such an opportunity by changing public perceptions that allow leaders to take steps toward rapprochement (Akcinaroglu, DiCicco, and Radziszewski 2011). Though public hardening in rivalries often ties the hands of leaders, disasters with high fatalities change public attitudes toward accommodation. Surprisingly, we found that the post–Cold War era, as a systemic shock, has not made it possible to reduce tensions once the crisis is triggered. We also did not find regime change to be a predictor of escalation. Regime change may bring radical institutional improvements that make it possible for new leaders to alleviate informational and commitment issues in rivalries. The type of regime change thus may matter.

FUTURE EXTENSION

Future research could focus on three main areas to improve our understanding of crisis bargaining in the context of rivalry. First, many of the existing factors that complicate conflict de-escalation fall under some form of informational problems. Questions about resolve and the credibility of threats, and the inability to objectively update information, contribute to bargaining failures. Considerably less attention, however, has been given to commitment problems, which are especially relevant for rivalries given that in this context states have a high expectation of future threats and hostile interactions. As such concern about rivals' opportunity to renege on agreements in the future presents an obstacle for peaceful settlement of crises. The main focus currently is one key issue at stake, specifically territory that increases commitment problems and is linked to less durable peace (Werner 1999). Other factors that could change the opportunity structure for the enemy to abandon the deal in the future and that its rival might be concerned about should be considered. For example, rivals that have multiple trade linkages could have more capabilities in the future as these linkages might not only foster trade but also possible assistance of such actors in case of a dispute. Even if trade linkages are nascent, the potential for trade connections to grow could create more opportunities for the enemy to challenge the terms of an existing settlement and be more successful at it. This might suggest that commitment problems in crisis bargaining could be more severe when one or both adversaries are economically connected to many other actors. On the other hand, if rivals increase trade with each other, the dependence on the other could limit the incentive to increase territorial disputes. This is because settled borders secure property rights and bring policy certainty that reduces the costs associated with economic transactions (Simmons 2005).

Second, there is a need to focus on the evolution of the crisis after the initial attempts to de-escalate. Democratic rivals may benefit from greater informational clarity due to the presence of audience costs (Colaresi and Thompson 2002; Tomz 2007a), while shocks can create an opening for more sustained rapprochement (Akcinaroglu, DiCicco, and Radziszewski 2011). Rivals occasionally abandon coercive strategies in favor of making positive gestures toward the enemy. Sometimes this happens outside of a specific dispute. For example, rivals have been known to provide foreign aid to other rivals that experience domestic instability to reduce the risk that such instability might bring more uncertainty to the rivalry itself (Uzonyi and Rider 2017). Sometimes, however, rivals make concessions during a crisis situation outside of the democratic dyad context and in the absence of shocks. When these positive gestures are studied the focus is on why a rival state might pursue such strategies and when (Akcinaroglu, Radziszewski, and Diehl 2014; Akcinaroglu and Radziszewski 2017). The outcome is usually considered in the context of a multistep effort to end the entire rivalry (Kupchan 2010). Future research should examine how a positive gesture directed by one rival toward another in the midst of an international crisis situation might lead to different outcomes.

One expectation is that since such positive gestures are likely to be on the lower spectrum of concessions (Akcinaroglu, Radziszewski, and Diehl 2014), they must be carried out in a series of gestures to signal credibility. Once the gestures signal credible intent to de-escalate, the enemy should respond with positive gestures in turn, following the logic of tit for tat. Another possible expectation is that the enemy abandons the logic of reciprocity and the coercive shadow of rivalry takes over, in which case the enemy interprets the rival's positive gestures as weakness and exploits that weakness by pressing for more demands. A coercive response to accommodative gestures would likely leave the concession-making rival no choice but to abandon its positive overtures or face political costs for being too soft in response to coercion. This dynamic would likely escalate the crisis. Empirical analysis of these competing outcomes across various types of rivalries would shed more light on the consequences of positive gestures in specific crises rather than on how they impact the termination of the entire rivalry.

The analysis could also explore the different types of de-escalatory patterns that such gestures might generate. Existing literature in general refers to de-escalation as the phenomenon when a dispute does not end in war. More specific outcomes, such as de-escalation with a partial agreement, de-escalation with a settlement of the dispute, or de-escalation with no agreement, should be studied to broaden our understanding of the relationship between accommodation and the nature of de-escalation.

Finally, existing research on crisis bargaining focuses predominantly on the dyadic context, the adversaries themselves, with limited concentration on the broader developments surrounding the crisis at hand. Colaresi and Thompson (2002) have shown that a crisis involving multiple actors has a greater chance of escalation, and Akcinaroglu, Radziszewski, and Diehl (2014) show that when a state faces not one but multiple severe threats, it is likely to offer a series of positive gestures to the state that is most militarily capable. As some rivals must balance conflicts with multiple enemies, developments in one rivalry can impact crisis bargaining in another. Conducting network analysis to consider a variety of linkages between rivalries can shed light on the extent to which states might adopt different strategies in crisis bargaining depending on the actors' broader connectivity in the system (Akcinaroglu and Radziszewski 2017). This in turn could have significant implications for understanding why some and not all crises escalate.

APPENDIX

In what follows we list the variables we used in the analysis, and we describe how we measure them.

Dependent Variable

Escalate: If any of the actors escalated in the crisis, this is coded as 1. Escalation means important or preeminent violence as compared to minor or none and is coded from the Actor Level-International Crisis Behavior, ICB1 (v13) Project. We used Cenvio, the centrality of Violence in ICB1, for this variable.

International and Dyadic Factors

Trigger Weak Actor, CINC Ratio: This is coded as 1 if the actor that triggered the crisis is the weak one in the dyad. We use the Composite Indicator of National Capability (CINC) index from the National Material Capabilities (v5.0), Correlates of War (COW) data set to code the weaker actor in the dyad. The CINC index is based on six variables recorded in the data set: military expenditure, military personnel, energy consumption, iron and steel production, urban population, and total population. We use the Trigent variable from the Actor Level-International Crisis Behavior, ICB1 (v13) to code the actor triggering the crisis. If the triggering actor is also the weaker actor, then this is coded as 1.

The CINC ratio is the ratio of the weaker actor to the stronger one in the dyad. It measures the degree of asymmetry in the capabilities of the two actors. The farther this ratio is from 1, the more the power relationship is asymmetric in the dyad.

Number of Crisis Actors, Number of Actors: These are both taken from the System Level-International Crisis Behavior, ICB1 (v13) project. The two variables report the number of crisis actors involved in an international crisis and the number of states perceived by the crisis actors to be involved, including the crisis actors themselves. For previous crises, we coded this as 1 if the actors in the dyad had crises before this event. For this, we looked at whether the same dyad had any observations in ICB1 before the crisis year.

Min Simultaneous Hostile Threats, Min Rivals: Both are taken from the Dyadic Militarized Interstate Dispute Dataset MID, (v4.3). These variables denote the total number of strict hostile threats (HIGHHOST, coded as 4 = use of force and 5 = interstate war) the actors have with any rival in the same year. We then take the minimum number of strict hostile threats in the dyad. We also code the minimum number of rivals each actor in the dyad has in that year.

Superpower Involvement: This is taken from the System Level-International Crisis Behavior, ICB1 (v13) project. It denotes superpower involvement by the United States and the USSR in post–World War II crises.

Domestic Factors

Minimum Audience Costs: We use Uzonyi et al.'s (2012) Audience Cost capacity index to measure regime sensitivity to audience costs. This index ranges from 0 to 3 and is based on two dimensions, openness of executive recruitment (xropen) and restrictions on political participation (parcomp), which are taken from the Polity IV Project. We take the minimum score in the dyad.

Democratic Dyad: If both actors in the dyad score 5 or above in the Polity IV Project, this variable is coded as 1.

Leader Change, Min Leader Tenure, Irregular Change, and *Regime Change* are coded from Archigos (Goemans, Gleditsch, and Chiozza 2009) data, which comprise a data set on political leaders. We code leadership change preceding or during a crisis. If there is such a change in any of the actors in the dyad, this variable is coded as 1. Irregular entry is coded as 1 if the leader exits in an irregular way such as a coup, revolt, or assassination. *Min Leader Tenure* is the minimum leadership tenure of any of the actors in the dyad. We code the leadership tenure of each actor using the leadership entry date from Archigos and the trigger date of the crisis.

Leader Changes That Are Not Successors: Leader changes that are not successors are coded from Change in Source of Leaders Support, CHISOLS (Leeds and Mattes, Leeds, and Matsumura 2016). We code any change in the source of leader support in the crisis year (solschangedummy in CHISOLS) by any actor in the dyad. It is coded as a dummy.

Systemic and Domestic Shocks

Post–Cold War: This is coded as 1 for all years succeeding 1990.

Natural Disasters: This is coded from EMDAT, International Disasters Database. We took the total deaths in natural disasters for both actors in the dyad in the crisis year. We categorized the fatalities as 1 (< 100), 2 (100 ≤ and < 1,000), and 3 (> 1,000) to denote the severity of the consequences of the disasters. We then summed the categorical score of both actors. This ranges between 1 and 6. For example, 6 denotes that both states had a severe natural disaster in the crisis year.

Regime Change: Coded from CHISOLS, this is coded as 1 if any regime change occurred in the dyad in the crisis year.

Individual Factors

Previous Crisis: This is taken from the System Level-International Crisis Behavior, ICB1 (v13) project. This variable is coded as 1 if the actors in the dyad had a crisis before this event. For this, we looked at whether the same dyad had any observations in ICB1 before the crisis year.

*Territory, Contiguity, Contiguity*Territory*: Territory is direct contiguity (separated by land or by a river border) between the actors in a dyad. It is taken from Direct Contiguity (v3.2) in the COW project. Territory is taken from the Actor Level-International Crisis Behavior, ICB1 (v13). We use Gravity = 3 in ICB1 to denote the territorial threat. Contiguity*Territory is the interaction of the two variables.

Intangible Territorial Claims: This is coded as 1 if there is a territorial threat to any of the actors in the dyad and the issue is coded as ethnic in System Level ICB1 data (Ethnic = 1 or 2).

Multiple Issues: This is coded from System Level ICB1 data (Issues variable). This variable is coded as 1 if the principal issue for the crisis actors is composed of two or more issues.

Table 5.A1. Logistic Model on Crisis Escalation

Escalate	*Model 1*	*Model 2*
International and Dyadic Factors		
CINC ratio	2.86(0.87)***	3.60(1.10)***
Weak state as trigger	0.74(0.16)	0.62(0.14)**
Number of actors	1.06(0.02)***	1.04(0.01)***
Number of crisis actors	1.11(0.05)**	
Superpower involvement	1.02(0.05)	1.06(0.05)
Minimum hostile threats		1.07(0.13)
Minimum rivals	0.82(0.07)**	
Domestic Factors		
Minimum audience costs	0.51(0.11)***	
Democratic dyad		0.39(0.11)***
Leadership change	2.25(0.40)***	3.02(0.57)***
Irregular leadership change		2.79(0.55)***

Escalate	*Model 1*	*Model 2*
Individual Factors		
Territory		0.47(0.20)*
Contiguity		0.43(0.10)***
Contiguity*Territory		5.08(2.60)***
Intangible territorial claims	2.02(0.60)**	
Multiple issues		0.75(0.14)
Previous crisis	1.12(0.10)	1.16(0.10)*
Systemic and Domestic Shocks		
Post–Cold War	2.18(0.59)***	1.65(0.44)*
Natural disasters	0.76(0.09)**	0.66(0.08)***
Regime change	0.51(0.26)	0.44(0.23)
Constant	1.43(0.57)	1.10(0.42)

* $p < 0.1$, ** $p < 0.05$, *** $p < 0.01$. Odds ratios are reported. Constant estimates baseline odds. N1 = 721, N2 = 758. We report odds ratios, defined as the probability of success and the probability of failure. Ratios less than 1 coincide with a negative coefficient and mean the odds of failure are greater.

NOTES

1. While Diehl and Goertz (2000) focus on the number of militarized disputes and the time span within which these disputes occured in order to identify and classify rivalries, Colaresi, Rasler, and Thompson (2007) rely instead on interpretive analysis of states' perceptions of each other as competitors and enemies whose threats can become militarized in order to identify what they refer to as strategic rivals.

2. Militarized disputes are defined as cases of conflict where a state engages in the threat, display, or use of military force short of war toward another state (Jones, Bremer, and Singer 1996, 163).

3. Senese and Vasquez (2005) show that this relationship is curvilinear, with the chances of war decreasing after twenty-eight militarized interstate disputes. However, their findings are not exclusively focused on international rivals.

4. This mechanism may be weaker in authoritarian states where propaganda can shape public perceptions to a greater extent and make domestic audiences more supportive of leaders' empty threats (Weiss and Dafoe 2019).

5. We include all the variables we mentioned in Table 5.1 in our analysis except for four. First, we were unable to find any data sets to code hawkish or dovish behavior. Second, when we tried to use the two variables, claim salience (ICOWsal) and the total number of militarized disputes over the issue that occurred during any dyadic claim (midsiss) from the Issue Correlates of War (ICOW) data set, we had a lot of missing observations. Third, we lost six of seven observations partly because ICOW is coded until 2001 and partly because the two data sets are incompatible. Fourth, no data set can capture public perceptions of the rivalry except for country-level surveys.

6. The data are in dyad format for each crisis.

Chapter Six

Nuclear Weapons

Matthew Fuhrmann

Nuclear weapons are different than any other military technology. They are unique because of the speed with which they can kill people and destroy property, as well as their production of radioactive fallout upon use. The detonation of a 150 Kt nuclear weapon—the size of the bomb North Korea tested in 2017—over New York City would kill an estimated 728,220 people.[1] Because of their destructive potential, nuclear weapons have been a centerpiece of world politics for more than seventy-five years.

The United States first detonated a nuclear explosion in the New Mexico desert in July 1945. One month later it dropped nuclear weapons on the Japanese cities of Hiroshima and Nagasaki. The Soviet Union then obtained nuclear weapons in 1949. The American and Soviet nuclear arsenals featured prominently in numerous Cold War–era events—most notably crises over Berlin from 1958 to 1961 and the 1962 Cuban Missile Crisis. The collapse of the Soviet Union on December 25, 1991, seemingly relegated nuclear weapons to the back burner of international relations, as the prospect of nuclear war now appeared exceedingly remote. But it soon became clear that nuclear weapons were still relevant in the altered international landscape. India and Pakistan built nuclear weapons and openly tested them in May 1998. North Korea followed suit, carrying out a total of six nuclear tests from 2006 to 2017. Some policymakers in the United States fear that Iran may soon build nuclear weapons as well.

In total ten countries have built nuclear weapons since World War II (see Table 6.1). Many others has seriously tried or at least considered the option. Another group of countries has hosted nuclear forces on their soil that belonged to an ally. For example, the United States continues to station fifty nuclear weapons at an airbase

Table 6.1. Nuclear Weapons States and Others That Tried

Nuclear Powers	*Countries That Hosted Foreign Nuclear Weapons*	*Select Countries That Considered Getting Nuclear Weapons*
China	Belarus	Algeria
France	Belgium	Argentina
India	Canada	Australia
Israel	Cuba	Brazil
Pakistan	Czechoslovakia	Egypt
North Korea	Cyprus	Iran
Russia	Denmark	Iraq
South Africa	East Germany	Italy
United Kingdom	Germany / West Germany	Libya
United States	Greece	Nazi Germany
	Hungary	Norway
	Italy	Romania
	Kazakhstan	South Korea
	Malaysia	Sweden
	Mongolia	Switzerland
	Morocco	Syria
	Netherlands	Taiwan
	Philippines	West Germany
	Poland	Yugoslavia
	Singapore	
	South Korea	
	Spain	
	Taiwan	
	Turkey	
	Ukraine	
	United Kingdom	

in Turkey—even as relations between the two countries have soured recently (see Fuhrmann and Sechser 2019).

How has the spread and possession of nuclear weapons influenced international relations? This chapter draws on a large body of scholarship to distill what we know about the political effects of nuclear weapons.[2] It focuses on two ways in which nuclear weapons shape international conflict behavior: (1) by deterring armed conflict and (2) by enabling revisionist foreign policies.[3]

With this information in hand readers can consider whether the spread of nuclear weapons is destabilizing or desirable.[4] Bear in mind that there are two ways to think about this issue. One approach is to consider how the spread of nuclear weapons influences the international system as a whole. Another is to ask how nuclear proliferation affects the foreign policy of an individual country, like the United States. What is good for one country might not necessarily be good for the world, and vice versa.

DETERRING ARMED CONFLICT

Deterrence is about shaping the behavior of another actor. The goal of the actor engaged in deterrence, who is typically called the *defender*, is to prevent someone else from taking an undesirable action. Nuclear deterrence can work through either punishment or denial. The leader of a nuclear-armed country could threaten to inflict a massive amount of pain on an adversary by razing one or more of its largest cities. To illustrate, in an apparent attempt to discourage Pakistani aggression, Indian defense minister George Fernandes said in December 2001, "We could take a strike, survive and then hit back. Pakistan would be finished" (quoted in Black 2010). The goal in nuclear deterrence by denial, by contrast, is usually to convince an adversary that an invasion or some other act of aggression would not succeed. Threatening to use nuclear weapons on the battlefield, for instance, might dissuade an adversary from carrying out a land invasion.

There are different types of deterrence, as summarized in Table 6.2.[5] General deterrence occurs when a state attempts to preserve the status quo outside of a crisis. Immediate deterrence takes place if general deterrence has failed and the defender seeks to restore the status quo ante. To illustrate, in 1962 the Soviet Union surreptitiously placed nuclear missiles in Cuba. This was a failure of general deterrence since President John F. Kennedy had previously warned Soviet leader Nikita Khrushchev not to place offensive missiles on the island. Once Washington discovered the missiles, the United States practiced immediate deterrence by implicitly threatening the Soviets with nuclear war unless Khrushchev dismantled the missile bases. Deterrence becomes *extended* if a nuclear power seeks to protect an ally rather than itself.

Does nuclear deterrence work? The answer is complicated. At the most basic level deterrence works when the potential aggressor believes that the expected costs of aggression exceed the benefits. If this is not the case, a nuclear threat may lack credibility. Research on nuclear deterrence focuses on four main things that

Table 6.2. Nuclear Deterrence Typology

		Who Is Being Protected?	
		Yourself	*An Ally*
Has There Been An Initial Provocation?	No	*General deterrence* Example: North Korea preventing a US invasion	*General extended deterrence* Example: the United States protecting Western Europe from Soviet invasion during the Cold War
	Yes	*Immediate deterrence* Example: Cuban Missile Crisis	*Immediate extended deterrence* Example: the United States defending Taiwan after Chinese shelling of offshore islands in 1954–1955

influence threat credibility: (1) the type of conflict the defender seeks to deter, (2) capabilities and nuclear superiority, (3) strategy and posture, and (4) the nuclear taboo—a normative opprobrium associated with nuclear use. I discuss each of these in turn.[6]

What Type of Conflict?

A country might face a full spectrum of military threats. Nuclear strategist Herman Kahn (1965) famously created a ladder of escalation with forty-four rungs ranging from the perception of a crisis to what he referred to as *insensate war*. What exactly does having nuclear weapons allow a country to deter?

One school of thought suggests that nuclear weapons allow countries to deter virtually any military threat, even low-level conventional conflict. The reason is that disputes that start off small could eventually escalate to the nuclear level. A rational actor, based on this line of thinking, would consider the possibility of escalation and avoid even small military provocations against nuclear powers (Jervis 1984, 13; Mearsheimer 2001, 129; Horowitz 2009, 251). Jervis (1984, 151) explains, for example, that US military responses to Soviet actions in Angola and Ethiopia during the Cold War would not have prompted an immediate nuclear attack. "But what would have happened had such steps been taken was impossible to foresee, and the chance of mutually undesired costs, if not complete devastation, weighed heavily on American decision makers."

This view often assumes that nuclear weapons loom in the background of any potential crisis even if nuclear use seems unlikely in the present moment (Jervis 1984, 150). The fact that nuclear weapons *could* be used shapes interactions with a nuclear power and induces caution (Kroenig 2013, 142). Paul Nitze (1956, 195) nicely characterized this perspective: "[W]hether or not atomic weapons are ever again used in warfare, the very fact of their existence, the possibility that they could be used, will affect all future wars."

The majority of studies that have evaluated this claim found that nuclear weapons do not reliably decrease—and may increase—the likelihood of low-level conflict (Kugler 1984; Huth, Bennett, and Gelpi 1992; Gartzke and Jo 2009; Sample 2002; Sobek, Foster, and Robison 2012; Gibler, Rider, and Hutchison 2005; Bell and Miller 2015; Fuhrmann and Tkach 2015). However, a couple of recent studies report a negative correlation between nuclear weapons possession and the onset of low-level military conflict, particularly if the challenger is nonnuclear (Narang and Mehta 2019; Lee et al. 2020). Beardsley and Asal (2009a) find that nuclear weapons reduce a country's vulnerability to violent crises, some of which might be classified as "low level." But they do not find evidence of deterrence benefits when it comes to the onset of nonviolent crises.

A second school of thought makes a more limited claim: nuclear weapons deter major war but may not deter lower-level conflict. In this view nuclear threats lack credibility against uses of force short of war. For more serious confrontations, though, nuclear weapons induce stalemate because nuclear use might be on the table. Low-level disputes do not become wars where nuclear threats might be credible, according to this line of thinking, because countries can effectively control escalation. This notion is at the heart of a perspective known as the theory of the nuclear revolution (TNR) (Jervis 1989). The first implication of TNR, according to Jervis (1989, 23–24), "is that military victory is not possible. From this it follows that if statesmen are sensible, wars among the great powers should not occur." Theorists such as Kenneth Waltz (1981) and Charles Glaser (1990) have echoed this view.

The TNR perspective holds that nuclear weapons contributed to the absence of major power war in Europe after 1945 (see Gaddis 1987). At the same time, according to a recent list of wars compiled by Dan Reiter, Allan Stam, and Michael Horowitz (2016), nuclear powers have been targeted in wars on five occasions: Russia and China fought against US forces during the Korean War, Egypt attacked Israel during the 1969–1970 War of Attrition, Egypt and Syria targeted Israel in the 1973 Yom Kippur War, Argentina instigated the 1982 Falklands War with the United Kingdom, and India and Pakistan fought the Kargil War in 1999. These disputes were possible because countries believed that they could prevent nuclear weapons from entering the fray by limiting their war aims or seeking aid from third parties, among other strategies (Avey 2019). Even within wars—which are typically defined as conflicts with at least one thousand battle-related deaths—there is thus a variation in the degree to which nuclear threats are credible. For proponents of the TNR perspective, nuclear arsenals are most useful for controlling escalation and preventing large-scale conflicts that would threaten the country's survival.

Studies seeking to statistically assess whether nuclear powers are less war prone have produced conflicting findings. Some have found that two nuclear-armed states are less likely to fight wars compared to other pairs of countries (Sample 1998; Rauchhaus 2009). Others show that nuclear powers are not unambiguously less likely to experience wars (Geller 1990; Gibler, Rider, and Hutchison 2005; Bell and Miller 2015).

A third perspective holds that nuclear weapons are useful for deterring one thing only: nuclear use. Using nuclear weapons in response to any other action, people in this camp argue, is simply implausible, so potential aggressors can dismiss the possibility of nuclear escalation in those cases. Former secretary of defense Robert McNamara (1983, 79) perhaps best captures this line of thinking. "[N]uclear weapons serve no military purpose whatsoever," he argued. "They are totally useless—except only to deter one's opponent from using them."

This argument is the least controversial of the three. It is also the most difficult to test using historical evidence because nuclear weapons have (thankfully) only been used twice in war—and not since August 1945. While it is true that a nuclear power has never suffered a nuclear attack, neither has a nonnuclear-weapons country since World War II. The traditional way to test predictions about nuclear deterrence—by comparing the rates at which nuclear and nonnuclear states experience some event—is not feasible in this context.

Allies versus the Homeland

There is general consensus that extended deterrence is more difficult than protecting the homeland (Schelling 1966). Indeed, it can be difficult for a nuclear power to convince adversaries that it would fight to defend an ally—especially if nuclear weapons could become involved. To cite one classic example, former French leader Charles de Gaulle doubted that the United States would sacrifice New York in order to save Paris.

Fewer studies have sought to empirically assess the viability of extended nuclear deterrence. Among those that have, there is evidence that allies of nuclear powers are less likely than other states to be targeted in military disputes. Huth (1990) finds that nuclear weapons contributed to extended deterrence success in an examination of fifty-six crises from 1885 to 1983. Fuhrmann and Sechser (2014) show that having a defense pact with a nuclear power lowers a country's risk of being targeted in military disputes that result in at least one fatality. Alliance commitments are effective because they communicate information to third parties about the nuclear power's resolve to fight (see Morrow 1994; Fearon 1997). Reneging on an alliance promise would harm the nuclear power's reputation (Gibler 2008), so it would not provide security guarantees unless it intended to uphold them. Potential third-party attackers recognize this and take security guarantees more seriously when they are codified in a formal treaty. However, support for the nuclear deterrence hypothesis in the context of extended deterrence is not universal. At least one study—Huth (1988a)—finds no relationship between the defender's possession of nuclear weapons and extended deterrence success.

Capabilities and Nuclear Superiority

The amount of nuclear firepower a country needs to successfully deter attacks has been a source of debate in scholarship for decades. This debate remains lively today.

One perspective suggests that states can engage in nuclear deterrence even if they do not possess assembled nuclear warheads (Schell 1984; Mazarr 1997; Fuhrmann and Tkach 2015). This could work in one of two ways. First, as a country gets close to a bomb, its rivals might fear that it *might* be nuclear armed, leading the adversaries to exercise restraint. For example, as India and Pakistan approached the weapons threshold during the 1980s the possibility that an attack would trigger nuclear retaliation induced caution (Hagerty 1995; Fair 2014). Second, potential aggressors might know for certain that a state is currently nonnuclear but realize that it could build a bomb quickly in the event of a crisis. Scholars refer to these states as *latent nuclear powers*. According to a database compiled by Fuhrmann and Tkach (2015), thirty-three countries have developed sensitive dual-use nuclear technology needed to build weapons. If a latent nuclear power is attacked, it could quickly assemble a nuclear bomb and potentially use it against the aggressor. There is some evidence that latent nuclear powers are less vulnerable to military disputes than states that lack the key building blocks for a bomb (Fuhrmann and Tkach 2015; Spaniel 2019, chapter 4). Yet many scholars are skeptical that latent nuclear deterrence can work (Waltz 1997; Mehta and Whitlark 2017).

Other scholars argue (or assume) that the deterrence benefits of an arsenal kick in once a state openly tests a nuclear device or assembles its first nuclear weapon (Gartzke and Kroenig 2009; Rauchhaus 2009; Horowitz 2009; Fuhrmann and Tkach 2015). In that situation an opponent may not be able to rule out the possibility that a military attack would result in nuclear escalation—even if the nuclear power's arsenal was potentially vulnerable.

Another perspective, exemplified by TNR, suggests that countries need a second-strike capability in order to deter (Waltz 1981; Jervis 1989). This means that they could retaliate with a nuclear strike after suffering an initial attack by an adversary. In that case mutually assured destruction (MAD) would hold. The challenger knows that a nuclear attack would result in the catastrophic destruction of its own territory and population. Based on the TNR perspective, MAD is therefore a highly stable situation in which the catastrophic effects of nuclear war induce caution and deter war. On the other hand, nuclear weapons are useless for deterrence if the adversary knows that it could wipe out the nuclear power's ability to retaliate in a disarming first strike. That kind of situation could be unstable, as it incentivizes a country to strike first during a crisis.

Based on this line of thinking, a country does not necessarily need a large nuclear arsenal in order to deter. Because nuclear weapons are so terrifyingly destructive, the prospect of a "little" nuclear war is enough to make any rational leader think twice before mounting an attack. A small nuclear arsenal is therefore sufficient for deterrence to hold as long as the potential attacker knows that it could not eliminate the state's nuclear capabilities in an initial strike (Waltz 1990, 736).

The second-strike school offers differing opinions about the level of certainty required for deterrence (for a more complete discussion, see Lieber and Press 2019, chapter 3). One perspective suggests that the mere chance of nuclear retaliation can deter conflict because the costs are so large. These scholars differ on exactly how likely retaliation must be, but they are united in the view that it does not have to be certain (Brodie 1946; Freedman 1988; Waltz 1990). Ask yourself this: if you were the American president, would you attack a nuclear-armed country if there was a 50 percent chance that doing so would lead to a single nuclear attack against New York? What if there was a 25 percent chance? Or 10 percent? Some would respond in the negative because the consequences are simply too great to justify the chance. Others disagree. These scholars argue that nuclear retaliation has to be certain (or near certain) in order for deterrence to work (Jervis 1984; Glaser 1990).

Another line of thinking suggests that a country needs to have nuclear superiority over its adversary in order to reap deterrence benefits. A nuclear monopoly—when one state has a nuclear arsenal and its adversary does not—is the most extreme form of nuclear superiority. When studies find that nuclear weapons yield political benefits, they tend to show that those advantages weaken considerably if the opponent is also a nuclear power (Beardsley and Asal 2009b; Narang and Mehta 2019; Lee et al. 2020).

In situations where two adversaries possess nuclear weapons, some scholars argue, deterrence depends on having the more powerful arsenal. Proponents of larger nuclear arsenals make three main claims. First, even if a nuclear exchange would be devastating for both countries, it would be worse for the state with the inferior arsenal. Second, having a nuclear advantage helps a country prevail in a crisis by increasing its resolve relative to that of its adversary. As Matthew Kroenig (2018) puts it, "military nuclear advantages increase a state's willingness to run risks in international conflicts … in a game of chicken we might expect the smaller car to swerve first even if a crash would be disastrous for both." Third, achieving a reliable second-strike capability is more difficult than the TNR perspective implies (Long and Green 2015; Press and Lieber 2019). Technological advancements and counterforce targeting make it easier to destroy an adversary's nuclear capabilities than is commonly assumed. Even submarines—which many analysts assume are survivable—can be located and targeted in ways that are not always fully appreciated (Long and Green 2016). As a result Press and Lieber (2019) conclude that "a small, vulnerable nuclear force did not give the Soviet Union adequate protection" during the early Cold War period. This view does not necessarily imply that states need massive nuclear arsenals or dramatically more weapons than their opponents in order to deter, but it does suggest that small arsenals may not be sufficient for deterrence. It implies that China, which has historically maintained a small nuclear arsenal, will continue modernizing its forces and seek greater capabilities to deter the United States (Talmadge 2019b).

Strategy and Posture

The preceding discussion about nuclear superiority leads to a more general question: what capabilities and doctrine are necessary to achieve relevant national security objectives? A large body of work considers this question. It considers how a country operationalizes its nuclear forces—which is known as its *nuclear posture*—and what implications this has for deterrence and peace.

Work on nuclear posture begins with a basic idea: the effects of nuclear weapons on international stability depend in part on how a country tries to use its nuclear arsenal. To understand the effects of nuclear arsenals, then, we need to have a deep appreciation of a country's goals and strategy. Much of the work on this topic has focused on the US-Soviet rivalry during the Cold War. More recently scholars have examined regional nuclear powers and the potential for nuclear escalation in US-China interactions (Cunningham and Fravel 2015; Talmadge 2017).

Vipin Narang (2009/2010, 2014) identifies three postures that regional nuclear powers could pursue: assured retaliation, asymmetric escalation, and catalytic. The goal in assured retaliation is to deter nuclear use and blackmail. China and India have adopted this posture. In asymmetric escalation, a strategy pursued by France and Pakistan, the principal objective is to deter conventional conflict as well as nuclear use. States with catalytic postures, like South Africa, rely on their arsenals to influence third-party states. The goal here is to motivate powerful actors, often the United States, to intervene on a state's behalf by threatening nuclear escalation in a regional conflict. Superpowers such as the United States could pursue one of these strategies, but they also have the means to chase more ambitious goals. For example, they could try to protect allies from aggression by placing them under a nuclear umbrella.

Nuclear strategists debate what countries must do in order to meet their relevant political objectives. Much of the discussion centers on five specific issues.

The first is about delivery systems. Nations have historically utilized three main ways to deliver nuclear weapons: bombers, missiles, and submarines. The United States relies on all three—the so-called *nuclear triad*. By contrast, the United Kingdom currently relies just on submarines to launch its nuclear missiles. Some research seeks to explain variation in the adoption of platforms (Gartzke, Kaplow, and Mehta 2014). Whether all three delivery systems are necessary for deterrence has been a source of debate, especially in recent years. Some have argued that US intercontinental ballistic missiles (ICBMs) are unnecessary and may actually invite instability (Talbot 2018). Missile silos are relatively vulnerable to a first strike because they are stationary, whereas bombers and submarines can be put in constant motion (missiles can also be made mobile via road or railway but the United States does not currently do this). Others counter that the land-based ICBM force is central to the deterrence mission (Kroenig 2018).

A second issue has to do with targeting (Sagan 1990). Counterforce targets are military sites or capabilities that could inflict damage on a country in retaliation for an initial attack. Countervalue strikes, by contrast, target an adversary's cities or industrial centers. Scholars and policymakers have debated what combination of counterforce and countervalue targets is necessary to achieve deterrence. The evolution of former secretary of defense Robert McNamara's views on targeting underscore this debate. In June 1962 he took a strong stand against targeting Soviet cities with nuclear weapons, arguing that American nuclear strategy should hinge on "the destruction of an enemy's forces, not his civilian population" (quoted in Scoblic 2009). However, he later emphasized the importance of countervalue targeting. McNamara argued that the best way to deter was by threatening to inflict "unacceptable damage" on an adversary. He defined this as the capacity to destroy between one-fifth and one-fourth of the Soviet population and one-half of its industrial base (Office of the Secretary of Defense, n.d.). Thus, based on the Soviet population in the mid-1960s, the United States believed that it needed to kill no fewer than 40 million civilians in order to deter Moscow. Some scholars have challenged this notion. Waltz (1981), for example, calls this requirement "absurdly high."

A third and related debate has to do with the development and deployment of tactical nuclear weapons, which are meant for battlefield use as counterforce tools. This debate focuses largely on US tactical nuclear weapons deployed abroad. During the Cold War the United States deployed tactical nuclear weapons on the territory of strategically significant countries such as Taiwan, South Korea, and the United Kingdom. Washington still maintains B-61 nuclear gravity bombs in Belgium, Germany, Italy, the Netherlands, and Turkey (Fuhrmann and Sechser 2014). Some have argued that these weapons bolster deterrence by serving as "tripwire" forces (Schelling 1966). They would be used early in an armed conflict before they could be eliminated in a disarming attack, based on this logic, thus guaranteeing that the United States would fight to defend its allies. Because the Soviet Union understood this, its leaders would be less likely to launch an attack. Moreover, these weapons were thought to bolster deterrence by denial. If the Soviets tried to invade, large numbers of their troops would be killed, and it would be difficult for forces that were not immediately hit to traverse the territory without suffering potentially fatal radiation poisoning.

Yet there is not strong empirical support for this idea. Fuhrmann and Sechser (2014) find that, controlling for the effects of a military alliance, countries that host land-based foreign nuclear weapons are not less likely to be targeted in violent military disputes than those that do not. This evidence suggests that stationing nuclear forces on an ally's territory does not bolster extended deterrence by making nuclear threats more credible. The presence of a robust defense commitment is usually sufficient to protect allies, regardless of where a nuclear power's assets might be deployed (Fuhrmann and Sechser 2014, 932).

No first use (NFU) of nuclear weapons is a fourth topic that has attracted attention in both scholarship and policy circles. An NFU pledge means that a country would not be the first to use nuclear weapons in war. The United States has traditionally not held nor does it currently have a declaratory policy of NFU. Some have argued that this is a wise policy since it enhances Washington's ability to deter chemical or biological attacks, as well as large-scale conventional conflict (Quester 2000). In this view creating ambiguity about the conditions under which a state would use nuclear weapons is good for deterrence. Critics of NFU also suggest that declaratory policy is largely irrelevant because pledges are hollow (Quinlan 2007). The Soviet Union had an NFU pledge during the Cold War, for instance, but it is not clear that this shaped American perceptions of Moscow's intentions in any meaningful way. Others counter that an NFU policy would enhance peace and stability by reducing the role of nuclear weapons in world politics, which would bolster global nonproliferation efforts and lower the risk of nuclear war (Sagan 2009; Gerson 2010).

A fifth prominent debate is about the command and control of a nuclear arsenal (Bracken 1983; Nolan 1989; Feaver 1992). Who should have launch authority over nuclear forces? And what is the ideal relationship between civilian leaders and military commanders when it comes to the management of a nuclear arsenal? There have been many close calls involving nuclear weapons. For example, during the Cuban Missile Crisis US personnel at a base in Minnesota mistook a bear—yes, a bear—for a saboteur and sounded an alarm, leading forces at a nearby base in Wisconsin to conclude that nuclear war with the Soviet Union had begun (Sagan 1994, 3). Pilots rushed to their nuclear-armed planes and prepared to take off but were stopped at the last moment. Tight controls on nuclear use may prevent accidents or unintended disasters (Feaver 1992, 165). At the same time, drawing on organization theory, Sagan (2004, 74) shows that extra security measures can paradoxically undermine nuclear security. Others have argued that tighter control over a nuclear arsenal can weaken deterrence by undermining threat credibility. From the standpoint of deterrence, a nuclear power wants its adversary to conclude that a provocation would result in certain and near-immediate nuclear retaliation. The optimal policy, therefore, might be to pre-delegate launch authority to local commanders in the event of a crisis—meaning that they could fire nuclear weapons without explicit authority from the civilian leadership.

This debate is playing out in the United States today. The president has the sole authority to order a nuclear attack; he does not need confirmation from the secretary of defense or anyone else. This policy is designed to bolster deterrence, but it has led to concerns that an erratic leader could carry out a catastrophic attack with essentially no institutional constraints. For some, the bar for nuclear use is too low. Members of Congress introduced legislation in 2017 that would prohibit the president from using nuclear weapons unless Congress had issued a formal declaration of war (Mecklin 2017).

The Nuclear Taboo

Nuclear weapons have not been used in war since 1945. Nina Tannenwald (1999, 2007) has argued that this trend is due to the evolution of a "nuclear taboo."[7] This taboo, she contends, "is associated with widespread popular revulsion against nuclear weapons and widely held inhibitions on their use" (Tannenwald 2007, 8).[8] In this view nuclear weapons are unlike any other military technology. They cannot be used the way that a country could, for example, use bullets, tanks, or drones to attack an adversary. Nuclear attacks are seen as morally repugnant due to the scale of the destruction they would cause. Normative prohibitions on nuclear use could weaken the effectiveness of nuclear deterrence by making a potential aggressor question whether it would suffer nuclear retaliation.

Other scholars refer to the nonuse of nuclear weapons since 1945 as a "tradition" (Sagan 2004; Paul 2009). They argue that material considerations rather than normative ones help explain leaders' reluctance to use nuclear weapons. Nuclear use could invite military or political blowback (see Sechser and Fuhrmann 2017). It could, for instance, invite widespread nuclear proliferation as countries look for ways to protect themselves from blackmail or nuclear-backed aggression in the future (Sagan 2004). Based on this line of thinking, leaders generally refrain from thinking seriously about a nuclear attack except in dire circumstances, because they want to avoid the price tag of nuclear use. If the material costs of nuclear use were low and the benefits high, however, these scholars are not convinced that the taboo would fully restrain nuclear use.

Recent experimental studies show that the nuclear taboo is weaker than one might think. Press, Sagan, and Valentino (2013) find that Americans are willing to use nuclear weapons in situations where doing so provides a military advantage. They also find that those who oppose nuclear use worry mostly about precedent setting and other material considerations rather than moral factors. In a related study Sagan and Valentino (2017) show that the majority of Americans are willing to support nuclear attacks that could kill millions of foreign civilians if that option protects US forces and better enables Washington to achieve its war aims. When strategic factors call for nuclear use, then, it is not clear that the taboo would restrain the American public.

These studies have prompted other articles on this topic—some of which reach different conclusions. Rathbun and Stein (2019), for example, find individual-level variation in support for using nuclear weapons. People with a particular moral attitude—specifically, support for retribution—are willing to support nuclear use in situations where others are not. Other recent research has broadly supported the notion that the nuclear taboo is fragile (Gibbons and Lieber 2019).

ENABLING REVISIONIST FOREIGN POLICIES

The discussion in the previous section considers the defensive functions of nuclear weapons. The focus is on whether nuclear weapons help countries deter military threats with greater ease. However, having nuclear weapons may also help countries change the status quo. Scholarship has looked at three main ways in which nuclear weapons could facilitate revisionist foreign policies: (1) conflict instigation, (2) coercive diplomacy, and (3) preventive war.

Conflict Instigation and Nuclear Shields

If nuclear weapons provide deterrence benefits, can nuclear powers mount military provocations with impunity? The answer would appear to be in the affirmative, since nuclear powers could use the threat of a nuclear attack to deter retaliation. B. H. Liddel Hart wrote in 1954 that a nuclear arsenal—particularly the development of the hydrogen bomb—"reduces the likelihood of full-scale war, it increases the possibility of limited war pursued by widespread local aggression" (quoted in Krepon 2004, 1). Glenn Snyder (1960, 31) echoed this view: "When the strategic balance is stable—when both sides have the capacity to strike back powerfully after absorbing a first strike—the tactical balance tends to become unstable because limited attacks can be undertaken and limited wars can be carried to fairly high levels of intensity without serious danger that either side will decide to initiate all-out strategic warfare." The idea that nuclear weapons may decrease the risk of total war while simultaneously increasing the likelihood of lower-level conflict became known as the *stability-instability paradox*.

There is some evidence to support the view that having a nuclear arsenal makes a country more likely to initiate military disputes short of total war (Beardsley and Asal 2009a; Rauchhaus 2009). Proponents of this notion often point to Pakistan as a case that supports the theory (Kapur 2007). After testing nuclear weapons for the first time in 1998, Pakistan instigated the 1999 Kargil War and supported violent non-state actors that carried out attacks against Indian interests with greater frequency. However, other studies do not find strong support for the nuclear emboldenment hypothesis. Bell and Miller (2015) find no evidence of a greater conflict risk in dyads with two nuclear-armed states. They do find, however, that nuclear states are more likely to instigate disputes against nonnuclear countries. These disputes tend to occur against new adversaries, which Bell and Miller interpret as evidence that states' interests expand once they obtain nuclear arsenals.

Coercive Diplomacy

Scholars have also considered whether nuclear weapons facilitate coercive diplomacy. This body of work asks whether nuclear powers engage in *compellence*—the use of military threats to change the status quo—more effectively.

Nuclear compellence could work, according to one school of thought, for the same reason that nuclear deterrence does: these weapons are so destructive that countries will do just about anything to avoid nuclear punishment. As Robert Pape (1996, 38) puts it, "Even if the coercer's nuclear resources are limited, the prospect of damage far worse than the most intense conventional assault will likely coerce all but the most resolute defenders." In his examination of nuclear blackmail Betts (1987, 218) finds evidence consistent with this view: "Attempts to exploit nuclear leverage in the past seem useful at best and not costly at worst, unless a country is operating from the inferior position." He reaches this conclusion after examining the most serious Cold War nuclear confrontations.

Some statistical studies show that nuclear powers can change the status quo with greater ease than their nonnuclear counterparts. Gartzke and Jo (2009) find that having nuclear weapons increases a country's diplomatic status. Beardsley and Asal (2009b) show that nuclear powers are more likely to prevail in crises against nonnuclear states. Kroenig (2013) finds that, in crises with two or more nuclear powers, the state with the nuclear advantage is more likely to win. The mere potential to build nuclear weapons can also aid coercive diplomacy, according to a recent analysis by Volpe (2017), since states can threaten to build nuclear weapons if their demands are not met.

Another view suggests that nuclear weapons have little utility for things other than deterrence. The TNR perspective holds that nuclear weapons cannot easily be used to change the status quo (Jervis 1989, 29–30). Expanding on this idea, Sechser and Fuhrmann (2017) introduce *nuclear skepticism theory*. They argue that coercive nuclear threats usually lack credibility because, in coercive diplomacy (as opposed to deterrence), the stakes are typically low relative to the costs of using nuclear weapons. The target therefore does not fear nuclear punishment if it fails to comply with the coercer's demands. In 2013, for example, North Korea made nuclear threats in an apparent attempt to compel Washington to lift economic sanctions and end joint military exercises with South Korea. Air Force commander Ri Pyong-chol said, "Stalwart pilots, once given a sortie order, will load nuclear bombs, instead of fuel for return, and storm enemy strongholds to blow them up" (Choe 2013). Yet there is no evidence that North Korea was able to extract concessions from the United States during this episode. President Obama did not appear to worry much about nuclear punishment: "We won't allow North Korea to create a crisis and elicit concessions," he said (BBC 2013).

Nuclear Brinkmanship

Nuclear strategists have identified a possible solution to the credibility problem in nuclear blackmail (this applies to deterrence as well). They call it *brinkmanship* or the *manipulation of risk* (Schelling 1966). The basic idea is that countries can turn unbelievable nuclear threats into credible ones by taking dangerous actions that raise the possibility of some unintended disaster. To illustrate how this might work, Schelling (1966) asks us to imagine two mountain climbers who are tied together. One might coerce the other by saying, "Do as I say, or else I'll jump." This threat lacks credibility since carrying it out would cause both climbers to be killed. But what if the climber, after making the threat, walks to the edge of a cliff and appears to be off balance? There is now the possibility that he could slip and fall, generating a disaster for both climbers that neither wants to materialize. The other climber might now comply with the demand even though he or she believes that his or her counterpart would not intentionally jump. Countries can engage in brinkmanship with their arsenals by pre-delegating launch authority or alerting nuclear forces to actions that could result in nuclear use even in situations where a leader knows that doing so would be suicidal (see Sagan 1985; Narang 2014).

A classic case of nuclear brinkmanship occurred during the Cuban Missile Crisis. During the crisis the United States moved to defense readiness condition (DEFCON) 2—one step away from nuclear war. In doing so, it gave pilots and local commanders launch authority over nuclear weapons. Shortly thereafter a U-2 spy plane accidently ventured into the Soviet Union after the pilot became disoriented. Worried that the Soviets might perceive this as the first stages of an attack—the two countries were, after all, in the midst of the most serious nuclear crisis they ever faced—the United States scrambled fighter jets to guide the U-2 back home. Because of the nuclear alert, these fighters were armed with nuclear weapons. At the same time, the Soviets had deployed MiG fighters to intercept the American aircraft. It is not hard to imagine that this incident could have provoked a catastrophic clash (see Sechser and Fuhrmann 2017, 36). Brinkmanship theory expects that leaders should anticipate this kind of danger and ultimately back down. In a crisis with two nuclear actors, the side that ultimately wins a game of brinkmanship will be the one that has the higher resolve—that is, more to lose (Jervis 1984). According to another perspective, which I highlighted earlier, nuclear superiority will enable a state to push harder in brinkmanship situations, thereby increasing the odds that it will prevail (Kroenig 2013).

Others are less convinced that brinkmanship offers a clear solution to the credibility problem in nuclear coercion. Sechser and Fuhrmann (2017) identify three problems with brinkmanship theory. First, it requires leaders to take dangerous actions that raise the possibility of catastrophe, such as alerting their nuclear forces. But leaders may wish to maintain control in a crisis rather than cede it.

Second, attempts to manipulate the risk of nuclear war may not be detected or correctly interpreted by the other side. Sagan and Suri (2003) document, for instance, President Nixon's attempt to end the Vietnam War on terms favorable to the United States by alerting American nuclear forces in 1969. The Soviets detected the alert but they were not sure what to make of it. No senior official in Moscow appears to have connected the alert with Vietnam, making it impossible for the Soviets to comply with the US demand. Third, even in the presence of brinkmanship, targets of nuclear coercion may still see threats as unbelievable.

Preventive War

Many have argued that pursuing nuclear weapons is destabilizing because it provides incentives for preventive war (Sagan and Waltz 2003, chapter 2). Rivals of a potential proliferator may seek to attack it in order to destroy its nuclear infrastructure, thereby delaying (or perhaps ending) the proliferator's nuclear ambitions. In 1981 Israel bombed an Iraqi nuclear reactor that could have aided Saddam Hussein's ability to build nuclear weapons. Israel took similar action in 2007, destroying a reactor in Syria that was being built with North Korean assistance. And the 2003 Iraq War was sold in part as necessary in order to end Iraq's nuclear weapons program. Fuhrmann and Kreps (2010) identify ten other cases—some of which extended for multiple years—in which a state seriously considered preventive strikes against a rival's nuclear program but ultimately opted against it. This includes the US consideration of force against China in the 1960s and North Korea in the 1990s.

Scholars have explained when and why nuclear proliferation-induced preventive attacks occur or are considered. One perspective suggests that these events happen when the anticipated costs of nuclear proliferation are high. Prior violent conflict, a highly autocratic potential proliferator, and divergent foreign policy interests all increase the likelihood of attacks or considered attacks against nuclear programs (Fuhrmann and Kreps 2010). Whitlark (2017) argues that threat perceptions vary across individual leaders. Movement toward preventive strikes happens, she claims, when heads of government believe that the spread of nuclear weapons induces instability and that the potential proliferator may be undeterrable. Other studies highlight the role of information and intelligence. Bas and Coe (2016) show that peacetime consideration of preventive strikes becomes more likely when intelligence estimates reveal that a program is on the cusp of yielding a bomb. There may be uncertainty about a state's nuclear-related progress, since countries can conceal relevant facilities, materials, or research (Fuhrmann 2018). In the presence of large power shifts ambiguity about a potential proliferator's intentions can lead to war (Debs and Monteiro 2014). In addition, as Spaniel (2019, chapter 8) demonstrates, uncertainty about whether the nonproliferator is committed to taking preventive action can also lead to conflict. Commitment problems can lead to nuclear-induced preventive wars as well, especially if the potential proliferator is conventionally weak (Schub 2017). Finally, the costs of war shape an attacker's enthusiasm for preventive military action. Ludvik (2019) shows that a credible threat of conventional retaliation deters attacks against states that may harbor bomb-related ambitions.

The political effects of preventive attacks have also been debated in scholarship. Some scholars contend that these strikes aid nonproliferation by destroying critical infrastructure (Kroenig 2013). In addition, preventive strikes can make foreign suppliers more reluctant to assist the targeted country and convince the proliferator to change strategies in ways that delay its programs (Kreps and Fuhrmann 2011). On the other hand, attacking a country in the name of nonproliferation can backfire in the long run by increasing the target's resolve to get nuclear weapons (Reiter 2005; Braut-Hegghammer 2011).

CONCLUSION: DIRECTIONS FOR FUTURE RESEARCH

This chapter has summarized what we know from scholarship on how nuclear weapons influence military conflict. As is hopefully clear from the preceding discussion, seventy-five years into the nuclear age we still lack consensus on some important questions about the political effects of nuclear weapons. Much more work is needed to resolve existing debates and address understudied questions. I focus on three potentially promising directions for future research.

The first is to use creative methodological tools to answer old questions. The standard approach for studying nuclear deterrence is to compare the conflict rates of nuclear and nonnuclear states. This strategy can be fruitful but it presents researchers with a well-known challenge. Identifying the causal effect of nuclear weapons on conflict based on real-world events is difficult because nuclear weapons are not randomly assigned. Countries seek and obtain nuclear weapons for particular reasons. These reasons—rather than the nuclear weapons themselves—could account for trends that we observe in historical data. Scholars typically account for this by controlling for confounding variables, factors that affect getting nuclear weapons and the likelihood of conflict, in their statistical models. We know, for example, that nuclear powers tend to have more powerful militaries than nonnuclear states and conventional military capabilities also bolster deterrence. It is therefore important to include some measure of conventional military power in a statistical model of nuclear deterrence, and many scholars do this. But some confounding factors are difficult to measure or observe, so we cannot easily account for them in a statistical model. If unmeasured confounders exist, the resulting conclusions could be misleading.

This does not imply that we should abandon observational studies of nuclear deterrence (on the virtues and pitfalls of this method, see Fuhrmann, Kroenig, and Sechser 2014). Yet supplementing this approach with other methods could be fruitful. Scholars have successfully used survey experiments to study interesting topics in nuclear security, particularly in the context of the nuclear taboo debate (Press, Sagan, and Valentino 2013; Sagan and Valentino 2017). Future work could extend this research by looking at other questions about the political effects of nuclear weapons where public perceptions are relevant, such as support for military interventions based on the target's nuclear status. Researchers are also beginning to use historical war games played by government officials (Pauly 2018) and experimental war-gaming (Reddie et al. 2019) to better understand nuclear dynamics. Scholars could further advance knowledge by considering other novel methodological approaches to nuclear security.

Second, it would be productive to devote more attention to the problem of nuclear latency. As discussed previously, there are many more latent nuclear powers in the world than actual weapons states. These threshold states have been neglected in scholarship historically. Fortunately, this issue has received more attention over the past decade (Sagan 2010; Fuhrmann and Tkach 2015; Mehta and Whitlark 2017; Volpe 2017; Spaniel 2019). Yet there is still much that we do not understand about how latent nuclear capabilities influence peace and stability. Additional work in this area would be welcome.

Third, technological innovations may be changing what we know about nuclear weapons and international conflict. The advent of sophisticated cyberwar capabilities and other emerging technologies could upend the traditional understanding of nuclear deterrence and coercion. Scholars have recognized this possibility (Press and Lieber 2019; Sechser, Narang, and Talmadge 2019) but there is still a lot to be learned at the intersection of nuclear weapons and emerging technologies.

NOTES

1. This is based on the simulator available at http://www.nuclearsecrecy.com/nukemap/.
2. For an earlier attempt that makes some similar points, see Berkemeier and Fuhrmann (2017).
3. This chapter focuses on mainstream international relations research from the rationalist and constructivist traditions. There is also significant work on nuclear weapons that is broadly associated with critical theory. In a feminist analysis of nuclear deterrence, for example, Cohn (1987) discusses how the language associated with nuclear war is gendered.
4. This is the question posed in a classic debate between Sagan and Waltz (2003).
5. I thank Kelly Kadera for suggesting a table along these lines. For literature discussing these distinctions, see George and Smoke (1974), Morgan (1983), Huth and Russett (1984), and Danilovic (2002). See Kahn (1965, 282) for another deterrence typology.
6. Much of the scholarship described in this chapter assumes that leaders behave rationally. For critiques of deterrence rooted in psychology, see Jervis, Lebow, and Stein (1989).
7. This is an influential work rooted in the constructivist tradition. For other constructivist-based scholarship about nuclear weapons, see Rublee (2009) and Harrington de Santana (2009).
8. A taboo may also operate in the realm of chemical weapons. See Price (2007).

Chapter Seven

Outcomes and Consequences of War

Stephen L. Quackenbush

Much of the literature on war and international conflict more broadly focuses on its causes. In comparison, the literature examining what happens once war starts is much less extensive. Scholars have nonetheless learned a lot about the evolution of war, including not only the outcomes and consequences of war—the focus of this chapter—but war duration and war termination as well.

This knowledge is of great importance because understanding the factors driving war outcomes is a vital element of understanding international conflict more generally (Biddle 2007; Quackenbush 2015). Furthermore, war has tremendous consequences for domestic and international politics, so understanding the nature of those consequences is also of vital importance. If leaders expect positive outcomes and consequences, then they should be more likely to go to war, while the reverse should also hold true.

In this chapter I examine what we know about the outcomes and consequences of war. While these topics are related, they are distinct. Explaining war outcomes requires us to understand the factors that affect the likelihood of victory, defeat, or achieving a draw. The study of war consequences entails identifying how war changes the status quo. After discussing the war outcomes and war consequences literature I consider important directions for future research before drawing overall conclusions.

EXPLAINING WAR OUTCOMES

Scholars have developed a number of explanations of war outcomes—winning, losing, or achieving a draw—over the years. The traditional understanding was that power is the most important and perhaps the only determinant of war outcomes.

Other studies identify a variety of other realpolitik or domestic factors that drive war outcomes. Further studies have examined the outcomes of insurgency and counterinsurgency campaigns.

Power

Traditional explanations of war outcomes focus on relative power of the combatants as the primary explanatory variable, arguing that the more powerful side is more likely to win wars. Despite this widespread agreement that power matters, there is a great deal of disagreement over which aspects of power matter most, or how power is best measured (Rothgeb 1993).

Some argue that wealth is the most important aspect of power. Rosen (1972) examines the relationship between wealth and victory and finds that the wealthier side won 80 percent of wars. Rosen argues that wealth is more important than cost tolerance, which he calls the willingness to suffer. Organski and Kugler (1978) also focus on wealth as the central element of power, although they modify wealth by accounting for the ability of governments to extract resources from society. Importantly, a foundational assumption they make is that if we can measure "the power of nations accurately enough," then we can predict the outcome of wars "with reasonable assurance" (Organski and Kugler 1978, 141).

Another common measure of power is the Composite Indicator of National Capability (CINC) (Singer, Bremer, and Stuckey 1972). This measure is comprised of six separate indicators encompassing military, economic, and demographic factors. Wayman, Singer, and Goertz (1983, 497) seek to determine which of these components is most important for determining the outcomes of wars and militarized interstate disputes. They find that "an advantage in industrial capabilities is more strongly associated with victory than is an advantage in military or demographic capabilities."

In contrast to this literature arguing that wealth is the key to power and in turn war outcomes, others focus on military capabilities. In particular Henderson and Bayer (2013) reexamine this issue, trying to determine whether wealth or military capability has a stronger impact. They conclude that while both make victory more likely, military capability is more important.

Despite this focus on power as the primary or even the exclusive determinant of war outcomes, the stronger side wins only about 60 percent of the time, depending on what measure of strength is used (Biddle 2004). This has led some scholars to examine why weaker states emerge victorious in war as often as they have. Rosen (1972), Mack (1975), and Maoz (1989b) provide early examinations of this question, although Sullivan and Arreguin-Toft have provided tremendous advances in our ability to explain these paradoxical conflict outcomes.

Sullivan (2007, 2012) argues that understanding the objectives for which war is fought provides a key reason that weak states sometimes defeat more powerful ones. She distinguishes between brute force objectives, which can be achieved through military victory, and coercive objectives, which can only be achieved through bargaining. Because brute force objectives do not necessarily require target compliance and coercive objectives do, the latter are more difficult to achieve. However, Sullivan contends that strong states are more likely to pursue coercive objectives, which can lead to defeat when they are unable to force the weaker state to comply with their demands.

Arreguin-Toft (2001, 2005) focuses on strategic interaction to explain why stronger states lose wars against weaker opponents. He distinguishes between two basic strategies for each side: direct attack versus barbarism for the strong actor, and direct defense versus guerrilla warfare for the weak actor. Arreguin-Toft argues that strong states are advantaged when both sides use the same basic approach (i.e., direct-direct or indirect-indirect), while weak states are advantaged when the two sides employ opposite approaches (i.e., direct-indirect or indirect-direct).

Realpolitik Factors

Stam (1996) greatly advanced the study of war outcomes in political science by systematically theorizing about the factors driving war outcomes and conducting a broad quantitative analysis of them. In contrast to common conceptions of war outcomes in terms of either winning or losing, Stam argues that there are three possible outcomes of a war: win, lose, or draw. The Korean War is one of many examples of a draw.

Unlike previous studies that typically focused exclusively on power to explain war outcomes, Stam accounts for a wide range of material and nonmaterial factors at both the international and domestic levels. In particular, Stam (1996) highlights the importance of military strategy, as represented in a simple classification of maneuver, attrition, and punishment strategies. Whereas a maneuver strategy is focused on the movement of forces to achieve advantageous positions against the enemy, an attrition strategy focuses on defeating the enemy by killing personnel and destroying material. In contrast, a punishment strategy focuses on increasing the enemy's costs, with two basic types: strategic bombing and guerrilla warfare. Stam's results indicate that strategy is the most important driver of war outcomes, with maneuver strategies being particularly successful.

Stam (1996) also accounts for terrain. Dense terrains such as jungles, mountains, and urban areas are more difficult for armies to operate in, particularly with armored vehicles. On the other hand, open terrain such as plains and deserts make it much easier for armies to maneuver and bring firepower to bear. There is also an important interaction between strategy and terrain because different strategies work best on different types of terrain. For example, maneuver strategies are best suited to open terrain types, whereas punishment strategies generally depend on rough terrain in order to be effective. States using a strategy that is appropriate for the terrain in question are more likely to win.

Another realpolitik factor that multiple scholars have focused on is war initiation. Initiators are generally expected to be more likely to win because they can dictate the course of events early in the war, and they are expected to initiate only when their expected utility for doing so is beneficial (Bueno de Mesquita 1981;

Morrow 1985). Further, Gartner and Siverson (1996) argue that most wars remain bilateral because initiators are less likely to attack a country if they anticipate that others will come to the target's aid. Thus, states will avoid initiating what they expect to be a difficult war.

Wang and Ray (1994) examined the fate of initiators in wars involving great powers dating back to 1495. They find that the relationship between initiation and winning has grown stronger over time, as initiators won only about half the time between 1495 and 1799 but have won about two-thirds of the time since 1800. One important caveat to their findings is their focus on great powers; war between major powers has become less common over time, just as initiators have become more likely to win.

The distance between a state's home territory and the location of the fighting in the war also matters (Stam 1996). The loss of strength gradient makes it difficult to project power over long distances (Boulding 1962). This has an important impact on war outcomes because as distance increases, it becomes harder to inflict costs on the opponent. However, distance is less of an obstacle for stronger states—particularly major powers—because they are better able to provide logistical support to their forces across long distances.

While the focus of this chapter is on political science literature examining war outcomes, there are many studies of war outcomes in other fields as well (e.g., Overy 1995). Further, theorists of military strategy such as Sun Tzu and Carl von Clausewitz are relevant to a broad understanding of the field. Quackenbush (2016) tests Clausewitz's (1984) argument that centers of gravity are the key to victory in war. In the process, Quackenbush (2016) finds that capturing the enemy's capital city, destroying the enemy's army, and knocking enemy allies out of the war all significantly increase the likelihood of victory.

Morey (2016) examines the impact of military coalitions on war outcomes, distinguishing between different types of multilateral wars. Sometimes states fight on the same side in a war, but they do not actively coordinate. He classifies these cases, such as Egypt, Jordan, and Syria in the Six-Day War, as fighting a war in parallel. In contrast, other cobelligerents have a unified command structure, a joint command, or develop a joint plan of battle, in which case he classifies them as a coalition. Morey (2016) finds that fighting as a coalition significantly increases the chances of victory in war.

Domestic Factors

Another popular focus has been the impact of domestic factors on war outcomes. Lake (1992) was one of the first to note that democracies are significantly more likely to win than non-democracies, finding that democracies win 81 percent of their wars while non-democracies win only 43 percent of theirs. He argues that this is due to the difference in rent-seeking behavior in democracies and non-democracies. Because they are constrained from earning rents, democracies will "possess greater national wealth, enjoy greater societal support for their policies, and tend to form overwhelming counter-coalitions against expansionist autocracies" (Lake 1992, 32), and thus be more likely to win.

Others focus on different mechanisms through which democracies are more likely to win wars. Reiter and Stam (1998) examine the interaction between regime type and initiation. They find that while democracies and initiators are always more likely to win than non-democracies or targets, democratic initiators are particularly likely to win. However, Bennett and Stam (1998) find that democracies need to win quickly or their advantage disappears.

Choi (2003, 2004) argues that the key impact of democracy on war outcomes is that democracies cooperate well with others, so they make better coalition partners than non-democracies. Therefore, fighting a war alongside more democratic partners increases the likelihood of victory, while increasing the number of one's non-democratic partners makes winning less likely. This is particularly significant since democracies are consistently more likely to ally with other democracies than with non-democracies (Siverson and Emmons 1991; Gibler and Sarkees 2004).

The most comprehensive examination of the impact of regime type on war outcomes was conducted by Reiter and Stam (2002). They do not find much support for previous arguments that democracies are more likely to win either because they tend to have strong economies and are thus able to overwhelm opponents through massive production of weapons and equipment, or because they tend to ally together and fight. Instead, Reiter and Stam (2002) find that two different factors are the most important ones that make democracies more likely to win. First, armies of democracies tend to encourage soldiers at all levels of the chain of command to take individual initiative, and second, democracies are better at selecting winnable wars to fight. Reed and Clark (2000) point out that if Reiter and Stam are correct that democracies are better at selecting winnable wars to fight, then the resulting selection effects make evaluating the ability of democracies to win wars very difficult. The finding that democracies are better at choosing wars they can win is well supported in the literature, even by studies not specifically focused on war outcomes (e.g., Reed 2000; Bueno de Mesquita et al. 2003).

While much of the literature examining domestic determinants of war outcomes focuses on democracy, not all of it does. Murdie (2012) examines the impact of civil-military relations on the outcomes of international crises. She finds that medium levels of civil-military conflict can actually be useful because there is an inverse U-shaped relationship with crisis success. Narang and Talmadge (2018) also focus on civil-military relations, but they move beyond simply viewing them as either conflictual or harmonious to examine them in more depth. While there are a number of different ways to account for civil-military relations, they look at whether a regime is plagued by coup concerns and whether a country's military focuses on internal tasks rather than fighting war. They find strong and consistent evidence that states either plagued by coup concerns or with militaries focused on internal tasks are less likely to win wars.

A final domestic factor to consider is political repression—whether political opposition to the government is allowed within a state, or whether it is silenced through censorship, imprisonment, or other means. Stam (1996) argues that repression can have an important effect on war outcomes because it allows leaders to hide costs in an effort to maintain legitimacy. Repression works differently for democracies, however, because high levels of political repression undermine democracy.

Insurgency and Counterinsurgency

In addition to these studies seeking to explain war outcomes in general, there is a large body of literature focused on the outcomes of counterinsurgency campaigns. This has become more prominent in the twenty-first century as the United States has struggled against insurgencies in Iraq and Afghanistan and other insurgencies have captured world attention in Nigeria, Syria, and elsewhere.

There are two primary approaches to counterinsurgency: direct and indirect (Nagl 2005; Kilcullen 2010). A direct, or attrition, approach focuses on trying to eliminate the insurgent forces directly through some combination of killing and imprisonment. Direct approaches to counterinsurgency stem from arguments that "a war is a war is a war" (Summers 1989, 27). Some scholars argue in favor of such direct approaches (e.g., Porch 2013). However, most counterinsurgency theorists argue in favor of indirect approaches, which focus on the hearts and minds of the population in order to eliminate support for the insurgents. For example, Findley and Young (2007, 396) develop a computational model to compare the two and find that "a hearts and minds approach is likely to be far more successful than an attrition approach." Economics is also important for indirect approaches. Berman, Shapiro, and Felter (2011) examine the political economy of counterinsurgency outcomes, arguing that increasing provision of services can reduce insurgent violence, chiefly because it increases satisfaction with the government. Essentially, hearts and minds can be bought, and evidence of US reconstruction projects in Iraq supports their theory.

Part of the problem with an attrition strategy for counterinsurgency is that it tends toward indiscriminate violence. Downes (2007) examines the effectiveness of indiscriminate violence and finds that it is generally ineffective. Relatedly, Kocher, Pepinsky, and Kalyvas (2011) find that the use of aerial bombing by the United States in South Vietnam was counterproductive for American efforts at counterinsurgency because it led to indiscriminate violence against civilians, regardless of their support for the insurgents. This leads to greater support for the insurgents within the population, growing the ranks of the insurgents by creating what Kilcullen (2009) labels accidental guerrillas. In contrast, Lyall (2009) finds that indiscriminate Russian artillery fire in Chechnya seemed to either decrease insurgent attacks or have no significant impact.

One reason that states might prefer direct approaches to counterinsurgency is that they are often expected to work better for conventional warfare. Based on this, Quackenbush and Murdie (2015) examine the relationship between conventional war and counterinsurgency. Given the manifest differences between the two types of warfare, some have argued that states are hampered in their ability to fight one type of war because they are prepared for the other type, which they have experience with. However, Quackenbush and Murdie (2015) find that experience fighting conventional wars does not negatively impact a state's success in counterinsurgency, or vice versa. Perhaps states that start fighting an insurgency with a direct approach could fruitfully switch to an indirect one. Enterline, Stull, and Magagnoli (2013) examine this question of whether changes in strategy in the midst of a counterinsurgency campaign can alter the chances of victory. They find that strategy change can make an important difference, but the shift in strategy needs to come fairly early in the war to really matter.

While a number of studies have focused on the difference between direct and indirect approaches, others have sought to examine a broader range of factors affecting counterinsurgency outcomes. Lyall and Wilson (2009) find that highly mechanized armies are less effective at fighting insurgencies. They argue that mechanized units tend to patrol the area in their vehicles, resulting in substantially less interaction with the local population than light infantry forces patrolling on foot. Because of this need to interact with the population, counterinsurgency places greater emphasis on manpower. Using the same data, Lyall (2010b) finds that regime type has no impact on the outcomes of counterinsurgency wars, in contrast to previous arguments that democracies make inferior counterinsurgents.

Friedman (2011) finds that the larger the number of personnel used by the state, the more likely counterinsurgency is to succeed. However, he finds that there is no optimal threshold of force density, contrary to many previous arguments (Galula 1964; Thompson 1966; Department of the Army 2007). Furthermore, he also argues that the key to victory is how the forces are used, not how many there are. Successful counterinsurgency requires understanding the local context rather than simply applying a set blueprint across the board (Greenhill and Staniland 2007).

Part of what makes counterinsurgency so difficult is that the local context of insurgencies can vary greatly, so studies have sought to determine how context matters. For example, Lyall (2010b) finds evidence that soldiers that share the same ethnicity with the population are more effective counterinsurgents than others. This has important ramifications for outside powers fighting against insurgents in another state, and for governments of multiethnic states.

EXPLAINING WAR CONSEQUENCES

A distinct but related area of inquiry is the consequences of war, or the ways that a war changes the underlying context in the future relations between states. Unfortunately, this is one of the least studied topics in international politics, particularly from a scientific perspective. There are a number of difficulties in studying war consequences because they could be just temporary, they might not be clear right away, they can be indirect, and they are not necessarily negative.

Additionally, the consequences of a war can be different for different groups, such as participants versus nonparticipants or winners versus losers. These multiple types of effects of war create serious obstacles to studying and understanding the consequences of war (Thompson 1995).

While larger wars tend to have greater consequences, sometimes small wars can have large consequences as well. For example, Bueno de Mesquita (1990, 28) shows that the Seven Weeks' War in 1866 had tremendous consequences, particularly because it "fundamentally changed the international order by providing the foundation for German hegemony on the European continent." This happened despite the small size and short length of the war. He concludes that "great consequences may result from small causes" (50).

To illustrate the multifaceted consequences of war, I look at the Franco-Prussian War (Howard 1981; Wawro 2003). Although it is not comprehensive, Figure 7.1 diagrams some of the primary consequences that followed German victory in the war. Domestically the war led to the collapse of the French Second Empire and the unification of Germany. The French collapse led to the creation of the Third Republic, which led to further consequences for French society. German unification led to rising German power and German colonization, which were important at the international level. Key direct international consequences

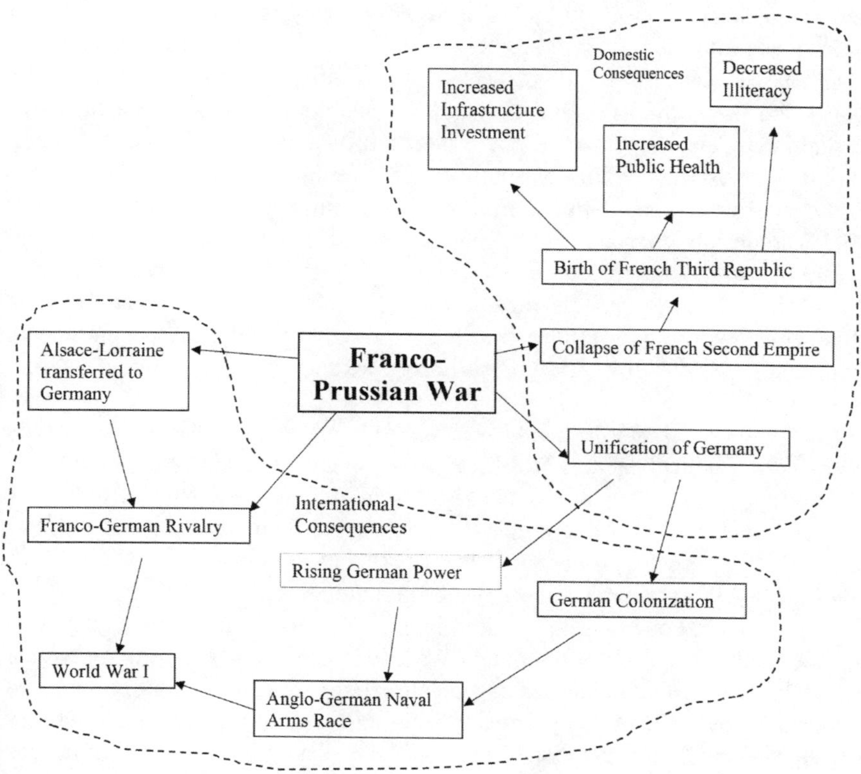

Figure 7.1. Consequences of the Franco-Prussian War.

of the war were the transfer of territory to Germany and the growth of the Franco-German rivalry. The rise of Germany fed into the naval arms race with Britain and eventually led to World War I. This case shows how a war can have a wide range of consequences, including regime stability, territorial holdings, arms races, and interstate rivalries, potentially increasing risks for future wars.

I now turn to examining specific consequences that have been identified in the literature. The first section examines domestic consequences, while the second focuses on international-level consequences. The third section examines recurrent conflict and rivalry.

Domestic Consequences

The Franco-Prussian War is one of many wars that led to significant domestic political changes. Similar to Germany, war plays a crucial role in state formation and expansion (Tilly 1975; Rasler and Thompson 1989; Spruyt 2002). Indeed, it was crucial for the development of the modern nation-state system as we know it. "War is an important factor in the origin of states and their subsequent expansions in territorial and functional terms" (Thompson 1995, 168). War can lead to permanent raises in state expenditure levels (Peacock and Wiseman 1961), although permanent increases are much more likely if the war is global. Rasler and Thompson (1989) conduct a broad empirical analysis of the impact of war on state expenditures and find that its effect is mostly temporary, although the effect of global wars is abrupt and permanent. More recently, Besley and Persson (2009, 2010) find strong evidence that external wars lead to increases in state capacity.

Similar to the French experience, interstate war can impact the likelihood of intrastate conflict within a country. States usually attempt to extract more resources from the population during war in order to pay for its high costs. Tilly (1978) argues that this can lead to internal conflict if segments of the population resist this extraction of resources and the government reacts with force. Although the details of their arguments vary, similar ideas abound in the literature (Stein 1980; Gurr 1988). Rasler (1986), however, shows that regimes that accommodate demands of various domestic groups during warfare are able to avoid internal conflict.

Even without leading to a civil war, war has important consequences for the survival of political leaders in office. In particular, war defeats increase the probability of regime change. The existence of such a connection between war and regime change provided the foundation for other theories about the effects of regime type, such as selectorate theory (Bueno de Mesquita et al. 2003) or Goemans's (2000) argument about war termination.

One of the first quantitative studies of the effect of war on regime change was by Bueno de Mesquita, Siverson, and Woller (1992). They sought to determine what factors affect the likelihood of violent regime change, such as through revolutions and coups d'état. They find that winning a war—for either the initiator or the target—greatly decreases the likelihood of violent regime change compared to the initiator losing. A loss by the target state also decreases the likelihood of violent change compared to the initiator losing, but not as much as a victory. Thus losing a war greatly increases the likelihood of violent regime change, particularly for initiators.

In a follow-on study, Bueno de Mesquita and Siverson (1995) seek to further clarify the effect of war on regime change through a survival analysis of the hazard that political leaders face of being removed from office. They find that each increase in the prewar tenure for a nondemocratic leader increases their expected duration in office after the war, while for democratic leaders, the effect is indistinguishable from zero. More costly wars (in terms of battle deaths) increase the risk of removal from office following war. On the other hand, winning the war increases the chances of survival in office, exactly as we should expect. They find, however, that this effect is smaller than the effect of prewar tenure for autocratic leaders. An autocratic leader who has a long tenure in office thus has a higher probability of maintaining power following a war loss than a democratic leader who wins a war.

Chiozza and Goemans (2011) conduct a broader analysis of the relationship between leaders and international conflict, looking not only at the impact of war on regime change but also at the role that leaders play in the outbreak of international crises and wars. They find that removal from office, whether forcible or through regular procedures, is always more likely following losses than wins. Democratic leaders, however, are continually at a greater risk for removal from office than nondemocratic leaders. In addition, they find that leaders' decisions about whether or not to initiate conflict are influenced by their expectations about the likelihood of removal from office. If leaders anticipate regular removal from office (such as through elections), then they have little to gain and much to lose from international conflict. In contrast, leaders who anticipate forcible removal from office (such as by coup or revolution) have little to lose and much to gain from international conflict. This increased likelihood of removal from office drives democratic leaders to be more selective about which wars to fight, as discussed earlier.

War can also have important consequences for the domestic economy. One positive effect that war can have on economic conditions within a country is through the acceleration of technological innovation. Given their vast scale and long length, both World War I (Hartcup 1988) and World War II (Hartcup 2000) led to a wide variety of technological innovations. For example, World War II led to a great acceleration of the development of nuclear energy (Rhodes 1986). In addition, electronic computers were invented during World War II, being used most notably for code breaking and in the development of the atomic bomb. The war also led to great advances in medical technology, particularly through the widespread use of penicillin. The German V-2 rocket developed during the war is not only an early version of modern ballistic missiles; it also paved the way for missions to the moon and elsewhere in space.

Several scholars have conducted broad, quantitative studies of the impact of war on economic growth within countries. Wheeler (1975, 1980) examined the consequences of war for industrial growth in sixty cases from 1816 to 1965 and found that the effect of war was significant nearly 75 percent of the time. The direction of change, however, varies greatly. In some cases war has a positive effect on growth, in others it has a negative effect, and in some cases war has only a temporary impact on industrial growth. Rasler and Thompson (1985, 1989) focused on the economic growth of major powers from 1700 to 1980. They also find that less than three-quarters of wars exert significant effects on economic growth, with those split nearly evenly between positive and negative impacts. More recently, Koubi (2005) examined the consequences of both interstate and intrastate wars in a broad sample of countries from 1960 to 1989. She found that war has a consistently positive impact on postwar economic performance, which grows even stronger following wars of greater duration and severity.

By definition wars lead to deaths so, rather obviously, wars can lead to an increase in mortality rates. The increase in mortality rates is not necessarily confined to military personnel, but can also come from civilian casualties as the result of bombings and battles, in addition to increases in premature death due to deteriorating wartime environments. Valentino, Huth, and Croco (2010) find that democracies minimize the number of casualties their country suffers in war. Further, democracies shield both their military personnel and their civilian populations from human costs. This effect compounds when interacted with conscription, with democratic volunteer armies suffering far fewer battle deaths than others (Horowitz, Simpson, and Stam 2011).

Fazal (2014) shows that advancements in modern military medicine have led to decreasing battle deaths, even though overall casualty rates have not necessarily decreased. This has ramifications for our classification of cases as wars or not, which relies on reaching a battle death threshold.

Wars, particularly if they are lengthy, also tend to lead to declines in marriage rates and childbirth, although there can also be a baby boom after the war as families are reunited. All of these factors, driven by the number of people mobilized for the war, can cause great strains on family relationships. Between the increased mortality rates and decreased birth rates, war can lead to major shifts in the age and sex structure of populations. These war-induced demographic changes are stronger the larger the war and the extent to which the home front quality of life deteriorates, and are smaller the greater the distance between the country and the combat zone (Thompson 1995).

War can also have a large impact on social change within societies (Marwick 1974). One undoubtedly positive consequence of war has related to women's rights. Since war fighting has been a historically male activity, war has led to a variety of changes in gender roles and expectations in society. When large numbers of men went off to fight during the world wars, many women started working outside the home—in factories and other positions previously held by men—for the first time. This impact of women on the war effort during World War I contributed to women's right to vote finally being granted in the United States in 1920. World War II saw even more women entering the workforce in the United States and elsewhere. This gave them greater power and voice in society, and contributed to a greater level of gender equality.

Finally, war can have important impacts on societal learning patterns. Because war participation is costly in terms of lives lost and wealth expended, one would expect that greater war costs would create greater reluctance on the part of leaders and societies to repeat the experience. This leads to the idea of war-weariness, where countries are expected to be less likely to fight again soon after a war (Farrar 1977; Levy and Morgan 1986). Since societal memories tend to fade with time, war-weariness is expected to only have a temporary effect. Early studies found little to no empirical support for the war-weariness hypothesis (Stoll 1984a; Levy and Morgan 1986). However, Garnham (1986) argues that democracies are more likely to grow weary of war than others. Further, Pickering (2002) finds that war-weariness exists for states that suffer a series of defeats, but states that keep winning wars keep fighting them.

International Consequences

As the Franco-Prussian War illustrates, the international consequences of a war can be immense and can lead to changes within dyads and the international system.

Wars often lead to territorial changes. If you compare the map of Europe in 1914 to the map today, the differences are dramatic. Some of the changes came peacefully, such as the breakup of Czechoslovakia into the Czech Republic and Slovakia in 1993. Most territorial changes, however, came as the result of war. Goertz and Diehl (1992b) find that territorial changes are likely to lead to a militarized dispute within five years, particularly if the exchange is perceived as illegitimate. If, however, the territorial changes lead to the resolution of territorial issues between the states, then future conflict is much less likely (Gibler 2012).

War can also affect the balance of power between states, as the outcome may reduce or augment a state's relative capabilities. The United States emerged from World War II as the most powerful country in the world by far. Siverson (1980) argues that the extent of a war's consequences is directly related to the changes produced in power relationships between states; the more power is changed by a war, the greater its consequences will be.

War has played a major role in colonization and decolonization in history (Maoz 1990). A number of extra-state wars occurred as European states fought against indigenous peoples to conquer territories, particularly in Africa and Asia, which were then held as European possessions for many years. Decolonization occurred in two major waves. Following World War I Germany lost all of its colonies, Austria-Hungary was broken up into several new states, and Russia lost parts of its empire (particularly Poland and the Baltic States). In the decades following World War II all of the colonial powers—particularly Britain, France, Portugal, and Belgium—lost nearly all of their remaining colonies. This wave of decolonization was a direct consequence of World War II, although it often took a number of years and a variety of intermediate events before the former colonies were finally granted independence.

Wars can also lead to a number of changes in alignment patterns within the international system, including shifts in formal alliances. For example, in World

War II the United States, Britain, France, and the Soviet Union were allies fighting against Germany, Italy, and Japan. After the war Germany, Italy, and Japan all became firm allies of the United States (along with Britain and France), while the Soviet Union became the primary enemy. Unfortunately there has been little systematic research on these effects.

In addition to the domestic economic impacts, war can have a large impact on international economic conditions. In particular there is strong evidence that war creates inflationary surges in the international economy (Hamilton 1977; Thompson and Zuk 1982). However, there have been very mixed results regarding the consequences of war in other economic areas such as prices and production (Thompson 1995). Long cycle theory (e.g., Modelski 1987; Goldstein 1988) argues that there is a cyclical relationship between major power wars and global economic conditions.

Central to the economic consequences of war are the costs related to it. Keynes (1920) argued that the gap in economic fortunes between the winners and losers would increase in the near future following the war but the economic devastation of the losers would then bring chaos to the entire international economic system, causing the winners of the war to fall down to the losers, making the economic gap between the winners and losers disappear. A second perspective is that all states would experience permanent economic loss following a major war (Angell 1911; Nef 1950).

Organski and Kugler (1977, 1980) introduce a third perspective, the phoenix factor. The basic idea of the phoenix factor is that economic consequences of war are temporary. The losers of a major war are expected to lose more economically than the winners. They do not, however, remain down permanently. Rather, like the phoenix of Greek mythology, the losers are expected to rise from the ashes of their defeat and recover relatively quickly. Organski and Kugler (1977, 1980) find strong support for their expectations, but they only look at the consequences of the two world wars. Kugler and Arbetman (1989), however, add the Franco-Prussian War to the analysis and rule out an additional alternative explanation. In addition, Koubi (2005) analyzes the economic performance of a large sample of countries following both interstate and intrastate wars from 1960 to 1989 and finds further empirical evidence supporting the phoenix factor. Kugler et al. (2013) build on these studies by looking at both demographic and economic consequences of conflict.

Recurrent Conflict and Rivalry

The Franco-Prussian War also illustrates how conflict can lead to further conflicts between the same states. Recurrent conflicts are quite common among pairs of states that have fought previously, and many studies have sought to explain them. I group the literature into four theoretical approaches: bargaining, enduring rivalries, conflict management, and deterrence.

The first major theoretical perspective focuses on bargaining, and is connected with the bargaining model of war more broadly (Fearon 1995; Powell 2002).

Bargaining explanations of recurrent conflict have made predictions about the effect of outcomes, settlements, changes in relative power, and third-party involvement on durations of peace following conflict.

The importance of outcomes for explaining recurrent conflict stems from the bargaining model's assumption that power is central to explaining international conflict. For example, Blainey (1988, 293) argues that "wars usually begin when two nations disagree on their relative strength, and wars usually cease when the fighting nations agree on their relative strength." Decisive war outcomes are expected to be stabilizing because they clarify the relative power of states.

Bargaining theorists argue that the mode of settlement used to end a dispute is not particularly important in determining the duration of peace following a conflict (Werner 1999; Werner and Yuen 2005). Rather, they argue that incentives to renegotiate the settlement drive recurrent conflict. This argument stems from the bargaining model's implication that the terms of settlement agreed to reflect the belligerents' mutual expectations about the consequences of continued fighting (Werner 1998). As long as those expectations remain the same, peace should last. However, if those expectations change—i.e., if either belligerent anticipates that it would fare better in a new conflict than the last—then they have an incentive to demand renegotiation. This may lead to a new conflict. Werner (1999) argues that the primary indicator of these incentives are shifts in power: as power shifts in one state's favor, that state has an increased incentive to challenge the settlement and a recurrence of conflict becomes more likely.

The final key prediction of the bargaining perspective focuses on third-party involvement. The bargaining model identifies commitment problems as a primary reason that bargaining breaks down, and thus as an important cause of war (Fearon 1995; Powell 2002). Walter (1997, 2002) argues that third-party involvement—either in the form of security guarantees or intervention with troops on the ground to enforce peace agreements—provides a way through which commitment problems can be overcome. Thus third-party involvement is expected to be an important way to increase the stability of peace following a conflict. Walter (1997, 2002) examines the effect of third-party involvement in the settlement of civil wars, although the bargaining perspective expects them to be pacifying following interstate conflicts as well (Fortna 2004b).

The second major theoretical perspective focuses on enduring rivalries. The enduring rivalries literature (e.g., Diehl and Goertz 2000; Colaresi, Rasler, and Thompson 2007; Valeriano 2013) focuses on pairs of states that view each other as enemies and fight with some regularity. The study of rivalries covers much more than just recurrent conflict, as discussed in chapters 5 and 11 of this volume.

Rivalry theorists agree with bargaining theorists that outcomes play a central role in explaining recurrent conflict. For example, Hensel (1994) looks at patterns of recurrent interstate disputes by examining the effects that outcomes have on future conflict. Analyzing contiguous Latin American dyads, Hensel finds that 93 percent of disputes characterized by a decisive outcome are followed by a later dispute, compared to only 85 percent of stalemates and 73 percent of compromise outcomes. However, he also finds that the post-dispute stability of decisive outcomes is (on average) over two years longer than compromise outcomes, and over six years longer than stalemate outcomes.

The third primary theoretical perspective has focused on conflict management. This literature has examined the effects of factors such as third-party intervention, UN peacekeeping missions, and regime type on the likelihood of reaching negotiated resolutions to conflict and stable peace following international conflicts.

The primary forms of conflict management that this literature focuses on are third-party involvement and peacekeeping, which are expected to lead to more stable peace following conflicts. Gartner and Bercovitch (2006) find two contrasting influences at work in the relationship between mediation and post-dispute peace. First, they contend that the effects of mediators are positively related with stability; clearly, this is the purpose of mediators' involvement in the first place. However, there is a selection effect that is negatively related to stability. Conflicts that attract the involvement of mediators tend to be conflicts that are inherently difficult to settle peacefully. If the sides in a dispute could peacefully resolve a dispute by themselves, there would be no need for a mediator. Beardsley et al. (2006) advance a similar argument, but also consider whether the principal actors engage in another crisis within the five-year period following a dispute.

Frazier and Dixon (2006) explore the utility of varying conflict management techniques and the contexts in which they are most likely to succeed. Their study focuses on the efficacy of different mediators—intergovernmental organizations (IGOs), states, and coalitions—and the conflict management techniques available to them, including verbal offers, mediation attempts, adjudication, and military intervention. Overall, they find that IGOs are the most effective managers while military interventions are the most effective conflict management type. Frazier and Dixon are interested in which combination of mediator and technique is most likely to bring about a negotiated settlement.

The conflict management literature assumes that negotiated settlements are inherently desirable compared to other settlement types, and thus that negotiated settlements lead to greater durations of peace after a dispute (e.g., Butterworth 1978; Dixon and Senese 2002; Frazier and Dixon 2006). Accordingly, imposed settlements—where the terms are dictated by one side to the other—are expected to produce less stable relations following conflict. For the same reasons, victor-imposed regime changes are expected to be less stable.

The final theoretical perspective focuses on deterrence to explain recurrent conflict. Senese and Quackenbush (2003) argue that settlement type is an important driver of recurrent conflict, and imposed settlements are the most stable type of settlement. This is because states can rely on unilateral deterrence to prevent recurrent conflict following imposed settlements, while they must resort to mutual deterrence following other types of settlements. That is a significant difference, as Zagare and Kilgour (2000) demonstrate that unilateral deterrence is more stable than mutual deterrence. Senese and Quackenbush (2003) find strong support for their expectations in an analysis of recurrent militarized interstate disputes between states from 1816 to 1992.

In subsequent research Quackenbush and Venteicher (2008) again find that imposed settlements are the most stable. Furthermore, they "find that the type of outcome reached has no direct bearing on post-dispute stability; its primary importance is that decisive outcomes make imposed settlements possible" (740). Settlement type thus matters more than outcome for recurrent conflict, in contrast to other studies claiming that outcomes are more important. Furthermore, Quackenbush (2010) finds that imposed settlements have a stronger impact on recurrent conflict than territorial issues, although territory is an important driver of recurrent conflict. Peterson and Quackenbush (2010) find an important interaction between settlement type and international trade in driving recurrent conflict, with imposed settlements paving the way for trade to have a pacifying effect.

Relatedly, Fortna (2003, 2004a) develops a theory of agreements, arguing that cease-fire agreements can help maintain peace by altering the incentives for war and peace, reducing uncertainty, and helping prevent or manage accidents that could lead to war. Furthermore, the stronger these agreements are, the more durable the peace following them is. Like the mode of settlement, her analysis highlights agreement strength as another important political factor characterizing the termination of the previous conflict.

A final argument explained by the deterrence approach to recurrent conflict is that conflicts terminated with a victor-imposed regime change are followed by longer durations of peace than other conflicts (Lo, Hashimoto, and Reiter 2008). By establishing a new regime in the defeated state, the victor is able to create more friendly relations between the states, thereby making enforcement of the settlement easier. This argument also augments the primary argument of the deterrence perspective that imposed settlements are more pacifying than other types. Although such regime changes can help with international stability, Bueno de Mesquita et al. (2003) argue that states have incentives to impose non-democratic regimes so that they can act in accordance with the imposer's wishes without having to worry about domestic political competition.

DIRECTIONS FOR FUTURE RESEARCH

While we know a great deal about the outcomes and consequences of war, there is also a great deal that we still need to learn. In this section I consider directions for future research in these areas.

War Outcomes

Many promising and important paths exist for future research on war outcomes. One clear need is connecting battle outcomes with war outcomes. Wars are made up of a series of battles and campaigns. As we have seen, scholars have identified a number of factors that drive war outcomes. Many studies have focused on explaining military effectiveness and the outcomes of battles (e.g., Biddle 2004; Grauer and Quackenbush 2020). However, we do not have a clear understanding of the connection between the two. Certainly there is a basic expectation that winning battles helps a state win the war. But we just as certainly know that winning battles does not automatically translate into winning the war.

Another promising avenue for future research is studying the impact of geography on war outcomes. At this point we know some basics regarding the relationship between geography and war outcomes. For example, Stam (1996) shows that the roughness or openness of terrain matters greatly, particularly because of its interaction with military strategy. Quackenbush (2016) shows that capturing the enemy capital makes victory more likely, and geography plays an important role in determining the vulnerability of capital cities to capture. But there is clearly much more to learn about the role of geography. Studies of civil wars have employed spatial analysis to examine the movement of rebel groups, placement of peacekeeping forces, and sundry related issues (e.g., Townsen and Reeder 2014; Reeder 2018). Spatial analysis has also been used to study the geographic spread of interstate conflict (Braithwaite 2006). The study of interstate war outcomes would do well to catch up, particularly because geography might be a key for linking battle and war outcomes.

Most studies of war outcomes have focused on one or two key variables of interest, while controlling for a few common variables. Unfortunately the result is that we never quite see how all of these different variables compare to one another head to head. This could be dealt with at least in part by creating a general model of war outcomes. Bennett and Stam (2004) employ an approach they call comparative hypothesis testing, in which they pit a variety of theories about the causes of war together to determine which ones enjoy the most robust empirical support. The closest thing that exists to this on the topic of war outcomes is Stam (1996), but the field has advanced in a variety of ways in the decades since his study.

A final key area to explore in future research is looking for possible temporal patterns related to war outcomes. Studies generally analyze the full range of data available, which could be 1816–2007 for the Correlates of War data (Sarkees and Wayman 2010), dating back to 1495 when using Levy's (1983) great power war data, or some shorter time period if other data are used. While some studies have explored the possibility of differences in findings between different periods of time, most have not.

War Consequences

Given the paucity of systematic research on the subject, there are plentiful directions for future research on the consequences of war. One promising area is to examine the connection between war consequences and the outbreak of war. From a basic rational choice understanding of conflict, leaders go to war because they expect to get something from it. So an expectation of positive consequences (at least compared with the alternative) will make war more likely. The development of a general model through which consequences of war could be forecasted would be tremendously useful in this regard.

Another concern that future research should address is whether our understanding of war consequences is overly dependent on certain cases. For example, many stylized facts are about the consequences of World War I and/or World War II. But while those wars had tremendous consequences for the world, they are not necessarily representative of other wars. Similarly, we need to ensure that our understanding of war consequences is not driven by the experience of the United States or other great powers. For example, Thies (2005) suggests that Tilly's (1975) ideas about war driving state building, based on the European experience, do not directly translate to Latin America.

Future research should also examine the consequences of war for women. Some recent research shows that women are commonly targeted for rape and sexual assault during peacekeeping operations (Karim and Beardsley 2016) and civil wars (Cohen 2013). However, these and related subjects that disproportionately affect women have not been studied systematically in the context of interstate war.

Pushing our understanding of recurrent conflict further would be another useful avenue for future research. Given that the different theoretical perspectives lead to different expectations—such as whether imposed settlements or negotiated ones are more stable—it would be useful to determine which approach is best supported by logic and evidence.

Further study of learning and war-weariness would also be useful. Do leaders and publics learn over time that there are more profitable ways to pursue national interests than war? Or do they learn that war can be profitable? There have been a variety of observations that war is becoming less frequent (e.g., Goldstein 2011; Gleditsch et al. 2013). Perhaps that is the ultimate consequence of war.

CONCLUSION

In this chapter I examine what we know about the outcomes and consequences of war. War outcomes depend on power, military strategy, regime type, and a myriad of other factors. War consequences shape domestic and international politics in a variety of ways, including leading to regime change, recurrent conflict, and a variety of economic consequences. Future research should build on the foundation provided by these studies.

A number of related subjects deserve our attention, and are important parts of a comprehensive understanding of international conflict. In particular studies of war duration (e.g., Bennett and Stam 1996, 2006) examine why some wars end quite quickly while others continue on for years. Relatedly, studies of war termination (e.g., Goemans 2000; Reiter 2009) seek to explain what brings states to stop fighting when they do. Making connections across this literature will further advance our understanding of the consequences of war.

Part II

FACTORS THAT PROMOTE PEACE

Chapter Eight

The Liberal Peace

Michael Mousseau

The fall of the Berlin Wall in 1989 serves as a milepost in the emergence of the concept of the "Liberal Peace": the idea that democracy, trade, and international governmental organizations can affect relations among nations. During the Cold War realist thinking dominated the study of international politics (Vasquez 1979), a view that informs us that domestic factors like democracy can hardly matter in foreign affairs. International trade, if it was thought to have any effect on nations' foreign policies, was largely viewed as a core cause not of peace but of wars of imperial conquest and exploitation (Wallerstein 1974; Choucri and North 1975; cf. Polachek 1980). Most studies reported little or no relationship of international organization membership with peace (e.g., Domke 1988). Yet within a decade of the end of the Cold War, the Liberal Peace had established itself as the foremost research program in the study of international conflict.

This chapter documents the rise of the Liberal Peace, with special emphasis on the causal mechanisms of the various explanations for it. It will be seen that more than a generation after the fall of the Berlin Wall a paucity of evidence remains for most theories of how democracy, trade, and international governmental organizations can affect relations among nations. This does not mean all these theories are wrong, however: the state of evidence could reflect a deficiency of effort in testing for causal microfoundations. Most of the empirical work can be characterized by what Mearsheimer and Walt (2013) call "simplistic hypothesis testing," meaning tests derived not from precise, theoretically deduced expectations but from general correlational speculations or deductions that ultimately can be interpreted in myriad ways. Some of the theoretical work seems disconnected from historical reality as we know it. This chapter is a call to arms, so to speak, for field researchers to give more attention to causality and history if we wish to make progress in understanding the causes of war and peace among nations.

I start by addressing the emergence of each pillar of the Liberal Peace—democracy, trade, and international governmental organization (IGO) membership—while assessing the state of the evidence at the dyadic level. Many studies have investigated the liberal variables at the national (monadic) and global (systemic) levels of analysis, in addition to the dyadic (pairings) level. For instance, democratic nations may be more peaceful than others at the monadic level (Ray 2000), and the proportion of democracies in the world could affect the behavior of non-democracies at the systemic level (Mitchell 2002). However, the Liberal Peace emerged foremost at the dyadic level, meaning relations between states that evince Liberal characteristics. As such, most theoretical and empirical work has been at this level, and space limitations make ancillary investigations at other levels largely beyond the scope of this chapter.

After reviewing the story of each pillar of the Liberal Peace, I offer directions for future research that, by examining the purported causal linkages more directly, have the promise of furthering progress in our understanding of it. I then identify two likely true real-world facts that have been largely overlooked and that probably matter for understanding the Liberal Peace: American hegemony has greatly reduced the level of anarchy over large swaths of the globe; and economic development, which may well cause democracy, trade, and IGO membership, may also cause the peace. I conclude with my appeal for more attention to causality and history.

THE DEMOCRATIC PEACE

Coined by Dean Babst (1964), the term "democratic peace" is the observed historical pattern that nations with democratic governments fight each other less often than other types of nations, even as they seem to fight other nations about as much as everybody else. This pattern appears across an array of definitions of what we might mean by "fighting": from the relatively minor act of threatening to use military force to sustained war. The often overlooked fact of the democratic peace is that it is only an observed historical correlation. Observations do not explain themselves, and we do not know what causes them. This section briefly chronicles the discovery of the democratic peace and the difficulties that have beset those who have sought to explain it.

With democracy broadly defined as states with elected legislatures that limit the power of executives, it was only toward the end of the Cold War that the historical absence of wars between democratic states began to be widely noticed (Doyle 1986; Levy 1988; Russett 1990). While today we can say that the dearth of militarized conflict between democracies probably began after World War II, if not World War I (Jenke and Gelpi 2017, 2275), it is possible that widespread attention had to wait for two developments in the study of war. The first is the fall of the Berlin Wall, which drew interest in the possibility that democracy may act as a united and powerful force in global politics, thus undermining the veracity of the realist tenet that domestic factors cannot affect patterns of war and peace among nations.

The second development was a range of advances in the study of war that coalesced by the late 1980s. These include a revolution in conceptual maturity and data generation that can be credited to J. David Singer and the Correlates of War (COW) project (Small and Singer 1982); wider realization of the need to analyze interstate pairings, or dyads, of nations (Most and Starr 1989), without which we could not draw confident conclusions on relations *between* democratic nations; and greater appreciation of the need for large representative samples along with controls for candidate confounding variables (Bremer 1992).

By the early 1990s it became increasingly clear that democratic nations were statistically less likely than others to have militarized confrontations with one another (Maoz and Russett 1992; Bremer 1993). Prior reports of democratic peace had appeared, but they either did not analyze dyads (e.g., Chan 1984; Weede 1984), relied on limited samples (Rummel 1979), or did not control for geographic contiguity (Small and Singer 1976; Doyle 1986; Maoz and Abdolali 1989). Critics of the statistical outcomes surfaced (Spiro 1994), but were convincingly challenged (Russett et al. 1995). By the middle of the decade the democratic peace could reasonably be described as "the most replicated research program in the modern study of international politics" (Maoz 1997, 162).

Today the main challenge regarding the democratic peace is in explaining it. There are two kinds of accounts: those that identify the cause as coming from democracy, and those that suggest some other factor may cause both democracy and peace.[1] Regarding how democracy may cause the democratic peace, an influential study by Doyle (1986) noted early on Immanuel Kant's eighteenth-century thesis that if people choose their leaders—that is, the very people who will be doing the fighting, dying, and suffering in foreign wars—they will tend to constrain their leaders from fighting wars that do not serve their interests (Kant 1939 [1795], 14). This view was reinvigorated more than a century later by US president Woodrow Wilson at the time of World War I. Maoz and Russett (1993) (see also Russett 1993) called this view the "structural" explanation, as distinguished from what they called the "normative" account, which emphasizes democratic norms and values of tolerance and compromise (Dixon 1994).

The most vexing challenge for anyone seeking to explain how democracy can cause the democratic peace has been in accounting for how democracies can avoid fighting each other, even as most studies find democracies fight other nations at normal rates.[2] To solve this quandary, both structural and normative explanations assume that leaders of democracies make foreign policy decisions with consideration of their perceptions of other nations' regime types: that democratic leaders consciously choose to behave differently with other democracies than they do with non-democracies. In structural selectorate theory the perception that a regime is democratic provides information that its leaders have public support in their foreign policy positions. Thus democracies consider each other more formidable adversaries than they otherwise would be, which has the effect that they avoid fighting each other (Bueno de Mesquita et al. 2004). In structural audience cost theory democratic leaders are assumed to believe that voters do not like their leaders backing down after having made a militarized threat.

Democratic leaders can thus use this fact to send and receive costly signals regarding their intentions to fight if pressed, which has the effect that they avoid fighting each other (Fearon 1994a). In the normative explanation democratic leaders are thought to apply the democratic norms of negotiation and compromise with nations they perceive as democratic, but less so with nations they perceive as nondemocratic (Dixon 1994).

Despite the crucial nature of regime perceptions in the causal chain for all mainstream explanations for how democracy can cause the democratic peace, we have little evidence that leaders actually consider the regime types of nations in their foreign policy decision making. A number of quasi-experimental studies have sought to gauge the effect of another state's democratic status on public preferences to use military force in the dyad, to my knowledge all reporting supportive results (Mintz and Geva 1993; Rousseau 2005; Johns and Davies 2012; Tomz and Weeks 2013). However, subjects in these experiments may understand "democracy" as meaning something else, including that a country is a friend. Tomz and Weeks (2013) sought to address this problem by manipulating factors that might be associated with democracy, such as alliance. But as with correlational studies, there is no way to know in this design if every candidate confounding factor is manipulated. Further, all of these experiments drew on subjects from three countries: the United States, Britain, and Israel. As a result causation is hardly isolated, since other factors these cultures may share—such as value change associated with economic development—can easily account for why respondents may value democracy or associate it with friendship. In fact, cross-cultural experimental work has shown that subjects in societies with higher levels of market integration act differently than subjects from other societies (Henrich et al. 2005).

Like most central questions of conflict processes, the field remains divided as to the cause of democratic peace. This could be the result, I have suggested (Mousseau 2019a, 1), of no theory of how democracy could cause the peace having predicted enough new facts to generate widespread support. A number of studies have corroborated various outcomes we would expect to see if the normative explanation is correct: democracies appear more likely than other countries to negotiate their differences (Dixon 1994), use or promote third-party mediators (Raymond 1994; Mitchell 2002), compromise (Mousseau 1998), and cooperate with one another (Mousseau 1997). However, none of these predictions is unambiguously novel enough to discount with confidence other factors from explaining them after the fact.

The structural explanation appears to have attracted the greater attention from field researchers, even as some meta-analyses report it as having a weaker predictive record than the normative explanation (Ungerer 2012). The most important prediction from structural accounts has been that democracies are more likely than other states to win their wars, largely because they make better decisions and have greater resolve than other countries (Lake 1992; Stam 1996; Reiter and Stam 2002; Bueno de Mesquita et al. 2004). However, it now appears that democracies may not be more likely to win their wars, as the pattern is not robust to minor alterations in measurement and design (Desch 2003; Downes 2009). Neither is war-winning necessarily structural: Reiter and Stam (2002) identify cultural reasons why democracies might win their wars. As is known today, democratic war-winning, if it does occur, is limited to the subset of Western democracies (Henderson and Bayer 2013), suggesting a cultural, not structural, effect of democracy, or something related with democracy.

While the structural explanations have had other successful corroborations (Schultz 2001a; Bueno de Mesquita et al. 2004), they have also accrued some anomalies. If democracies do better at selecting into easier wars, then they should tend to fight shorter ones (Reiter and Stam 2003), but they do not (Henderson and Bayer 2013). Neither are leaders of democracies more likely than autocracies to be punished from defeats in war (Chiozza, Giacomo, and Hein E. Goemans, 2004b). In fact, the democratic tendency to compromise in domestic politics can mean that democratic leaders make poorer rather than better decisions compared with their autocratic counterparts (Blagden 2019). Tomz and Weeks (2013) found that democratic publics are more concerned with value issues than issues of cost in decisions to escalate foreign conflicts, contrary to most structuralist theorizing. A number of studies have emerged casting doubt on whether the core fact of one of the leading structural hypotheses, audience costs (Fearon 1994), even exists (Snyder and Borghard 2011; Downes and Sechser 2012; Trachtenberg 2012).

As mentioned, a correlation of democracy with peace does not imply democracy causes it: some other factor could explain both democracy and the peace. Early on Farber and Gowa (1995) suggested common interests, rooted in alliance making, may account for the democratic peace. However, most studies have consistently shown that democracy remains a significant force for peace after controlling for alliance (Bremer 1992; Maoz and Russett 1992). To my knowledge, as of this writing there are two confounding factors with standing evidence in the literature: market-oriented development (Mousseau 2000, 2019a, 2019b) and territorial settlements (Gibler 2012; cf. Park and James 2015), each of which could cause both democracy and peace. Only the former theory, however, can be described as "liberal" and fits into the Liberal Peace, and therefore is discussed in this chapter. Before then the following sections examine developments in the next two pillars of the Liberal Peace: trade and international organization.

THE TRADING PEACE

Like the democratic peace, the idea that trade can cause peace did not become a mainstream issue in the literature until after the end of the Cold War, and probably for the same reason: the dominance of realist thinking, which discounts a role for trade in causing peace. Unlike the democratic peace, however, the trading peace was also delayed by the preceding two centuries of history, which contained numerous wars waged explicitly for the purpose of extracting resources and labor from weaker countries. Prevailing thinking during the Cold War is perhaps best epitomized by Wallerstein (1974) and Choucri and North (1975), popular tomes that argue that a core cause of international conflict is the search for raw materials and markets.

There are three kinds of trading peace arguments. They are often flung together in ambiguous ways, but they are distinct causal claims and should be treated as such. I therefore discuss each in turn.

Gentle Commerce

The idea that trade reduces conflict became commonly accepted among eighteenth-century Classical Liberals such as Baron de Montesquieu, David Hume, and Adam Smith. Hirschman identified the origins of this thinking in the veneration of the pursuit of self-interest that largely defines the Liberal thinking of the period. He quotes Samuel Ricard in 1704:

> Through commerce, man learns to deliberate, to be honest, to acquire manners, to be prudent and reserved in both talk and reason … so as not to arouse any adverse judgement … for fear of damaging his credit rating. (Hirschman 1992, 109)

Game theoretic applications have corroborated the Classical Liberal insight linking honesty and self-interest using the famed prisoner's dilemma: we know that among self-interested actors thinking rationally, in repeated interactions and with expectations of future ones, trust and cooperation emerge (Axelrod 1984). However, the gentle commerce thesis has a core challenge: it works only in communities small enough that individuals know each other, at least by reputation. The prisoner's dilemma does not predict trust and cooperation to emerge among strangers, meaning among those with no prior interactions or when there are no expectations of future ones. Merchant networks can be developed to extend individual reputations for honesty (Milgrom, North, and Weingast 1990, 8), but the depth of these networks is necessarily limited within communities and cannot be applied to the long-distance trade of international commerce (see also Clague et al. 1999, 187).

Opportunity Costs

Trade interdependence theory largely originates with Polachek (1980), who argued that mutual dependency on trade can raise the cost of conflict for both nations in a dyad, the higher costs being the presumed loss of trade with the adversary. Oneal and Russett (1997) set the measure widely used nowadays, with country *i*'s dependency on trade with country *j* the percentage of country *i*'s economy dependent on trade with *j*; *inter*dependence is gauged as the level of dependency of the less dependent state in the dyad. Today the opportunity cost thesis has been recorroborated about as often as the democratic peace correlation, and appears for foreign direct investment as well as trade (Lee and Mitchell 2012). While challenges have emerged (Barbieri 1996; Green, Kim, and Yoon 2001), the thesis is supported in most studies (King and Zeng 2001; Gleditsch 2002; Xiang, Xu, and Keteku 2007).

However, the opportunity cost thesis faces several vexing challenges. One is the crucial assumption that nations in conflict bear the costs of the loss of trade with their adversaries. It seems at least sometimes they do not: the northern states of the United States, for instance, traded with Canada during the War of 1812 (Barbieri and Levy 1999). While evidence generally supports that trade drops during militarized conflict (Anderton and Carter 2001; Li and Sacko 2002; Keshk, Pollins, and Reuveny 2004; Kim and Rousseau 2005; Long 2008), it is hard to have confidence in the extent to which this happens because data on trade tend to be unreliable during wartime (Levy and Barbieri 2004, 7). In any case Barbieri and Levy (1999) have shown that trade normally returns to prewar levels soon after wars end, suggesting that losses in trade from fighting may not be large enough to inhibit the pursuit of the initial goals over which nations fight.

The second quandary for the opportunity cost thesis is varying elasticity and substitutability of trade goods (Li and Reuveny 2011). An impending loss of access to bananas is less likely to constrain a nation otherwise willing to fight over an issue than an impending loss of access to capital goods, food, or oil. Similarly, a nation that can substitute some lost resources for others, or substitute resources from one foreign nation with those of another, will be less constrained than those less able to do so. For the opportunity cost thesis to work, we have to assume that elasticity and substitutability vary randomly with trade interdependence in dyads. This is an untenable assumption given that wealthy and industrialized nations are probably less dependent on inelastic goods than others, and are better positioned to substitute resources from one nation with those from another. It follows that if opportunity cost theory is correct, then the wealthy industrialized nations might be the least constrained and thus the most coercive of all types of nations. However, this inference goes against the spirit of Classical Liberal thinking, which treats the wealthy industrialized nations as archetypical Liberal states (Rosecrance 1985; Weede 1996).

The third challenge for the opportunity cost thesis is perhaps the most pressing: the direction of causality between trade and conflict. In Liberal interdependence theory, the expected loss of trade inhibits the resort to militarized coercion, as nations care about losses in trade more than they care about other issues of international politics—at least sometimes with some regularity. However, the observed correlation of trade interdependence with peace does not necessarily mean nations are sacrificing political goals for economic ones: it could reflect a reversed causal path of nations sacrificing economic goals for political ones. A number of studies report evidence for this reversed path of causation (Reuveny 2001; Long 2008).

Still, the reversed direction of politics affecting trade is not inconsistent with the Liberal opportunity cost thesis because in this theory there must be some expectation that nations do not fight and trade at the same time. At issue, then, according to Keshk, Pollins, and Reuveny (2004), is whether trade predicts peace above and beyond any impact of peace on trade. Keshk et al. report no evidence for such an impact, and Kim and Rousseau (2005) report identical results. However, Hegre, Oneal, and Russett (2010) report that—with consideration of certain controls—an impact of trade on peace is significant after consideration of the reverse impact of peace on trade.

Some have suggested that the opportunity cost thesis is ultimately not about opportunity costs but about the opening these costs bring for nations to show their resolve in conflict. Interdependence can facilitate costly signaling in that a dependent state that escalates a dispute is able to demonstrate that it is willing to sacrifice the gains from trade over the issue at hand, yielding information that its adversary can use to reassess its options and choose to back down (Morrow 1999; Gartzke, Li, and Boehmer 2001). However, even if signaling could work in conjunction with the opportunity cost thesis, it does not appear to supplant it: Polachek and Xiang (2010) have shown formally that opportunity costs can affect peace in the absence of signaling.

Economic Modernization

As mentioned, varying elasticity and substitutability in trade goods in dyads suggests that the wealthy industrialized nations should be the least constrained by the loss of trade in conflict, and thus the most coercive of all types of nations. Yet the third strand of trading peace literature asserts the opposite: that the costs of war increase with economic modernization. There are three main modernization arguments.

First, Rosecrance noted that "land prices have been steeply discounted" in the "wealthiest industrial countries" (1996, 49). As a result, "developed countries would rather plumb the world market than acquire territory" (46). However, Rosecrance was not particularly clear about how a devaluation of land can cause a preference for trade over plunder: land is not the only commodity that can be plundered.

The second way modernization can make trading cheaper than fighting is that it brings with it developed infrastructure, such as roads, railways, and ports, which has the effect of reducing the costs of trade (Rosecrance 1985). However, developed infrastructure also reduces the costs of war. I have argued that developed infrastructure is essential to "pay for, design, build, and maintain the ships and planes considered necessary to move troops overseas and maintain their support over large distances" (Mousseau 2005, 69). I am not aware of any reason to think that development's war-reducing effect is greater than its war-inducing effect. In fact, Britain's early industrialization was key to its military occupation of much of the world, which was only made possible—at least profitable—by its industrial mass production of the Gatling gun.

The third argument for how modernization increases the cost of war is from Norman Angell (1910, 49–50), who argued that war cannot be profitable because "the act of military confiscation upsets all contracts," and property titles become "waste-paper." Scholars often overlook Angell's definition of modernization, which is not about trade or industry but property and contract rights: *if* the conqueror respects the property rights of the conquered, there can be no profit from the war; *if* the conqueror confiscates the property of the conquered, it will suffer economic blowback because "the common prosperity of all of us come[s] to depend upon the reliance which can be placed on the due performance of all contracts" (61). Scholars often depict Angell as predicting the end of war, but Angell's core thesis of note was far more narrow: he asserted only that states with property rights would not fight each other *for profits*, and then *only if leaders understood* that such wars are unprofitable (304).

THE INTERNATIONAL ORGANIZATION PEACE

Drawing on Immanuel Kant's eighteenth-century proposal for a world federation of democratic republics, Russett and Oneal (2001) placed the idea of international organization firmly into the Classical Liberal peace with democracy and trade. Traditional arguments for how international governmental organizations (IGOs) can promote peace among nations recognize that the international system is formally anarchic, and therefore there is nothing to stop a nation from using coercion against another. In this environment international institutions are believed to promote peace in two primary ways. First, they facilitate information sharing, which can smooth coordination and assuage fears of cheating and relative gains seeking (Keohane and Martin 1995); they can also act as active mediators in conflicts (Russett and Oneal 2001; Mitchell and Hensel 2007). For nations having something to fight about, the information from IGOs can enhance the odds of a member state correctly appraising the other's level of resolve, thus boosting the chances of a bargained, or nonviolent, solution (Boehmer, Gartzke, and Nordstrom 2004).

The second way IGOs can promote peace in an anarchic environment is by increasing the cost of war. Presumably nations join IGOs because they yield some benefits; otherwise nations would not join them (Keohane and Martin 1995, 46). Assuming that the benefits of institutions are sacrificed in conflict, shared IGO membership must increase the opportunity cost of war, and thus sometimes constrain states from fighting when they otherwise would (Russett, Oneal, and Davis 1998; Ikenberry 2001). There is some debate, however, on the extent to which IGOs can constrain nations from fighting: Mearsheimer (1994) has voiced that IGO memberships can be reflective of state interests.

All of these arguments assume an international competitive environment, and that IGOs abet in keeping this competition from escalating to militarized behavior. It could be, however, that IGOs do more: that they reduce the competitive nature of international relations. Drawing on constructivist ideas, Bearce and Bondanella (2007) argue that IGOs act to socialize member states out of competitive anarchy (see also Russett, Oneal, and Davis 1998). In analyses of interstate dyads they find a strong association between shared IGO membership and shared preferences. This is a noteworthy finding in that it suggests that IGOs affect member states' foreign policy preferences, causing them to become more aligned with one another. However, the authors do not gauge actual preferences: they gauge "revealed" preferences, meaning behavior observed after the fact that might reflect commonality of preferences, drawing on data on voting in the United Nations General Assembly. Reliant on a proxy gauge of preferences, the causation could easily be reversed, with common preferences causing shared IGO memberships. Bearce and Bondanella (2007) lag their measure for IGO membership five years behind their dependent variable of revealed preferences, but this step is not adequate for inferring with confidence the direction of causation between preferences and IGOs.

To date the evidence linking shared IGO membership with peace in interstate dyads is mixed and the direction of causality far from certain. Most older studies have tended to report little or no relationship (e.g., Domke 1988); some even report an increased tendency for members of IGOs to fight (Gartzke, Li, and Boehmer 2001). More recent work suggests that well-institutionalized and security-oriented interventionist IGOs are associated with less militarized conflict (Boehmer, Gartzke, and Nordstrom 2004; Mitchell and Hensel 2007). Others report that the effect of IGOs may be indirect: it seems states with indirect IGO ties (Dorussen and Ward 2008) and commonality in IGO networks (Kinne 2013) are less likely to fight each other than other states. Other studies are reporting variance over time, with the peace-inducing effect of IGOs possibly a twentieth-century phenomenon (Maoz et al. 2006), or perhaps an artifact of the Cold War (Anderson, Mitchell, and Schilling 2016). Overall, the state of knowledge remains as Russett and Oneal (2001) described it two decades ago, with far less confidence that IGOs have some association with peace among nations, causal or not, compared with the other two pillars of the Liberal Peace, democracy and trade.

TURNING THE SPOTLIGHT ON CAUSALITY

While we know a lot more today than we did a generation ago, we have seen how, as with many big questions of global politics, the field has not been able to coalesce around any leading explanations for the Liberal Peace. One reason could be that most studies in the Liberal Peace are correlational, and many of the correlation findings accord with multiple explanations. While more refined measures can help, as would statistical models guided by more precise causal mechanisms, three types of noncorrelational designs could move our understanding of the Liberal Peace forward: survey experiments, process tracing, and structured interviews.

As can be seen in rows 1–3 in Table 8.1, most explanations for democratic peace rely crucially on the assumption that democratic leaders make decisions according to their perceptions of other nations' regime types. Yet we have negligible evidence that leaders actually do this. As mentioned, most survey experiments have limited their subjects to the populations of democracies with shared histories and developed economies. While assignments to experimental groups are randomized, the populations tested in these studies are not. We thus need experiments with subjects that include the populations of older democracies outside of the economically developed West. Survey experiments that can isolate concern for the democratic status

Table 8.1. Evidence for Theories of Liberal Peace

Row	*Theory*	*Democratic Peace*	*Trading Peace*	*International Organization Peace*	*Primary Weaknesses*
1	Democratic norms	X			Regime perceptions
2	Democratic resolve	X			Regime perceptions
3	Audience costs	X			Regime perceptions
4	Gentle commerce		X		Not tenable
5	Economic modernization		X		Theoretically incomplete
6	Property rights & trade/FDI		X		Not examined
7	Opportunity costs		X	X	Evidence: Can be reversed
8	Information			X	Not examined
9	Constructivist			X	Evidence: Can be reversed
10	Bargaining models	X	X	X	Evidence: Not promising
11	Economic norms theory	X	X	X	More research needed

of adversaries in the populations of nations such as Colombia, India, and Turkey would go a long way toward making the public perceptions idea convincing.

Second, if democratic foreign policy decision makers actually act on their perceptions of other nations' regime types, we should be able to see this with in-depth process tracing of leadership decision making in international crises. Layne (1994) has documented that this did not happen during the Fashoda Crisis of 1898. Peceny (1997) has argued persuasively that popular perceptions of Spain as a non-democracy in the United States before the Spanish-American War were based on Spain's behavior in Cuba, not on the nature of the Spanish regime. Oren (1995) documents a similar pattern regarding US perceptions of Germany during World War I. Other crises need to be equally investigated. Among the most obvious are a series of crises between Ecuador and Peru, Greece and Turkey, and India and Pakistan. For more recent crises, structured interviews of surviving foreign policymakers could be invaluable.

In addition, the underlying causal mechanism of selectorate theory (Bueno de Mesquita et al. 2004) can be tested with interviews of ex-leaders of democratic dyads that have had conflicts. In all three democratic dyads mentioned earlier in this chapter—Ecuador and Peru, Greece and Turkey, and India and Pakistan—democracy has varied over time. In structured interviews foreign policy planners can be asked if they altered their calculations of their adversary's resolve as it changed its regime status. Regarding audience costs theory, foreign policy planners can be asked if they took notice of the adversary making a clear public threat, and whether they considered it more credible than they otherwise would have because they perceived the adversary as democratic.

Regarding the trading peace depicted in Table 8.1, we have seen how the gentle commerce and economic modernization theses in rows 4 and 5 are incomplete: in current formulations neither is equipped to account for the trading peace. More promising is Norman Angell's thesis in row 6, which we have seen predicts an interaction of property rights (in both states in a dyad) with trade or financial integration. While this thesis is more than a century old, to my knowledge it has not been investigated in a large-N study.

Next in Table 8.1 is opportunity costs, which applies to both the trading and IGO legs of the Liberal Peace. This remains a promising thesis, yet to my knowledge no one has examined the crucial assumption that leaders and publics (at least sometimes) value the opportunity costs of trade or IGOs more than they do other values, such as security. Regarding the trading peace, survey experiments with subjects drawn from poorer countries as well as richer ones could detect if people are willing to sacrifice security for wealth. In-depth process tracing of leadership decision making in international crises should inform us whether foreign policy decision makers consciously decide to sacrifice security (or other values) for wealth. If we cannot see wealth as more important than security in publics or their leaders, then it is likely that the causation is reversed, with peace facilitating the interdependency, as critics of the Liberal Peace suggest (Keshk, Pollins, and Reuveny 2004; Kim and Rousseau 2005).

Regarding opportunity costs and the IGO peace in row 7, we need survey experiments to learn if publics value the opportunity costs of international organization more than they do other values, such as security. Regarding the information role of IGOs in row 8, process tracing can seek to learn if leaders *ever* draw on information made available from IGOs to monitor cheating and relative gains seeking (Keohane and Martin 1995). If there is little evidence of concern for opportunity costs or reliance on IGO-generated information, then more attention should be shifted toward cultural or constructivist accounts (row 9). Perhaps most needed are tests for reverse causality: do IGOs predict revealed preferences above and beyond any impact of revealed preferences predicting IGOs? Such a test promises substantial progress on the debate of interests and IGOs initiated a generation ago by Mearsheimer (1994), and can go a long way toward supporting constructivist predictions of the peace-inducing effects of IGOs.

As can be seen in row 10 of Table 8.1, bargaining models have been applied to all three legs of the Liberal Peace. Bargaining models apply only to cases where parties are already in competition over some resource, as they explain militarization as "an equilibrium phenomenon when information about the behavior of the parties is imperfect" (Greif, Milgrom, and Weingast 1994, 758). Applied to international politics, these conflicts are said to involve demands for territory, tribute, or some sort of trade inequity (Fearon 1995, 389–90). Bargaining models thus assume that Liberal nations make these kinds of demands on one another about as often as other kinds of states, with militarized conflict normally averted because democracy, trade, and IGO memberships facilitate costly signals that, among them, yield more accurate calculations of each other's resolve or credibility.

To assess the viability of bargaining models for the Liberal Peace, we need to know if liberal states make demands for territory, tribute, or trade inequities among each other at the same rate as other states. If liberal states make fewer such demands on each other than other states, then we would know that something other than costly signals and bargaining is needed to fully account for the Liberal Peace.

In fact, bargaining models would seem to offer at best weak explanans for the Liberal Peace. Between 1948 and 2001 the COW militarized interstate dispute (MID) data identify 1,269 new militarized confrontations between nations. During this same period states rarely made demands on one another for tribute or trade inequities, and the ICOW data set (Hensel and Mitchell 2007b) identifies only eighty-eight new territorial or maritime demands made by one previously existing state on another. Since it seems most militarized confrontations do not arise from failed bargaining circumstances, bargaining models may not be up to the task of accounting for the Liberal Peace.

WHAT LIBERALS OVERLOOK: HEGEMONY AND DEVELOPMENT

At this point we cannot know if the failure of any Liberal Peace theory to prevail from myriad investigations so far is a consequence of all of them being inaccurate depictions of the real world, or of not enough testing. Two major issues, for me, raise somber misgivings about the efficacy of many of these theories.

First, most of the Liberal Peace theories ignore the fact of US hegemony since World War II. In March 1947 the greatest power on earth announced with the Truman Doctrine that it will oppose any nation that makes a coercive demand on another for territory, tribute, trade inequity, or any kind of servile status. In all likelihood the frequency of these kinds of demands dropped sharply at this time; certainly nations that wish to stay in the good graces of the United States refrain from making these kinds of demands. In this way US hegemony since World War II has increased the cost of war for all nations, not simply democratic trading ones that join IGOs. Since we know as an objectively true fact that this hegemony promotes democracy, trade, and international organization, US hegemony has the potential to explain all three pillars of the Liberal Peace.[3]

The second problem the Liberal theories share is that they largely overlook the proverbial elephant in the room: economic development. There are a number of reasons to think turning our attention toward economic development has research promise. First, democracy is not exogenous in nature: something causes it (Thompson 1996). Theorists of democratic peace have almost entirely ignored an enormous body of research on the causes of democracy. In this body there is a near consensus on two key issues. First, economic development, or something related with economic development, causes democracy, rather than the other way around (Burkhart and Lewis-Beck 1994). Second, democratic culture related with economic development is a crucial part of the explanans for democratic longevity (Dahl 1997).

It is also obvious that economically developed democratic states are more likely than others to be trading states that initiate and join IGOs.

The second reason to consider economic development as an explanans of the Liberal Peace is that Ungerer (2012) has shown that the normative explanations for the democratic peace have a stronger corroborative record than the structural ones. An enormous literature—again almost entirely ignored by theorists of Liberal Peace—has yielded overwhelming evidence that links economic development with value change (for a review of this literature, see Gilman 2007). Might economic development produce democratic values, peace, trade, and international organization among nations?

PUTTING THE PEACES TOGETHER

There remains thus far one theory of Liberal Peace not yet discussed: my own thesis of how market-oriented economic development can produce peace and all three pillars of the Liberal Peace—democracy, trade, and international organization (Mousseau 2000, 2019a, 2019b). In brief, a rise in contractual economic flows among strangers in a society heightens demand for a state that enforces contracts. A rising demand for contract enforcement in turn yields a value for democracy, as the democratic rule of law is the only way to render a state's impartiality in the enforcement of contracts credible enough for many in a society to regularly take the chance of trusting strangers in contract. Since contractualist societies are trading and law-oriented societies, contractualist democracies are trading states that easily form IGOs to facilitate economic cooperation among them.

There is much more to this story, which can be examined elsewhere (Mousseau 2000, 2019a, 2019b): what matters here is its promise for the Liberal Peace. Drawing on Table 8.1, the theory does not assume that leaders perceive and act on each other's regime (or economic) types. Thus the theory does not suffer from the primary weakness of the three main competing theories of democratic peace listed in rows 1–3.[4] Because the theory relies on a strong state for the enforcement of contracts, it does not face the gentle commerce limitation of being workable only in small communities (row 4). Yet, because market-oriented development means a contract-intensive economy, it can account for the long-observed association of economic modernization with peace, property rights, and trade (rows 5–6).

Just as the developed democratic state is principally constrained to safeguard the equal protection of the rule of law at home, it is principally constrained to safeguard the equal protection of the rule of law abroad. This deduction is not idealist: these states are not giving up their interests. Rather, contractualist states have greater interests than others in the self-determination of all nations, as the essential foundation for a stable and profitable global marketplace. The theory thus predicts behavior by modeling interests rather than behavior from assumed interests, and additionally does not assume an inherently competitive world. Therefore, opportunity costs (row 7), information (row 8), and bargaining (row 10), play no role in the theory. Yet the theory is not susceptible to the constructivist assumption that

IGOs matter (row 9), because it does not assume that values and interests originate in actor interactions or IGOs; in the theory state- and individual-level values and interests originate in domestic economic structure.

The theory can explain why the United States decided after World War II to enforce a global order based on the principle of self-determination, why most of the developed democracies chose to ally with it, and why they continue to do so even as the relative supremacy of the United States has declined dramatically since 1945 (Mousseau 2019b). It seems in any given year about half of all democracies do not have contractualist economies (179), and these non-contractualist democracies are not in the statistical democratic peace (Mousseau 2000). In fact, the correlation of democracy with peace is reported to disappear once the contractualist economy is included in standard regression models (Mousseau 2019a). Neither do the non-contractualist democracies tend to win their wars, at least to the extent that Henderson and Bayer's measure of "Western" (2013) serves as a proxy for contract-intensive economy. The theory has also been corroborated in multiple domains, including the areas of democratization (Aytac, Mousseau, and Orsun 2016), international terrorism (Meierrieks 2012; Boehmer and Daube 2013; Krieger and Meierrieks 2015), civil wars (Mousseau 2012), military coups (Powell and Chacha 2016), state capacity (Enia 2017), and human rights (Wright and Moorthy 2018). Still, one weakness of the theory as an explanation for the Liberal Peace is that few researchers have, for whatever reason, subjected it to rigorous testing in the trading and IGO pillars.

CONCLUSION

This chapter has sought to document the Liberal Peace, with special emphasis on assessing the state of evidence for the various theories of how democracy, trade, and international organization can cause peace among nations. As summarized in Table 8.1, eleven mainstream causal mechanisms or approaches have been applied to the three pillars of the Liberal Peace, with some applicable to two or more of the pillars. It is my hope that this chapter not only offers an introductory summary of the Liberal Peace and the state of evidence behind it but also serves to help readers reassess the comparative promise of the competing theories and to identify the most hopeful directions going forward.

Starting with the democratic pillar of the Liberal Peace, we have seen that all mainstream explanations for it are susceptible to the veracity of one bold assumption: that leaders or publics of democracies accurately perceive and act on their perceptions of other nations' regime types. We have seen too that the state of evidence supporting this assumption is weak. To further progress, I suggested experimental surveys with subjects in older and poorer democracies and in-depth process tracing that might document if democratic leaders have ever considered and acted upon their perceptions of the regime status of other nations. Structured interviews of foreign policymakers in rival dyads where democracy has varied over time can determine if leaders varied their calculations of their adversaries' resolve according to their regime status over time. They can also inquire if democratic leaders ever gave notice to an adversary's making of a clear public threat and understood such threats as more credible than they otherwise would have precisely because the adversary was democratically elected, and then whether they accommodated the adversary's demands as a result.

Regarding the trading peace, we have seen how the gentle commerce and economic modernization theories are incomplete in present formulations. More hopeful is the opportunity costs thesis, which applies also to the international organization peace. However, we have little evidence of leaders choosing to sacrifice national interests, whatever they may be, for the benefits of trade and IGO memberships. We thus need survey experiments, with at least some subjects drawn from developing countries, to determine if people are willing to make these kinds of sacrifices. Process tracing of crises can determine if leaders have ever chosen to sacrifice national interests for the benefits of trade and IGO memberships. Until such evidence is reported, the fairer conclusion is of reverse causality: that leaders sacrifice the benefits of trade and IGO memberships as they pursue national interests.

As seen in the bottom two rows in Table 8.1, only two mainstream theories or approaches apply to all three pillars of the Liberal Peace: bargaining approaches and economic norms theory. Bargaining models have attracted the most attention in the literature, but we have seen that bargaining failures can seem to account at best for only a small portion of militarized conflicts. Using the ICOW data, research is needed to determine if liberal states make territorial demands on each other as much as other states: if they make fewer such demands, then we would know that something other than bargaining is needed to fully account for the Liberal Peace.

Economic norms theory departs from all other mainstream Liberal explanations in that it identifies the relationships of all three pillars with peace as spurious. A fourth factor—economic modernization in the form of contract-enforcing states—is identified as the cause of peace, all three Liberal pillars, and the US-led Liberal hegemony since World War II. More research is needed, however, to determine the strengths and weaknesses of this theory regarding the trading and IGO pillars of the Liberal Peace.

More precise causal studies could revive perceptions-based theories of how democracy can cause peace and prop up the opportunity cost theories. Meantime, it is worth thinking about why these and other suggested studies have largely not materialized. One reason could be a lack of interest among field researchers in understanding real-world causality (Mearsheimer and Walt 2013). The dearth of causal studies could also reflect the malady that negative results are often hard to publish. This is especially so if findings go against popular trends, as reports of the Liberal Peace once went against popular trends during the Cold War. Yet, until we have more precise evidence, we will continue to remain in a state of theoretical turmoil, with new generations of scholars having to select one guidepost or another seemingly at random. Without more rigorous causal testing we will, as we should, continue to have little confidence in most theories of how democracy, trade, and IGO membership can cause peace among nations.

NOTES

1. Another possibility is the reverse causation of peace causing democracy (Gates, Knutsen, and Moses 1996; Thompson 1996). While there may be some reverse causation (James, Solberg, and Wolfson 1999), it appears too weak to nullify the impact of democracy on peace (Mousseau and Shi 1999; Reiter 2001; Rasler and Thompson 2005).

2. While some early studies reported the monadic finding that democracies are more peaceful overall than other types of nations (Maoz and Abdolali 1989; Chan 1984; Rummel 1979), most studies today report that democracies are about as conflict prone as other states; they are largely in peace only with each other (Buhaug 2005; Quackenbush and Rudy 2009; cf. Ray 2000).

3. Some early studies speculated or implied some role for US hegemony in the democratic peace (e.g., Farber and Gowa 1995; Rosato 2003, 600). Lemke and Reed (1996), drawing on power transition theory, suggested that democracy could cause a state to ally with the American hegemony, explaining the democratic peace. However, McDonald (2015) reports evidence that alliance with the hegemon, not democracy, explains the democratic peace, as many democracies do not ally with the United States and these non-allied democracies are not in this statistical peace.

4. Ray and Dafoe have described my theory as dependent on the assumption that nations act on their perceptions of other nations' economic types (Ray and Dafoe 2018, 194–95, 197). They provide no supporting citation for their claim, and they are incorrect (Mousseau 2019a, 7).

Chapter Nine

The Territorial Peace

Current and Future Research

Douglas M. Gibler and Steven V. Miller

INTRODUCTION

Territorial peace theory (Gibler 2012) argues that external threats to the state over the distribution of territory affect bargaining among groups in society. These threats create both pressure and opportunity for state leaders. Territory is a salient issue for citizens, who have learned over time that displacement from territory is an existential threat. Citizens want state leaders to provide for their security, which creates an opportunity for state leaders to centralize power in their position for discretionary decision making on deterring the external threat. Deterring the external threat impels leaders toward mobilizing large, land-based armies for territorial security. This augments the leaders' position vis-à-vis potential regime dissidents because large armies decrease the costs associated with violent repression against redistributive demands from the general public or challenges to the leader's office from other elites. Autocracy emerges from territorial threat. Territorial peace, defined as the presence of stable borders and absence of external threats to territory, makes democracy more likely.

This argument cuts across the traditional divide between international and comparative politics scholarship. For example, by focusing on the effects of territorial threat in regime development, the theory links endogenous theories of democracy that emphasize societal-level barriers to pluralism and economic group competition, to exogenous theories of democracy that rely on geographic diffusion and threat environments. Domestically, territorial threat leads to centralized institutions, bargaining among societal groups, and changes in individual attitudes receptive to authoritarian leadership; these centralized polities are also more likely to repress their citizens. Internationally, prior territorial threats are likely to condition the state by making it more likely to engage in international conflict, and, of course, the absence of threat leads to peace or at least a highly selected set of conflicts. Thus states at territorial peace are more likely to negotiate compromises to their disputes and are also more likely to win disputes they choose to escalate because territorial peace provides states more opportunity for selection into conflicts that are easier to negotiate and easier to win. This selection effect largely explains the well-identified association of democracy with compromises and victories.

In this chapter we take stock of this argument and the scholarship that has followed from it. Our review suggests the core claims of the territorial peace argument are well supported across a wide variety of applications. These include analyses of a rich array of data sources, like the World Values Survey, the Women in Parliament data, and the Political Instability Task Force, as well as our most accessible conflict data sets. Analyses that seek to test the core claims of the territorial peace theory produce results broadly in line with its core claims while also illuminating new mechanisms to support the territorial peace. Critiques of the territorial peace often misrepresent the theory's scope and argument, certainly as it pertains to the supposed peace among democracies.

Our review proceeds as follows. First, we provide a condensed version of the book-length argument introduced in Gibler (2012). Here we take care to emphasize that the territorial peace is an argument whose foundation is the importance of disputed territory to international conflict, but its core is an argument of state development under these conditions. Thereafter, we review the scholarship that has followed from it, elaborating on how scholars working with the book's main argument have produced intriguing and important analyses to support its claims of territory's importance to individual-level political attitudes, domestic political processes, and interstate conflict patterns. We conclude with a discussion of where we believe future research should proceed regarding the territorial peace argument.

THE TERRITORIAL PEACE ARGUMENT

The territorial peace argument emerged amid growing scholarship linking territorial issues and all patterns of interstate conflict (see Hensel and Goemans, chapter 1 of this volume, for a review of this scholarship). The early work on territorial conflict was a challenge to structural realism, arguing that the most common approach to security studies remained limited when most interstate conflict is dyadic and localized to particular territorial issues. Meanwhile the democratic peace research program emerged simultaneously as a regime-based challenge to structural realism complaining about the important variation in conflict patterns by government type, especially the peculiarities of democracies. Gibler (2007) stands out as the first articulation of a territorial conflict research program that has direct implications for the democratic peace. Gibler's analysis suggests that the democratic peace is epiphenomenal to the presence of peaceful borders. As such, Gibler breaks new ground in showing that prioritizing the importance of borders and territorial conflict is not only a cogent response to structural realism but also a cogent response to the democratic peace that emerged as the most prominent critique of structural realism in security studies. Both lack the explanatory power of an issue-based paradigm that draws attention to disputed territory and peaceful borders.

Gibler (2012) expanded this analysis into an argument of "territorial peace," a fuller case for the importance of disputed territory to all matters of state development and interstate conflict. The foundation of the territorial peace argument is rooted in scholarship that had accrued to that point on why territory is so important to the conduct of international politics. Briefly, no issue in international politics has the salience of territory, and external threats to territory matter more than external threats over other issues.

Disputed territory's importance has a clear material component. In primitive societies, for example, individuals learned over time that land was the primary source for food and shelter and that displacement from territory was an existential threat. There is an intangible component to territory's importance as well. Occupation of territory creates symbolic and psychological attachments to the land itself. Territory is a means to self-categorization and identity. Over time both processes of territory's material and symbolic importance aggregated from primitive societies to states.

This is why, for example, Iran and Iraq fought an eight-year war whose focal point was control of the oil-rich province of Khuzistan, east of the strategically valuable Shatt-al-Arab waterway that served as Iraq's outlet to the Persian Gulf. It is also why the partition of the British Raj created an immediate strategic rivalry between India and Pakistan whose focal point is Kashmir. Kashmir is unique because the territory has no known precious commodity or even speculated commodity of interest to India or Pakistan. Even Indian leaders at the time of the partition never suspected that Kashmir had material value but chose to pursue it anyway (Hamid 1986, 274) because the territory holds primarily symbolic value for both sides in the rivalry. Taken as a whole, territory is highly salient for individuals, and threats to territory constitute both material threats to livelihood and symbolic threats to a sense of self that is connected to a piece of territory as "homeland."

The importance of disputed territory is the foundation to the territorial peace argument but the core of the argument is one of state development. External threats to territories have played an outsized role in how states formed and developed. The domestic political processes that Gibler (2012) outlined are multiple and need not be exactly linear or identical in all cases, though the development of states under territorial threat share these characteristics.

First, the salient nature of disputed territory and the external threat to it becomes a socialization mechanism for individuals. This part of the territorial peace argument is well traveled with clear origins in the early social psychology literature (e.g., Simmel 1955; Coser 1956). Coser (1956, 38) communicates this point nicely, if broadly, by noting that conflict serves to establish and maintain the boundary lines of societies and groups, along with the identities associated with them. Thus external territorial threats make national divisions salient for individuals, cleaving individuals to "in-groups" whose attachments are to the state under threat.

Two processes follow from this increasing attachment to the state as an in-group. Classic identity scholarship has long cautioned that cohesion with an in-group is not a frictionless process. Individuals who cleave together around a common identity in response to an external source of threat are likely to become intolerant in the process. The sources of their intolerance include the source of threat, which becomes a salient "out-group" under these conditions, as well as other would-be dissidents and deviants whose failure to conform to majority group norms is interpreted as a sign of weakness when shows of strength under threat are paramount.

The second process under these conditions is the emphasis that individuals afford to their security under conditions of external territorial threat. Individuals fear the costs of potential displacement from territory and gravitate toward state leadership to provide for their security. This is a sort of "rally effect." Society rallies around the state leader, who becomes more a symbol of national unity in the face of external threat (cf. Lee 1977, 253).

Territorial threat creates an opportunity for state leaders as these processes of group attachment and fear create conditions favorable for the consolidation of power for the position of the state leader. A society that cleaves together as a show of resolve against a source of external threat tilts the bargaining position in favor of the leader against rival elites. It becomes politically untenable for these rival elites to challenge the state leader's authority, providing more regime security for the state leader. Further, citizens who prioritize their security under conditions of external threat are ultimately permitting more discretionary power for the state leader on how to best provide for their security. A threat to territory leaves little patience among citizens for any kind of political obstacle that would be easily construed as hindrances to territorial defense.

Large, land-based armies are of course indispensable for deterring territorial threats. The goal for challenging states is territorial occupation and annexation, for which an army is necessary. For targets of territorial threat, a comparable army capable of maintaining occupation is imperative. States whose homeland territory are directly threatened by neighbors are thus likely to mobilize large, land-based armies. Those states without such territorial threats are unlikely to raise such large, land-based armies. The United States is an illustrative case here. The US military saw little in the way of a professional, permanent standing army during its formative years. Instead, the US military—and the army, in particular—was best characterized by rapid mobilizations and large-scale demobilizations around situational conflicts/emergencies prior to the start of the Cold War (e.g., Sparrow 1952). The geographical endowment of the United States, characterized by two friendly neighbors to its north and south and by its separation from Asia and conflict-prone Europe by two vast oceans, can account for this.

The shift toward large, land-based armies under territorial threat augments the opportunity for state leaders to consolidate power in their position. Large, land-based armies serve dual purposes for state leaders. They are useful for deterrence against external threats and provide the optimal means for territorial security. These armies are also useful instruments of repression against potential regime dissidents. State leaders confronted with challenges to their office can pursue either peaceful negotiation or violent repression. Territorial threat decreases the cost of violent repression, making it a more attractive option for state leaders while also making dissidence prohibitively more expensive for regime outsiders. Centralization of power is a natural consequence of territorial threat as states pursue territorial defense. However, autocracy is more likely to emerge under conditions of territorial threat and stable borders. Conversely, the absence of territorial threat decreases the need to mobilize large, land-based armies and makes repression by state leaders more expensive relative to negotiations. Democracy is more likely to emerge as a result. From this perspective it should be unsurprising that a country like Switzerland, with its natural mountain buffers, is one of the longest-running decentralized democracies in the world.

The territorial peace argument links territorial issues to state development as a challenge to the primacy that democratic peace scholars have traditionally afforded to understanding international politics by reference to regime type. The causal arrow in democratic peace scholarship treats regime type as exogenous to the conflict patterns that regime type may explain. It then assembles a body of scholarship that, at its core, contends pairing democracies together will lead to the absence of war between them. The territorial peace argument contends this argument mis-specifies the exogeneity of regime type. Instead, regime type is endogenous to the presence of stable borders and territorial peace. Democracies are more likely to emerge under conditions of territorial peace. Autocracies are more likely to emerge under conditions of territorial threat. The democratic peace, per the territorial peace argument, is a function of omitted variable bias. It should be no surprise then that the zones of peace and democracy that emerged in post–World War II Europe (cf. Gleditsch 2002) began when major countries in Western Europe settled their borders following the war.

Territorial threat keeps foreign policies more local and directed toward the immediate nature of the threat on the border. These conflicts should be difficult to resolve peacefully, are resistant to compromise or negotiation, and are more likely to escalate toward war. Conversely, territorial peace creates new foreign policy opportunities for states. The presence of stable borders and territorial peace makes conflict unlikely with neighbors. This allows state leaders to pursue military development that shifts from large, land-based armies toward other means of power projection across the globe, like navies and air forces. States at territorial peace can become more selective in whom they fight and where. This makes noncontiguous conflict more likely for states at territorial peace, albeit over issues of lesser salience. States are more likely to pick winnable conflicts or those that are more amenable to negotiation and compromise. This is why the natural territorial security of the United Kingdom as an island in Europe or the United States' aforementioned geographical endowment permitted both a more democratic regime relative to mainland Europe and an ability to select conflicts in more remote parts of the globe.

THE EMPIRICAL SUPPORT FOR THE TERRITORIAL PEACE

Studies that have followed Gibler (2012) have offered wide-ranging support for many of his core claims. We review these explications of the territorial peace argument at the multiple levels in which they operate. The first and perhaps most intriguing level of explication is at the individual level. Numerous works have done well to explicate the individual-level mechanisms that underscore disputed territory's salience.

The second area of research supporting the territorial peace argument looks at the effect of disputed territory on domestic political processes—namely, regime type—to test the mechanism by which state leaders use territorial threat to centralize power and consolidate more autocratic regimes. The third strand of this supporting research looks at variations in conflict patterns that successfully elevate the territorial peace over democratic peace explanations for interstate conflict. The fourth strand of research we review looks at those works that have tried to argue for the continued importance of the democratic peace in light of the challenge posed by Gibler's (2012) territorial peace argument.

Individual-Level Support

The most intriguing extensions of the territorial peace argument, especially for the state development component, have been at the individual level. Studies leveraging external territorial threat indicators with survey data have done well to illuminate the micro-level foundations of the state development hypothesis in the territorial peace argument. Here much of the important work on this front has been done by Gibler, Hutchison, or Miller, either together or in their solo works.

Hutchison and Gibler (2007) are the first to empirically demonstrate the importance of disputed territory to individuals, emphasizing Coser's (1956, 38) claim that conflict is an in-group socialization mechanism. Fearing the nature of territorial threat, citizens in these threatened states cleave together as a show of resolve against the source of the territorial threat while becoming intolerant of groups they perceive to deviate from in-group norms. Their analysis links territorial conflict indicators in the Correlates of War (COW) militarized interstate dispute (MID) data with the World Values Survey data to demonstrate that allowing these least-liked groups political liberties would be seen as introducing weakness into the political system at a critical time when a show of strength against the external territorial threat is necessary. Multilevel models of World Values Survey data show territorial threat increases political intolerance toward least-liked groups.

Gibler, Hutchison, and Miller (2012) offer a direct test of the in-group identification component of the territorial peace argument. They use Afrobarometer and World Values Survey data to show that citizens in states targeted by territorial threats are more likely to self-identify with their country whereas citizens in states that routinely initiate territorial revisions are more likely to self-identify with their ethnicity. The argument suggests threatened states identify with their country as a natural response to external threat but that territorial revisionist leaders often stake their territorial claims on ethnic grounds, which is broadly consistent with the territorial peace argument. The most intriguing component of the analysis is the within-case study they do of Nigeria, which demonstrates the spatial variation of identity attachments contingent on proximity to disputed territory.

Elsewhere, Hutchison and Miller dedicated much of their early scholarship to exploring the different processes of disputed territory's salience at the individual level. Hutchison's interests focus on the mechanics of government trust and approval as functions of territorial threat. His analyses suggest an interesting heterogeneity in citizen attitudes toward governments under territorial threat. Territorial threat allows governments to better mobilize their citizens on terms advantageous to the government (Hutchison 2011b). This claim has a lot of intuition from works in American politics (e.g., Rosenstone and Hansen 1993), but generalizes in a manner consistent with the territorial peace argument about state development under territorial threat. The main findings from this set of analyses is that territorial threat allows governments to more easily mobilize citizens. An increase in government approval follows even if declining trust in government emerges as a side effect (Hutchison 2011a).

Miller's analyses look more at citizen attitudes toward what type of authority the government should have, directly testing the territorial peace argument that the salience of territory for individuals makes conditions amenable for autocracy. His analyses of cross-national survey data are broadly consistent with this claim. Miller (2017) links the territorial peace to terror management theory (e.g., Landau et al. 2004), offering an argument that disputed territory constitutes a salient threat to individual livelihood that draws individuals toward a strong leader unencumbered by legislatures or elections in order to provide for their security. Citizens are happy with the changes around executive authority that follow territorial threat, and are happier overall, provided the government is pursuing defense of the coveted territory (Miller 2013). Further, S. Miller (2018) bridges the cohesion and state development hypotheses in showing that territorial threat allows for greater permissiveness of government corruption while decreasing tolerance of private corruption in society.

Other scholarship has accrued that supports the territorial peace at the individual level as well. For example, Tir and Bailey (2018) suggest responses to disputed territory emphasize more "masculine" values of militarization and centralization around traditional hierarchies, decreasing women's welfare and enlarging the gap in the relative welfare between men and women. Kim (2019b) uses World Values Survey data to show that individuals are more willing to fight for their country if they live in a country with competing territorial claims.

Complementing Gibler, Hutchison, and Miller's (2012) within-case analysis of Nigeria, Tanaka's (2016) survey experiment in Japan shows respondents farther from disputed territories are more likely to view the Liancourt Rocks dispute through a nationalist lens. Justwan and Fisher's work brings survey experiment data to bear on territorial peace hypotheses with a unique focus on the territorial disputes of India (e.g., Justwan and Fisher 2017, 2018, 2019). Their findings are multiple and unpack the nexus of trust and disputed territory, at least in India's particular territorial disputes.

All told, the territorial peace argument enjoys some of its most important support at the individual level. Here multiple scholars have bridged an argument of international politics and state development with comparative political behavior insights to better illuminate the individual-level mechanisms of the argument.

Domestic Political Processes and Regime Type

Territorial threat scholarship has also illuminated Gibler's (2012) centralization argument. Gibler (2010) first showed that disputed territory constitutes a salient external threat that creates strong incentives for institutionalized opposition to support the head of state. This is a rally effect argument that Gibler (2010) uses to show how the presence of territorial threat allows state leaders to remove potential veto players from the regime to better consolidate the state leader's position to secure tenure. This argument is consistent with another argument by Tir (2010) about the incentives for unpopular leaders to salvage domestic support and regime control through a territorial diversion. Rally effects and diversionary theories of war have long plagued scholarship on leaders and war since few arguments are so intuitive in theory with so little empirical support for them (see Wolford, chapter 14 in this volume). However, territorial threat scholarship may offer some clues as to what issues lead to these rallies in a manner consistent with the territorial peace argument. The decision making leading to the Falklands War, the canonical case of rally effects and diversionary conflict, may well illustrate this.

Nam Kyu Kim's more recent work, both solo and with coauthors, has produced results consistent with the territorial peace argument of state development. For one, territorial threat creates conditions favorable for the emergence of a particular kind of autocracy: military rule. External territorial threats increase the capacity of the military to both deter external rivals and intervene in politics (Kim 2019a). Indeed, dictators in territorial rivalries are particularly vulnerable to coup attempts from the very agency they must empower against an external threat (Florea 2018).

However, the nature of the military regime that follows is typically more "collegial" than "strongman," greater resembling the military junta in Greece than, say, Idi Amin's military rule in Uganda. Women's representation in parliament typically declines as well. Tir and Bailey (2018) demonstrate why this operates at the individual level, but Kang and Kim (2020) are able to show this with the cross-national Women in Parliament data. Finally, Hong and Kim (2019) unpack the repression component of the territorial peace argument, at least as it pertains to mass killing as a particularly severe form of repression. Their findings do establish a link between external territorial threats and mass killings by the state, but offer some caveats about its scope conditions. Typically mass killing is more likely to follow in states whose leaders commit themselves to more exclusionary ideologies. This should follow naturally from the territorial peace argument about the salience of disputed territory and territorial conflict as an in-group socializing mechanism, but the connection here is made explicit in the analysis.

One particularly intriguing explication of the domestic political processes of territorial threat comes from Gibler and Miller (2014). They note that enough interstate territorial conflict scholarship had accrued to that point, linking territorial threat with changes in domestic political institutions as well as domestic political attitudes, but no analysis had tried to bridge interstate territorial conflict with domestic political conflict. The salient nature of external territorial threat should have dampening effects on militarized challenges to the central government's authority. This is a logical implication of the territorial peace argument: large, land-based armies that emerge under conditions of territorial threat should increase the government's capacity to extract resources from its citizens while also increasing the costs of mobilizing an insurgency campaign against the government. Their analysis shows that external territorial threat both increases state capacity and decreases the likelihood of armed conflict and civil war over control of the central government. Importantly, they find lingering effects of territorial threat for periods as long as thirty years after territorial settlement.

Territory, Democracy, and Conflict Processes

The territorial peace argument is primarily one of state development and the importance of territory to domestic political processes. However, it has major implications for the democratic peace, which to that point had produced an assembly of facts about (1) how democracies avoid war with each other and (2) how democracies perform in the conflicts they choose to fight. The territorial peace argument contends that democracy is not exogenous to conflict patterns when democracy is itself endogenous to territorial peace. A slew of findings have challenged much of the democratic peace scholarship on this topic.

Gibler's (2007) analyses, later extended in his 2012 book, directly challenge the main democratic peace finding about the ability of joint democracies to avoid a war with each other. Levy (1988, 662) famously noted that democratic peace emerged as "the closest thing to an empirical law" in all of political science, but no democratic peace scholarship to that point had taken serious inventory of the territorial conflict scholarship and included issue-type indicators in their statistical models. Gibler's analyses show that, had they done so, the results would be much less supportive of the core claims of this purported empirical law.

Indeed, democratic transitions are often a result of peaceful territorial transfers (Gibler and Tir 2010) and are more likely to cluster in zones of peace following territorial settlement (Gibler and Tir 2014). Gibler and Braithwaite (2013) leverage conflict "hot spot" data (cf. Braithwaite 2010) to cast further doubt on the universality of democratic peace claims, elevating the importance of regional stability following territorial peace to explain conflict patterns typically afforded to the democratic peace. Miller and Gibler (2011) offer further evidence that core empirical regularities about the pacific nature of democracies follow from omitted variable bias. They show jointly democratic disputes are no more or less likely to end in a negotiated compromise after controlling for territorial threat indicators.

Gibler and Hutchison (2013) also offer evidence skeptical of an audience cost component to democratic peace. Some scholarship rooted in the democratic peace (e.g., Schultz 2001; Slantchev 2006) had offered an audience cost component to the research program, contending that democratic publics are better at sanctioning democratic leaders for foreign policy failures. This would underscore why democracies avoid costly wars with each other and fare better in the conflicts they do fight. However, democracies rarely have territorial disputes and the disputes they do have are unlikely to generate the kind of audience costs implied by audience cost scholarship. Gibler and Hutchison (2013) show that democratic leaders do not have the kind of institutional advantage that democratic peace scholarship suggests. For the subset of democracies with territorial issues, democracies are more likely to back down, to have a higher number of incidents, and to take escalatory steps when compared to other regimes. All would be inconsistent with democratic peace theory's argument that democracy's unique institutional designs relative to autocratic systems make democracies better at signaling resolve in conflict.

The other territorial peace challenges received wisdom about where and under what conditions democracies choose to fight. Democratic peace scholarship had assumed that the conflicts democracies choose to fight will be quicker and more easily won at a lower cost to a democratic government, given electoral pressures from a general public that has more say in selecting a leader and punishing a leader for bad foreign policies.

However, Gibler and Miller (2013) show that much of these claims follow from omitted variable bias. Democracies rarely have territorial disputes with neighbors and, controlling for that, there is no independent effect of democracy to support these claims. Democracies are no more likely to have quicker disputes or emerge victorious in their disputes once controls are added for the threat environment.

CRITIQUES AND CHALLENGES TO THE TERRITORIAL PEACE

To date, most scholarship following the publication of Gibler (2012) has produced findings that broadly support the argument's central claims while further explicating some of the processes in the argument. However, not all scholarship has been receptive to the claim that territorial peace precedes and is ultimately exogenous to democratic peace. We review some of those critiques of these stronger claims of the territorial peace.

Park and Colaresi (2014) were the first to directly challenge the territorial peace argument that there is no democratic peace separate from a territorial peace. Their review of Gibler's (2007) replication data contended there were several inconsistencies in the data he described and the statistical model he presented. Park and Colaresi argue that correcting those inconsistencies and reestimating the data restores the statistical significance to the core hypothesis of democratic peace that jointly democratic dyads avoid conflict with each other. Importantly, they suggest the findings are stronger for democratic peace than the territorial peace argument. However, Gibler (2014) notes this critique misses the point of the territorial peace argument. The territorial peace argument is primarily an argument of the importance of disputed territory to contiguous dyads while the support for Park and Colaresi's (2014) critique leans heavily on noncontiguous dyads. A focus on just contiguous dyads, per the territorial peace argument, shows no statistically discernible effect of joint democracy on conflict avoidance.

Owsiak's recent research presents the argument that these competing paradigms could be better integrated and complement each other. His work over time on the border settlement data suggests that border settlement increases peace among contiguous dyads, nondemocratic or democratic (Owsiak 2019). Further, democratic contiguous dyads do not experience such a pacifying effect in the absence of border settlement. Democratic peace claims, he finds, work better among noncontiguous dyads even as peace typically requires both democracy and border settlement in his analysis. This would endeavor to put territorial peace on equal footing with the democratic peace.

However, further exploration of the border settlement data provides more evidence to elevate territorial peace claims over democratic peace claims. Owsiak and Vasquez (2019) find that democratic dyads may handle their disputes more peacefully because they have settled their borders prior to becoming democratic. This removes one of the most dangerous issues from their mutual foreign policy agenda. Indeed, Gibler and Owsiak (2018) find that border settlement precedes democratization, which makes a conjoint explanation meaningless.

Direct challenges to the territorial peace argument misrepresent the scope of the argument and unnecessarily elevate the importance of the democratic peace of noncontiguous dyads. Efforts to better integrate territorial peace arguments with democratic peace arguments to put both on equal footing will still emphasize the explanatory power of the former over the latter. Overall, with the exception of a recent contribution from Hegre, Bernhard, and Teorell (2019) that brings the new V-Dem data into this debate, no study has emerged that could adequately restore the significance of the democratic peace after controlling for the insights from the territorial peace.

WHAT'S NEXT AND WHAT'S MISSING IN THE TERRITORIAL PEACE RESEARCH PROGRAM

Gibler (2007) broke new ground in directly challenging the democratic peace as a function of omitted variable bias, offering the first articulation of disputed territory's primacy over regime type in understanding international conflict. Gibler (2012) later expanded this earlier analysis into a fuller argument of territorial peace. The foundation of this argument is the importance of disputed territory as a salient issue in international conflict, but the core of this argument is one of state development. External threats to territory create conditions that permit the centralization of authority in the state leader's position and create conditions amenable to the consolidation of autocracy.

Territorial peace, conversely, reduces the state leader's ability to consolidate power through executive centralization and the mobilization of large, land-based armies. Democracies are more likely to emerge under conditions of territorial peace. This makes democracies endogenous to territorial peace, challenging the democratic peace scholarship on conflict patterns as a function of omitted variable bias.

A slew of scholarship has accumulated since Gibler's (2007) first analysis and the publication of his 2012 book that offers further support for the main argument. Indeed, there is no shortage of a diverse array of scholarship that both directly supports the argument's core claims while providing new information about scope conditions to better clarify the theory. The main argument is well supported across multiple applications. In this section we discuss what we believe would be welcome additions to the territorial peace research agenda.

We believe there is still plenty of opportunity to explore some of the book's core arguments and that these would be welcome additions to this already thriving body of scholarship. For one, not a lot of scholarship has emerged that directly discusses one of the core features of the territorial peace argument: Whereas the argument is primarily one of state development, it makes considerable emphasis of the importance of large, land-based armies to states under territorial threat. Conversely, states at territorial peace should be less likely to need or have these armies and their militaries should be more likely to emphasize navies or air forces. Gibler (2012, e.g., 32–33) makes this point explicit at various points in his book. However, the evidence in support of this argument is only indirect. For example, North and Weingast (1989) argue that the British monarchy felt more pressure to compromise with parliamentary rivals because the bulk of the military force for the island country was naval and invested abroad.

Downing (1993) has a more historical argument about what exposure to warfare in early modern Europe did for democratic development. In his analysis states that were more exposed to warfare and with armies more domestically mobilized for it, like France, became more autocratic and those states that were less exposed and with armies less mobilized for warfare, like Sweden and the Netherlands, deviated less from their early constitutions. They were in a better position to become more democratic. The territorial peace argument no doubt has ample inspiration from the likes of Tilly (1985), Boix (2003), and Acemoglu and Robinson (2006), but direct analyses of the heterogeneity of militaries as a function of territorial peace have been lacking to date. Analyses of military personnel data (cf. Singer, Bremer, and Stuckey 1972) will only indirectly address this part of the territorial peace argument.

Some clarification of the repression component of the argument would be a welcome addition to this research program. A lot of work in this area has been subtly inconsistent with each other. For example, Gibler's (2012) argument is fairly direct in contending that the centralization of power of the state leader and the mobilization of large, land-based armies makes repression cheaper and more attractive for state leaders who would like to better secure their position and tenure. However, repression is endogenous to dissent. Other parts of the argument no doubt mention regime dissidents and deviants from in-group norms, but outward displays of dissent should become unappealing to those minorities. Repression is theoretically both cheap for leaders and unappealing for dissidents to incur. Elsewhere Miller (2013) suggests that citizens are mostly happy when the leader takes aggressive measures toward territorial security and also suggests that individuals prefer the kind of leader who could more easily squash regime dissidents (Miller 2017). Citizens, at least from this perspective, are happier with government outputs on the balance but prefer a leader who could more easily repress them at the leader's discretion. Wright (2014) and Hong and Kim (2019) try to directly answer this question, but both results do not complement each other well. Wright (2014) argues territorial conflict for quasi-democracies makes repression more attractive while autocrats expend more resources on territorial conflict and are less able to repress at home. Hong and Kim (2019) offer the strongest results that show external territorial threats increase mass killing of regime dissidents. However, mass killing is a particularly severe form of repression and it is still unclear how external territorial threats explain other forms of repression. As with any good theory, more and better clarifications of the theory's component parts will both clarify the theory and bring forward many more research questions to answer.

Chapter Ten

The Peace Puzzle

Understanding Transitions to Peace

Andrew P. Owsiak, Paul F. Diehl, and Gary Goertz

In 1942 Quincy Wright produced an impressive volume that chronicled and analyzed the causes of interstate war.[1] As Karl Deutsch succinctly noted in the second edition, Wright's volume had "one purpose: the understanding and control of war and its eventual prevention" (Wright 1965 [1942], xiii). Most other scholars approach the study of war with similar motivations (see Vasquez 1993, 2009). The thinking goes that, if only the field—and, more important, policymakers—could understand why and how wars occur, then the causes of war might be mitigated to create a more peaceful world.

From this perspective conflict below the threshold of war signifies greater peace, but significant variation exists in lower-conflict relationships. Consider for example, the following pairs of states or dyads—Iran-Iraq (1934–2005), the United States-Egypt (1978–2015), Belize-Guyana (1981–2015), and Sweden-Denmark (1952–2015)—as representative of so-called peaceful relationships. None of these dyads was at war all of the time, if ever. Nevertheless, their relationships differ markedly. Perceptions of threat, enmity, and competition (i.e., rivalry) dominated one of these dyads (e.g., Iran-Iraq). Others cooperated substantially (e.g., Belize-Guyana), even harmonizing policy and rendering war unthinkable (e.g., Sweden-Denmark). Finally, remaining dyads lie between these two extremes (e.g., United States-Egypt).

Existing work does not adequately explain this variance in peaceful relationships. Conflict trends indicate that the world has become more peaceful over time (e.g., see Goldstein 2011; Pinker 2011 on general conflict trends, see Pettersson, Hogbladh, and Oberg 2019: and chapter 16), but only if we define peace as "not fighting." Such a narrow conceptualization—called "negative peace," or the absence of violence (Galtung 1959)—rests on flawed logic. It equates one type of peace (negative peace) with all forms of peace, even though we know this to be misleading, if not false. In addition to negative peace, for example, Galtung (1959, 1996) conceptualizes positive peace, "a cooperative system beyond 'passive peaceful coexistence' that can bring forth positively synergistic fruits of the harmony. In a set of states this leads to a continuum from total separation, dissociation, to total association, union" (Galtung 1996, 61). States at positive peace have integrated, institutionalized, and harmonized relations; they are close friends (Diehl 2019).

If one accepts Galtung's argument, then the widely held belief that the causes of peace are simply the inverse of the causes of war—that is, causal symmetry exists among the causes of war and peace—is not only misguided but also myopic.[2] First, although much research focuses on how to stop conflict-prone dyads from fighting (i.e., rivalry termination; e.g., see Bennett 1996; Owsiak and Rider 2013), its inquiry always ends when the fighting stops. Once nations have achieved negative peace, discussions of how to produce greater peace dissipate. Second, scholars who track the expansion of negative peace do so in limited ways. The evidentiary basis for claims of widespread (negative) peace rest largely on the debatable decline in battle deaths from war (for counterviews, see Braumoeller 2019); they do not consider cooperation or other peaceful behavior in the international system. Moreover, the theoretical explanations for peace that scholars advance often lack coherence or empirical tests. Pinker (2011), for example, offers multiple explanations at once, including the presence of nuclear weapons, the feminization of society, greater economic prosperity, and many others; with so many mechanisms—and none tested empirically—the overall explanation becomes imprecise and, in scholarly parlance, overdetermined. Goldstein (2011), in contrast, gives almost exclusive causal weight for greater peace to the United Nations, overstating the effect of that institution in a way that does not map well with global peace trends.

In our conception any expansion of global peace necessarily entails an aggregate shift in the peacefulness of dyadic *relationships* (i.e., relationships between pairings of individual states). Entertaining that idea requires a general inquiry into transitions away from escalation, war, and hostility—as well as a specific inquiry into the factors that encourage individual dyadic relationships to move toward greater peace beyond the negative variety. We begin with a simple question: is peace expanding in the international system? Documenting this trend shows the need to explain two critical relationship transitions (Goertz, Diehl, and Balas 2016). First, how do pairs of states move from rivalry to negative peace? Second, how do dyads transition from negative to positive peace? After reviewing the existing, limited research that addresses these questions, we conduct an original empirical analysis to uncover the factors that facilitate more peaceful relationships during the period 1946–2015.

THE CONTINUUM OF INTERSTATE RELATIONSHIPS

For a dyad to be classified as peaceful, there needs to be sufficient interactions among the dyad members to judge whether the interactions are hostile, peaceful, or somewhere in between. In conflict research "politically relevant dyads" typically identify this group; because members of politically relevant dyads are geographically proximate or include at least one major state, they have (in theory) a greater ability and desire to fight one another. This sample works well in conflict studies, for it identifies dyads with a high potential for fighting. It fares worse, however, when studying peace. Sufficient "nonmilitary" interaction might occur among noncontiguous, nonmajor states (e.g., Poland-Estonia after 2004). Dyads with a "potential for peace" therefore differ, at least partially, from those with a potential for war.

Accordingly we require a broader conceptualization of the dyads in which peace is possible. To identify these pairs, we track numerous factors: major state status, contiguity, militarized encounters, colonial relationships, and common membership in security and economic institutions. Each indicates significant government-to-government ties (Goertz, Diehl, and Balas 2016; Diehl, Goertz, and Gallegos 2019), and together they move us beyond the potential for militarized encounters as the defining feature of interstate interactions.

Applying this conceptualization yields a unique, limited set of cases: dyadic *interstate relationships* (n = 2,631 during the period 1900–2015). Through studying the interactions within these relationships, one can evaluate whether a given relationship is hostile, harmonious, or somewhere in between at any moment in time. Indeed, Goertz, Diehl, and Balas (2016) categorize all interstate relationships along a five-point continuum—a "peace scale"—that runs from extremely hostile (i.e., rivalries; see also chapter 5) to highly integrated and friendly (i.e., security communities (see also Davenport, Melander, and Regan 2018; Diehl, Goertz, and Gallegos 2019). That continuum appears in Figure 10.1, and a relationship's precise placement along it depends broadly on myriad considerations: violent interactions, the issues under dispute (and how states handle these issues), communication and transnational ties, diplomacy, and areas and degrees of cooperation, among others.

Scholars and policymakers have given the hostile (or rivalry) side of the continuum a tremendous amount of attention (for overviews, see Dreyer and Thompson 2011; Diehl and Goertz 2012). *Severe rivalries* exist when the involved states perceive one another as threats, enemies, and competitors (e.g., India-Pakistan 1947–present; see Colaresi, Rasler, and Thompson 2007). Unresolved salient issues drive such sentiments, thereby encouraging the rivals to pursue their issues

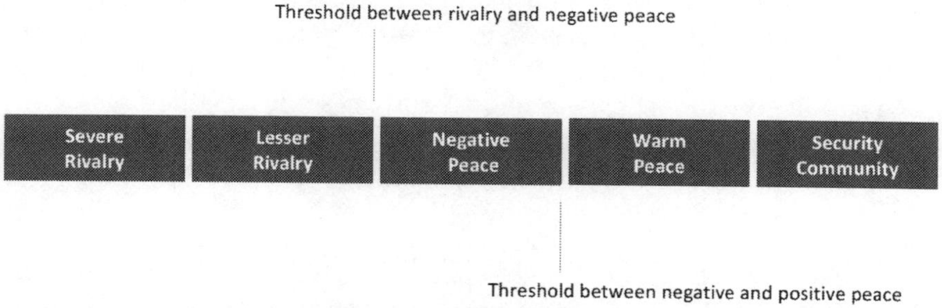

Figure 10.1. Interstate Relationships and Key Transition Points.

via frequent and intense violent episodes. These episodes fail to resolve the issues with finality, recur, and therefore build a hostile history of interaction. Rivals eventually expect that violent interactions over the disputed issues will continue for the near future. In response, they plan their foreign policy around confronting the rival—a step that institutionalizes, or "locks in," the rivalry for an extended period, making it subsequently more difficult to escape.

Lesser rivalries exhibit traits similar to but less extreme than those of severe rivalries (e.g., Colombia-Venezuela during the period 1841–1982). The hallmarks of rivalry (i.e., threat, enmity, and competition) linger, along with unresolved issues. These characteristics fuel violence and diplomatic hostility, yet lesser rivalries differentiate themselves from severe rivalries in the frequency and intensity of this hostility. They present fewer and less severe violent interactions, with many crises handled nonviolently.

Violent episodes disappear almost entirely within *negative peace* relationships (see also, inter alia, Bayer 2010; Kupchan 2010).[3] Dyads at negative peace possess a number of characteristics. They might maintain war plans or issue statements that suggest conflict remains possible; despite such possibilities, however, they almost never threaten or fight one another militarily (i.e., they experience few militarized interstate disputes [MIDs]). This occurs because they have largely resolved their major disputed issues, leaving only unresolved issues of low salience. Moreover, states at negative peace recognize one another diplomatically, communicate officially with one another, and engage in peace negotiations with one another—perhaps even signing peace agreements. At negative peace, then, states are neither close friends nor bitter enemies (e.g., Egypt-Israel in the period after 1989), a description that matches the majority of dyadic relationships in the world at any given time (see Table 10.1).

Moving further right along the relationship continuum in Figure 10.1 pushes states into "positive peace." Dyadic relationships clear a high bar to qualify for such a categorization. Extensive military cooperation (e.g., alliances) and private economic ties (e.g., trade) alone are insufficient. Common alliance members (e.g., Turkey and Greece within the North Atlantic Treaty Organization) and trade partners can be extremely hostile toward one another, even if they share some security or economic interests. The United States and China, for example, trade extensively with one another but remain fierce military and political competitors. What then distinguishes the dyads at positive peace from other dyads that merely cooperate? The key characteristics include mechanisms for—and expectations of using—peaceful conflict management to address disputes, as well as the understanding that military force for such a purpose would be "unthinkable" (i.e., has a zero probability). Beyond this, dyads in positive peace relationships coordinate and integrate government policies more deeply across a range of issue areas. They build institutions to assist with conflict management, develop greater functional interdependence, and remain satisfied with the status quo vis-à-vis one another.

Two relationships lie underneath the broader "positive peace" label: *warm peace* and *security communities*. The latter constitute the most peaceful relationship type. They might manifest conceptually when two distinct political entities

Table 10.1. Peace in the International System, 1945–2015

	Peace Scale Level					
Period	*Severe Rivalry*	*Lesser Rivalry*	*Negative Peace*	*Warm Peace*	*Security Community*	*Total Relationships*
1900	9%	7%	84%	0%	0%	1,095
1905	9%	7%	84%	0%	0%	1,164
1910	10%	8%	82%	0%	0%	1,194
1915	11%	12%	77%	0%	0%	1,387
1920	6%	8%	85%	1%	0%	1,800
1925	5%	6%	89%	1%	0%	1,835
1930	4%	5%	90%	1%	0%	1,867
1935	6%	6%	88%	1%	0%	1,895
1940	9%	11%	79%	1%	0%	1,580
1945	8%	9%	79%	5%	0%	1,600
1950	10%	9%	70%	9%	1%	1,902
1955	10%	11%	67%	10%	3%	2,193
1960	10%	8%	72%	7%	2%	2,984
1965	10%	8%	75%	6%	2%	3,384
1970	9%	6%	76%	7%	2%	3,705
1975	8%	5%	77%	8%	1%	4,449
1980	7%	6%	74%	12%	1%	4,989
1985	6%	5%	73%	14%	1%	5,404
1990	5%	3%	77%	10%	5%	6,082
1995	4%	2%	80%	6%	7%	7,754
2000	3%	2%	81%	6%	9%	8,830
2005	1%	2%	75%	5%	17%	10,292
2010	1%	2%	74%	5%	18%	10,579
2015	1%	2%	74%	5%	18%	2,128

Notes: A row entry begins a five-year period (e.g., 2010 covers 2010–2014). The final row covers only one year because the data set ends.

Source: Diehl, Goertz, and Gallegos 2019, 9.

merge; in practice, though, states within secure communities retain substantial sovereign independence (e.g., dyadic relations among European Union [EU] members). This independence, however, exists alongside intense policy coordination and harmonization. Extensive communication links, transaction flows (e.g., in people and goods), and security interests tie security community members together (Deutsch, Burrell, and Kann 1957). The mutual benefits that result can then create shared values and identities, and render war an unthinkable mechanism for addressing disagreements that arise.

As with the severe/lesser rivalry distinction, warm peace relationships display characteristics similar to security communities, leading the two categories to differ more in degree than in kind. The greatest difference concerns the level of policy integration and harmonization. The United States (US)–United Kingdom (UK) relationship (1941–present) offers an example. While a member of the EU (1992–2020), the UK held security community relationships with the other EU members. In contrast, the US-UK relationship—although peaceful, positive, and close—never achieved the deep policy integration and harmonization that defined EU membership. It therefore qualifies as a warm peace relationship.

At any moment in time, a dyad's relationship falls into one of the five categories just outlined. Nevertheless, dyadic relationships also change over time. We focus on two such changes—or transitions—in this chapter. The first involves a shift from either rivalry level (i.e., severe or lesser) to the continuum's midpoint (i.e., negative peace). The second concerns a shift from that midpoint to either positive peace level (i.e., warm peace or security community). In both cases we explore the conditions that promote the expansion of more peaceful relationships in the international system.

INCREASING PEACE IN THE INTERNATIONAL SYSTEM

Before evaluating how and why interstate peace expands in the international system, we first document that trend. Table 10.1 describes how the distribution of interstate, dyadic relationships across the peace scale changes over time (Diehl, Goertz, and Gallegos 2019). Negative peace relationships are consistently more common than the alternatives—nearly 67–90 percent of dyadic relationships fall within this category at any given time.[4] Most relevant for our purposes, the overall distribution of relationships changes markedly over time. Rivalries—both lesser and severe—peak at about 25 percent of dyadic relationships in the early twentieth century, but constitute only 3 percent of such relationships by 2015. Where do these relationships go? Rivals typically resolve their outstanding, disputed issues and transform their relationship—often moving (next) to negative peace (e.g., Mozambique-South Africa).[5]

At the other end of the continuum we see an increase in warm peace and security community relationships over time. These manifestations of positive peace were largely unknown before 1920 and generally rare through 1950. Thereafter they jumped dramatically upward, a trend that continues into the twenty-first century. The EU's eastward expansion drives much of this trend, creating new relationships and transforming others (e.g., from negative to positive peace) as new members join.

The year-to-year probability that a given interstate relationship shifts from one category to another on the peace scale is low. Nevertheless, we observe more than seven hundred such transitions in the data over two centuries, most commonly between adjacent categories along the peace scale (Diehl, Goertz, and Gallegos 2019). Roughly 70 percent of the transitions move dyads in a more peaceful direction, but some differences exist over time. Prior to 1945 relationships transition equally toward and away from greater peace. After 1945, however, the system marches steadily—with few perturbations—toward more peaceful relationships. Shifts from rivalry (i.e., severe or lesser) to negative peace (e.g., Uganda-Kenya) explain much of this trend, yet many other factors reinforce it. Once reaching positive peace, for example, reversals toward less peaceful relationships rarely occur (e.g., see members of the

Council for Mutual Economic Assistance [COMECON] after the Cold War ends). Many new dyadic interstate relationships that form after World War II similarly qualify as positive peace relationships (e.g., EU expansions)—yet another indicator of peace expanding throughout the international system. Collectively, then, the international system trends toward greater levels of peace, even with some backsliding (see Israel-Iran; Diehl, Goertz, and Gallegos 2019).

SCHOLARLY WORKS ON TRANSITIONS TO PEACE

What accounts for the changes in dyadic relationships that produce the systemic trend toward greater peace? How do dyadic relationships transition from less to more peaceful? The answers require us to consider two separate transitions: rivalry to negative peace and negative peace to positive peace. Existing research addresses these transitions separately, both theoretically and empirically.

Transitions from Rivalry to Negative Peace

One way, albeit indirect, to consider transitions from rivalry to negative peace is to explore how rivalries end. In theory a rivalry might transition to either negative or positive peace when it ends (see Figure 10.1). In practice, however, rivalry termination leads most immediately to negative peace (e.g., Israel-Egypt after Camp David; see Diehl, Goertz, and Gallegos 2019). A subsequent transition might then take the dyad from negative to positive peace (e.g., France-Germany), but only through several intermediate steps. Thus, even though research on rivalry termination does not specify the type of relationship that former rivals develop, we can generally infer that they move to negative peace after their rivalry ends. As a result, research on rivalry termination sheds light on transitions from rivalry to negative peace.

A substantial body of research examines how rivalries end. Many of its arguments propose that significant political events, or "shocks," facilitate rivalry termination. The punctuated equilibrium model that Diehl and Goertz (2000) advance, for example, posits that shocks disrupt rivals' established patterns of hostile interaction (see also Cornwell and Colaresi 2002). Some of these shocks originate at the system level (e.g., wars or a major shift in power distribution; Diehl and Goertz 2000), change the security environment, and prompt a response (e.g., a new alliance, sometimes even with an old rival). Other shocks occur within a rival state. Civil wars, for example, encourage states to terminate a rivalry in order to deal with more pressing, internal problems, while regime change potentially alters governments' policy preferences, thereby expanding the opportunities available for ending a rivalry. Any major political event that affects states' perceptions of external or internal threat, or that influences states' capabilities, thus could cause rivalries to end (see also Rasler, Thompson, and Ganguly 2013). The magnitude of the event required to end a given rivalry will depend on how entrenched the rivalry's dynamics and expectations are. Nevertheless, shocks will be (near) necessary—but not sufficient (i.e., they could instead reinforce a rivalry)—conditions for rivalry termination (see also chapter 5).

Rivalries also end when the involved states can no longer sustain them. Rival states orient their foreign policy to confront their counterpart—a salient threat, enemy, and competitor. Too many rivalries distract foreign policy attention and drain state resources. Indeed, Cox (2010) argues that multiple rivalries cause domestic policy failures, as states divert resources from domestic to foreign policy needs. At some point, which varies with the state's available resources, a state enters a "strategic predicament," in which it reevaluates its various enemies (see Kupchan 2010). It might then end one rivalry in order to pursue another—perhaps new—one (Bennett 1996; Fravel 2008).

The existence of multiple rivalries, however, does not necessarily lead a state to end all its rivalries; states strategically decide which rivals to continue pursuing, which to accommodate, and which to end. A state facing several rivalry threats is more likely to moderate or end its competition with rivals that have strong economic connections to others in its broader network; in contrast, a threatened state maintains its competition more often with a rival allied to its other rivals. Major powers and regime type (i.e., democracy) alter these general behaviors somewhat, producing more accommodation than would otherwise occur (Akcinaroglu and Radziszewski 2017; see also Akcinaroglu, Radziszewski, and Diehl 2014).

A third international factor concerns borders. "Unstable borders" fuel many rivalries (Rider and Owsiak 2015). Resolving interstate borders—or territorial disputes more broadly—removes this fuel, lessening hostility and fostering rivalry termination in the process, especially in contiguous dyads (Owsiak and Rider 2013). Owsiak, Diehl, and Goertz (2017) find direct support for this logic. When neighboring states fully delimit their mutual border, the likelihood that the dyad transitions from rivalry to negative peace rises, while the threat of backsliding from negative peace to rivalry falls. Civil war, however, undercuts the stability and predictability that border delimitation brings. Activities associated with civil war breed border instability and unpredictability: cross-border rebel movements, military support, and refugee flows. As a result dyads without civil war have greater border stability; this stability in turn increases the odds of a dyad being at negative peace and prevents it from sliding into rivalry.

Alongside international factors, domestic factors also promote rivalry termination. As noted earlier in this chapter, Cox (2010) interprets domestic policy failure as a primary instigator of the rivalry termination process. Policy failures cause preferences to change, which potentially expand the terms that the involved governments prefer to continuing a rivalry. If preference change has such an effect, however, then *any* significant domestic preference change will create an opening for rivalry termination. When do these preference changes arise? The most obvious moments involve regime change or a new government coming to power. Not all such moments will do, though. If one hawkish leader replaces another, the effect on rivalry will be minimal (Vasquez 1993, 2009). Thus states need a leader more willing to make concessions to and reach compromises with an "enemy" (i.e., a more "moderate" or "dovish" one). The best way to capture this dynamic might be when a government's support coalition shifts significantly—something that happens more frequently when regimes or governments change (Rasler, Thompson, and Ganguly 2013).

The scattered individual effects discussed so far might fit within a broader, more coherent argument.[6] Goertz, Diehl, and Balas (2016), for example, integrate many of them and supply more precise information on where relationships move after rivalry (e.g., negative or positive peace). Their study primarily examines trends in the international system (as the unit of analysis), but what it theorizes and empirically confirms applies to individual dyadic relationships as well. It argues that the issues over which states fight have shifted markedly over time, with drastic consequences for international peace. By 1945 states had sidelined the most salient, violence-prone issues on their foreign policy agenda, territorial disputes (Vasquez 2009; see also chapter 1), largely through the development of four interrelated norms. The first and broadest norm outlawed gaining territory through conquest (i.e., military force; see also Hathaway and Shapiro 2017). The remaining norms then managed the creation of new territorial entities. A norm against secession prevented the fragmentation of existing states, without the mutual consent of both the rump and newly emerging states. The decolonization norm similarly dictated that colonies gain independence through a peaceful transfer of power from parent to colonial state. Finally, the *uti possidetis* norm prevented any new states from widely contesting the ("existing") placement of territorial borders; it encouraged new states instead to adopt existing administrative borders as international boundaries.

Despite these norms, states advance(d) competing territorial claims over large tracts of land (e.g., see Frederick, Hensel, and Macaulay 2017). The system demanded, however, that they handle these claims differently than in the past. Military force would no longer be permissible as a conflict management tool (see also Schenoni et al. 2020); states had to employ nonviolent tools. Indeed, Goertz, Diehl, and Balas (2016) chronicle an upsurge in mediation, arbitration, and adjudication, particularly after 1990—largely concentrated within rivalries.

These various developments altered international relationship patterns in two ways. First, they prevented new rivalries from forming. Territorial conflict sits at the heart of many rivalry competitions (Rasler and Thompson 2006; Rider and Owsiak 2015). It is a highly salient issue, one over which states disproportionately resort to violence. Among neighbors with relatively equal power, this leads to cycles of fighting and stalemate; hostility builds and fighting recurs while the issue remains unresolved. Feelings of threat, enmity, and competition then form (i.e., rivalry) as states perceive a threat to their authority for the foreseeable future. Interrupting the cycle of violence, stalemate, hostility, and perception formation causes the probability of rivalry formation to fall. International norms facilitated such an interruption after 1945, discouraging states both from reopening previously settled territorial questions (Owsiak 2012) and from using violence to address any unresolved territorial disagreements.[7]

At the same time, the norms helped existing rivalries end. If territorial disputes fuel many rivalries, then removing—or controlling—these disputes promised to destroy rivalries' raison d'etre. The norms supplied the recommended means (e.g., mediation, arbitration, or adjudication) and ends (e.g., settlement terms through focal points) for resolving many territorial disputes. Moreover, they reduced open hostility generally (e.g., suppressing violence), thereby raising the odds that any peaceful conflict management among "enemies" would succeed. In this way territorial norms pushed relationships in the system toward negative peace. The decline of war and the rise of peace therefore go together, but not because of pure causal symmetry. A third factor, international norms, drove both trends.

Studies of rivalry termination say little about relationship transitions directly, for they miss half the equation (i.e., what post-rivalry relationships look like). Nevertheless, because we can infer that rivals move to negative peace when rivalry ends, we gain insight from this research into the transitions that interest us here. Border stability, as well as various international and domestic shocks, often encourage relationships to transition from rivalry to negative peace. As the persistence of rivalries underscores, however, these factors are not sufficient to end rivalries. They also cannot explain transitions to positive peace. That requires a different set of factors.

Transitions from Negative to Positive Peace

A less voluminous body of research addresses transitions to positive peace. The research discussed so far does not apply well to these transitions, which involve a different process than rivalry termination. To end rivalries, states must resolve issues of contention (Owsiak and Rider 2013). Moving from negative to positive peace, in contrast, requires a series of affirmative, cooperative actions. Involved states go beyond merely "not fighting" to working together.[8]

Goertz, Diehl, and Balas (2016) indicate that negative peace is probably a prerequisite for any (subsequent) movement to positive peace—that is, interstate relationships rarely jump from rivalry to positive peace directly. An explanation for shifts to negative peace will therefore only create the permissive conditions for subsequent positive peace transitions. Nevertheless, Goertz, Diehl, and Balas suggest how the latter might unfold. They speculate, for instance, that geographic proximity could be a necessary condition for positive peace relationships to develop. Neighbors interact more with one another—positively and negatively—on myriad dimensions, creating greater opportunity for collaboration and conflict. Geographic proximity also brings common problems that might be best addressed through coordination and cooperative institutions.

Strong and legitimate institutions—both domestic and international—offer a second condition. At the domestic level, Goertz, Diehl, and Balas (2016) propose that positive peace results not from *any* democracy, but rather from *high-quality* democracy. The latter meets the "minimum standard" for democracy (e.g., elections), but also scores highly on multiple dimensions of democratic characteristics (e.g., political participation, transparency, and rule of law, among others). Diehl,

Goertz, and Gallegos (2019) extend this logic further, arguing that high-quality *negative peace* might be most conducive to positive peace transitions. High-quality negative peace incorporates high-quality democracy and adds institutionalized mechanisms for resolving disputes, making the probability for war (or rivalry) extremely low. Because border settlement (Gibler 2012; Gibler and Owsiak 2018; see also chapter 9) and capitalist economic development may be a precursor to high-quality democracy (Mousseau 2013; see also chapter 8), these factors also could contribute to the foundations necessary for positive peace relationships.

At the international level positive peace often grows through strong institutions and organizations, most notably regional economic institutions (REIs) (e.g., the EU or the North American Free Trade Agreement [NAFTA] and its successor). Institutions formalize their members' commitment to coordinate interactions—a key characteristic of positive peace—and, over time, even cause member interests to converge (Bearce and Bondanella 2007). They typically do this across a wide range of issues (e.g., trade, communication, and other policies), providing extensive rather than isolated cooperation that deepens over time. Cooperation consequently expands beyond the members' original integration, thereby creating the potential to transform warm peace relationships into security communities.[9]

Kupchan (2010) provides an alternative story, one that theorizes the transformation of interstate relationships from one end of the peace scale to the other in a series of steps. A "strategic predicament"—a state being confronted with multiple rivalries—motivates the first step: a reconciliation with, conciliatory gesture toward, or concession offered to an enemy (i.e., "unilateral accommodation"). A positive response from the enemy (i.e., "reciprocal restraint") toward these overtures might next set a cooperative "tit-for-tat" sequence in motion. At this second step diplomacy takes over. Any disputes resolve peacefully without jeopardizing overall relations (i.e., similar to negative peace), and a series of diplomatic gestures and actions (often over many years) reinforce the march toward more peaceful relations.

The process shifts from diplomats to the public in the third step. Here cooperation extends to various political, economic, and social realms (i.e., "societal integration"). These cooperative actions increase in depth and number until "warm peace" develops. Finally, the fourth step brings attitudinal change, in which people within the former rival states come to view one another as friends rather than enemies (i.e., the "generation of new narratives and identities"). This mirrors the central characteristic of a security community.

In the end transitions to positive peace remain understudied. Some analysts speculate about these transitions (Goertz, Diehl, and Balas 2016), whereas others confine their analysis to a few case studies in which relationships transition across the entire peace scale in multiple steps (Kupchan 2010). Because relationships rarely pass from one end of the peace scale to the other, however, the generalizability of any case study research remains potentially limited. Most relationships that transition to positive peace involve pairs of states at negative peace with no history of rivalry. We consequently know less about positive peace transitions than about how rivalries end.

TOWARD A SYNTHESIS AND SORTING OUT DIFFERENT CLAIMS

Which factors are associated with transitions toward more peaceful relationships? In the sections that follow we undertake a preliminary examination of peaceful transitions during the period 1946–2015 to investigate this question (n = 248)—with a focus on transitions both from rivalry (i.e., either severe or lesser) to negative peace (n = 164) and negative peace to positive peace (i.e., either warm peace or security community) (n = 84; see thresholds in Figure 10.1).

Research offers limited guidance about the factors that theoretically affect the peaceful transitions we study here. Nevertheless, we derive three sets of potential factors from it, related to border stability (e.g., settled borders or the absence of civil war), international cooperation (e.g., shared membership in a regional economic institution), and domestic politics (e.g., democracy or changes in a leader's support coalition). To these we add various realpolitik concerns that scholars typically associate with conflict and its escalation (e.g., recent wars, power [a]symmetry, and power shifts). We suspect that peace and war involve different processes (Diehl 2016), making a causal symmetry assumption untenable. An empirical test needs to (in)validate that suspicion, though, and including realist factors allows us to learn whether they can account for transitions to peace. The specific variables we include in our analyses appear in Table 10.2. Variables all accept a value of (1) if present or (0) if absent.

We conduct two distinct analyses. A static analysis, what Goertz (2017) calls an "absolute test," first presents the percentage of cases in which each individual variable is temporally proximate at the time of a transition. Based on the initial results, we examine each relationship transition to determine the combinations of variables that exist most frequently at transition points. Because not all combinations appear frequently, we present only the most common combinations (see Table 10.4). We stratify each of the two analyses according to the type of transition—to negative and positive peace, respectively—as these likely involve different processes and factors.

EMPIRICAL PATTERNS

The first set of analyses—presented in Table 10.3—considers a variety of individual conditions associated with peaceful transitions. We begin our discussion with transitions from rivalry to negative peace (first column, Table 10.3). Two main factors stand out. First, stable borders matter significantly for transitions to negative peace—more so than any other factor. More than 87 percent of dyadic relationships that transition out of rivalry do so after the dyad members have settled (i.e., fully delimited) their borders. Noncontiguous dyads share no common borders, of course, so settled borders only affect transitions within contiguous-dyad relationships (75 of 164 transitions; see denominators in Table 10.3). Nevertheless border stability is an important factor that carries across both contiguous and

Table 10.2. Factors Affecting Peaceful Transitions

Category	*Variable*	*Operationalization*	*Lag (in Years)*	*Data Source*
Borders	Settled borders	The dyad has fully delimited its borders.	1	Owsiak (2012)
	Absence of civil war	No dyad member experiences a civil war.	10	Gleditsch et al. (2002); Pettersson, Hogbladh, and Oberg (2019)
Cooperative International	Dyad contiguous to another positive peace relationship	At least one dyad member is land contiguous to another state in a positive peace relationship.	1	Diehl, Goertz, and Gallegos (2019)
	Shared membership in regional economic institution	The dyad members both belong to the same regional economic institution.	0	Diehl, Goertz, and Gallegos (2019)
Domestic	Joint democracy	Both dyad members score 6 or higher on the Polity Autocracy-Democracy Index.	1	Marshall and Jaggers (2019)
	Change in leader support	The government in at least one dyad member state experiences a shift in its supporting coalition.	10	Mattes, Leeds, and Matsumura (2016)
Realist	Interstate war	Dyad members go to war.	10	Sarkees and Wayman (2010)
	Power parity	The ratio of the stronger to the weaker state's capabilities is less than 1.5:1.	1	Singer (1987)
	Shift to power asymmetry	The dyad shifts from power parity to non-parity.	10	Singer (1987)
	Shift to power symmetry	The dyad shifts from power non-parity to parity.	10	Singer (1987)
	Outside rivalry added	At least one state gains a rivalry relationship (excluding the dyad in question).	10	Diehl, Goertz, and Gallegos (2019)
	Outside rivalry ended	At least one state loses a rivalry relationship (excluding the dyad in question).	10	Diehl, Goertz, and Gallegos (2019)

Notes: We lag the variables to establish causal order. A lag will not be appropriate for the regional economic institution variable, as states often join such institutions coterminous with or following a relationship transition.

Table 10.3. Dyadic Characteristics at Relationship Transition, 1946–2015

	Transition from Rivalry to Negative Peace	*Transition from Negative Peace to Positive Peace*
Border Stability		
Settled borders (contiguous dyads)	87.67% (62/75)	90.00% (36/40)
Absence of civil war	54.88% (90/164)	97.62% (82/84)
Cooperative International		
Dyad contiguous to another positive peace relationship	50.00% (82/164)	85.71% (72/84)
Shared membership in regional economic institution	23.17% (38/164)	86.90% (73/84)
Domestic		
Joint democracy	9.56% (13/136)	83.11% (64/77)
Change in leader support	68.15% (107/157)	87.34% (69/79)
Realist		
Interstate war	4.38% (7/160)	0.00% (0/79)
Power parity	13.66% (22/161)	11.39% (9/79)
Shift to power asymmetry	9.94% (16/161)	7.59% (6/79)
Shift to power symmetry	11.80% (19/161)	6.33% (5/79)
Outside rivalry added	38.41% (63/164)	15.48% (13/84)
Outside rivalry ended	53.66% (88/164)	33.33% (28/84)
Total number of transitions	164	84

Notes: Variation occurs in the denominators because of differences in the variables' temporal domains.

noncontiguous dyads: civil war. Earlier we argued that civil war has deleterious consequences for regional stability and relations, particularly among neighbors (see also Owsiak, Diehl, and Goertz 2017). Roughly 55 percent of transitions out of rivalry occur in dyads that have *no* civil war during the ten years prior to the transition. Rivals need borders to stabilize for (negative) peace to follow.

Second, changes in leadership support coalitions appear in the ten years prior to 68 percent of transitions from rivalry to negative peace. All governments respond to the policy preferences of their key supporters. When these preferences shift (e.g., the group of supporters changes significantly, occasionally bringing a new leader to power) the government policy more likely changes too.

Opportunities for concessions or policy moderation—even to enemies—then may open. Haiti offers an illustration. It experienced numerous leadership coalition changes during 1991–2004, a period in which it had a (lesser) rivalry with the United States. After President Aristide left office in 2004, however, the government's support coalition changed again and the relationship moved to negative peace. Other relationships frequently follow a similar pattern. Leadership support coalition changes, however, will not be sufficient for relationship transitions; some coalition changes do not transform relationships. Nevertheless, support coalition changes create an opportunity for states to redefine their relationships.

A congruent theme emerges in transitions to positive peace (see second column, Table 10.3). Stable borders retain their importance. In relationships that make such transitions, 90 percent have settled borders prior to the transition—a figure similar to the one associated with transitions from rivalry to negative peace. In addition, whereas 55 percent of transitions out of rivalry occur in dyads that have no civil war during the ten years prior to the transition, more than 97 percent of transitions from negative to positive peace include this condition. The absence of civil war implies greater government stability and strength, a characteristic that permits states to work toward improved relations and greater integration with others (i.e., focus outward).

At the international level, nearly 86 percent of dyads share membership in a regional economic institution (e.g., the EU) at the time of transition. Roughly as many dyads (87 percent) have at least one member contiguous to another positive peace relationship when their relationship transitions to positive peace. These two traits often work in tandem. States typically institutionalize cooperative commitments through regional organizations. As the organizations succeed, they then expand membership—usually via geographic contiguity—to nearby neighbors. Positive peace therefore clusters and spreads (Goertz, Diehl, and Balas 2016).

A similar dynamic affects domestic institutions. Approximately 87 percent of all transitions from negative to positive peace occur when one of the involved states has a change in leadership coalition during the ten years preceding the transition. As the coalition supporting a government changes, opportunities for redefining interstate relations open as well. These changes often occur when new governments come to power—more common in democratic states. Not surprisingly, then, in addition to support coalition changes generally, more than 83 percent of dyads that transition from negative to positive peace relationships contain two democratic states.[10] Why would (joint) democracy matter more at this phase, as opposed to the transitions from rivalry to negative peace? Most notably, democracies tend not to be rivals with one another. Exceptions exist (e.g., Greece-Turkey 1975–1979, 1983–2013), but most often for short spells, in which rivalry and joint democracy overlap. After that either the rivalry ends or a reversal away from democracy in one of the states occurs. Joint democracy therefore promotes peace once a relationship achieves negative peace—perhaps after borders settle (Gibler and Owsiak 2018; Owsiak and Vasquez 2019). In addition, prominent institutions encourage both democracy and positive peace, usually with the former preceding the latter. The EU, for example, requires new members to obtain a minimum level of democracy *before* admission. This causes democracy to cluster and spread alongside positive peace.

Regardless of which transition we consider, Table 10.3 also shows that realist variables perform poorly—a sign that the assumption of causal symmetry does not hold (i.e., the causes of conflict and peace are not two sides of the same coin). Interstate wars, power parity, and shifts to greater power (a)symmetry coincide with few transitions, either to negative or positive peace (i.e., less than 10 percent of cases each). Shifts in rivalry portfolios perform somewhat better, particularly with regard to transitions from rivalry to negative peace. At least one dyad member gains a rivalry in the ten years prior to 38 percent of such transitions (e.g., the United States' relationship with Iraq improved in 1984, after the United States developed a rivalry with Iran in 1979), while at least one dyad member loses a rivalry in the ten years prior to 54 percent of such transitions (e.g., Argentina moved its relationships with Brazil and Chile to negative peace in 1980 and 1984, respectively). These figures halve when we consider transitions from negative to positive peace instead (i.e., 15 percent and 33 percent, respectively). Regardless, we find some support for a central argument in the research on rivalry termination: changes in a rivalry portfolio often encourage states to reorient their foreign policy, which includes ending ongoing rivalries.

COMBINATION PATTERNS

In the second analysis we consider the combinations of characteristics that coincide most frequently with relationship transitions, rather than individual characteristics alone. In doing so, we emphasize the strongest factors uncovered in the previous analysis: border settlement, the absence of civil war, a change in a leader's support coalition, joint democracy, and a change in rivalry portfolios. We do not include shared membership in regional economic institutions (REIs) or contiguity in the list for theoretical and empirical reasons. States often join REIs coterminous with or after a relationship transition to positive peace, making it difficult to ascertain how they might integrate into a causal story of relationship transitions. Contiguity to a positive peace relationship, in contrast, largely results empirically from EU expansion. Although they produce strong effects (Table 10.3), we therefore set these latter two factors aside.[11]

Table 10.4 reports the five strongest combinations for each transition type. The data suggest two main findings. First, the combinations that coincide with transitions to negative peace are weaker than those that coincide with transitions from negative to positive peace. The "best-performing" combination—settled borders and a change in leader support (e.g., Chile-Argentina in 1984; see also Schenoni et al. 2020)—accounts for only 50 percent of transitions out of rivalry. This might occur because no single, predominant pathway from rivalry to negative peace exists. Rivals form disputes over many issues, and link myriad issues together as the rivalry develops (Vasquez 2009; Dreyer and Thompson 2011). Escaping hostile rivalry relationships may therefore vary widely as well.

Table 10.4. Most Common Factor Combinations at Relationship Transitions, 1946–2015

Transition from Rivalry to Negative Peace	*Frequency(in %)*
Settled borders *and* change in leader support	56.00% (42/75)
Settled borders *and* absence of civil war	42.67% (32/75)
Change in leader support *and* outside rivalry ended	39.49% (62/157)
Settled borders *and* decrease in rivalry ended	34.67% (26/75)
Absence of civil war *and* change in leader support	34.39% (54/157)

Transition from Negative Peace to Positive Peace	*Frequency(in %)*
Settled borders *and* absence of civil war	90.00% (36/40)
Absence of civil war *and* change in leader support	87.34% (69/79)
Absence of civil war *and* joint democracy	83.11% (64/77)
Settled borders *and* absence of civil war *and* change in leader support	80.00% (32/40)
Settled borders *and* change in leader support	80.00% (32/40)

Notes: Variation in the denominators occurs because of differences in the variables' temporal domains.

Second, our combinations not only account for more transitions from negative to positive peace but also approximate a singular transition path. The top five combinations each account for more than 80 percent of transitions to positive peace.[12] The absence of civil war, for example, appears in four of the five most common combinations that accompany such transitions. A change in leadership support coalition and settled borders each similarly feature in three of the top five combinations. Variation consequently exists across the combinations but is less than what we observe in the combinations that accompany transitions out of rivalry. This implies that fewer pathways carry states from negative to positive peace.

ALTERNATIVE PATHS TO PEACE

Although we focus on relationship movement along the peace scale continuum, this is not the only way for states to develop positive peace relationships. In 368 cases a dyad with no prior relationship *began* its relationship at positive peace (e.g., East Germany-Romania after World War II, or Slovakia with numerous states after joining the EU). These relationships experienced no "transition" per se; rather, positive peace resulted from a different process.

An examination of these cases reveals similar characteristics to those we identify in positive peace relationship transitions: settled borders, the absence of civil wars, and changes in leadership support coalitions. In addition, at the time the relationship forms members of these dyads nearly always (97 percent) share membership in an REI and overwhelmingly (90 percent) have at least one contiguous neighbor already in a positive peace relationship. This suggests that positive peace spills over into additional relationships, facilitated through geographic proximity and key institutions—as the EU illustrates best.

CONCLUSION AND FUTURE RESEARCH DIRECTIONS

Is peace expanding in the international system? If so, how and why? In this chapter we first documented aggregate trends in the distribution of dyadic interstate relationships. These have become less hostile (i.e., rivalries decline) and more peaceful—with a notable rise in the number of positive peace relationships. We reviewed existing studies to find an explanation, but research remains less developed here than on the causes of war. As a result we supplemented our review with exploratory analyses to determine the factors most associated with dyadic relationship transitions toward greater peace. Our analysis uncovered numerous findings:

- Peaceful transitions share a number of characteristics—most notably, stable borders and changes in leadership support coalitions.
- Border stability is a near necessary condition for peaceful transitions, particularly among neighboring states. To the extent that states pass through negative to positive peace, these factors may also increase the likelihood that a dyad's relationship qualifies as a candidate for (a subsequent) transition to positive peace.
- The characteristics associated with transitions from rivalry to negative peace differ from those associated with transitions from negative to positive peace—particularly those related to international cooperation and domestic institutions.
- Factor combinations do not produce a single, unique transition path from rivalry to negative peace. They do, however, imply that fewer paths carry states from negative to positive peace than from rivalry to negative peace—largely because the factors involved link theoretically in the former combinations (e.g., border settlement, the absence of civil war, joint democracy, and a change in a leader's support coalition).
- Realpolitik factors perform poorly, underscoring that war and peace are causally asymmetrical.

As a relatively new research area, future studies could advance our understanding of peaceful transitions in numerous ways. A first step would be developing a theoretical framework that accounts for peaceful transitions. We identify a series of factors that coincide with peaceful transitions. How do these factors fit together? Through what causal process do peaceful transitions progress? Only theory can answer these questions, and when it does, the limited research on the topic, especially with regard to the establishment of positive peace, will likely advance significantly.

Second, scholars need to consider the influence of factor combinations and sequences. The steps-to-war model offers a successful illustration of how to approach the task (Senese and Vasquez 2008); it outlines four factors that each increase the probability that an interstate conflict will escalate to war (i.e., territorial disputes, rivalry, alliance formation, and arms races). Scholars might develop a similar "steps-to-peace" model. Neither our empirical assessment nor existing research offers a clear formula for what that combinatory model might look like. In some ways, though, this differs little from the steps-to-war model. It not only began inductively, but research shows that only 45 percent of war cases include all its factors (Sample 2018a).

Sequencing provides a related avenue for future exploration.[13] The distinction between underlying (e.g., settled borders) and proximate (e.g., change in a leader's support coalition) causes suggests a temporal ordering of some effects (Vasquez 2009). Contiguous dyads overwhelmingly possess settled borders (i.e., fully delimited borders; Owsiak 2012) before their relationship transitions from rivalry to negative peace (87 percent) and before negative peace relationships move to positive peace (90 percent). Having stable borders therefore seems to be a key underlying condition for peaceful transitions—a point that the absence of civil war underscores. Stable borders will be insufficient for peaceful transitions; many dyads possess settled borders and see no civil war, and yet they do *not* undergo such a transition. Nevertheless, any proposed sequence might place stable borders near the front of the sequence, as laying groundwork for subsequent factors. A similar formulation might apply for the roles contiguity and joint democracy play in transitions from negative to positive peace.

Our analysis remains less clear on the proximate causes of peaceful transitions—with one exception. Changes in a leader's support coalition not only affect relationship transitions strongly but also appear as the "last event" before almost every relationship shift. This suggests that domestic politics significantly alters interstate relationships (e.g., see Colaresi 2005). They and domestic political shocks more generally (i.e., other than civil war and regime change) deserve greater scrutiny.

Finally, we may not yet have all the pieces. A factor we did not identify could play a key role in peaceful transitions. Moreover, some of our conditions might need refinement. Diehl, Goertz, and Gallegos (2019), for example, specify high-quality negative peace as a necessary condition for the transition to positive peace. Although a promising notion, it is conceptually unrefined, and scholars have not yet examined it empirically.

In a volume on war why does the study of peace matter? Scholars and policymakers often study the causes of war in order to prevent it and transform hostile relations into more peaceful ones. That goal is laudable. Peaceful relationships not only lessen—perhaps eliminate—the probability of war, but also carry numerous additional benefits (e.g., greater trade, communication, security, etc.). Understanding how these transitions unfold therefore deserves significantly more attention, for we cannot simply negate the causes of war and expect greater peace. To expand and (perhaps) accelerate international peace, the explicit study of peace needs as much attention as the study of war. On that score we still have a long way to go.

NOTES

1. We thank Joshua Jackson, Yahve Gallegos, George Williford, and Chad Clay for their research assistance. Sara Mitchell, John Vasquez, Jaeseok Cho, Michael Mousseau, Krista Wiegand, and other contributors to this collection also offered valuable feedback on earlier drafts of this chapter, for which we are grateful.

2. On causal (a)symmetry, see Goertz and Mahoney (2012).

3. As indicated in what follows, "negative peace" means more than the absence of war here. For an overview of different conceptions of peace, see Diehl (2019).

4. Heterogeneous cases comprise this category, including former rivals (e.g., China-South Korea), as well as states that interact sufficiently but that are neither well integrated with nor openly hostile toward one another (e.g., Moldova-Ukraine).

5. The raw number of interstate rivalries has also declined.

6. See also Schenoni et al. (2020), who argue that peace comes through territorial settlement, which is itself a function of three individually necessary and sufficient conditions: attention, significantly altered preferences, and third-party assistance.

7. States still contest territory—sometimes violently (e.g., see Crimea in 2014). These rare cases do not reverse general trends, although they underscore that actors can violate norms.

8. The cooperation involved extends beyond the limited cooperation, or "islands of agreement" (Blum 2007), that rivals experience. Unlike rivals, members of positive peace relationships have effectively managed or resolved the issues that precipitate armed conflict, leading to more fundamental, deep, sustained cooperation.

9. See also the research on reciprocity, including, inter alia, Goldstein et al. (2001) and Goldstein and Pevehouse (1997).

10. A few non-jointly democratic dyads transition to positive peace. These cases typically involve a powerful state coercing cooperation (e.g., the Soviet Union and members of COMECON/Warsaw Pact or Vietnam-Cambodia). They are also the rare cases that backslide from positive to negative peace.

11. The latter two factors largely move together and will be near necessary conditions for a transition from negative to positive peace in Europe.

12. Why would one combination account for 90 percent of cases and another 87 percent? Some variables, particularly settled borders, carry spatial limitations (e.g., require contiguity). This alters the subset of cases to which each combination applies. Moreover, many factors that facilitate positive peace correlate. This need not, however, undercut their individual importance as necessary conditions for transitions to positive peace.

13. Senese and Vasquez (2008) explicitly argue that sequencing does not matter in the steps-to-war model, although their discussion also indicates sequential ordering between some factors (e.g., arms races appear last). Sample (2014, 2018a) investigates potential sequencing in that model, and concludes both that some cases fit the model and that alternative paths to war exist.

Chapter Eleven

Conflict Management of Territorial and Maritime Disputes

Emilia Justyna Powell and Krista E. Wiegand

INTRODUCTION

States get involved in disputes over all sorts of issues from trade to nuclear proliferation to territory. Territorial and maritime disputes are the most common type of dispute in the past century. They are burdensome for more than the parties involved; these conflicts more often than not affect regional and international security. In most instances territorial and maritime disputes are very costly, not only with regard to military defense or threats of hostile activities but also with regard to a lack of cooperation. In these disputes actual, tangible losses experienced by both parties are accompanied by lost profits resulting from damaged bilateral relations. These types of disputes are different in their resolution because of their scale and their long-term consequences compared to change in leadership or a trade dispute.

In this chapter we briefly talk about the similarities and differences between territorial and maritime disputes, and we describe the political and legal methods of peaceful resolution. We also review the international relations (IR) scholarship that asks, "what drives states to seek different peaceful resolution (PR) methods?" focusing on the selection of dispute resolution methods.

TERRITORIAL AND MARITIME DISPUTES

Territorial and maritime disputes both stem from unsettled concerns about sovereignty. Indeed, issues of land and those of sea boundary reside at the center of sovereignty (Gamble 1976). States also have river disputes, which tend to be less salient than territorial or maritime disputes (Hensel et al. 2008). However, it is crucial to recognize that the underlying conception of the law of the sea is that the land dominates over the sea. Indeed, the general rule of international law is that a state cannot exist without a territory. At the same time many states do not have access to the sea and are landlocked. In contrast to maritime areas, land is arguably the most rudimentary feature of a state, and state identity is deeply associated with historically occupied land. Thus "the land territorial situation constitutes the starting point for the determination of the maritime rights of a coastal state" (Shaw 2017, 410). Maritime boundary disputes therefore frequently evolve from antecedent disputes over land. Also, the conceptualization of states' rights associated with maritime jurisdictional zones has evolved over time. This reality suggests that many maritime disputes are usually more recent than territorial disputes, and they continue to crystalize as states expand their maritime claims (Lathrop 2014).

Recurring pressure on the part of coastal states to expand the reach of territorial sea and other jurisdictional zones is reflected in international conventions regulating maritime law. The 1982 Convention on the Law of the Sea (UNCLOS) constitutes the most important treaty regulating this area of international law.[1] The Convention explicitly clarifies the extent and scope of the different maritime zones. Importantly, patterns of dispute resolution—in particular methods most frequently employed—frequently differ for territorial and maritime disputes. In short, states seem to prefer different PR methods depending on the nature of their claim. This is partly due to the fact that UNCLOS offers its state members a variety of specific dispute resolution methods. The proportion of the law of the sea disputes in relation to disputes dealing with other issue areas has grown rapidly. Reflective of these patterns, a substantial part of the UNCLOS treaty deals with dispute resolution. To summarize, though maritime and territorial disputes deal with issues of sovereignty, the nature and patterns of dispute resolution tend to exhibit some idiosyncrasies.

PEACEFUL SETTLEMENT OF DISPUTES IN INTERNATIONAL LAW

States involved in a territorial or maritime dispute have several options when seeking peaceful resolution. Article 2(3) of the Charter of the United Nations (UN) stipulates that states should "settle their international disputes by peaceful means in such a manner that international peace and security, and justice, are not endangered."[2] The particular PR methods—negotiations, good offices, inquiry, conciliation, mediation, arbitration, and adjudication—are listed and regulated in Chapter 6 of the Charter, which is specifically devoted to peaceful dispute settlement. However, states cannot be forced to use any particular method, and every attempt at resolution must be preceded by both parties' consent. Therefore, for a specific venue to be employed, both disputants must agree to the conflict management process in question. In other words, "Their preferences have to converge to some extent" (Powell 2020, 168).

International law does not compel states to resolve their differences. Indeed, some contentions remain unsettled for decades. Thus the scope and function of international PR methods differ from dispute resolution in domestic legal systems. Throughout the life span of a dispute, states usually employ more than one method of settlement, and each party proposes its own most preferred venue hoping that the other side will consent.[3] In many instances disputants start with less formalized methods, and then when these methods prove unsuccessful, states attempt more formal methods. However, resorting to less formalized methods is not a precondition for legalized methods. Naturally the lack of a genuine compulsory jurisdiction, corresponding to what is the standard in domestic courts, suggests that contesting states may never resort to a court. In this regard, dynamics of territorial and maritime disputes tend to be similar to those of other types of disputes. As a general rule states usually try to push back against pressures to adjudicate a dispute, and more often than not they prefer to use less formal settlement venues such as negotiations or mediation.[4] Also, in many situations disputants are bound via existing commitments to try a specific PR method. States may additionally choose to maintain the status quo—continue the dispute—and take no foreign policy option to handle the contention.

It is useful to conceptualize PR methods on a continuum, a scale of legality/formality (Powell and Wiegand 2010). On one side of this continuum are negotiations, as the method least laden with rules and procedures. The nonbinding methods—good offices, inquiry, conciliation, mediation—are located in the middle of the scale. Though each of these methods entails a set of specific rules, in all of them the disputants maintain partial control over the contention and may reject suggestions of the intermediary. Arbitration and adjudication are situated at the other end of the spectrum. Both of these methods yield binding decisions reached as a result of relatively formal proceedings. One can perceive this scale in terms of legality since formality in the context of international dispute resolution entails increased presence of international law, both procedural and substantive, in the resolution process (Powell and Wiegand 2010).

Bilateral negotiations involve direct talks between the disputants with no third-party intervention. This PR method is much more political than legal because it involves bargains, concessions, and compromises on territorial sovereignty or maritime areas. Simply put, the disputants try to resolve all their differences by themselves. Nonbinding third-party methods—good offices, conciliation, commission of inquiry, and mediation—are quite different from bilateral negotiations. All these methods entail many more rules and procedures that make dispute resolution much more formal. Rather than talking through issues themselves, the disputants engage with an intermediary, who—depending on the specific method—has more or less prerogative with regard to the resolution process. For instance, a good officer's task is relatively straightforward: to motivate the disputants to begin communicating and maintain dialogue throughout the process of settlement.

In contrast with mediation and conciliation, a good officer does not actively participate in the construction of settlement terms (Collier and Lowe 1999, 27).

Conciliation entails the intermediary producing a proposal or a series of proposals that set out the terms of the settlement. In a way, conciliation has some aspects of mediation and inquiry and constitutes a relatively fluid framework within which numerous options are feasible. What distinguishes conciliation from mediation is that a conciliator usually offers proposals officially and based on independent investigation. In contrast, a mediator typically informally suggests settlement on the basis of documentation provided by the parties (Merrills 2017, 26).

Via arbitration and adjudication, the two types of binding third-party mechanisms, disputes are settled according to international law. Decisions of arbitral panels and international courts are binding. Whereas adjudication entails the submission of a dispute to a permanent court with largely fixed composition, in arbitration the disputants may select arbitrators. Both these methods entail formal procedures, although in general arbitration is more flexible than adjudication. By way of illustration, states have a high degree of flexibility in the composition of an arbitral tribunal. Furthermore, it is in the hands of the disputants to set the arbitration procedure and frame the issues for the arbitrators to consider. Additionally the parties also decide how the tribunal is to acquire evidence, whether arbiters can issue separate opinions, and if the final award will be published. Much more formalized, adjudication entails the submission of a dispute to an international court, such as the International Court of Justice (ICJ) or the International Tribunal for the Law of the Sea (ITLOS), a court that specializes in maritime disputes. International courts usually have a fixed composition of judges and strictly adhere to international law in the process of adjudicating (Powell 2020).

Taken together, instances of arbitration and adjudication in territorial and maritime disputes are significantly outnumbered by nonbinding methods such as negotiations and mediation. Interestingly, with time states have become more acceptant of international courts and arbitral tribunals. Since the early 2000s the ICJ has had a relatively busy docket, a pattern that is quite likely to continue. The number of ICJ judgments in contentious cases has substantially increased, despite the fact that several recent territorial contentions have ended up in arbitration, or less formal methods such as mediation. The twentieth century brought an unprecedented proliferation of international courts, and the UN's principal judicial organ, the ICJ, has become increasingly active. With the gradual development of international law, states see more and more value in resolving their maritime and territorial contentions with the help of an unbiased third party. Arguably this trend "is reflective of the rule of law and the growth of international cooperation" (Shaw 2017, 848). The increased presence and activity of international adjudicators has substantially contributed to the evolution of international law, as courts' decisions clarify existing rules. Yet states seem to resort to arbitration more so than before. This is particularly true for maritime disputes, since arbitration prominently features in the UN Convention on the Law of the Sea.

In contrast with adjudication, arbitration offers disputants a higher degree of control over the proceedings, the composition of the tribunal, and the scope of contention.

Additionally, arbitration tribunals are better suited to hear cases of a technical nature that may necessitate scientific and technical expertise. The increased need for specialized knowledge beyond that of international law has caused the evolution of international courts. By way of illustration, in certain circumstances ITLOS may ask technical or scientific experts to sit with it.[5] Expanding flexibility in dispute resolution is likely to continue as interstate disputes become more and more complex. This is particularly true for maritime disputes, since they frequently deal with overlapping, complicating claims.

As mentioned before, a substantial part of UNCLOS is devoted to dispute resolution. More specifically, Article 287 enables state members to choose one of four compulsory procedures if an issue of interpretation/application of the Convention arises: the International Tribunal for the Law of the Sea (ITLOS), the ICJ, and two types of arbitration. States may choose these options a priori and they may specify these procedures' rank order. ITLOS, the ICJ, and arbitration are employed if states cannot come to agreement through other peaceful conflict management strategies such as negotiations. If states do not make an explicit choice, the default procedure is arbitration. The specific setup of dispute resolution in UNCLOS explains why arbitration is employed more often in the context of maritime disputes when compared with territorial disputes.

WHAT WE KNOW ABOUT PEACEFUL DISPUTE RESOLUTION

International relations scholars have recognized for some time that interstate dispute resolution is as important as the causes of armed conflict. This genre of research has primarily focused on identifying conditions under which disputants will attempt peaceful resolution versus engaging in militarized conflict. Scholars have written about a variety of aspects of territorial and maritime disputes, including the probability of resolution, effects of territorial changes and peaceful territorial transfers, types of peaceful resolution method, and timing of third-party intervention (Goertz and Diehl 1992b; Hensel 1996a, 2001; Huth 1996; Huth and Allee 2002; Tir and Diehl 2002; Tir 2003, 2006b; Allee and Huth 2006b; Hensel et al. 2008; Gibler and Tir 2010; Powell and Wiegand 2010).[6]

Several factors are linked to states' selection of specific peaceful settlement methods, ranging from military balance and alliance membership to domestic factors such as regime type and type of domestic legal system. Scholars have studied how these dyadic factors and characteristics of states may affect states' PR preferences, but arguably the most significant research has been on the characteristics of the disputed territory and maritime areas, especially how salient or valuable they are. In what follows we examine the three distinct categories of factors influencing dispute resolution: (1) characteristics of the disputed territory, (2) relations between the disputing states, and (3) characteristics of disputing states.

CHARACTERISTICS OF THE DISPUTE: SALIENCE OF TERRITORY

Probably the most significant findings about the dynamics of peaceful dispute resolution are based on how the disputing states value the disputed territory or maritime areas. States seem to care about disputed territory more than other contentious issues (Goertz and Diehl 1992b; Vasquez 1993, 2009; Huth 1996; Hensel 2001; Senese and Vasquez 2003, 2008; Owsiak and Mitchell 2019). In fact, states most frequently engage in war over issues of territorial sovereignty (Goertz, Diehl, and Balas 2016). When compared with maritime disputes, territorial disputes have a higher degree of salience, and they are significantly more likely to be subject to peaceful settlement. However, despite the high salience of territorial disputes, attempts at peaceful resolution have smaller chances of success (Hensel et al. 2008). River disputes are more likely to be resolved through nonbinding third-party methods, and maritime disputes are more likely to experience settlement through formal legal methods.

Several studies examine the types and levels of salience/value associated with the disputed territory. Territory has tangible value—also referred to as intrinsic value (Diehl and Goertz 2002)—if it includes economic resources, has strategic benefits, or constitutes a mainland territory, and has a permanent population (Huth 1996; Hensel 2001; Hensel et al. 2008). Intangible or relational value (Diehl and Goertz 2002) is signified by ethnic links to the land, homeland territory status (compared to dependency status), or symbolic, nationalist value based on lost territorial autonomy or feelings of attachment to the territory (Diehl 1992; Hensel 2001; Diehl and Goertz 2002; Hensel and Mitchell 2005; Wiegand 2011).

Hensel (2001) shows that higher cumulative levels of territorial salience increase the likelihood of bilateral negotiations. At the same time salience does not necessarily affect the choice of third-party resolution methods. Gent and Shannon (2010) find that increased salience discourages leaders from resorting to binding resolution methods. In a subsequent study Gent and Shannon (2011b) show that as the value of the disputed territory increases, nonbinding third-party PR methods are less likely. Interestingly, several studies provide somewhat contradicting findings to Gent and Shannon (2011b) and demonstrate that higher salience levels seem to be associated with attempts at third-party PR methods (Allee and Huth 2006b; Mitchell, Kadera, and Crescenzi 2009). An interesting observation is that, with the exception of Allee and Huth (2006b), all of the aforementioned studies use the same data, the Issue Correlates of War (ICOW). Thus it seems that empirical choices associated with research design are likely influencing the divergent findings. This debate should provoke future research.

In addition to the overall territorial salience, scholars have focused on specific types of salience mentioned earlier in this chapter. Tangible value has been found to encourage cooperation and settlement between adversaries (Rosenau 1967; Mansbach and Vasquez 1981; Vasquez 1993). Huth and Allee (2002) demonstrate that when a territory has strategic value as opposed to economic value, peaceful settlement attempts are more likely. However, at the same time, such resolution attempts are less likely to be successful if they involve exchange of territorial concessions between adversaries. States are not eager to turn decisions about sovereignty over to a third party in such cases. In the context of territorial disputes, Powell (2020) shows that when the disputed land has economic value, Islamic states prefer less legalized PR methods over nonbinding venues.

Whether a territory has intangible value also influences states' choices of PR methods. By way of illustration, Hensel and Mitchell (2005) find that intangible value including ethnic similarity, homeland status, and lost territorial autonomy makes peaceful settlements more likely. With regard to specific resolution methods, Wiegand (2014) shows that higher levels of intangible salience encourage mediation. Huth and Allee (2002) show that ethnic value in particular increases the likelihood of legally binding PR methods. Yet it seems that this relationship may hinge on regions or on the domestic culture of dispute resolution. For instance, Powell (2020) finds that Islamic states dealing with ethnically valuable land avoid adjudication and arbitration in favor of bilateral negotiations. In sum, there is a considerable disagreement in the scholarship with regard to how the different aspects of territorial salience affect states' behavior in territorial and maritime contentions, and this disagreement calls for additional research.

RELATIONS BETWEEN DISPUTING STATES

Huth's (1996) book on territorial disputes stands out as a leading study of how relations between disputing states affect the trajectory of territorial disputes. Huth examines a variety of traditional realist factors against the domestic politics and issues at stake. The overall results indicate that several factors related to a dispute's international context, such as the balance of military forces, influence the likelihood of dispute settlement. The primary finding is that military capabilities have a negative effect on the likelihood of dispute settlement, while values of territory and democratic norms have a positive influence on dispute settlement (Huth 1996). A more recent study examines military balance as a factor that could influence the type of settlement method, specifically in the context of democratic dyads (Ellis, Mitchell, and Prins 2010), showing that bilateral negotiations are twice as likely to occur when disputing states have power parity compared to power asymmetry.[7]

Several studies have examined the role of military alliances in shaping states' preferences toward PR methods (Huth 1996; Frazier 2006; Owsiak and Frazier 2014) between disputing states and the effect on dispute resolution rather than just as a control variable. This literature suggests that disputants are less likely to attempt binding resolution forums when they share common security interests compared to when they do not. Thus shared membership in military alliances encourages bilateral negotiations (Huth 1996; Leeds, Long, and Mitchell 2000; Allee and Huth 2006b). Huth (1996) demonstrates that when disputing states have common security ties, the likelihood of peaceful settlement increases by nearly 17 percent.[8]

Past relations between the disputants—a history of failed or successful settlement, armed conflict, and defeat in armed conflict, as well as past experience using specific dispute resolution methods—have received considerable scholarly attention. The disputing states may be in greater need of third-party help if they share a conflictual past. Indeed, most attempts at mediation occur in hostile relationships or rivalries.

The intra-dispute dynamics also shape states' choices of PR methods. The disputing sides become more open to legalized settlement methods once negotiations fail to deliver a result satisfactory to both sides. Hensel (1999) finds that a history of failed settlement attempts in a particular dispute increases the probability of subsequent third-party assistance, while a history of successful settlement attempts increases the probability of bilateral negotiations. For example, Bahrain and Qatar failed at resolving their territorial and maritime dispute on multiple occasions prior to both states agreeing to submit their dispute to the ICJ in 1991.[9] The contention was over the Hawar Islands, the island of Janan/Hadd Janan, the shoals of Qit'at Jaradah and Fasht ad Dibal, and a townsite of Zubarah (Powell 2020, 174).[10] For many years Bahrain and Qatar proposed a range of PR methods, and all of them proved unsuccessful in resolving the contention. Finally the dispute was resolved at the ICJ. The 2001 judgment successfully ended the dispute by awarding both sides a portion of the contested territory.

Similarly, several studies confirm that third-party methods such as mediation or arbitration are most likely to be employed by adversaries with a history of failed agreements (Hensel 2001; Hensel et al. 2008; McDowell 2018).[11] More generally, in the context of 1816–1986 territorial claims, Goertz et al. (2016) demonstrate that states treat binding third-party methods—arbitration and adjudication—as a last resort. In a way, legalized dispute settlement is exceptional in state relations. Another way that past relations affect dispute settlement is the role of past territorial settlement treaties, which are found to have no effect on the likelihood of future territorial changes (Hensel, Allison, and Khanani 2009) and less armed conflict (Owsiak 2012).

A record of past militarized conflict can also influence the likelihood of peaceful dispute resolution efforts. One explanation for this relationship is that states might have learned that peaceful settlement is less costly. Additionally, settling a dispute in a peaceful manner offers greater prospects for success than militarized conflict (Hensel 2001). Similarly, Hensel et al. (2008, 127) find that states are more likely to pursue peaceful settlements in general "when they have a history of recent militarized conflict over the same issue and/or a history of recent failed attempts to settle the same issue peacefully," compared to when they lack such shared experiences. Ellis et al. (2010) argue that a history of militarized conflict encourages democratic states specifically to seek third-party dispute resolution. In fact, these authors show that democratic dyads are 50 percent less likely to pursue third-party assistance in disputes in which the disputants have been involved in no or little past militarized conflict. Nonetheless, the above-mentioned studies embrace a relatively narrow approach to the concept of past experience.

Powell and Wiegand (2014) and Wiegand and Powell (2011) take a more comprehensive approach and analyze states' past experiences with different

PR methods in all territorial disputes. These authors argue that disputants learn from all of their past interactions with each settlement method, and subsequently engage in forum shopping. The outcomes of resolution attempts in all territorial disputes thus play a vital role in states' subsequent choices of a settlement method. This is particularly true with regards to legally binding methods. Wiegand and Powell (2011) examine how challenger states use their own win/loss record and the win/loss record of the target state to estimate the probability of winning in a PR forum. This research shows that in the context of territorial disputes, challengers with a positive win/loss record in binding PR methods, arbitration and adjudication, are more likely to pursue the same methods again. At the same time such positive past experience with binding PR methods dissuades challengers from pursuing bilateral negotiations.

An example is the 2003 agreement by Malaysia to submit its case against Singapore to the ICJ in the *Sovereignty over Pedra Branca/Pulau Batu Puteh, Middle Rocks and South Ledge (Malaysia/Singapore)* case.[12] Arguably the chief motivating factor behind Malaysia's decision to resort to adjudication was Malaysia's victory in the ICJ case against Indonesia, *Sovereignty over Pulau Ligitan and Pulau Sipadan (Indonesia/Malaysia)*. The ICJ judgment was largely in favor of Malaysia. The day after the Court issued the judgment in the *Indonesia/Malaysia* case on December 17, 2002, the Malaysian deputy prime minister announced publicly that winning the case against Indonesia had encouraged Malaysia to submit its dispute with Singapore to the ICJ: "Pulau Batu Puteh will be next … following the positive outcome of arbitration by the International Court of Justice (ICJ) on the dispute between Malaysia and Indonesia over Sipadan and Ligitan" (Bernama News Agency 2002). According to observers of the Malaysia-Singapore case, between 1998 and 2002 "there was no movement towards ICJ resolution. However, Malaysia's victory in its Sipadan dispute with Indonesia last December [2002] reignited its media's interest in Pedra Branca," prompting Malaysia to agree to ICJ resolution of its dispute with Singapore on February 6, 2003, only six weeks after Malaysia won its case against Indonesia (Lim 2003; Powell and Wiegand 2014). Interestingly, Powell and Wiegand (2014) find that levels of domestic rule of law, together with the win/loss record, shape states' preferences toward PR methods. In particular states with high levels of domestic rule of law are more likely to return to binding methods in other disputes only if they had positive experiences with binding methods in the past. In contrast, states with low domestic levels of rule of law seem less concerned with their past win/loss record when selecting a resolution method.

CHARACTERISTICS OF THE STATE

Domestic factors, in particular regime type, constitute the key factors shaping states' views of PR methods. Much of this literature is rooted in the democratic peace argument (James, Park, and Choi 2006), or the territorial peace argument (Gibler 2012; Gibler and Owsiak 2018; see also chapter 9, herein). Research based on the democratic peace examines how the normative and structural aspects of democracy influence states' strategies in resolving territorial and maritime disputes. We know that democratic dyads are "more likely to adopt compromise solutions to problems as a matter of course" (Ellis et al. 2010, 374). Extending the democratic argument to the systemic level, Mitchell (2002) finds that the higher proportion of democracies in the international system is associated with more frequent usage of third-party PR forums by nondemocratic dyads. Relatively high numbers of democracies at the systemic level enable the spread of democratic norms. Thus, non-democracies are more likely to consider resorting to third-party PR methods. Taking a normative approach, Dixon (1994) finds that democratic dyads are more likely to seek third-party dispute resolution methods, though this study is not specifically about territorial disputes. Democracies' respect for in-court proceedings should arguably carry over into international dispute resolution.[13]

Interestingly, there is both a theoretical and empirical disagreement about which type of resolution method is more likely among democratic states: bilateral negotiations or third-party methods. Some scholarship also shows that the effects of democracy disappear in the midst of territorial claims with high salience (Park and James 2015). James, Park, and Choi (2006) find limited support for the democratic peace argument, showing that democratic dyads are not very likely to seek peaceful resolution in territorial disputes. Instead, issue salience has more of an effect. Hensel (2001) argues that states involved in disputes over contentious issues like territory prefer to have more control over the outcome of negotiations. This particular study shows that democratic dyads are much more likely to use bilateral negotiations for peaceful dispute resolution, while they are much less likely to pursue third-party methods. Shannon (2009) finds that democratic dyads are no more or no less likely to attempt third-party resolution methods compared to other dyads. Gent and Shannon (2011a, 2011b) argue that binding dispute resolution attempts can be just as costly to domestic audiences since leaders can be punished for giving up control to an arbiter or international court. Some studies show that the effects of democracy disappear in specific contexts, such as in territorial disputes of Islamic states (Powell 2020). It seems that "democratic norms may work quite differently in societies where there are strong alternative norms favoring informal dispute resolution" (188).

Similarly, in their analyses of peaceful attempts to resolve territorial disputes, Powell and Wiegand (2010) and Wiegand and Powell (2011) find that joint democracy has no statistical impact on states' decisions to choose binding methods. This scholarship is consistent with the findings of Mitchell et al.'s (2009) study, which demonstrates that democratic dyads are not more likely to choose third-party resolution methods. In fact, most of their models of third-party settlement attempts indicate that joint democracy decreases the likelihood of resorting to a third party. This is consistent with the high number of current and former authoritarian states, such as Libya, Qatar, Bahrain, El Salvador, Honduras, and Burkina Faso, seeking dispute resolution at the ICJ. Delving further into the relationship between democracy and PR methods, Ellis et al. (2010) find that democratic dyads seek bilateral negotiations rather than third-party help, specifically when the disputants have not fought each other militarily, the issues at stake are salient, and military balance is closer to power parity. The mixed findings about democracy suggest that other factors such as major power status, issue salience, and past experience have stronger effects on the selection of peaceful dispute resolution. For example, despite the strong democratic status of the United States, the US government is often unwilling to bind itself legally at the ICJ in dispute resolution based on its major power status. We know that states with high power asymmetry like the United States are less likely to pursue legally binding resolution methods, based on a significant amount of research (Hensel 2001; Mitchell 2002; Simmons 2002; Allee and Huth 2006b; Hansen, Mitchell, and Nemeth 2008; Hensel et al. 2008; Shannon 2009; Gent and Shannon 2010).

Going beyond the normative argument, scholars argue that in some circumstances democracies are drawn to legally binding third-party PR methods due to structural characteristics of democratic domestic institutions. Leaders in democracies—with larger winning coalitions—are more accountable to their constituents. Thus, to remain in office (or have the party remain in office), the democratic leaders are more cautious about their policy decisions (Huth and Allee 2002). Consequently, democratic leaders are hesitant to engage in any unpopular policies that may threaten their reputation, and subsequently increase the likelihood of domestic punishment. In the context of territorial disputes this implies that democratic leaders will likely avoid settlement options that contradict discourse about the disputed territory. Pursuing peaceful settlement—particularly when it involves potential concessions or compromises—can be a costly strategy. Within this genre of literature Choi and Chiozza (2003) examine the role of individual leaders, focusing on their vulnerability to domestic audiences. They find that dispute resolution is more likely when new democratic leaders are in power, since these leaders are less likely to be punished having just received endorsement from their constituents. In contrast, authoritarian leaders are more likely to attempt peaceful resolution as they spend more time in office since they have accumulated political experience and are therefore less constrained by domestic audiences.

Huth and Allee's work (2002, 2006a) contributes interesting insights into the relationship between democracy and the resolution of territorial disputes. The primary finding of the 2002 piece is that policy successes "may help deter political opposition, strengthen a democratic leader's hold on office, and increase the stock of political capital upon which leaders can draw to advance their broader policy agendas" (Huth and Allee 2002, 71). Thus, as time passes since an election, democratic leaders are less likely—and authoritarian leaders are more likely—to seek negotiations for settlement in order to avoid potential domestic punishment. Moreover, when the disputed territory is valuable and therefore controversial to domestic audiences, democratic leaders are less likely to offer territorial concessions. Focusing specifically on legalized dispute settlement, Allee and Huth (2006a and 2006b) show that the likelihood of legal dispute resolution triples when the disputing states possess democratic political institutions. The argument is based on the notion that legalized binding resolution methods can act as more legitimate political cover for leaders who anticipate potential opposition to dispute resolution attempts (Allee and Huth 2006a, 286). Simmons's (2002) study dovetails nicely with these arguments asserting that international arbitration and adjudication provide democratic leaders with useful means to settle a dispute when domestic political opposition is likely to block a negotiated solution. Further work by Huth, Croco, and Appel (2011) confirms that democratic leaders with a clear need for domestic cover are more likely to pursue adjudication as the means of peaceful settlement, specifically when states have weak legal claims to the disputed territory.

Moving beyond normative and structural aspects of regime types, several studies focus on the role of domestic legal traditions—civil, common, and Islamic—on states' preferences toward PR methods. In short, the three major legal traditions embrace divergent beliefs about dispute resolution, and these beliefs—supported by societal preferences—are projected on the international realm (Mitchell and Powell 2011). Using an affinity explanation, Powell and Wiegand (2010) argue that states select dispute resolution methods that are most similar to those embraced by their domestic legal systems. This study shows that civil law states are more likely to seek binding third-party dispute resolution (especially adjudication), Islamic law states are more likely to seek nonbinding third-party resolution, and common law states are more likely to seek bilateral negotiations. Powell (2015 and 2020) examines the differences and similarities between the Islamic legal tradition and international law, focusing in particular on peaceful dispute resolution. She demonstrates that the relationship between Islamic law states and PR methods "is shaped by sub-categorical legal features, specific legal traits that characterize a domestic legal system" (Powell 2020, 273). Some Islamic law-based legal features in Islamic law states' legal systems constitute a natural bridge with international nonbinding third-party binding venues. Overall, the main contribution of this genre of research is that it goes beyond regime type and recognizes that other institutional state-level characteristics influence states' choices of PR methods.

In addition to regime type, level of domestic accountability, and domestic legal tradition, states' memberships in international organizations (IOs) can influence the likelihood and choice of PR methods. This is an important consideration since IOs often play the role of mediators or arbiters in peaceful dispute resolution (Abbott and Snidal 1998; Bercovitch and Schneider 2000). Overall these studies show that increased IO membership—especially those that call for peaceful dispute settlement among their members—encourages disputing states to seek third-party assistance. This scholarship argues in general that being part of many peace-promoting treaties is a credible signal of a state's "increased embeddedness in the global promotion of peace" (Powell 2020, 189). Disputing states that sign and ratify a higher number of such agreements are more likely to resort to third-party PR methods (Hensel 2001; Mitchell et al. 2009). Focusing specifically on territorial, maritime, and river disputes, Hansen et al. (2008) find that regional and global IOs constitute more effective intermediaries when the IOs are highly institutionalized and include higher numbers of democratic members with converging interests. However, the impact of membership in peace-promoting IOs may have important regional aspects. In the context of territorial disputes, Powell (2020) does not find that the variance in regional and global commitments helps explain why some Islamic law states are more accepting of specific PR methods over others.

CONCLUSION

It is likely that a discussion of the empirical research presented in this chapter would have not been possible before the twenty-first century. The relatively new focus on the active attempt to resolve territorial disputes has now gained traction largely due to the emphasis on positive peace. The main factors that influence peaceful resolution of disputes include (1) characteristics of the disputed territory or maritime areas (issue salience/value), (2) relations between disputants, particularly past experiences between the disputants, and (3) characteristics of the state, mainly regime type and domestic legal system.

Because there are multiple foreign policy options, not only the methods of peaceful resolution but the use of force as well, scholars continue to debate how these outcomes should be conceptualized and coded. Conceptually, the question is first whether governments make decisions to pursue a foreign policy strategy and then once they make this decision, whether they select to use force or peaceful resolution mechanisms, or whether the decision is made in a combined sense so that leaders decide to pursue resolution through force (or peaceful resolution) at the same time. This question is important because the researchers must decide whether they must deal with selection effects.

It is also crucial to recognize that dispute resolution constitutes a dynamic process over time with many decision points along the way. Indeed, throughout the life span of a dispute, the parties usually propose a multiplicity of PR methods. Moreover, maritime and territorial disputes may span many years or even decades. This reality should shape the way that scholars conceptualize dispute resolution processes as unfolding in distinct—though intricately interrelated—stages (Jones and Metzger 2018; Wiegand, Powell, and McDowell 2020). Though the final PR method that the disputants agree to certainly matters, previous exchanges of resolution proposals reveal the disputants' changing preferences (Wiegand and Powell 2011; Wiegand et al. 2020).

A related question about the policy outcome is whether armed force and peaceful resolution are treated as exclusive strategy options. Some scholars (Huth 1996; Huth and Allee 2002) treat this policy option as three distinct choices that could occur in any given year, while others, particularly scholars involved in the ICOW data project, include both policy options in the same observation (Hensel et al. 2008). Other scholars view using force as an option that could occur at the same time as peaceful attempts for resolution, or that both policy strategies could be pursued multiple times in any given year. This problematizes the explanations for this critical dependent variable given that the primary question about territorial disputes is about the likelihood of armed conflict and the likelihood of peaceful settlement. These are questions that scholars of territorial and maritime disputes will need to continue discussing.

Future areas of research should also focus on the dynamic, evolving, and flexible nature of settlement methods offered within the framework of international law. How does the structure of each PR method juxtapose with solutions adapted by different regions of the world and different legal cultures? The newly emerged field of comparative international law demonstrates that domestic context—comprising localized laws, understandings, and norms—shapes how states approach international treaties and institutions (Roberts 2017; Roberts et al. 2018; Powell 2020).

Peaceful settlement methods offered by international law are thus likely to be used as a flexible set of arrangements or legal guidelines. These broad guidelines allow states of different regions and different legal cultures to express their own settlement preferences driven by domestic societal preferences. In an important way, "dispute resolution is what states make of it" (Powell 2020, 290). States with different beliefs about conflict management can tailor PR methods to suit their own specific preferences. Equally important to the scholarship is framing international dispute resolution not only as a political choice that states make but also as a legal phenomenon that takes place in a broader context of international law. Thus, in order to generate meaningful insights, the scholarship has to draw equally on the international relations literature and the international law literature. After all, conflict management—whether domestic or international—is largely a legal enterprise. Overlooking the numerous contributions provided by either discipline is therefore not a constructive route forward for the scholarship.

NOTES

1. Convention on the Law of the Sea, December 10, 1982, 1833 U.N.T.S. 397.
2. The UN Charter website: www.un.org/en/charter-united-nations/index.html.
3. See Table 10.2 in Greig, Owsiak, and Diehl (2019) for a detailed outline of the characteristics of different types of peaceful resolution methods.
4. Literature on this topic is extensive. See, among others, Ratner (2006, 821); Goertz et al. (2016). The scholarship has been prolific in trying to explain why and under what conditions states use legal mechanisms. See Posner and Yoo (2005); Allee and Huth (2006b); Gent and Shannon (2010); Mitchell and Powell (2011); Davis (2012); Huth, Croco, and Appel (2013); and Powell and Wiegand (2014).
5. Article 289 of UNCLOS.
6. A growing number of studies has focused particularly on the selection of PR methods (Mitchell 2002; Simmons 2002; Allee and Huth 2006b; Gent and Shannon 2010, 2011a, 2011b; Huth, Croco, and Appel 2011, 2013; Wiegand and Powell 2011; Lefler 2014; Powell and Wiegand 2014; Powell 2015, 2020; Owsiak and Mitchell 2019; Wiegand, Powell, and McDowell 2020).
7. A number of other studies focusing on other key factors, but controlling for ratio of military balance, find that power relations between the disputing states influence the

choice of resolution method (Hensel 2001; Mitchell 2002; Simmons 2002; Allee and Huth 2006b; Hansen et al. 2008; Hensel et al. 2008; Shannon 2009; Gent and Shannon 2010). Focusing specifically on the selection of legal dispute settlement, these studies show that arbitration or adjudication are more likely if the disputing states have relative power parity. The underlying logic is that the more powerful state in a dyad is unlikely to trust and accept third-party judgments. For example, in the Article VII arbitration over maritime features in the South China Sea, not only did militarily powerful China ignore the 2016 ruling but China also refused to participate in the arbitration brought by the militarily weak Philippines in 2013. In contrast, disputants who are relatively equal in power are more willing to trust and accept third-party binding resolution outcomes. For example, Simmons (2002) shows that arbitration is less likely when there is greater power asymmetry between disputing states, while Mitchell (2002) similarly demonstrates that third-party involvement in dispute resolution decreases as power asymmetry increases. Powell (2020) demonstrates that Islamic states involved in territorial disputes are more likely to negotiate in situations of power asymmetry, avoiding binding venues.

8. Interestingly, focusing specifically on mediation, Frazier (2006) finds that when a third party is allied with one of the disputing states, this relationship does not influence the likelihood of third-party mediation in territorial MIDs (militarized interstate disputes).

9. Maritime Delimitation and Territorial Questions between Qatar and Bahrain (*Qatar* v. *Bahrain*), ICJ Judgment of 16 March 2001, ICJ Rep 2001, 40. The dispute concerned several other territories, including the island of Janan/Hadd Janan, the shoals of Qit'at Jaradah and Fasht ad Dibal, and Zubarah, a townsite on the northwest coast of Qatar (see Schulte 2004).

10. Maritime Delimitation and Territorial Questions between Qatar and Bahrain, Merits, ICJ Judgment of 16 March 2001, ICJ Rep. 2001, 40.

11. Interestingly, a study by Gent and Shannon (2010) shows that previous successful settlement attempts reduce the likelihood that decision makers will seek binding resolution methods.

12. Sovereignty over Pulau Ligitan and Pulau Sipadan (Indonesia/Malaysia), Application for permission to Intervene, ICJ Judgment of 23 October, 2001, ICJ Rep 2001, 575.

13. Many studies build on, alter, and expand this argument, and examine more closely the causal relationship between democracy and formal PR methods, arbitration and adjudication (Simmons 1999, 2002; Hensel 2001; Huth and Allee 2002; Mitchell 2002; Allee and Huth 2006a; Mitchell and Hensel 2007; Mitchell et al. 2009; Shannon 2009; Ellis et al. 2010; Gent and Shannon 2011b).

Part III

EMERGING TRENDS IN INTERSTATE WAR RESEARCH

Chapter Twelve

Cyber War

Brandon Valeriano, Ryan C. Maness, and Benjamin Jensen

INTRODUCTION

Grasping what the international relations and security studies community knows about war also means grappling with how the conduct of wars will evolve in the future. Emergent technologies will play a role in reshaping the meaning of war as the character of conflict evolves over time. New tools and methods to wage battle and conduct statecraft below the level of armed conflict will inevitability alter what we know about the process of war, as well as the relations between adversaries in times of peace.

The current new technological advancement that has coercive potential is represented by the development of cyber operations. The challenge is that these new tools often come with vastly inflated potential and generally fail to alter the course of warfare. As some propose, cyber tools offer a technology posed to reshape war and even the international system (Kello 2013, 2017; Clarke and Knake 2014).[1] Cyber conflict is part of the present and will be a factor in the future of war, but this statement comes with many caveats.

Cyber conflict will not reshape war and will only move things at the edges, potentially increasing the ability of states to signal discontent. Cyber tools are unsure, limited in reach, mostly nonlethal, costly to develop, require a confluence of events to make them work, and generally are unable to shape the dynamics of battle (Valeriano and Maness 2014, 2015; Kostyuk and Zhukov 2019). Cyber tools are poor means of compulsion and deterrence (Valeriano, Maness, and Jensen 2018). They can offer an additive value in combat if offensive cyber operations can take out the command and control (C2) of the opposition or disable weapons platforms while in use, but the utility of the offense is dependent on the failure of the defense.

Although not as destructive in the physical realm, perhaps the relevance for cyber tools is found in operations in the information environment (OIE). If true, cyber-enabled information operations may be the force multiplier that states are seeking, where covert operations such as espionage and psychological warfare become critical in wearing down the enemy's will to fight or reaching the domestic population. However, these are old tactics being utilized by a new technology, not a game changer for how states start or conduct war.

The shape of war remains resilient and new tools only reshape the contours of war, not the nature of the fight. Both offensive and defensive options based on emergent technologies offer great promise, but they often inevitability crumble when confronting the true brutality of war. The tank enables great maneuverability but is limited in production quantities, needs vast amounts of fuel, and remains a costly investment. Stealth bombers remain vulnerable to detection despite their name and are being utilized in sparing quantities in conflicts. Unmanned vehicles (UAVs) reduce the need for manned aircraft, but their main utility is loitering, reconnaissance, and not taxing the resources of the state deploying these forces.

The conjecture for some is how war can decline when future technologies like cyber options—malicious digital weapons—will increase lethality and offer a means to continue the fight in new ways. This chapter explores this question and reviews what we know about cyber conflict, covering the nature of war, dynamics of coercion, escalation, and the possibility that these tools can limit the onset and exacerbation of conflict. Findings to this point suggest a limited potential of cyber technologies to transform what we know about war, but this does not mean that other technologies cannot reshape the future potential for warfare.

CYBER OPERATIONS AND THE ONSET OF WAR

For many, cyber security is a top national security threat and a challenge to the stability of society that requires a reorientation of national strategy. The mystery surrounding cyber operations shapes the perception that we are vulnerable to digital violence. The fear of the unknown animates many projections about the future of war. The potential for cyber operations to reshape war mystifies pundits and observers alike, but this threat is now more than thirty-five years old. Cyber tools are not new weapons—the challenge is that the ubiquity of the Internet expands potential attack surfaces, giving the opposition more targets of interest. The other issue is that dependency on digital communication can now endanger C2 facilities, making it impossible for leadership to communicate.

The utility of cyber operations fortunately seems to be generally confined to fiction as presented in science fiction, such as *Battlestar Galactica* (Dykstra et al. 2003), or popular fiction such as the novel *Ghost Fleet* (Singer and Cole 2015). Fiction has animated our beliefs about cyber operations, a key event being the release of the blockbuster film *Wargames* (1983), which prompted President Ronald Reagan to establish the first task force to examine our national cyber vulnerabilities (Kaplan 2016).

For this exercise, we define cyber conflict as "the use of computational technology for malevolent and destructive purposes in order to impact, change or modify diplomatic and military interactions between states" (Valeriano and Maness 2015, 21). Many have suggested our future will be dictated by cyber conflict (Kello 2013; Clarke and Knake 2014), while others question the utility of cyber operations for war (Rid 2012; Gartzke 2013; Valeriano and Maness 2015). If we define war as a crisis event between nation-states with one thousand battle deaths (Small and Singer 1982), there is little probability of "cyber war" because cyber options have yet to be the direct cause of even one death.

It is possible that death and destruction can occur via cyber operations, but these sorts of dramatic attacks represent a massive escalation in intensity between states that would likely only happen if there is an ongoing major war. In fact, the only cyber operations that could provoke kinetic responses occur during ongoing hostilities (Borghard and Lonergan 2019). What is more likely is that cyber operations will be utilized for sabotage (Rid 2012) or deception (Gartzke and Lindsay 2015), but even the utility of these operations is dubious because of the dynamics of cyber tools' coercive impact.

What is different about cyber tools is that the code or malware utilized in operations can be readily repurposed. There is extensive evidence of operations spreading beyond the initial attack zone, with some even suggesting one of the most famous cyber operations (the Stuxnet attack against Iran) was repurposed to enemies of the United States for their own ends. A malicious group named Shadow Brokers released code supposedly developed by the US National Security Agency (NSA), and America's adversaries have utilized this code to attack the country that developed the code, such as North Korea's Wannacry ransomware campaign in 2017 (Mohurle and Patil 2017). This process is akin to turning an opposition missile right back at the attacker.

It is still early in the overall development of cyber operations and learning their optimal strategic utility. While the tool is not new, it is certainly not mature. Defenses and offenses still need to balance to meet the threat, artificial intelligence will pose another challenge to cyber operations, making defenders more capable, and information operations in the cyber domain have led to widespread disinformation and political warfare. States with little concern for ethics, morals, and responsibility could utilize cyber operations for destructive purposes against civilians, but these grand fears are unlikely to materialize. This brings us to our key question: has the character of war evolved with the onset of the cyber era?

THE EVOLVING CHARACTER OF WAR

When studying war every academic will inevitably come across Carl von Clausewitz's dictum on the nature and character of war. The nature of war can change slightly, but it is wholly subordinate as an enterprise to the "original violence of its elements, hatred and animosity" (Clausewitz 1984, chapter 1: 7). The conditions of humanity mean for many that the essential components of violence and horror behind war will never change. The nature of war remains the same, but the character of war can change because "from the character, the measures, the situation of the adversary, and the relations with which he is surrounded each side will draw conclusions by the law of probability as to the designs of the other, and act accordingly" (Clausewitz 1984, chapter 1: 8).

In short, the nature of war is dependent on the characteristics of humanity itself, but the character can change with the situation, the politics, the technology, and the culture of the moment. If those in control of the course of war sense a change in the opportunity for war because new technologies alter the landscape, the character of war has inevitably changed. Almost directly tied to the concept of strategic culture (Snyder 1977; Johnston 1995), the course of strategy can be observed through behavior, but generally these behaviors do not change much through time.

The problem for those who promote a revolutionary view of cyber technology (Kello 2017) is that cyber tools have not changed the nature or character of war despite predictions to the contrary. Instead, cyber options only reinforce the patterns of old. States do not fight over technology but rather over traditional causes such as territorial issues (Valeriano and Maness 2015).

As we discuss in this chapter, cyber technologies do not alter the coercive patterns of conflict, the dynamics of escalation, and the lust for total victory. Instead, cyber operations just reinforce what has come before. As tools of the strong, cyber options serve to subjugate the weak. As tools of the weak, cyber options will cause disruption and serve notice to adversaries about capabilities, but they are unlikely to alter the strategic balance of conflict and give the initiative to the weak.

The cyber conflicts of the present are the result of the rivalries of the past, from the United States versus Russia (Maness and Valeriano 2015), to North Korea versus South Korea, and to even the United Arab Emirates versus Qatar (Valeriano, Maness, and Jensen 2017). Prior diplomatic failures and aggressions shape the cyber conflicts of the future. We have yet to see an international dispute caused solely by a strategic cyber action, and we are unlikely to witness such an event because the character of conflict has not changed with the advent of cyber conflict.

The case of Iran is illustrative of the typical dynamics of cyber conflict. There was an observed uptick in cyber actions by Iran after the United States pulled out of the nuclear agreement signed by President Obama (Joint Comprehensive Plan of Action 2015). Expressing dissatisfaction through cyber conflict is a typical outgrowth of cyber options, but Iran is not seeking to shape the behavior of its adversary through direct conflict because the strength of the United States serves as a deterrent of sorts. Instead it responds with attacks against Saudi Arabia and its oil giant, Aramco (once in 2012 in response to Stuxnet and again in 2019 in response to JCPOA and the War in Yemen, Baezner 2019).

The counter to this narrative is the constant Russian bombardment of cyber-related tactics in Ukraine, the supposed testing ground for future cyber operations (Greenberg 2019). The problem is there is no evidence that cyber operations against Ukraine, while it was directly involved in a war with Russia, were out of the norm for a combat operation. The crucial question is whether the use of cyber operations changes anything. For Kostyuk and Zhukov (2019), the answer is definitively no during the Ukraine-Russia war. Cyber operations did not change the flow of the battlefield, and even seemingly devastating attacks such as the Not-Petya operation against Ukraine's financial services sector and the Estonia attacks in 2007 had a limited impact.

Taking a sobering look at the possibilities of digital technology is not meant to dismiss the possibilities of the future; rather it serves to remind us that strategic calculations and the base politics of the past do not change with the fostering of new weapons leveraged to fight. Cyber operations and other modern technologies such as nuclear weapons might alter the bargaining patterns of states, but the patterns of conflict remain the same (see chapter 6 of this volume). The pull of the past remains strong, and the importance of this volume on the causes of war blends the past with the new to bring us to new understanding about the present.

THERE IS NO CYBER WAR

Despite many claims to the contrary (Kello 2013, 2017; Buchanan 2016), cyber operations have not fundamentally revolutionized international relations. Instead, they have simply been added to the low end of coercive options short of war (Nye 2017; Valeriano, Maness, and Jensen 2018) that states can utilize to compel their adversaries to moderate their behavior, and many of these options are information related, limiting physical destruction.

Thomas Rid was instrumental in pushing the field of cyber security quite early with the conjecture that cyber war will not take place (Rid 2012). The main idea behind this statement was that death is unlikely to result from cyber war—in a pure definitional sense, war is unlikely. This does not mean that cyber conflict is unlikely or rare. Rid was clear early on suggesting that "cyber sabotage is easy," and goes further discussing the use of Russian active measures as a method of information warfare (Rid 2020).

At the same time Valeriano and Maness (2012) started to advance the position through data and evidence that cyber conflict is a relatively underused tactic with limited ability to affect the battlefield. Cyber war is an inflated threat incapable of changing the facts on the ground. They also demonstrate that Russia's ability to cause mischief in international affairs is relatively toothless, since it depends on limited techniques such as cyber conflict. Even the US presidential election hack of 2016 resulted in minimal success in changing voters' minds about the candidates (Bail et al. 2020).

The debate between Lindsay and Kello (2014) in the pages of *International Security* also pushed the field in new directions, placing scholars into corners with Langø (2016) labeling the divide as a split between cyber skeptics and cyber revolutionaries. Kello advances the revolutionary perspective, suggesting the context and content of international relations will change with cyber options being incorporated into national defense strategies (Lindsay and Kello 2014). Lindsay and Kello (2014) are skeptical of the potential of cyber factors to alter the international discourse and instead argued they would be added to the limited portfolio of options states can utilize during conflict, and perhaps more useful in times of peace and below the level of armed conflict.

The reality is that a more moderate path needs to be carved out. Scholars are not skeptical of cyber operations, only of the dramatic proclamations of change and difference cyber options introduce. Threat inflation dominates the field and news as cyber operations have been used sparingly so far, giving us limited examples of their utility. In the absence of evidence, base fears color the shape of the fear.

Gartzke and Lindsay's (2015) article on offense, defense, and deception was critical for the development of the field of cyber security. It pushes our aperture away from war and toward the idea that cyber conflict is primarily what might be called an intelligence game (Rovner 2019). Valeriano, Jensen, and Maness (2018) demonstrate empirically that the great majority of observed cyber conflicts are what might be categorized as espionage. The rise of cyber operations might coincide more closely with the evolution of espionage and other operations in the information environment rather than the evolution of war.

The perspective of restraint is critical for cyber security. This perspective is advanced by Valeriano and Maness (2015), who argue that restraint is the outcome we witness in cyber operations because of the limitations of the weapons, the vulnerability within each state, and the general inability of the options to alter the cost-benefit calculus of the adversary. States will refrain from more harmful and damaging cyber operations, using more low-level tools to keep the adversary confused and at bay.

The stability-instability paradox introduced by Snyder (1965) has been instrumental in articulating a view that the instability that seems evident in cyber space is stabilizing. "We should expect to see a lot more creative exploitation of global information infrastructure, but threat actors have strong incentives to restrain the intensity of their exploitation" (Lindsay and Gartzke 2016: 176). Restraint can be seen as an outcome of the limited coercive value of cyber operations, but it can also be a choice or a strategy given the logical limitations of technology to provide an advantage to the attacker.

Figure 12.1 shows the number of cyber incidents over time for the years 2000–2016. Data were extracted from the Dyadic Cyber Incident and Campaign Dataset (DCID), version 1.5 (Maness, Valeriano, and Jensen 2019). These data record all known state-initiated cyber incidents and campaigns between rival states (Klein, Goertz, and Diehl 2006) at the dyadic level. Incidents capture variables including method (vandalism, denial of service, network intrusions, network infiltrations), target type (private sector, civilian government, military government), strategic/coercive intent (disruption, short- or long-term espionage/manipulation, degradations), presence of a following information operation (yes or no), and severity score (1–10, with a score of 1 being minor probing and 10 being massive death as a result of a cyber operation), among others.[2]

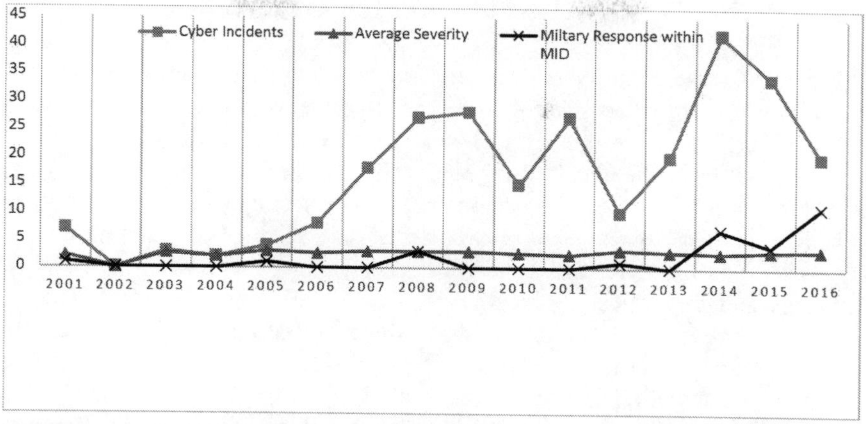

Figure 12.1. State-initiated Cyber Incidents, Severity Average, Military Responses: 2000–2016.

Figure 12.1 shows that cyber incidents have been increasing over time but remain around the same severity score through the duration of the 2000–2016 period. States have found cyber incidents useful for a foreign policy tactic, but have yet to raise the stakes with more severe cyber operations that could lead to escalation. The stability-instability logic seems apparent in international cyber conflict.

Figure 12.1 also shows that military responses to cyber operations have been rare throughout the timeline, and only recently have military responses to cyber operations been on the rise. Data for military responses were compiled from three separate data sets: the Integrated Crisis Early Warning System (ICEWS) (Boschee et al. 2018), the Correlates of War (COW) Militarized Interstate Dispute (MID) data set (see Ghosen, Palmer, and Bremer 2004), and the International Crisis Behavior (ICB) (Brecher and Wilkenfeld 1997) data set. The "military usage" ICEWS variable is coded within a three-month (ninety days) period after the cyber operation is either initiated (for non-espionage operations) or becomes public (most espionage variables). These variables are then overlaid with the MID variables (which only go to 2010) and the ICB variables (coded until 2015) to see if these military operations fall within the time period after the cyber operation.

Looking at Figure 12.1, we see that military responses increase after the year 2013, likely due to the rise in organized violence seen globally in recent years. This includes the Russia-Ukraine protracted conflict, the Syrian civil war, and the rise of China's use of cyber operations in its disputes with regional rivals over issues regarding the South China Sea. Russia has been infiltrating Ukrainian networks in tandem with its other grey zone or "hybrid" tactics, such as disinformation operations and arming dissidents, since the annexation of Crimea in 2014.

DISTINCT NATIONAL APPROACHES TO CYBER CONFLICT

The field of cyber security is dominated by conjecture because many suggest there is no way to collect data on cyber security events (Kello 2013). So many events occur in a covert domain that it would be impossible to collect a representative sample of the data in the system. This challenge was met by Valeriano and Maness (2014, 2015), who demonstrate that collecting data on cyber operations between rival states is not only possible but also revelatory and necessary for progress in the field. Since then, others, including the Council of Foreign Relations and the University of Maryland's START program, have collected data on cyber events, but the DCID data (Maness, Valeriano, and Jensen 2019) represent the only peer-reviewed data set that also considers the issues of internal validity and external reliability.

For too long single examples have been leveraged to make large claims about the process of cyber conflict. Most of these examinations would not even meet the standards of a true case study, but instead represent surface-level explorations of an event to advance a specific predetermined position. Early and credible case studies in the field have been few and far between. Lindsay (2013) and Barzashka (2013) explore the Stuxnet operation in detail, demonstrating US capability and intent in operations. While *precision* was the operative word that might describe the operation, Stuxnet likely had little overall effect on the ability of the Iranians to enrich uranium. Instead, to meet the challenge, Iran brought more centrifuges online and ended up enriching more uranium than before the attack.

Simple bivariate cross-tabulations (crosstabs) are utilized in the tables that follow to uncover relationships between two or more categorical variables. Measuring Pearson residuals gives us the key to where these significant relationships lie.[3] If the Pearson residual is above or below a score of two, statistical significance is present (at the 95 percent confidence level), where the utilization of a certain strategy or target type is more or less than expected, according to the normal distribution. In other words, if a positive significant relationship exists between variables (in bold and starred in the tables), that indicates that the preferred strategy, target type, or response is present. If a negative significant relationship exists between variables, it indicates that the preferred strategy, target type, or response is not present, and suggests that the other dependent variables are preferred over the significant negative highlighted variable.

Table 12.1 shows that the overall pattern of usage of cyber operations by the United States demonstrates a reliance of "degrade operations" that fit with the precision strike complex model of US strategy (Valeriano, Jensen, and Maness 2018). The data demonstrate that the United States uses these operations sparingly, although only four of the total cells are statistically significant due to data limitations. Usually these degrade operations require intelligence collection, and several of the espionage incidents are precursors to the eventual operations utilized against rivals. The United States has not used a disruptive strategy in the data set's current timeline; however, in 2018 U.S. CYBERCOM launched a several days-long DDoS (distributed denial of service) against Russia's Internet

Table 12.1. Crosstabs of Cyber Operations by Strategy Initiated by the United States and Its Four Major Adversaries 2000–2016

Initiator		*Disruption*	*Espionage*	*Degrade*
United States	Count	**0**	10	**11**
	Expected count	**5.47**	12.47	**3.07**
	Pearson residual	**–2.34****	–0.70	**4.53****
Russia	Count	19	35	11
	Expected count	16.92	38.58	9.50
	Pearson residual	0.51	–0.58	0.49
Iran	Count	8	21	4
	Expected count	8.60	19.60	4.82
	Pearson residual	–0.20	0.32	–0.37
China	Count	17	55	**2**
	Expected count	19.26	43.93	**10.81**
	Pearson residual	–0.52	1.67	**–2.68****
North Korea	Count	**13**	9	4
	Expected Count	**6.77**	15.43	3.80
	Pearson residual	**2.40****	–1.64	0.10

****$p < 0.05$, $n = 219$, Pearson chi-squared (8) = 46.24, $p = 0.000$****

Research Agency (IRA), the famed troll farm cited by Special Counsel Robert Mueller's team in 2019 (Nakashima 2019).

On the other hand, China focuses mainly on espionage operations directed at gathering information and capability for future power projection. China's cyber strategy is more long-term, where the development of its technology and military sectors is important to its rise and quest for parity with the United States. China also uses disruptive strategies, which are usually against its regional rivals when conventional disputes manifest in the South China Sea over territory.

Russia is a revisionist power and is attempting to achieve the global perception that comes with being a world power, and its specialized talent in cyber and disinformation campaigns is a big part of its foreign policy strategy. Russia typically utilizes espionage operations so as to enact cyber-enabled information operations against its rivals' domestic populations and to be able to sow discord and pursue its objectives more freely because of these disruptions. As Table 12.2 shows, Russia uses cyber operations against the private sector more than any other strategy when compared with the other four countries in the table, but these results are not statistically significant. Russia's strategy of stealing technology and sensitive information feeds into its revisionist strategy in cyber space to punch above its weight against richer or more powerful rivals.

China infiltrates the civilian sectors of rival governments more than any other country, and this reinforces its attempts to steal sensitive information, intellectual property, and government secrets as it continues its long-term rise. The United States targets the military sectors of rival governments, in line with its attempted adherence to the Law of Armed Conflict (LOAC) and precision-strike mentality, where collateral damage is limited.

Table 12.2. Crosstabs of Cyber Operations by Target Type Initiated by the United States and Its Four Major Adversaries

Initiator		*Private sector*	*Government nonmilitary*	*Government military*
United States	Count	**2**	7	**12**
	Expected count	**7.38**	10.07	**3.55**
	Pearson residual	**–1.98****	–0.97	**4.49****
Russia	Count	31	24	10
	Expected count	22.85	31.16	10.98
	Pearson residual	1.70	–1.28	–0.30
Iran	Count	15	16	2
	Expected count	11.60	15.82	5.58
	Pearson residual	1.00	0.05	–1.51
China	Count	17	**47**	10
	Expected count	26.02	**35.48**	12.50
	Pearson residual	–1.77	**1.93****	–0.71
North Korea	Count	12	11	3
	Expected count	9.14	12.47	4.39
	Pearson residual	0.95	–0.42	–0.66

****p < 0.05, n = 219, Pearson chi-squared (8) = 41.80, p = 0.000****

We know little of general cyber capabilities. It is difficult to examine the overall spending on cyber security because most budgets are "black" or unknown. Understanding weapons stockpiles is an even more difficult process because malware stockpiles are unknown and there is an intense debate regarding state assets. There is even a likely bigger challenge in the United States, with no overall coordination of malware or zero-day purchasing; US government organizations and allies may be in competition with themselves for overall access to "cyber weapons."

Slayton (2017) makes the point that much of the capacity for cyber power comes from the talent and skill of individuals in the space. Craig and Valeriano (2016) apply the traditional study of arms races to cyber competition. While many in the space assume there are cyber arms races, this is an unknown at this point since we have little grasp of overall capabilities for most states. Some states like North and South Korea are likely engaged in a cyber arms race, but we have no idea of the overall scale of the problem internationally.

Table 12.3 lists power measures for the ten most active cyber states, according to the DCID (Maness, Valeriano, and Jensen 2019). Latent Cyber Capacity is a compilation of six World Bank measures under two categories: infrastructure and knowledge capital. The infrastructure category records World Bank scores recording broadband subscriptions per one thousand people, secure servers per one million people, and percentage of high-tech exports out of total manufactured exports (World Bank 2020). The knowledge capital category records scores from the World Bank that include the number of Internet users per one thousand people, the number of science, technology, engineering, and mathematical (STEM) journal articles published, and the number of patent applications for each country

Table 12.3. Different Power Dynamics: Top Ten Most Active Cyber States

Country	*Latent Cyber Capacity (2016)*	*Economic Power (GDP billion $, 2019)*	*Military Power (Total billion $, 2019)*	*CINC (2012)*
United States	**6.82**	**20,490**	**750**	0.143291
China	6.43	13,400	237	**0.2181166**
S. Korea	6.22	1,531	44	0.0232826
Japan	5.86	4,970	49	0.0370358
Israel	5.32	351	20	0.0042498
Russia	5.03	1,578	48	0.0400789
Iran	4.53	440	19.6	0.0157625
India	4.35	2,597	61	0.0808987
Pakistan	4.11	305	11.4	0.0145536
N. Korea	4.01	40 (est.)	1.6	0.0132601

included in the DCID (World Bank 2020). These scores are normalized and then averaged to get the power scores listed in Table 12.3.[4]

The United States is the top cyber power in the international system, according to the Latent Cyber Capacity index, followed by China and South Korea. The United States has a robust technology industry and a growing high-tech infrastructure backbone, as well as many patents, STEM publications, and research universities. China's technology industry was kick-started by its espionage tech sector as well as its investment in modern infrastructure, and South Korea's homegrown technology sector put it near the top with the United States regarding these latent power scores. The rest of the columns show economic power measured in gross domestic product (GDP) per capita for 2019 (World Bank 2020), military expenditures per country (Global Firepower 2020), and the latest Composite Indicator of National Capability (CINC) scores for each country as of 2012 (see Singer 1987).

CYBER COERCION AND THE UTILITY OF CYBER OPERATIONS

In recent years the question of cyber coercion has dominated the literature in the cyber conflict field. The reason why is not difficult to uncover as coercion, or the ability of one state to influence the behavior of another, is the key question the field has generally ignored until recently. With many proclamations of the dramatic change brought on by cyber tools, few seemed to want to ask the basic question of how one would compel change with digital force. Answering the question of coercion directly impacts theories about the utility and effect of cyber operations, directly affecting the probability of war.

Coercion, under the construct advocated by Schelling (1980), can be divided into two forms, compellence and deterrence. Compellence is the ability to make an actor behave in a manner it otherwise might not. Deterrence is when an actor does not do what it otherwise might. Libicki (2016) was at the forefront in suggesting ways coercion may work in cyber space, but few have sought to investigate the question from a social science perspective until more recently.

Borghard and Lonergan (2019) suggest that cyber power as a tool of coercion will have limited effectiveness, because cyber tools often lack clearly communicated threats, (Libicki 2016) credibility, and reassurance. Valeriano, Maness, and Jensen (2018) focus on the empirical dynamics of coercion in their monograph. They find that compellence is limited, resulting in a change in behavior only 5 percent of the time. This rate is lower than most other forms of compellence, which traditionally sit above 44 percent on average.

Some countries are better at coercion than others—for example the United States succeeds at a rate of 38 percent, mainly because it focuses on the use of degrade campaigns that seek to change the behavior of the adversary. The United States also has the advantage of being the most powerful state in the system in terms of conventional capabilities, meaning it can back up its threats with action, ensuring compellent success more than usual.

Russia has been generally unsuccessful in leveraging cyber operations for effect. An examination conducted by Bail et al. (2020) suggests that the Russian use of information operations only really reached those already supporting Donald Trump. Changing behavior is not evidenced here—if anything, cyber-enabled information operations launched by Russia might have increased turnout and support for Trump, but the audience was ready to hear the message in the first place (Bail et al. 2020). Kostyuk and Zhukov's (2019) work is also instrumental in demonstrating the limits of Russian influence in cyber space. The scholars could identify no evidence of successful coercion during the war in Ukraine.

Deterrence is difficult to investigate empirically. Some suggest that the United States has deterred cyber conflict between adversaries above the threshold of armed conflict but failed below this zone. The challenge of course is that it is difficult to observe deterrence because successful deterrent moves will be unobserved, and successful deterrence entails nonaction. More critical is the reality that states have not mobilized for deterrence under the conditions where most scholars would accept its application (Brantly 2018). For true deterrence, the state must have defenses to forestall an attack in the first place, capabilities to respond, and credibility to launch retaliatory attacks, and it must also communicate what actions it wants to dissuade.

There is a possibility true deterrence can be implemented and tested in reality, but that would require a state to invoke real collaboration between the state and the private sector, ensuring that society overall is protected to forestall attacks in the first place. Layering deterrent principles in this way could increase the success of the construct, but once again we are left with a limited number of states with the ability to make this process work, suggesting the construct overall is limited.

The Cyberspace Solarium Commission (2020), in its comprehensive review of cyber policy and action, is advocating a policy of layered deterrence that mimics many of these stated panoplies (denial, cost imposition, credibility, entanglement, and a true public-private partnership). Achieving a whole-of-nation approach to cyber deterrence is critical, but also difficult to implement. Ensuring cooperation within the US government is difficult enough. Extending cooperation between the government and private entities is orders of magnitude more complicated, but also more critical given the tendency for the majority of targets to be in the private sector.

ESCALATION AND CYBER SECURITY

Escalation is defined as an increase in the nature or intensity of conflict. Extending escalation theory to cyber space would include situations in which "the target responds with more intense and costly cyber means (cyber escalation within the domain) or through breaching the cyber-kinetic threshold (cross-domain escalation)" (Borghard and Lonergan 2019: 123). Libicki (2016) simplifies cyber escalation into two factors: intensity (striking deeper, lasting longer) or more extensive (striking new targets), plus adds the consideration that an attack in cyber space can escalate to an attack on physical space.

Cyber escalation is then defined as a reaction with digital tools that increases intensity or aggression or that spreads the scope of conflict after an initial move in the digital domain. Theoretically the cyber domain can contain escalation because of the uncertainty introduced by cyber weapons (Buchanan 2016). There is also the aspect of civilian punishment, attacks that hit what might be deemed off-limits civilian systems, provoking escalation due to the violation in norms between two actors.

In practice, however, cyber escalation is rare. Stuxnet (2007–2010) is often cited as the prime example of cyber escalation yet, put in the context of the wider dispute between the West and Iran over the development of nuclear weapons, the Stuxnet attack is actually a de-escalatory move because the other options on the table at the time were conventional strikes that would have caused death and destruction.

Talmadge (2019a: 864) makes the point that technology itself is rarely a sufficient condition for escalation, "cast[ing] doubt on the idea of emerging technologies as an independent, primary driver of otherwise avoidable escalation." Technology became the mask for the processes that enable escalation, rather than the cause of escalation itself. It is not the domain that produces escalation, but the action in the domain that produces outcomes.

We have ample empirical evidence that escalation is limited in the cyber domain. Even in its simpler form, there is little retaliation, let alone escalation, in the domain or even out of the domain when cyber actions are the triggering events. When conflicts do escalate in cyber space, we observe limited engagement between the parties unless there is already an outright war. As Valeriano, Jensen, and Maness (2018: 18) note, "even when cyber exchanges between rivals escalate, they remain limited in scope outside of ongoing military conflict. That is, rivals may use cyber operations to probe the enemy, test their resolve, and signal the risks of significant escalation."

Data analysis supports these positions and is developed here from established data and taken from ongoing projects to support our background investigation into cyber escalation processes (Maness, Valeriano, and Jensen 2019). Table 12.4 shows the response patterns between the United States and its four major adversaries (Russia, Iran, China, and North Korea) in the cyber domain, as well as how each country responds to a cyber operation with a retaliatory cyber operation. Table 12.4 utilizes the cyber operation severity score from the DCID version 1.5 that measures the impact and national security importance of each state-initiated cyber operation between the years 2000–2016 between rival states (Maness, Valeriano, and Jensen 2019). The scale is interval and ranges from 1 to 10.

Table 12.4 shows that the United States is often on the receiving end of retaliation at a higher rate than expected. However, these responses to US cyber actions do not indicate within-domain escalation. The severity levels with a response score of 2 were retaliatory to US actions that were of a higher severity level. Of the seven actions at the severity level of 4, three represent a decrease in the

Table 12.4. Crosstabs of Response Severity to Cyber Operations Initiated by the United States and Its Four Major Adversaries

		Response Severity						
Initiator		*No Response*	*1*	*2*	*3*	*4*	*5*	*6*
United States	Count	**7**	1	**5**	1	**7**	0	0
	Expected count	**16.60**	0.10	**1.25**	0.96	**1.63**	0.39	0.10
	Pearson residual	**–2.35****	2.92	**3.36****	0.04	**4.21****	–0.62	–0.31
Russia	Count	52	0	6	5	1	1	0
	Expected count	51.35	0.30	3.86	2.97	5.05	1.19	0.30
	Pearson residual	0.09	–0.55	1.09	1.18	–1.80	–0.17	–0.55
Iran	Count	27	0	0	**4**	1	1	0
	Expected count	26.07	0.15	1.96	**1.51**	2.56	0.60	0.15
	Pearson residual	–0.20	–0.39	–1.40	**2.03****	–0.97	0.51	–0.39
China	Count	66	0	**0**	0	7	1	0
	Expected count	58.46	0.34	**4.39**	3.38	5.74	1.35	0.34
	Pearson residual	0.99	–0.58	**–2.10****	–1.84	0.52	–0.30	–0.58
North Korea	Count	21	0	2	0	1	1	1
	Expected count	20.54	0.12	1.54	1.19	2.012	0.48	0.12
	Pearson residual	0.10	–0.35	0.37	–1.09	–0.72	0.76	2.26

****$p < 0.05$, n = 219, Pearson chi-squared (24) = 76.50 p = 0.000****

Table 12.5. Crosstabs of Response Severity to Cyber Objectives

Cyber objective		Response Severity: No Response	1	2	3	4	5	6
Disruption	Count	58	0	**19**	4	2	3	0
	Expected count	64.99	0.65	**8.73**	4.20	5.50	1.62	0.32
	Pearson residual	−0.87	−0.80	**3.48****	−0.10	−1.49	1.09	−0.57
Espionage	Count	118	1	**4**	8	12	2	0
	Expected count	109.57	1.09	**14.72**	7.09	9.27	2.73	0.55
	Pearson residual	0.81	−0.09	**−2.79**	0.34	0.90	−0.44	−0.74
Degradation	Count	25	1	4	1	3	0	1
	Expected count	26.45	0.26	3.55	1.71	2.24	0.66	0.13
	Pearson residual	−0.28	1.44	0.24	−0.54	0.51	−0.81	2.39

****p < 0.05, n = 266, Pearson chi-squared (12) = 36.49, p = 0.000****

initial attack severity and four represent an increase by one tick in severity. The only other country that witnesses a statistically significant level of retaliation at a greater rate than expected is Iran, which is wholly due to US or Israeli operations. China's significant negative relationship with its severity score of 2 shows that it prefers higher levels of severity when it retaliates in the cyber domain. Many of the Chinese incidents involve entanglements with the United States, which is another great power with vast cyber capabilities. This propensity to use more severe attacks does not denote escalation, however. Escalation is rare in digital interactions as measured by rival states from 2000 to 2016.

Table 12.5 shows how responses are related to the overall cyber strategy of the initiating states. In terms of response severity by strategic objective, disruptive efforts by initiating states are usually met with retaliatory disruptions, further indicating that the cyber domain is, for the most part, non-escalatory. Espionage campaigns are also commonly met with cyber operations that either steal or signal capabilities or displeasure for the originating action, but do not lead to a tit-for-tat escalatory ladder, as indicated in Table 12.2. Only occasionally do we see disruptions or espionage operations escalate to the severity level of 5, with this happening only five times over the 2000–2016 period.

Results from war games have demonstrated the complicated empirical picture of the escalation landscape in cyber space (Schneider 2017). After examining war games from 2011 to 2016, Schneider finds that government officials were hesitant to use damaging cyber weapons. Most games only witness the use of cyber capabilities after the onset of conventional warfare, not before. For Jensen and Banks (2018), in the context of cyber options, escalation was the exception, not the norm.

To explore escalation when cyber options are present within the context of integrated options of national power, Jensen and Valeriano (2019) ran a series of war games on 259 participants, including members of the military, students, and policymakers. The simulation situated participants in a typical crisis that was highly likely to escalate given a rivalry situation over an ongoing territorial dispute when the crisis under examination was the third in a series over a five-year period.

The results demonstrated that escalation is not the norm. Based on a general baseline for all international conflict situations, most options fell below the 48 percent threshold. In fact, the only instance where escalation was the dominant option was when a cyber action started the crisis and the target had no cyber response options. This suggests there are implications of attacking a state with cyber options when they do not have the ability to respond within the domain. In most other situations, we witness few demands for escalation when cyber response options are on the table.

Overall, regardless of the situation, cyber escalation is usually not the dominant response. The reality is that even under dangerous conditions cyber response options can actually moderate crisis response patterns. Surveys demonstrate a great amount of fear in the cyber domain, but this does not motivate overreaction (Gross, Canetti, and Vashdi 2017). Figure 12.2 shows the results from a cyber campaign–directed dyads data set. What is measured in what follows is whether the combined cyber operation, which includes diplomatic, economic, and military variables extracted from ICEWS (Boschee et al. 2018), has escalatory responses from the target from all four domains recorded in the campaigns. For a cyber response to be recorded, the target responds within one year from the start date of the original cyber operation from the initiator for disruptions and degradations, and from the date the operation becomes public for espionage operations. For a diplomatic response, the time frame is one month (thirty days) after the cyber operation's initiation or public reporting, and for economic and military responses, the time frame is three months (ninety days) after the same criteria regarding the cyber operation. For cyber escalation, the severity score must go up at least one point regarding the cyber response. No cyber responses at the same severity score are included. For conventional responses, we use the Conflict and Mediation Event Observation (CAMEO) conflict-cooperation scores to measure escalation (Schrodt 2012). The CAMEO scale ranges from −10 to 10 where the more negative score, the more conflictual and the more positive score, the more cooperative the foreign policy action. If the conflict scores from each domain are lower (more negative) from the target state in retaliation for the cyber incident, escalation in each domain is recorded.

Looking at Figure 12.2, we find that only seven cyber operations result in a higher cyber response in terms of DCID severity scores from the target state. This indicates that cyber does not beget more cyber, and it is extremely rare that a cyber operation is met with a more sophisticated and damaging operation from the original victim. Diplomatic escalation, where the diplomatic response is more severe than the initiating state's diplomatic action during the combined cyber campaign, is found in about one-third of all 266 campaigns recorded in DCID version 1.5.

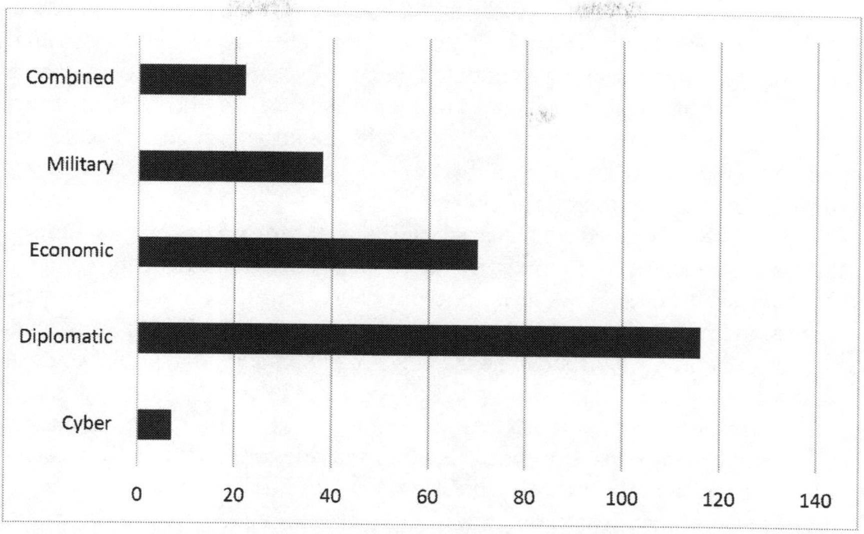

Figure 12.2. Escalatory Responses per Domain: Cyber Campaigns.

Economic escalation is found in 65 cyber campaigns, and this is a favorite response of the United States and its North Atlantic Treaty Organization (NATO) partners on less powerful states that launch cyber campaigns against Western powers. We see military escalation in nearly 40 cyber campaigns, and these are mostly embedded in ongoing conflicts in the post-Soviet space, the Middle East, and the South China Sea. Finally, we see combined escalation where the added totals of the diplomatic, economic, and military scores between the initiator and target are at a higher overall CAMEO score.

CYBER AS A TOOL OF REPRESSION

Overall, the field has missed the boat on cyber war. There is no "war," but there is a strong probability that cyber tools can be used to repress populations internally. Cyber weapons are better suited against the weak than the strong. Powerful states can fight back, and every state is vulnerable in cyber space—even North Korea, which has little dependence on the Internet. Yet individuals, activists, journalists, and members of the civil society community have no great ability to fight back or protect themselves against a committed state adversary. The individual actor has little recourse when paired in a fight against a system of state control.

Some jump to the conclusion that cyber war will be a reality between states, but Arquilla and Rondfeldt (1996) early on suggested that netwars will be a critical aspect of future cyber conflicts. Netwars will be fought between non-state or irregular forces, but the evolution of the domain will witness more conflicts between state and non-state actors, morphing netwar into something different and unforeseen.

As Valeriano and Pytlak (2016: 1) argue, "there has been a precipitous rise in malicious hacking, but it is not exhibited between states, rather it is from within them by governments seeking to maintain control over their populations. There is an increasing utilizing of cyber technology to silence dissent, often in direct contradiction with human rights law." This style of conflict can be termed "cyber repression," the use of digital tools by the state to repress, demean, and harass activists, journalists, and protectors within a state.

We can witness this effect by examining the pattern of Internet shutdowns historically. Gohdes (2015) finds evidence that Internet shutdowns coincide with wider state repression events, possibly increasing the ability of the state to control the population during times of turbulence. Gohdes (2015: 352) notes "that governments have the strategic incentive to implement Internet blackouts in conjunction with larger repression operations against violence opposition forces." It is not just that governments use Internet shutdowns as a form of repression, but that governments can also use the access they provide to the Internet to create a new style of targeted and precise repression (Gohdes 2020).

Instead of placing a fear of cyber war at the heart of upcoming interstate conflicts, its true place is likely a source of power by the state against adversaries during events of protest or rebellion. The utility of cyber weapons to control and subjugate a society are clear given the dependence many place on connectivity. Add to this the rise of the surveillance state and there are the makings for something more dangerous than the proponents of cyber war have ever dreamed. Limiting connectivity, repressing digital dissent, invading privacy, and surveilling domestic enemies are clear strategies for the state to prevail against internal adversaries and movements.

At the same time capable individuals can marshal resources and capabilities to challenge the state digitally in response to increased repression. They can attack critical infrastructure, the media, and even the ballot box. A series of aggressive events by the state against a disaffected population is only likely to provoke that population to respond in kind. The car bombs of the future might be digital.

WHAT WE KNOW ABOUT CYBER WAR

For many, cyber war is the future. Warfare is a persistent reality while war is a condition that exemplifies the escalation of hostilities. The general decline of wars between states leads some to believe that this trend will only be a blip in our history as new technologies reshape the global battlefield. The conjecture that future technologies will increase lethality has only met the devastating slap of reality.

The fear of the future shapes many visions of strategy, projecting a need to deal with an inflated threat now, before things get worse. This process has generally fed into the cult of the offensive (J. Snyder 1984) with many believing the best defense is a good offense in cyber space. The challenge is that evidence paints a much different picture—despite visions of a future filled with cyber conflict, we instead find evidence that cyber operations are limited as a coercive tool in international affairs. This is not to suggest that cyber operations will not be part of future battles. They will be adjunct and additive capabilities for all future fights. Cyber operations are not typically catalysts for war and may actually represent off-ramps away from war. Not all new capabilities need to be destabilizing; some technologies can instead increase the ability of states to signal displeasure and alter the course of conflicts.

Inherent in the analysis of cyber operations is the simple idea that cyber operations are not salient enough to spark wars. Too often scholars of technology and security forget to ask a simple question: what are they fighting over (Diehl 1992)? Those who ignore the findings of the field of international relations are doomed to repeat the errors of the past and find cause for wars in things that are not critical enough to spark international conflagrations. With the constant rise in cyber operations every year, however, states must be seeing a strategic utility to these options. Most cyber operations are usually below the threshold of armed conflict and can perhaps be considered a new type of political warfare, and a logical application of Clausewitz in times of peace (see Kennan 1948). States will try to gain bargaining advantages with their adversaries using digital tools in order to project power and cause asymmetries.

On the other hand, conflict processes scholars often fail to examine the strategic logic of their positions. If there is no inherent logic for the coercive power of a strategy, there is a dubious connection between that strategy, technology, or means of warfare and the onset of war itself. Technologies like cyber options, nuclear weapons, and chemical weapons are not reshaping the character of war; instead they are altering the bargaining landscape below the threshold for the use of force. This alteration of the international conflict landscape has not been noticed by the field at large. Moving forward, the field of conflict studies needs to be aware of the evolution of the conflict landscape and the reality of how emergent technologies reshape the perceptions of conflict. Cyber conflict is in our future, but cyber war will not come.

NOTES

1. *Weapon* is generally a poor term to describe digital packages with lines of code with unclear offensive and defensive abilities. Scholars are moving away from the term *cyber weapons*.

2. For a list and complete descriptions of the DCID variables, see https://drryanmaness.wixsite.com/cyberconflict.

3. Pearson residuals measure the distance between the expected value of a crosstab analysis and the observed value by the number of standard deviations. If a Pearson residual has an absolute value of 2 (either more than 2 or less than −2), then we can infer a statistically significant relationship exists between the two categorical variables for that specific cell. All significant relationships by cell in the following tables are in bold. All tables also have statistically significant chi-squared scores, indicating that all analyses to follow have significant relationships between the categorical variables and allowing us to move forward with cell-specific Pearson residual tests.

4. For a more detailed description of the Latent Cyber Capability index, see Valeriano, Maness, and Jensen (2018, chapter 3).

Chapter Thirteen

The Environment and Conflict

Water Wars

Sara McLaughlin Mitchell and Yufan Yang

In this chapter we consider how environmental factors influence the risks for interstate conflict.[1] We focus on renewable resources[2] (e.g., freshwater, fisheries) and climate change[3] (e.g., long-term changes or deviations in temperatures or precipitation).[4] In general the literature suggests that environmental risks for interstate wars exist, but they are often triggered by other conflict conditions (e.g., population growth) and that conflict escalation can be avoided through institutionalized cooperation (e.g., river treaties). We begin by reviewing the older literature connecting population growth, resource competition, and interstate conflict. This is followed by a review of research on water conflicts involving cross-border river and maritime areas and a discussion of how climate change influences interstate conflict. We then identify other theoretical conditions that intervene in the environment-conflict relationship such as economic development, domestic institutions, state capacity, and international institutions. We conclude with thoughts on challenges in this literature and we identify paths forward for improved understanding of how environmental factors influence conflict processes.

EARLY LITERATURE

Early work by scholars in this research area considered the potential for environmental-based conflict through the lens of resource scarcity induced by population growth and industrialization. Building upon Malthus's warnings of potential territorial conflicts driven by rapid population growth and declining food supplies, Hardin (1968) predicted that the exponential growth of the global population would severely increase resource competition. Choucri and North (1975) developed a theory of lateral pressure that connected population growth and resource demands, defining lateral pressure as "the process of foreign expansion of any activity that can develop national capabilities to meet the growing demand of a society" (16). The authors hypothesized that interstate conflict is more likely if two states face lateral pressure for securing resources from outside their borders and if they have a history of hostile relations. Focusing also on population growth as a driver of potential interstate conflict, Klare (2002; see also 2012) noted that the supply of renewable and nonrenewable resources faced intense demands due to increased population growth, spread of industrialization to a greater number of countries, and contested sources of resource supplies across international borders (e.g., freshwater in river basins).

While population growth seemed like a natural factor creating more environmental interstate conflict, early quantitative findings were mixed. Bremer, Singer, and Luterbacher found little relationship between population growth and interstate war: "Nation-years characterized by high density or rapid increases in density are followed by periods which are neither particularly high nor particularly low in the nations' war involvement" (1973, 340). Examining all state years between 1930 and 1989, Tir and Diehl (1998) found a positive link between population growth and militarized interstate dispute (MID) involvement (especially for states increasing military spending), although they found that population changes did not predict conflict initiation or escalation. Monadic analyses by Stalley (2003) from 1980 to 1992 on MID incidences show that increases in population density significantly raise conflict risks, even after controlling for other environmental factors such as soil degradation, water scarcity, and fish catches. A more recent monadic analysis by Reuveny and Barbieri (2014) on 123 countries from 1975 to 2001 shows stronger results linking population size and population growth to interstate war onset and war involvement, with population growth having the largest substantive effect of all variables.

However, one limitation of these studies is the focus on monadic research designs, which makes it difficult to see how joint population size/density/growth influences the risks for dyadic interstate conflict. A dyadic analysis of fatal MIDs and wars from 1885 to 2001 (Hegre 2008) shows that pairs of states that are geographically closer and larger in terms of population size and military capabilities are more likely to fight. Yet this analysis views interstate war through a gravity model lens, and thus population might be a risk factor for conflict because it creates more opportunities for fighting.[5] This is different from earlier studies in the environmental literature that viewed population growth as a risk factor for interstate war because of its effects on resource competition.

Many of the earlier studies connecting environmental resources and war stemmed from qualitative case studies, with the Soccer War, Ogaden War, Six-Day War, Chaco War, and Falkland War linked to land/timber scarcity, raw materials scarcity, and water scarcity (Reuveny and Barbieri 2014, 789). In the 1980s and 1990s many scholars and prognosticators warned of impending water wars between countries. In 1984 John Cooley argued that "the constant struggle for the waters of the Jordan, Litani, Orontes, Yarmuk, and other life-giving Middle

East rivers … was a principal cause of the 1967 Arab-Israeli war and could help spark a new all-out conflict" (3). In 1991 Joyce Starr described the potential for increased military clashes over the Jordan river basin, while identifying additional challenges in the Nile and Tigris-Euphrates river basins. Using information collected from multiple case studies, Thomas Homer-Dixon (1994) elaborated on the linkages between environmental scarcities and violent conflict within and between states, including conflict over water rights in the West Bank. Other case studies discussed conflict and tensions among states in river basins such as the Jordan (Lowi 1993), Nile (Swain 1997), Tigris-Euphrates (Daoudy 2009), and Ganges (Crow and Singh 2000). As seen in the next section, these warnings of water wars were not confirmed in many quantitative studies. In fact, research shows that cooperation is more common in river basins than war.

WATER WARS? RIVER CONFLICTS

Conflicts over international river basins are viewed as part of a broader class of issues involving resource conflicts (Klare 2002, 2012; Hensel et al. 2008; see also Hensel and Goemans, chapter 1 herein). Shared rivers can involve disagreements about borders, such as the Sino-Soviet border dispute in the late 1960s involving the Ussuri River and the ownership of Chenpao Island (Gleditsch et al. 2006, 365). Conflicts can also arise between upstream (e.g., Sudan) and downstream states (e.g., Egypt) over water quantity issues (e.g., the Jordan River [Lowi 1993] and the Syr Darya River [Dinar 2009]), water quality issues (e.g., the Rhine River [Haftendorn 2000]), and navigational rights (e.g., San Juan River).

As Gleick (1993, 84–85) notes, freshwater is a more salient security issue when there is significant resource scarcity, when states depend on other states for water supplies, when relative power in a river basin is asymmetric (see also Brochmann 2012), and when there are few alternative sources of water. Data collected by the Issue Correlates of War (ICOW) project record diplomatic conflicts over cross-border rivers, with salience captured through indicators measuring navigation (of people or commerce), fishing/resource extraction, hydroelectric power generation, irrigation, support of a permanent population, and whether the river flows through a state's homeland territory (Hensel and Mitchell 2017). In addition to river claims with higher issue salience having a greater likelihood of conflict, analyses of MID onset in claim dyad years also show that navigational rights are the most likely to spark military clashes (Hensel, Mitchell, and Sowers 2006). While ICOW analyses focus on linear relationships between salience and conflict, Dinar (2009) finds that the relationship between water scarcity and conflict is curvilinear, with river cooperation more likely for moderate levels of scarcity.[6]

When examining the relationship between shared rivers and interstate conflict using large-N statistical analyses, the findings are mixed. Some studies find that shared river basins increase the risk of dyadic militarized disputes (Toset, Gleditsch, and Hegre 2000), while other studies find that these results are weakened once controls for land contiguity are included (Brochmann and Gleditsch 2012). This stems in part from the high degree of multicollinearity between contiguity and shared rivers. For example, 80 percent of contiguous dyads share at least one river (Furlong, Gleditsch, and Hegre 2006). Diplomatic conflicts between states with shared river basins are more likely for dyads with longer rivers that cross borders and involve existing river treaties, while river claims are less likely to occur in basins with greater runoff (lower water scarcity) and in basins with more powerful downstream states (or hydro-hegemons) (Brochmann and Hensel 2009; see also Zeitoun and Warner 2006; Daoudy 2009). Jointly democratic dyads are less likely to witness river claims as well. Upstream/downstream dyadic configurations in river basins also face higher risks for militarized disputes than other types of river configurations, such as sideways or mixed river relationships (Furlong et al. 2006; Brochmann and Gleditsch 2012).[7] Dyads that share more rivers experience a greater likelihood of fatal MIDs (Furlong et al. 2006). As noted earlier in this chapter, other scholars emphasize variation in the salience of the contested river issues, with highly salient river claims in water-scarce regions like the Middle East being more prone to militarized conflict than river issues in water-abundant regions like North America (Hensel et al. 2006).

Even though studies find evidence of low-level militarized conflict in river basins, there are few instances of interstate wars or severe militarized interstate disputes. Of the 143 river claims identified by ICOW globally from 1900 to 2001, only 16 (11.2 percent) involve any militarized disputes over the claim, while only 5 (3.5 percent) experience fatal MIDs (Hensel and Mitchell 2017, 132). Territorial claims (41.8 percent) and maritime claims (27.3 percent) are militarized much more frequently. Aaron Wolf and his colleagues have categorized conflict and cooperation events for more than 260 international rivers, finding more frequent cooperation over rivers than conflict, including the signing of several thousand river treaties (Hamner and Wolf 1998; Wolf 1998). This reflects in part the active participation by regional organizations to resolve river claims as opposed to territorial claims handled with few institutional approaches and maritime claims addressed through a global organization (United Nations Convention on the Law of the Sea) (Owsiak and Mitchell 2019). Lee and Mitchell (2019) show that the presence of other (nonrenewable) resources can alter the likelihood of conflict in river basins. Situations where downstream states produce oil that can be traded with an upstream state (with no oil production) are more peaceful than situations where both states in the basin are energy producers.

WATER WARS? MARITIME CONFLICTS

When one thinks of modern water conflicts, they often involve clashes at sea over maritime boundaries, resource rights, and navigation through strategic choke points. China's maritime disputes with its neighbors over territorial (e.g., Senkaku/Diaoyu [Wiegand 2009; Fravel 2010], Paracel) and maritime (e.g., Spratly Islands, Gulf of Tonkin) areas highlight these modern "water wars," as well as the resurgence of conflict between long-time rivals such as Russia-Ukraine and Turkey-Greece over contested maritime (and territorial) spaces. Maritime issues have become central to states' security interests (Klein 2011).

While river issues have been managed primarily through diplomacy, maritime issues involve more frequent threats, displays, and uses of military force. Naval and coast guard ships often protect fishing vessels in contested waters (e.g., US Navy protecting fishermen off the contested coastal areas of Ecuador and Peru) and sail through contested areas to assert freedom of navigation rights. For example, the United States conducted several freedom of navigation operations in 2020 to challenge China's maritime claims around the Spratly Islands in the South China Sea (part of the nine-dash line). Greece mobilized its naval fleet in the Aegean Sea in 2020 to disrupt Turkish efforts to survey for oil and gas in a contested area of the Greek continental shelf. When maritime claims also involve territorial issues (e.g., ownership of an island like the Falklands), the chances for MIDs and wars are much higher (Mitchell 2020).

Of the 270 dyadic diplomatic maritime claims coded by ICOW from 1900 to 2010, more than 25 percent experience at least one militarized dispute (Hensel and Mitchell 2017; Mitchell 2020). Maritime claims involve competition over renewable (e.g., fisheries) and nonrenewable (e.g., oil, minerals) resources. The creation of offshore oil rigs in 1947 significantly increased the risks for MIDs over ICOW maritime claims with contested oil resources (Nyman 2015). Interestingly, however, studies have shown that renewable resources spark more severe interstate conflicts. Nemeth et al. (2014) examine different components of the ICOW salience scale for maritime claims[8] and find that migratory fish stocks increase the risks of MIDs more than other maritime salience factors. The movement of renewable resources across boundaries leads countries to defend resource rights up to and beyond exclusive economic zone (EEZ) boundaries. Canada, for example, fired upon a Spanish trawler in March 1995 fishing 218 nautical miles offshore in the halibut-rich Grand Banks (Song 1997), competing over a fish species that is highly migratory. Maritime conflicts between democracies are quite common and include maritime disputes between the United Kingdom and Iceland (Cod Wars), Canada and the United States (e.g., Gulf of Maine, Beaufort Sea, Juan de Fuca Strait/Dixon Entrance, Northwest Passage), and France and Spain (Bay of Biscay). Among all pairs of states with opportunities for maritime claims, jointly democratic dyads have the highest probability of experiencing diplomatic conflicts over maritime areas, creating a future flashpoint issue between democratic regimes (Daniels and Mitchell 2017).

Fisheries have been connected to conflicts in other studies as well (Bailey 1996; Asgeirsdottir 2007; DeSombre and Barkin 2011). Mitchell and Prins (1999) analyze all jointly democratic MIDs from 1946 to 1992 and discover that fishing resources are involved in 25 percent of the ninety-seven democratic disputes in this period. Fishing resources are also connected to smaller-scale conflicts, piracy, and civil wars. Poverty and favorable geographical conditions fuel "fish wars" in Southeast Asia, a region where "the distinctions between fishermen, smugglers, illegal traders and pirates can be blurred and the exercise of extreme violence is not uncommon" (Murphy 2010, 40). Declining small-scale fisheries catches increase conflict between different groups, especially between small-scale and industrial fishers (Pomeroy et al. 2007). Illegal, unreported, and unregulated (IUU) fishing has increased significantly as countries with large fishing fleets encroach on other states' claimed EEZ areas, sparking conflicts with local fishermen and governments protecting their sovereign resource rights. Somali fishermen saw their fish catches decline by 20 percent from 1992 to 2000, with at least ninety thousand tons of fish caught in Somali waters through IUU, encouraging Somalis to turn to piracy for their livelihood (Arky 2010, 8).

Interstate conflict over maritime zones and cross-border rivers can be explained by many of the factors that help us understand traditional paths to war, such as the steps to war in territorial disputes (see Hensel and Goemans, chapter 1 herein; Mitchell and Vasquez, chapter 19 herein). As Hensel et al. (2008) show, diplomatic issues that are higher in salience and that involve a longer history of conflict (rivalry) are more likely to experience MIDs, whether they are territorial, maritime, or river issues. They also find that joint democracy helps reduce the militarization of territorial and river claims, but it has no effect on maritime conflicts.

On the other hand, capability asymmetries (power preponderance) reduce the likelihood of MIDs for territorial claims but not for maritime claims, while naval capabilities (especially for the revisionist state) influence militarization of maritime claims but not territorial claims (Mitchell 2020). These results demonstrate the utility in studying environmental conflict across multiple diplomatic issues, as some of the factors known to cause war are present in some issue conflicts but not others.

Diplomatic disputes over maritime resources have seen more militarization than river issues, but few water issues have escalated to "water wars." Where are the water wars that were predicted decades ago? One answer is that water issues have been handled more successfully with conflict management institutions than territorial issues (Owsiak and Mitchell 2019). Rivers in particular have been governed by a large number of bilateral and multilateral treaties (Wolf 1998), and many of these agreements establish river commissions to monitor escalatory situations (Zawahri 2008). The river treaty literature shows that institutional design matters for the success of peaceful conflict management and effective resource management (Ostrom 1990; Tir and Stinnett 2012). The United Nations Convention on the Law of the Sea has also been a very successful multilateral institution for encouraging third-party peaceful settlement and preventing the onset of new maritime claims (Nemeth et al. 2014; Owsiak and Mitchell 2019). Another answer is that territorial issues are higher in intangible salience than water conflicts, involving historic homelands, ethnic groups, and sacred sites, which make division of contested territories more difficult (Hensel and Mitchell 2005). In this regard it is easier to find peaceful solutions to water conflicts than land conflicts. Finally, we may be looking in the wrong place for water conflicts, as clashes in the Tigris-Euphrates basin have shown. While Turkey, Iraq, and Syria have certainly experienced many interstate river claims and escalation to MIDs in a few cases, more conflict has occurred internally. Turkey's Greater

Anatolia Project (GAP) to build two dozen dams displaced thousands of Kurds inside the state. Decreased water supplies in the basin and severe drought issues contributed to the outbreak of the Syrian civil war (Beck 2014). Future research would benefit from a focus on how environmental conflicts influence interstate and intrastate violence and exploration of the transnational dimensions of such conflicts. Climate changes are projected to displace as many as a billion people by 2050 (Barnett and Webber 2009), for example, and these climate migrants could spark new interstate conflicts. Climate change will also put pressure on existing freshwater and fishery resources, change the baselines for existing EEZ maritime boundaries, and open up new sea lanes and resource areas in the Arctic that may spark conflicts. In the next section we discuss how climate change and volatility in weather patterns could act as potential triggers for interstate conflicts.

CLIMATE CHANGE AND INTERSTATE CONFLICT

Conflict scholars examine how climate change and climate variability influence the risks for interstate disputes and wars.[9] Climate change could lead to war through several paths. First, countries might become more territorially revisionist as resource stocks and productive agricultural lands are diminished by higher temperatures and less rainfall (Busby 2008). Second, rapid climate change can increase uncertainty about property rights, giving declining powers incentives to be more territorially revisionist (Gartzke 2012). Guo's (2007) survey of more than two hundred disputed areas since World War II finds support for this type of hostile lateral pressure, as resource scarcity at home prompts many states to make claims to neighboring states' territories. Third, greater volatility in weather patterns can create uncertainty about future resource stocks and encourage more revisionist behavior by countries to protect natural resources (Schmidt, Lee, and Mitchell 2021.). For example, Egypt has become more aggressive in its defense of water supplies on the Nile as upstream states like Ethiopia have constructed dams and as population growth and drought within Egypt have put greater demands on water supplies.

On the other hand, global warming over the past few centuries may have been associated with the decline of interstate warfare (see Braumoeller, chapter 16 herein), reflecting the economic efficiencies that globalization creates relative to territorial conquest (Brooks 1999). Zhang et al. (2007) examine climate change and conflict over a long historical period (1400–1900). Their initial research focused on China and found that warfare periods coincided with cold spells due to decline in agricultural production, forced migration, and violent competition between groups for food and other resources.

The authors discovered similar patterns in Europe, with periods of war highly correlated with periods of cold temperature anomalies. Tol and Wagner (2010) confirm these findings of colder periods experiencing more warfare (due to displacement from bad harvests) using rolling regression time series models of temperatures and conflict. However, they find the effects are getting weaker over time, such that it is difficult to infer that warmer climate periods of the modern period will avoid warfare. Analyzing the systemic relationship between global warming and war, Gartzke (2012) argues that it is important to control for economic development in these models, since growing interdependence could increase the costs of territorial conquest (relative to trade). Using a summed measure of MIDs in the system from 1816 to 2000 yearly and a measure of global temperature anomalies (from the 1951–1980 baseline), Gartzke's analyses show that temperature anomalies have little effect on fatal MIDs, while economic development levels better predict the decline in conflict over this period.

Schmidt et al. (2021) find similar patterns when studying the effect of climate factors on diplomatic conflicts (using ICOW data), showing that heat waves reduce the likelihood of issue claim onset or militarization (or the effects are insignificant). All these empirical studies suggest that higher average temperatures from long-term climate change are not likely to contribute to more frequent warfare, reflecting perhaps the improving adaptation of human society to climate change.

On the other hand, there is more empirical evidence that short-term and long-term changes in precipitation can fuel interstate conflicts. Devlin and Hendrix (2014, 28) point out that climate change can make some countries wetter or dryer on average, but also that climatic variability is greater, with more frequent floods or droughts and greater yearly variability in precipitation. They find that higher precipitation levels reduce the chances for MIDs, while greater variance in rainfall increases conflict risks.

Climate variability also increases the salience of contested issues and generates uncertainty about states' ability to comply with existing resource treaties (Schmidt et al. 2021). Climate changes in the Syr Darya river basin, for example, destabilized agreements between riparian states and increased diplomatic conflict (Bernauer and Siegfried 2012). Schmidt et al. find that precipitation deviations from long-term averages and greater short-term volatility in rainfall increase the likelihood of ICOW issue claim initiation, especially for challenger states (revisionists), and that precipitation volatility also increases the risks for claim militarization. Curvilinear patterns for climate volatility are also found in the intrastate violence literature, with conflict occurring more often with floods or droughts than average rainfall (Hendrix and Salehyan 2012).

CONDITIONS OF ENVIRONMENTAL WAR

Does environmental change always lead to interstate conflict? Literature suggests that when faced with environmental stress, the vulnerability of a state or a society could have conditional effects on the outcome (Kasperson et al. 2014). The vulnerability of a society to environmental changes, defined as "the extent to which a natural or social system is susceptible to sustaining damage from climate change" (IPCC 2001, 99),[10] is greatly shaped by its natural conditions, socioeconomic development, and capacity to cope with the stress. It depends on how sensitive a state is to experience insecurity due to environmental changes and how much a state could cope with it, usually referred to as susceptibility and adaptability of a state, respectively (Jones, Mattiacci, and Braumoeller 2017). Susceptibility is largely determined by a state's economic development and geographical conditions, while adaptability depends on economic development, domestic institutions, and international institutions.

Economic Development

Compared with developed countries, developing countries are more susceptible to environmental changes because agriculture is usually an important source of income in developing countries (Homer-Dixon 1994). Extreme weather and environmental degradation could greatly affect agricultural production and national income in developing countries (Miguel, Satyanath, and Sergeni 2004), slowing economic growth and increasing human insecurity. One may argue that free trade could mitigate the problem of resource scarcity, but developing countries are likely to be disadvantaged in international trade as well. For example, Tir and Diehl (1998) point out that poor countries may not have enough foreign currency for international transfer, or simply have little or nothing to trade for the resources needed. Hence, when faced with declining agricultural production caused by environmental change, developing countries are unlikely to benefit from international trade in the long term.

Environmental changes disproportionally influence developing countries and thus increase the level of economic loss both directly and indirectly. Directly, environmental changes result in reduced access to natural capital and thereafter national and household incomes; indirectly, developing countries, which are less able to provide welfare or aid to the affected population due to a lack of an effective social security system, can reinforce the problem of declining agricultural production (Barnett and Adger 2007). A sudden economic loss induced by environmental changes, which could lead to migration, population displacement, and competition over resources, could result in civil conflict (Koubi et al. 2012). Similarly, environment-induced economic loss could also lead to interstate conflict if transboundary migration occurs.

Reuveny (2007) identifies three cases where international migration leads to interstate conflict (Ethiopia-Somalia, El Salvador-Honduras, and Mauritania-Senegal), and he also finds that migrating to developed countries is unlikely to cause interstate conflict. An implication is that developing countries, being more susceptible to environmental changes, could produce more environmental refugees, which results in increasing tensions and potential conflict with neighboring countries if the neighboring countries are also resource constrained.

Domestic Institutions

Developing countries are more vulnerable to environmental changes, but whether the induced resource scarcity leads to conflict is also dependent on their adaptability. One important component of adaptability is domestic political institutions. Literature on domestic institutions focuses on state capacity and regime type, arguing that democratic institutions and strong state capacity could reduce the risk of environment-induced conflict (Koubi et al. 2012; Jones et al. 2017). Tracing back to Sen (1990), democratic institutions including regular elections and freedom of speech help to eliminate famines. Sen (1990) argues that in democracies with free media and active oppositions, the government would face severe punishment if it does not take necessary measures to prevent a famine. Similarly, the consequence of environment-induced resource scarcity, especially when it comes to life essentials such as food and freshwater, could be mitigated by democratic institutions. For instance, Koubi et al. (2012) find that climate-induced economic shocks are less likely to cause civil conflict in democracies.

Another important component of domestic institutions is state capacity. Jones et al. (2017) argue that states with coherent political institutions can mitigate the problem of resource scarcity through social redistribution and economic intervention. Autocracies are more likely to mobilize the military to deter potential unrest. Ignoring the role of government in managing resources and resolving potential conflict could lead to incorrect inferences (Salehyan 2008). In general, strong states are more capable of responding to environment-induced socioeconomic insecurity through either a social security system or repression. In the context of interstate conflict, democracies with strong state capacities are the least likely to experience environmental war because they could cope with environment-induced resource scarcity and the potential economic loss better. However, as we discussed earlier, some issues like maritime boundaries and resources can act as a flashpoint between democratic countries, in large part because they have greater levels of development (and thus larger fishing fleets) and demand for food (Daniels and Mitchell 2017).

International Institutions

Apart from domestic institutions, international institutions play an important role in international conflict management as well (see Powell and Wiegand, chapter 11 herein). Research on the interactive effect between international institutions and environmental war largely focuses on water issues. Overall, institutionalists believe that international institutions help promote cooperation and resolve disputes before they escalate into militarized conflict. In terms of international resource conflict, international resource management institutions can develop and evolve according to the changing resource conditions and their conflict management capabilities are further strengthened (Giordano, Giordano, and Wolf 2005). States' interdependence on river issues is likely to create the supply of international regimes, which increases the level of coordination (Dinar 2009). As noted earlier, disputes on shared rivers can arise from issues of navigation, fisheries, hydropower, and water quality, as well as agricultural usage. Shared rivers involve distribution and common goods problems, where international institutions could be helpful by providing more information transparency, lowering transaction costs, establishing enforcement systems, and introducing reputational concerns into states' utility functions.

Empirical evidence supports the argument that international institutions contribute to reducing water-related conflict. For example, Tir and Stinnett (2012) find that although water scarcity increases the risk of conflict, international agreements help lower this risk. Similarly, Brochmann and Hensel (2009) argue that river treaties significantly increase the likelihood of negotiation over river claims. Dinar et al. (2015) look at different institutional designs and contend that countries governed by highly institutionalized treaties are more likely to cooperate, and treaties with flexible and specific water allocation mechanisms are more conducive to international cooperation. To sum up, international institutions alleviate river disputes and promote cooperation, although some institutional designs could be more effective than others. The large number of cooperative treaties in international river basins helps explain why river issues are the least militarized issues coded by the ICOW project.

This section has illustrated that the relationship between environment and conflict could be conditional. The intervening factors can be socioeconomic development, domestic political institutions, and international institutions. Overall, low-income countries, non-democracies, weak states, and states less involved in international organizations are more vulnerable to environmental war. That said, some scholars argue that the real causal mechanisms linking environmental changes and conflict are not clear or robust. The following section briefly introduces some counterarguments of this field.

COUNTERARGUMENTS

A number of counterarguments, both theoretically and empirically, have challenged the relationship between environmental changes and war. Some scholars argue that environmental changes only have weak or indirect effects (Raleigh and Urdal 2007; Theisen, Holtermann, and Buhaug 2012; Koubi et al. 2012). On one hand, whether environmental changes cause conflict depends on socioeconomic conditions and political institutions, and thus there might not be an unconditional or direct relationship between environmental changes and conflict (Koubi et al. 2012). On the other hand, even if there is an indirect effect, this effect appears to be weak empirically (Sakaguchi, Varughese, and Auld 2017). Some scholars further suggest that in conflict research other factors such as socioeconomic development and state capacity are substantially more influential and should be of higher research priority (Theisen et al. 2011; Mach et al. 2019).

For example, Feitelson, Tamimi, and Rosenthal (2012) examine the tension over freshwater between Israel and Palestine and argue that a resource-induced conflict is unlikely. They point out that Israel, due to its strong state capacity and high economic development, is able to address the potential damage caused by climate change through desalination and other public polices and thus mitigate the problem of resource scarcity. They further suggest that although climate change could affect Palestinian farmers, urban populations are unlikely to be affected. Hence they conclude that climate change is unlikely to cause conflict between Israel and Palestine directly.

Similarly, Theisen, Holtermann, and Buhaug's (2013) research on drought and civil conflict in Africa reaches a similar conclusion, indicating that drought has no direct relationship on conflict onset. Instead, political exclusion plays a far more influential role. Likewise, Koubi, Bernauer, Kalbhenn, and Spilker (2012) question the causal pathways between climate change (deviations in temperature and rainfall), economic growth, and civil conflict.

Moreover, some scholars point out that the relationship between environment and war is not empirically robust or consistent (Buhaug 2010; Theisen, Gleditsch, and Buhaug 2013; Koubi et al. 2014; Wischnath and Buhaug 2014). Theisen et al. (2013) present a comprehensive review on recent studies on environment and conflict, including climate variability, natural disasters, and forced environmental migration. They find that empirical evidence in this field appears to be inconsistent and they argue that better theories and empirical modeling are needed. Similarly, Wischnath and Buhaug (2014), focusing on Asia, contend that there is little evidence supporting the climate-conflict argument. In their research empirical results are sensitive to research design, suggesting that the effect of climate variability on conflict found in other studies is not systematic or generalizable. Buhaug (2010) also challenges Burke et al.'s (2009) conclusion that warm weather leads to increased probability of civil war in Africa by applying a different research design, and he finds that Burke et al.'s (2009) results do not hold when alternative measures and model specifications are used. In short, a few scholars have pointed out that the effect of environmental changes on conflict is not empirically robust, and this field has not reached a consensus yet.

Civil war scholars have investigated many intervening variables between the environment and conflict, and they have done so in a much more systematic way than interstate scholars. Interstate conflict models that focus on the environment would benefit from the identification of factors that condition the relationships between resource competition, climate change, and war.

RESEARCH CHALLENGES

One reason that the relationship between environment and conflict is still debatable is that there are many research challenges in this field. One of the challenges is to disentangle the causal chain (Nordås and Gleditsch 2007). The most common causal chain, suggested by the literature of this field, starts from environment-induced resource scarcity, such as a shortage of food, freshwater, and other life essentials. Consequently, people who are affected by resource scarcity may fight over the remaining resources, or they have to leave the area. Migration could further result in resource competition in the receiving area as well as shift the demographic balance, creating conflicts between environmental migrants and host populations. However, testing this causal chain is extremely difficult.

First, many factors could interact with the damage caused by environmental changes and conflict. Some states are more vulnerable to environmental changes than others due to natural conditions and lower economic development, and whether environmental changes would lead to conflict also depends on domestic and international institutions. To draw conclusions on the underlying causal mechanisms between environment and conflict, one has to take these intervening variables into account. It is possible that environment does not have systematic or unconditional impact on conflict, but scholars need more comprehensive theories and better methods to incorporate interactive, contingent effects. Capabilities and institutions have different effects on escalation of diplomatic conflicts over rivers and maritime areas, for example, but we know less about how these factors interact with environmental changes. If climate changes and climate volatility generate greater uncertainty over resource stocks, this could increase states' incentives to capture neighboring territories and resources. By tracking environmental changes to the onset of diplomatic conflicts more carefully, we can learn more about the causal pathways by which the environment leads to war.

Second, though migration can potentially cause conflict, research finds that environment-induced international migration does not occur very often (Raleigh, Jordan, and Salehyan 2008). Gray and Wise (2016) also find inconsistent patterns of climate-induced migration across countries. This could be because the term *environmental refugee* is relatively new and the data are not fully collected, or because international migration is practically difficult for people affected by environment. Moreover, even if environment-induced migration does occur, it does not necessarily cause conflict. For instance, Reuveny (2007) examines thirty-eight cases of environment-induced migration across Asia, Africa, and America. He finds that nineteen out of thirty-eight cases exhibit conflict, among which only three cases are interstate conflict. In short, existing evidence suggests that more theoretical and empirical efforts are needed to demonstrate the causal mechanism between environment and conflict.

Third, another research challenge is methodological. As previously presented, empirical results of this field are mixed and sometimes contradict each other, and this situation could be partly attributed to methodological difficulties. Methodological difficulties, including endogeneity, nonlinear response, and poor controls (Burke, Hsiang, and Miguel 2015) could hinder scholars from observing real causal mechanisms. Among these difficulties, the endogenous problem probably receives the most attention because environment and human activity are hardly separable. This means that the causal pathways could go from conflict to environmental changes as well (see Miguel et al. 2004; Koubi et al. 2012). In fact, environmental changes, including environmental degradation, climate change, and extreme weather, are closely related to human activity, and environmental stress, militarized disputes, and political instability also reinforce each other (Salehyan 2008). Ignoring these intercorrelations and endogeneity implies that many models suffer from estimation bias and inaccurate forecasts.

On one hand, conflict can greatly lower the amount of available resources through accelerating environmental degradation such as the decline of arable lands and freshwater that results in further resource scarcity. Afterward the decline in available resources can cause a new round of conflict. On the other hand, state capacity and political stability also affect the quality of weather data collected from governmental stations, which is another potential source of bias. For example, in a recent research study, Schultz and Mankin (2019) argue that temperature data are usually collected from local governments, whose capacity has a great influence on the quality of the data. They find that measurement error of weather data widely exists in sub-Saharan Africa due to political instability and low state capacity. After correcting for measurement error, the results show that the effect of temperature anomalies on civil conflict almost doubles. Though measurement error in climate data due to political instability may be less severe in other regions of the world, this source of estimation bias cannot be completely neglected.

Another issue involves multicollinearity and the use of confounding controls, where the impact of environmental variables is absorbed by control variables such as agricultural production and population (Burke et al. 2015). These control variables could be affected by environmental changes and are therefore endogenously determined. Including these control variables in the model may lead to incorrect predictions as well. Another methodological difficulty is the possibility of nonlinear relationships. Nonlinear relationships are more commonly seen in research on water conflicts (Dinar 2009; Schmidt et al. forthcoming), but not fully accounted for in other research on environmental conflict. Hidalgo et al. (2010) demonstrate an invert-U shape relationship between rainfall deviation and agricultural income in Brazil, implying that the influence of climate change on economic growth may not be linear as some political scientists predict. However, most research on climate and conflict does not take nonlinear responses into consideration, which could also explain the inconsistent empirical evidence of this field.

FUTURE RESEARCH

Research on environmental causes of interstate conflict identifies several factors that increase risks for war, including resource scarcity, colder temperatures, and greater precipitation deviations from normal rainfall levels. Much of the interstate conflict literature focuses on water issues, such as water quantity or quality clashes in river basins or competition for oil, fish, and mineral resources in maritime zones. When these issues are highly salient, central to states' security concerns, and connected to territorial disputes, the likelihood for repeated MIDs and wars is high. Yet the lack of water wars over freshwater shows that countries prefer to find cooperative and equitable solutions for sharing a resource essential to all human life.

Future research could help address questions raised by this literature. First, we believe that interdisciplinary research would be beneficial for better theory building. Research on the relationship between the environment and conflict asks scholars to have a deep understanding of ecology, hydrology, and other social science disciplines. For example, a comprehensive analysis of water security issues requires profound knowledge from hydro-climatic aspects, eco-hydrological aspects, economic aspects, and sociopolitical aspects (Ludwig et al. 2011). Interdisciplinary cooperation could be a bridge connecting scholars of different fields, which contributes to scientific synergies. It allows researchers to share theories and data, develop innovative methods, discuss research agenda, and provide opportunities for novel perspectives. For policymakers, facilitating interactions between natural science and social science will also increase the adaptabilities to future challenges of environmental changes.

Second, the quality of empirical analysis could be significantly improved through fine-grained, georeferenced data. Geospatial data could provide more information at the local level, allowing for in-depth theory testing (Raleigh and Urdal 2007). This is important because environmental changes, such as soil degradation and water scarcity, usually happen at a local level instead of a state level. Using states as the unit of analysis may hinder scholars from observing subnational heterogeneity and lead to inaccurate research conclusions. In fact, geospatial data has benefited other international fields, such as civil war research, but it is largely neglected in environmental war research. This is partly because geospatial data on environment are very limited so far. Though raster data can be found from the United Nations Environment Program, which Raleigh and Urdal (2007) use, this data set is far from complete. It is important for researchers to collect high-quality, relatively complete, and recent geospatial data to conduct future research, and this task requires interdisciplinary cooperation as well.

Finally, environmental factors could be connected more explicitly to existing war theories. Do countries have diversionary incentives for war when facing several environmental shocks at home? When natural disasters occur in the context of rivalry, it could create an opportunity for a leader to divert the public's attention away from a poor government post-disaster response. How do relative capabilities influence the chances for war over environmental vs. nonenvironmental issues? The literature we reviewed shows that power preponderance may help prevent interstate conflict for some issues (territorial) but not others (maritime) and that different types of capabilities alter the chances for war distinctly (e.g. naval vs. air power). Why do democracies contend over some environmental issues (e.g. maritime) more than others (e.g. rivers) and how does this affect the prospects for democratic peace in the future? Why do regional and international organizations and treaties vary across different types of diplomatic issues and even among environmental issues? By making greater connections between existing theories of war and environmental variables, interstate war scholars can emulate the research by civil war scholars to better tease out the causal pathways by which resource scarcity and climate change may lead to violence. Given the accelerated pace of climate change, natural disasters, and resource scarcity for renewable (and nonrenewable) resources, such a focus in the war literature is much needed.

NOTES

1. For an excellent review of the earlier literature on environmental conflict, see Gleditsch (1998).

2. See Koubi et al. (2014) for a review of the literature connecting renewable and non-renewable resources to conflict.

3. Climate change refers to long-term changes in "mean conditions," like increases in global temperatures, changes in annual precipitation, or rising sea levels, and changes in the intensity or frequency of natural disasters (Barnett and Adger 2007, 640). Climate variability (Koubi 2019) reflects short-term deviations of weather patterns from long-run means (e.g., floods, droughts, heat waves).

4. Our review draws upon civil war research, but we do not provide a comprehensive analysis of that literature. See Barnett and Adger (2007), Nordås and Gleditsch (2007), Salehyan (2008), Theisen et al. (2013), Burke et al. (2015), Sakaguchi et al. (2017), and Koubi (2019).

5. Hegre argues that population should be a standard control variable in interstate conflict models like it is in civil war models because size effects alter the relationship between other variables (e.g., power parity) and war.

6. Dinar's dependent variable is a measure of river basin cooperation, thus the relationship he observes between water scarcity and river cooperation is an inverted U.

7. In mixed relationships states are both upstream and downstream on the same river or the river forms the border. In sideways relationships two states share the same river basin but no water flows from one state to another (Brochmann and Gleditsch 2012, 522).

8. This includes homeland territory (vs. offshore), fishing (migratory), resource basis for claim, oil, strategic chokepoint, and territorial claim (Hensel et al. 2008).

9. Much of this literature focuses on intrastate conflict, which we do not cover in detail. In addition to the sources listed in endnote 4, see also Hendrix et al. (2014) for a review of the environment-conflict civil war literature.

10. See the Intergovernmental Panel on Climate Change's 2001 Synthesis Report, https://www.ipcc.ch/site/assets/uploads/2018/05/SYR_TAR_full_report.pdf.

Chapter Fourteen

Leaders and War

Scott Wolford

Leaders, their incentives, and their personalities suffuse popular explanations for war. The War of 1812 was "Mr. Madison's War," the Mexican-American War "Mr. Polk's," and Woodrow Wilson decided almost unilaterally when and how the United States joined the First World War. Saddam Hussein attacked Iran in 1980 in hopes of destabilizing Ayatollah Khomeini's new theocratic regime. Popular accounts attribute Britain's response to Argentina's 1982 occupation of the Falklands to Margaret Thatcher's "iron" will. Bill Clinton's political opponents accused him of launching airstrikes on Iraq, Sudan, and Afghanistan to distract the public from domestic scandals. And it remains easy to debate whether a counterfactual President Al Gore would have launched the same Iraq War that George W. Bush initiated in 2003.

Popular narratives aside, the role of individual leaders in shaping patterns of war and peace has only recently received mainstream scholarly attention. This is due in large part to Waltz's (1959) influential conceptualization of three "images" of international politics: human nature, political institutions, and the international system. Waltz favored the third, which emphasizes strategic interaction in the absence of meaningful restraints on the use of military power. Yet how states leverage that military power, as well as what issues they pursue, against whom, and whether they ultimately go to war over those issues depends on how they are organized (i.e., their political institutions) and *who leads them*. This chapter surveys how the recent focus on leaders as units of both analysis and observation has advanced—and can continue to advance—our understanding of the causes of war.[1]

States change leaders more often than they change political institutions, neighbors, alliances, and rankings in the global power hierarchy (Chiozza and Goemans 2011, 200). Figure 14.1 plots the number and proportion of states experiencing at least one instance of leader change per year from 1875 to 2015.[2] In some years dozens of states see incumbents replaced, often as a result of war,

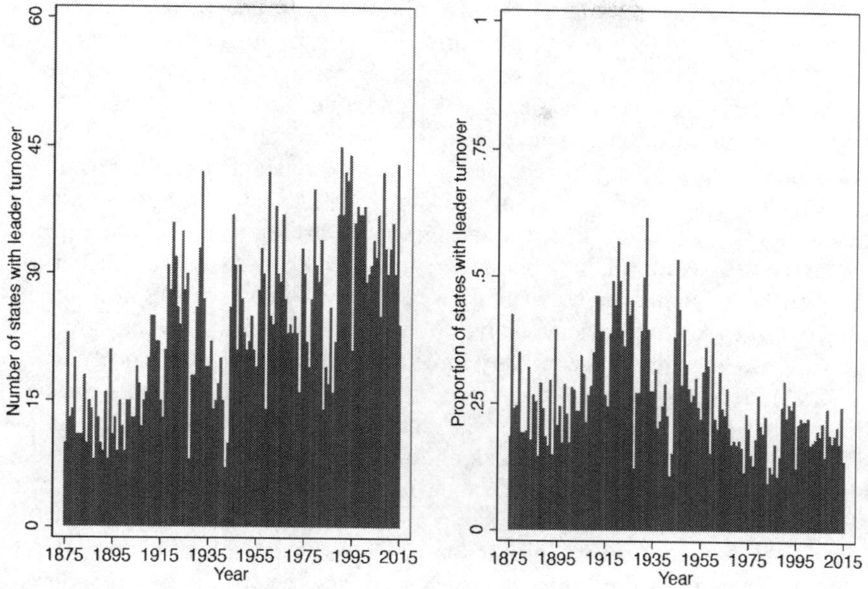

Figure 14.1. Number and Proportion of States with Leader Turnover by Year: 1875–2015.

whether at the hands of conquerors or their own constituents (Bienen and van de Walle 1991; Chiozza, Giacomo, and Hein E. Goemans, 2004b). On average, the international system has seen roughly twenty-four states experience leader turnover per year, which is good for about one quarter of the international system per year.[3] But without data of scope comparable to that available for system- and state-level concepts (see Bennett and Stam 2004), the relationship between leader-level variables and war could at best be measured in case studies. Small samples, however, bias analyses in favor of finding that leaders *do* matter, attenuating the metaphorical regression line, overweighting case-specific and idiosyncratic factors (like individuals' foibles and personalities) and underweighting systematic factors (like the distribution of power). This lack of data also limited the scope and stifled the growth of leader-level theorizing, as scholars looked to background, psychology, or ideology with limited examination of how leaders interact with their domestic and international political environments.

Recent years have seen this void filled by data describing (a) the means by which leaders enter and leave office (Goemans, Gleditsch, and Chiozza 2009), (b) leaders' domestic coalitions (Mattes, Leeds, and Matsumura 2016), and (c) their personal biographies (Ellis, Horowitz, and Stam 2015), allowing scholars to estimate models of conflict that place leader-level variables like time in office, political support, and personal background on the same footing as traditional state- and system-level variables like regime type and relative power. Leaders are no longer relegated to the error term of quantitative international relations (IR). Indeed, Chiozza and Goemans (2011, 201) show that leaders account for roughly one-third of the variation in yearly conflict outbreaks over the twentieth century—a larger share than variables measured on either countries or international systems.

These new data on leaders have also allowed us to choose the correct unit of observation for leader-centric theories, like those in the diversionary war tradition, that depend explicitly on divergences between the private and public costs and benefits of war (Chiozza, Giacomo, and Hein E. Goemans, 2004b, 604). Ensuring consistency between units of analysis and units of observation is no small thing: failing to account for leaders' private and public incentives can encourage misleading inferences about when and how leaders use war to ensure political survival (Chiozza and Goemans 2011) and about whether and where reputations form in international politics (Wolford 2007, 772–75; Wu and Wolford 2018). New theories have followed that try to make sense of how leaders constrained by institutions and relative power hope to stay in office, and how they are variously inclined to use military force to fit into the international system.

The modern literature on leaders and war has developed around two questions.[4] First, how do political survival motives influence leaders' decisions for war? Second, how are states' decisions for war shaped by which leader holds office? The first explores how political institutions shape the attractiveness of war for survival-minded leaders, abstracting away from individual differences. The second abstracts away from institutions to highlight how variation in personal attributes like personal experience, psychology, and ideology create variation in preferences over war and peace. Individual leaders are the units of analysis and observation in both approaches, but political institutions carry the explanatory weight in the first and individual preferences carry it in the second.

Differences between the two approaches are easily highlighted by how they handle risk acceptance: the political survival approach shows how domestic institutions can encourage otherwise cautious but survival-minded leaders to accept the gamble of war (Goemans and Fey 2009), and the personal biography approach explains why some leaders accept the risks of war regardless of political institutions (Horowitz, Stam, and Ellis 2015, chapter 1). That distinction aside, the same variables can represent different concepts in each approach; time in office can represent political security (Chiozza, Giacomo, and Hein E. Goemans, 2003, 2004a) in the political survival approach and either experience (Potter 2007) or accumulated information (Wu and Wolford 2018) in the personal attribute approach. This division is ideal-typical, and some work surely blurs it, but it's useful as an organizing principle.

In what follows I summarize and assess the contributions of the political survival and personal attribute approaches to our knowledge about the outbreak of war.[5] Both bodies of literature have generated more stylized facts than we can currently explain, and I highlight the need for theory to keep pace with the rapid production of empirical generalizations. I then describe a route forward for leader-centric research that integrates leader-level variables with coherent, strategic theories of war—several of which are available off the shelf. Adopting such a theoretical disciplining device, one that connects variables measured on leaders to well-understood concepts in theoretical models of war, is an important next step in showing not only *that* leaders matter but *why* they matter. I close with an argument that research on leaders and war will advance only to the extent that it heeds Waltz's (1959) warning that we not ignore the strategic interaction that defines the international system—a critical element in placing leaders' personal attributes and political incentives on sound substantive footing as factors explaining war and peace.

STAYING IN OFFICE

The political survival approach isolates tensions between a leader's private interest in retaining power and the public's interest in avoiding costly, destructive war.[6] If conflict offers leaders opportunities unavailable in peacetime to improve their chances of political survival, then war may not be inefficient—indeed, it can look like a consumption good—for the very leaders who make ultimate decisions over war and peace (Chiozza, Giacomo, and Hein E. Goemans, 2004b). Leaders are "interchangeable" in this approach (Carter and Chiozza 2018: 9), an important step for isolating the effects of political survival incentives on war from alternatives.[7]

Differences in leader behavior are a function of political institutions, which link political survival to international conflict by (a) identifying who can remove the leader from office, (b) determining the ease with which those who wish to replace the leader can do so, and (c) shaping the credibility of commitments not to kill, jail, or otherwise punish deposed leaders. When success bolsters their survival and defeat undermines it, and when losing office means losing life and liberty, the private benefits of retaining office can, under some conditions, encourage leaders to embark on wars for which state-centric approaches cannot account. However, one key premise—which institutions render political survival more sensitive than others to war outcomes—remains contested.

DIVERSIONARY WAR

That leaders might use force abroad to divert attention from trouble at home, whether by taking advantage of in-group/out-group biases or by demonstrating competence, is a venerable idea in both political science (see Levy 1989) and everyday opposition politics.[8] Yet empirical support for such a pattern is (at best) mixed across different models, data, and measurement strategies. Some studies find evidence for diversionary wars, others no evidence. There's no empirical consensus on the question the way there is over, say, neighbors being more likely to fight one another than non-neighbors (Vasquez 1995).[9] But with the advent of improved data and theoretical precision, recent work has shown that diversionary war is less a function of electoral politics in democracies, an original source of inspiration for the theory, than of the personal costs of losing office in nondemocratic countries. When the private stakes of retaining office are low—i.e., when losing office is not that painful—the costs and risks of war loom large, especially when cheaper means of boosting popularity—like economic policy—are available (Arena and Bak 2015). But when the private stakes are high, as they are when leaders can expect exile, jail, or death after leaving office, "fighting for survival" or "gambling for resurrection" can make sense (Goemans 2000; Chiozza and Goemans 2011). Yet high private stakes for officeholding are typically associated with *non*democratic countries, where losing office means losing one's freedom and access to rents or (at the extreme) one's life (Goemans 2008).[10] In democracies, by contrast, losing office is on average less painful, and diversionary incentives may be too obvious to the public (or, at least, a pivotal voter) to reap political rewards (Mitchell and Prins 2004).

The key explanatory concept in theories of diversionary war is an incumbent's risk of losing office, which pushes leaders to engage in diversionary behavior.[11] Most work measures diversionary incentives with indicators of economic troubles like unemployment, inflation, and the misery index (Mitchell and Prins 2004; Mitchell and Thyne 2010) or changes in gross domestic product (GDP) (Leeds and Davis 1997; Johnson and Barnes 2011); domestic unrest (Dassel and Reinhardt 1999; Haynes 2016); approaching elections (Stoll 1984b); and declining support among either the executive's co-partisans (Morgan and Bickers 1992; Meernik and Waterman 1996) or the public at large (Ostrom and Job 1986). Yet most of these studies take the state, not the leader, as the unit of observation, introducing potential error that leader-centric work has sought to correct. Chiozza, Giacomo, and Hein E. Goemans, (2003, 2004a) address this problem by estimating the contemporaneous risk of losing office on a sample of leader-years, finding that (a) conflict is more likely when leaders are secure (rather than insecure) in office, but (b) political insecurity *is* associated with the outbreak of interstate conflict for nondemocratic countries.[12] Sampling on leaders also allows the authors to show that conflicts launched from positions of political insecurity are associated with better political survival prospects in nondemocratic countries, adding evidence for the theory's explanatory value that could not have been recovered from a sample of states.[13]

Recent studies show that diversionary wars are rarest when the conventional wisdom most expects them, yet space remains for theoretical development. Good theories tell us both why things happen *and* where to look for evidence that they happen. Tarar's (2006) formalization of diversionary theory inside a standard bargaining model of war, which explicitly models strategic interaction, is illustrative. The analysis shows that the appeal of diversion, which he conceptualizes as a means of proving foreign policy competence, depends on the strength of the potential target; beating a weak opponent does not reveal much, while beating a stronger one signals competence but is less likely to come off successfully. If diversion is to happen, it has to be against a rough military equal—and not all states can just choose a target given that it's typically hard to project force outside one's immediate neighborhood (Lemke 2002, chapter 4). The occurrence of diversion should also be conditional on the anticipated costs of war, the value of the international status quo, and (critically for the personal attribute approach) who is expected to replace the incumbent.[14]

Finally, Tarar shows that strategic conflict avoidance, by which foreign states try not to run afoul of leaders with diversionary incentives (see, e.g., Smith 1996a; Leeds and Davis 1997), may not be all that prevalent as an explanation for the absence of a relationship between political insecurity and conflict. First, dissatisfied foreign opponents, like the rivals in Mitchell and Prins's (2004) story, can still profit from challenging an incumbent with diversionary incentives. And second, when the benefits of officeholding are sufficiently large, then no deal can prevent the incumbent from launching a war. As I discuss later in this chapter, exercises like Tarar's are useful for clarifying when and where we should be able to observe leader-specific processes in the historical record.

DOMESTIC INSTITUTIONS AND WAR

The political survival approach also promises an improved understanding of the link between domestic institutions and war. Assuming that leaders wish to stay in office, regime types determine both (a) the extent to which leaders are removed after losses and retained after victories, which we can call accountability or vulnerability, and (b) the credibility of commitments not to punish them after removal. The empirical patterns to be explained, at least those relating to conflict behavior, are well established, from the dyadic democratic peace (Bueno de Mesquita et al. 1999), to high rates of democratic military victory in interstate war (Reiter and Stam 2002), to dispute initiation rates among autocratic regime types, some of which are indistinguishable from democracies (Weeks 2012).[15] Yet this cluster of studies does not make a common set of assumptions about which institutions make political survival more sensitive to war outcomes, and therefore which leaders' decisions are driven by and which are free from concerns about retaining office. The challenge in moving this literature forward, I argue, lies in rethinking how we think about regime types and accountability.

The conventional story has leaders whose political survival is more sensitive to war outcomes giving battle with greater selectivity and making more substantial war efforts when they do fight, all in the service of avoiding political punishment—i.e., removal from office. Some scholarship identifies high levels of accountability with democracies (e.g., Bueno de Mesquita et al. 1999; Reiter and Stam 2002). Weeks (2008, 2012), however, argues that some autocrats—especially those in civilian regimes with strong institutions—are equally accountable. In analyses of conflict initiation, she finds that democrats and politically accountable autocrats are indistinguishable; personalist dictators and military regimes, though, appear uniquely violent. Debs and Goemans (2010) establish a similar set of patterns but explain them differently, assuming that autocrats face more painful post-tenure fates and that democrats' political prospects are *less* sensitive than autocrats' to war outcomes. Both sets of assumptions imply similar sets of facts—though Weeks (2012) is largely silent on the democratic peace, in contrast to Debs and Goemans (2010)—but they proceed from different assumptions about the sensitivity of democratic leaders' political survival to war outcomes. How can we adjudicate between them?

Scholars have attempted to answer this question by modeling survival prospects as a function of war outcomes. Challenging the standard account, Chiozza, Giacomo, and Hein E. Goemans, (2004b) find that *autocratic* leader tenures are more sensitive on average to war outcomes than democratic tenures. Debs and Goemans (2010) replicate this pattern and explain it by noting that the costs of removing democratic leaders—and thus their chances of survival after any war outcome—are always low, while defeat can change the costs of removing an autocrat more substantially. Croco and Weeks (2016), however, argue that analyses must account for both culpability for the war and vulnerability to removal, generating a set of patterns in which leaders in democracies, as well as autocracies with strong institutions, are equally sensitive to war outcomes on average. Autocrats overseeing weak institutions, meantime, are the least sensitive. Finally, Carter (2017) shows that whether observable differences exist between democratic and autocratic survival depends on the extent of military mobilization; only for war efforts large enough to distort politically popular domestic spending, like the social safety net, do we observe differences across regime type. And when we do, democracies come out the more sensitive in the empirical record.[16] Which regime types (if any) pose greater risks for leaders who lose wars or promise greater rewards for leaders who win them remains an open question.[17]

How we specify models of political survival shapes how we explain the relationship between regime type and leaders' decisions over war and peace. Given that competing theories make similar predictions about the democratic peace and the conflict behavior of autocratic regimes, it makes sense to look for places where predictions diverge. Yet several obstacles stand in the way of making such a judgment. First, Debs and Goemans (2010), Croco and Weeks (2016), and Carter (2017) each rely on different samples, definitions of war (and conflicts short of war), and operationalizations of conflict outcomes. These decisions entail assumptions that rest on equal logical footing, so it is not possible to tell *why* any two studies yield different results.[18]

Second, each analysis treats the occurrence of war as randomly distributed among the explanatory variables (regime type, culpability, vulnerability, mobilization, etc.), which no associated theory predicts to be the case. Croco (2011), for example, finds that culpable leaders do better on average in war and that non-culpable leaders do not get punished for losing, but the statistical model assumes culpability to be randomly assigned. Weisiger (2016), however, finds that autocracies facing worsening battlefield situations are more likely than democracies to replace culpable leaders with non-culpable leaders. This should make autocrats look *less* sensitive than they really are in observational data, indicating that selection bias may not work in the way that Croco and Weeks's (2016) interpretation requires, where democrats risk only defeats for which they will not be sensitive.[19] We might try to isolate close calls, where the outcome of the war was nearly a coin flip *ex ante*, but democratic advantage models (e.g., Bueno de Mesquita et al. 1999; Reiter and Stam 2002) predict that democracies should be underrepresented in "fair" fights. Unexpected or surprising war outcomes may also be useful (see Blimes 2009), but they would suffer from similarly biased distributions of regime types. Therefore, with endogenous rates of initiation shaping the distribution of war outcomes, it is not clear that extant work *can* recover each regime type's "true" sensitivity.

This is not to say that relative sensitivities are unimportant. But scholars can profit more from trying to *explain* relative sensitivities as part of the same process that explains the outbreak of war. The debate plays out over analyses of observational data, which all participating authors recognize to be plagued with strategic selection bias (cf. Schultz 2001a) even as they disagree over the direction in which it ought to work (Debs and Goemans 2010, 434; Croco and Weeks 2016, 593–94). Yet in whatever direction the bias works, the fact remains that each set of competing claims about political survival and conflict behavior predicts that evidence for relative sensitivity should be unobservable (i.e., off the equilibrium path).

Observed rates of conflict initiation and political survival are products of the same data-generating process. No participants in the debate dispute that, yet it should give us pause in taking observed rates of sensitivity as representative of underlying rates of sensitivity. It will prove more fruitful to develop competing theories further—say, by integrating them into explicit theories of war (is more later in this chapter)—to see where they differ in their observable implications, both of which would entail hypotheses about rates of initiation *and* rates of postwar survival. This approach would allow scholars to see which model's predictions do better across multiple outcome variables, including rates of both conflict *and* postwar political survival, giving us more information with which to update our beliefs about the link between political survival and war. It is time to let assumptions be assumptions and predictions be predictions.

WHO IS IN OFFICE?

The personal attribute approach abstracts away from institutions and political survival to isolate variation in leaders' individual biography, psychology, and ideology. Whether personal attributes render leaders willing to take greater risks (Horowitz et al. 2015), use force in general (Carter and Smith 2020), or stay the course through destructive conflicts (Dafoe and Caughey 2016; Kertzer 2016), they influence the leaders' war payoffs, which we can shorthand as their preferences.[20] If successive leaders of the same state evaluate war and peace differently, then leadership turnover can produce changes in foreign policy independent of political institutions. Therefore, it matters not only who leads but also how incumbent leaders' preferences differ from (a) their predecessors and (b) their likely successors (Wolford 2012, 2018).[21]

Leadership turnover can shape patterns of war and peace both after (*ex post*) and before it happens (*ex ante*). Leaders exercise wide agency in personal attribute research (Horowitz and Fuhrmann 2018, 2073), which helps explain variation within the same set of political institutions. Yet focusing on personal attributes also risks eliding strategic interaction in favor of idiosyncratic behavioral tendencies, a problem that looms large over studies of reputation, yet is only rarely addressed (Krcmaric, Nelson, and Roberts 2020).

LEADER CHANGE *EX POST*

Most personal attribute work isolates the *ex post* consequences of leadership turnover, explaining the onset of disputes and their escalation to war with reference to leaders' formative personal experiences (biography, e.g., Horowitz and Stam 2014); dispositions, pathologies, and personality traits (psychology, e.g., Yarhi-Milo 2018); and beliefs about how the world works and relevant political cleavages (ideology, e.g., Palmer, London, and Regan 2004; Saunders 2011; Bertoli, Dafoe, and Trager 2019). Unifying these approaches, Carter and Smith (2020) measure leaders' personal hawkishness as a function of biography, political preferences, and psychological traits with a Bayesian latent variable model.

Other personal attribute work explores the consequences of individual preferences in the context of leadership turnover, which brings to office new incumbents with incentives to bolster their reputations for either resolve or reliability (Wolford 2007; Gibler 2008; Dafoe 2012; Lupton 2018a, 2020; Wu and Wolford 2018). The first batch of studies has produced several stylized facts about which leaders are more likely than others to initiate militarized disputes, establishing that even if the content of foreign policy does not change when leaders take office, the means by which those policies are pursued manifestly does.[22] The second batch hints at a way to improve our ability to explain why particular leader attributes are associated with the outbreak of war.

Leader attribute research has established several empirical patterns. First, leaders see more conflict when they lack experience (Potter 2007; Bak and Palmer 2010), grow older in democracies and mixed regimes (Horowitz, McDermott, and Stam 2005), served in the military yet did not experience combat (Horowitz et al. 2015), are believed to be "mad" (McManus 2019), or participated in a rebellion (Colgan 2013; Horowitz et al. 2015).[23] Non-Western leaders who attended universities in Western democracies are also less likely to initiate military disputes than those who didn't (Barceló 2020). With respect to psychology, leaders are more likely to choose war when they value social capital (Yarhi-Milo 2018) or honor (Dafoe and Caughey 2016) and when they inherently seek excitement (Gallagher and Allen 2014).

Evidence for gender differences, which implicates both biology (McDermott and Cowden 2001) and *other* leaders' beliefs about gender (Reiter and Wolford 2019), is less consistent (Horowitz et al. 2015).[24] It's hard to tell if the apparent lack of difference between male and female leaders emerges because (a) there are too few observations of female leaders, (b) women who show traditionally masculine traits are the only ones to rise to power, or (c) women tend to come to power in violent international neighborhoods.[25]

Individual beliefs are also important; leaders representing right-leaning parties (Palmer et al. 2004; Bertoli et al. 2019) and those who see foreign threats as emanating from other states' politics (Saunders 2011) are uniquely willing to use force, and in the latter case expansive war aims, to realize their objectives.[26] Finally, Carter and Nordstrom (2017) leverage several measures of leader preferences to challenge the conventional understanding of how political survival incentives influence democracies' conflict behavior. When doves become lame ducks (i.e., legally barred from reelection), their rate of conflict initiation drops significantly. Hawks, though, are neither more nor less likely to initiate conflicts than lame ducks, suggesting that doves mask their preferences when they care about political survival, mimicking hawks in order to win elections (cf. Schultz 2005).[27]

The foregoing work focuses on *observable* elements of leaders' willingness to use force. Preferences, though, are subjective, *unobservable* to both other leaders and the analyst. That gives (a) leaders opportunities to misrepresent their preferences and

(b) opponents reason to doubt leaders' claims about their willingness to use force. Foreign states (as well as the analyst) can use observable factors like those identified earlier in this chapter to make guesses about who is likely to use force and who is not, but new leaders also come to power with private information over precisely how willing they are to wage war. Wolford's (2007) model of repeated crisis bargaining links this uncertainty over leaders' subjective estimates of the costs of fighting to the private information explanation for war (Fearon 1995, 395–401).

By tying reputations to leaders, not states, the model resolves puzzles about the apparent lack of connection between past actions and present outcomes (see also Dafoe 2012; Lupton 2018a, 2018b, 2020; Renshon, Dafoe, and Huth 2018). In stark terms, it shows that state-level work often looks for evidence with the wrong unit of observation (e.g., Press 2005). The key empirical implication for *ex post* turnover is that new leaders have incentives to fight, and their opponents have incentives to stake out aggressive bargaining positions that make new leaders more likely to fight—the former to cultivate a resolute reputation and the latter to force the revelation of information. This results in turnover-driven temporal cycles, such that disputes are most likely to escalate early in leaders' tenure and less so over time. Wu and Wolford (2018) recover the pattern empirically, finding it to be strongest among rival dyads and neighbors, precisely those states that expect sufficient future interactions to make reputation building worth the costs. The core insight about new leaders and uncertainty has held up across models of the duration of militarized disputes (Smith and Spaniel 2019), the onset of arms races (Rider 2013), and the imposition of economic sanctions (Spaniel and Smith 2015).[28] This empirical progress, it is worth noting, followed a process that linked leaders to a coherent, fully strategic model of war.

Caution is still warranted when it comes to interpreting the stylized facts generated by the *ex post* turnover literature. First, though the political survival literature is built around the idea that leadership turnover can be endogenous to conflict, personal attribute studies tend to assume that leaders come randomly into office—a potentially significant hurdle to inference when specific observable traits, as opposed to uncertainty around those traits, is a key explanatory factor. Krcmaric et al. (2020) identify several inferential hurdles posed by "personal biography" work, including (a) leader turnover and outcome variables caused by the same unmodeled factor; (b) exogenous changes that put new leaders into power *and* change the values of variables related to war, like the makeup of the domestic support coalition (Leeds, Mattes, and Matsumura 2016); (c) leaders with certain backgrounds being chosen by constituents in preparation for looming war; and (d) self-selection into socializing experiences, like military or educational institutions, that appear to shape the willingness to use force. Some studies are more vulnerable than others to these problems, but all should be wary of them.

Second, much of the empirical work on leader backgrounds models the initiation of low-level conflict, like militarized interstate disputes (Palmer et al. 2020), with leader backgrounds linked to a concept like risk attitude (Horowitz et al. 2015, 27–31). Yet absent a coherent theory of how disputes escalate to war—i.e., one that links leader biography, psychology, or ideology to either some bargaining friction or some identifiable parameter in such a model—it's not clear how to interpret these results. Hawkish leaders may initiate disputes, but are disputes initiated by hawkish leaders more or less likely than others to escalate to war? The answer is not straightforward: in an asymmetric-information crisis bargaining model (e.g., Fearon 1995), with one informed and one uninformed player, the probability of war decreases in observable elements of the informed player's resolve, yet increases in observable elements of the uninformed player's resolve, regardless of which side initiates the dispute.[29] This illustrates a simple yet important point:

> Individuals may be faced with uncertainty about the consequences of their choices, so their choices are not implied in any straightforward way by their preferences over final outcomes. (Wagner 2001, 5)

Different models give different answers to those questions based on the issues at stake and the bargaining frictions that cause war. Yet most empirical work in this branch of the literature has yet to take advantage of strategic theories of war, leaning instead on implicitly decision-theoretic accounts that elide the strategic interaction inherent to the conflict process. Answering these questions, specifying when certain attributes push leaders to war and when they do not, is the essential next step. The personal attribute approach will improve only to the extent that it can place the leader agency that is its strength inside theories of why leaders take their states to war.

LEADER CHANGE *EX ANTE*

Fewer studies consider the effects of anticipated leader change on the present chances of war. They are based on the very same insight that makes turnover important *ex post*—incumbent leaders are rarely bound by their predecessors to particular bargaining positions—yet they challenge our understanding of *how* leaders matter. Some even identify threats of omitted variable bias in studies of *ex post* leader change. Reputation work in the *ex post* literature recognizes that leadership turnover can create information problems, but the *ex ante* perspective highlights how anticipated leadership turnover can create commitment problems based on purely observable factors like those identified in the *ex post* literature. To wit, an incumbent's satisfaction with the status quo provides no reliable guide to their successor's satisfaction, and changes from one leader to another are both rapid (i.e., discontinuous) and potentially large enough to cause preventive war (Powell 2006; Wolford 2018). Ease of identifying the bargaining problem notwithstanding, this nascent branch of the literature suffers from obstacles to data collection and measurement. Intelligence agencies specialize in judgments over who is likely to succeed the leaders of rival states and how those successors are likely to behave (Wolford 2012, 517–18), but systematic data collection on the shadow cast by incumbents' successors does not exist, saddling this branch of the literature with some of the same problems of small-sample observation that plagued the literature before the arrival of extensive leader-level data.

Most work in this vein remains theoretical. Schultz's (2005) model of trust and cooperation identifies an incentive to keep leaders from dovish factions in office if it means forestalling the arrival of a hawk. Two related models by Wolford (2012, 2018), which like Schultz incorporate endogenous turnover, generate insights based on foreign states comparing a future with today's incumbents to future with their potential successor. First, when doves wait in the wings, foreign states may wait out a hawkish incumbent, refusing to negotiate in anticipation of an agreeable successor. Nikita Khrushchev's refusal to negotiate over Francis Gary Powers's freedom is a case in point; disinclined to give Vice President (and presidential candidate) Richard Nixon a foreign policy victory, he stalled until John Kennedy won the election of 1960 (Khrushchev 2000).

Second, when hawkish successors wait in the wings, two different outcomes are possible. When war can not lock in settlements, foreign states may grant ostensibly unnecessary concessions to keep doves in office, as the United States did to bolster Boris Yeltsin in Russia's 1996 presidential election (Wolford 2012, 523–25). But when war *can* insulate settlements against a hawkish successor's renegotiation, foreign states may attack dovish incumbents to seize what they can ahead of impending leader change, as the Balkan League did in the face of the Ottoman Empire's temporary weakness in 1912 (Wolford 2018, 178–82). Both Schultz (2005) and Wolford (2012, 2018) show that the shadow of an incumbent's successor can upend conventional expectations over the link between leader preferences and war. Doves, in particular, can find themselves embroiled in wars because of the shadows cast by hawkish successors. This poses a problem of omitted variable bias in models that link leader preferences to war, because increasing incumbent dovishness can appear to lead to increasing rates of conflict. Like reputation studies in the *ex post* turnover literature, these models tie leader-level variables to coherent theories of war, and they indicate that observed relationships between leader preferences and conflict can be misleading—unless analysts consider which bargaining problems plague negotiations between the leaders under observation.

The *ex ante* turnover literature has yet to see its empirical implications examined in large samples. For many cases, the data simply do not exist. One possibility for indirect measurement is to examine what happens to term-limited leaders once their successors become known—i.e., for those periods during which democracies have both presidents/prime ministers and presidents/prime ministers-elect.[30] Measures of incumbent hawkishness-dovishness can also be useful as they approach the limits of observed distributions; a very dovish leader is likely to be followed by a relative hawk, and vice versa.[31] Data limitations aside, the *ex ante* perspective sheds light on important historical cases, from Iraq's attempts to unseat Iran's new revolutionary regime in the early 1980s (Wolford 2012, 525–26) to the United States' gentle handling of Yeltsin's Russia, challenged electorally by both Communist reactionaries and virulent nationalists, in the mid-1990s.

The literature on domestic policy insulation may, however, offer some insights. Domestic factions are likely to engage in costly policy insulation, which has an obvious parallel in interstate war (Powell 2004), when they are sufficiently different from other parties and when they are unlikely to be in power again (de Figueiredo 2002). But until improved measures of the successor's shadow exist, the *ex ante* turnover literature will remain separated from the larger-sample *ex post* literature, even as it identifies grounds on which to challenge the latter's unconditional relationships between leader attributes and war.

CONCLUSION

New data have enriched the literature on leaders and war empirically, but the promise of integrating leaders into what we already know about war and peace has been only inconsistently realized. The development of theory has not kept pace with the production of stylized facts, and closing that gap should be a priority for future research. We have learned more about how variation in (a) the risks and consequences of losing office and (b) leaders' biography, psychology, and ideology is associated with variation in the onset of international conflict. Yet the lack of consistently—or at least clearly—applied theories of war looms large. Without clarity on when we are using which theories of war, it's difficult to compare predictions across theoretical models and to say what we can learn from our empirical models, whether regressions or case studies. In the political survival approach, for example, a clear theory of war might help us sort out which regimes (if any) are consistently more likely to punish leaders for poor war performance, by giving us a better sense of what to expect in the observational data. Yet the very theories we use imply that this question cannot (yet) be settled empirically. There is no "empirical question" until we have a theory to tell us whether, when, and why any such question *can* be answered empirically. Likewise, in the leader attribute approach, tighter links between leader attributes and either (a) war payoffs or (b) specific bargaining problems can give us a better sense of how these attributes relate not just to the willingness to initiate a dispute but also to the strategic processes by which disputes escalate to war. We are increasingly confident *that* political survival incentives and leader attributes matter for the onset of conflict, but we cannot be confident that we know *why* they matter without embedding them in a coherent theory of war.

Substantial empirical progress notwithstanding, patterns must still be organized, categorized, and ultimately explained. For that, we need not just conjectures about how leader-level variables might correlate with the outbreak of conflict, but also theories that tell us when, how, and why these variables are connected to each part of the conflict process. At the leader level or not, many of the concepts we care about when it comes to interstate war require a consideration of state-level and systemic factors, from political institutions to alliances to the distribution of power. Thankfully, coherent strategic theories that can give us both sharper predictions and better explanations are available off the shelf; Tarar (2006) shows that the bargaining approach to war can make sense of the empirically messy diversionary war literature, and Wu and Wolford (2018) show that it can do the same for questions of reputation and war.

Horowitz et al. (2018) claim that "the dominance of the bargaining model of war … limited the viability of leader-level arguments" (2181), but this is inaccurate. A charitable reading of their claim associates bargaining models with state-centric models, but this is not the case; the political survival literature is built on the idea that leaders bargain on behalf of states, and the attribute literature teaches us the most when leader attributes shed light on how they bargain. We cannot make useful leader-level arguments about the outbreak of war *without* a coherent theory of its origins, and the bargaining approach is uniquely suited to integrating multiple levels of analysis (e.g., Tarar 2006; Wolford 2007; Debs and Goemans 2010; McDonald 2015; Wu and Wolford 2018). Ironic though it may sound coming from someone who produces leader-centric research, fruitful advancement requires that the literature heed Waltz's (1959) injunction to remember that strategic interaction is the key to understanding international politics. His "third image" may represent the international system, but that system defines strategic interaction in ways that singular focuses on bad regimes (the second image) and bad people (the first image) do not—and that a singular focus on individuals to the exclusion of their domestic and international environments risks doing again. Strategic interaction gives substantive meaning to the otherwise apolitical first image, defining who leaders interact (and thus fight) with and the personal risks they face in waging war. We should lose sight of neither.

NOTES

1. Readers will note that I have omitted a great deal of work on leaders—e.g., that related to operational codes (Walker, Schafer, and Young 1998). I have done this to ensure I can meet space constraints, but also because it is natural to draw connections between the quantitative literature on disputes and war, where most work has been in the past few years, and explicit theories of the conflict process. These omissions are unfortunate in a descriptive sense, but they fit less neatly into the story to be told about the recent explosion of quantitative leader-level data.

2. Data from Goemans et al. (2009) and Gleditsch and Ward (1999).

3. Also notable is that, as the number of states (and democratic states at that) has increased and the number of interstate wars has fallen since 1945, so has fallen the rate of leader turnover.

4. Carter and Chiozza (2018) describe these two strands of literature as the "survival" and "personal attribute" approaches; Krcmaric et al. (2020) refer to the latter as "personal biography." I adopt these categories as they accord with my description of what factors each approach holds constant and allows to vary. Horowitz and Fuhrmann (2018) identify "institutional leadership" and "leader attribute" approaches, a distinction that is not as well suited to my purposes.

5. I am neglecting a growing literature on how leaders shape intrawar politics (see, e.g., Stanley and Sawyer 2009; Croco 2011; Weisiger 2016), but the underlying mechanisms—preferences, information, power, and political constraints—behave the same way before, during, and after conflict. Thus this chapter can still serve as a useful starting point for readers interested in those topics.

6. See Downs and Rocke (1994) for elaboration of this principal-agent mechanism, and Fearon (1995, 379n1) for a comparison to unitary-state models of war.

7. To be clear, I should note that no one who works in this tradition thinks that leaders do not have policy preferences; it is a matter of isolating and exploring the effects of survival incentives on behavior for later comparison to the effects of other incentives.

8. Chollet and Goldgeier (2008) document accusations from the opposition that Bill Clinton used force against Iraq, Afghanistan, and Sudan to distract from the Lewinsky

scandal and again against Iraq to boost his popularity before the election of 1996, among others.

9. Quite apart from disagreeing over whether diversion happens, scholars also disagree on what diversion *is*, as evidenced by the proliferation of proposed mechanisms, from out-group bias to evaluations of competence to opportunities for wartime repression.

10. An exception in democracies would be cases where leaders face (a) post-tenure prosecution for crimes committed prior to or during their tenure or (b) prospective losses of significant rents derived from unpunished corruption.

11. It is not clear why this ought to be the case, though. Leaders have informational advantages over their publics (Downs and Rocke 1994), and if they know that some event will soon happen to lower their popularity, then they have incentives to divert *before* such an event happens, like leaders calling elections in political systems with flexible election timing (Smith 2003).

12. Leaders' political prospects can also explain patterns of repression and dissent (Conrad and Ritter 2013) and choices of partners in military coalitions (Wolford and Ritter 2016).

13. Note that the authors' prediction here is about whether leaders improve their survival chances on the equilibrium path—that is, in observational data—making for a crucial distinction (discussed at greater length in what follows) with broader debates about which regimes render leaders more sensitive to war outcomes than others (Debs and Goemans 2010; Croco and Weeks 2016).

14. Mitchell and Thyne (2010), in fact, find that inflation is associated with the use of force only for high-salience issues—i.e., those for which the incumbent's state has a low valuation of the status quo.

15. Not all political survival stories focus on regime type differences. Potter (2013) shows that American presidents are more likely to authorize "substantial" uses of force after larger margins of victory than after small ones. Carter and Nordstrom (2017) show that leaders facing term limits in democratic systems are not more likely to use force than those under electoral constraints, and in fact dovish leaders are less likely to use force than they are with an active electoral connection. Mandate sizes and term limits are explicitly leader-level institutional variables, yet they have received less attention than regime types.

16. In a sample of elections in the United States, the United Kingdom, Israel, and India after the Second World War, Arena (2008) finds that war outcomes only affect incumbents' political survival if the opposition did not support the war, pointing to yet another obstacle to telling a straightforward story about regime type, war outcomes, and leader tenure.

17. Another possibility is that there may be no unconditional differences to uncover—i.e., that political institutions have no independent effect on survival rates, only interactive effects. I have more than a sneaking suspicion that this is the case.

18. That would require all assumptions to be the same but for the one we would like to compare.

19. Further, if victory demonstrates competence, then Tarar's (2006) results complicate the role of observed war outcomes still further; many may simply not reveal much about a leader's competence or lack thereof.

20. Preferences are defined over outcomes fixed by the analyst. Whatever features players enter an interaction with can be described as part of their preferences, even if those features relate to the processing and interpretation of information.

21. Readers will note that I characterize work on leadership turnover as part of the personal attribute approach, whereas Horowitz and Fuhrmann (2018) do not. They contend that for this work, the fact of turnover matters more than differences in leader preferences. But here I note that differences in leader preferences *make* turnover consequential. If lead-

ers could not differ in their preferences, then turnover would be relevant only to the extent that time in office shapes the risks of losing office—yet leader attribute work abstracts explicitly away from such differences. Therefore, when turnover matters solely as a function of differences in leader preferences (e.g., Wu and Wolford 2018; Smith and Spaniel 2019), the work properly falls into the personal attribute approach.

22. This is an important corrective to some views of how leaders might matter. Jervis (2013) contends that, if one were to remove proper names from histories of a country's foreign policy, one might conclude that leaders do not matter because policy goals would largely be the same; yet this places a burden on leaders to deviate from their predecessors in what issues are considered important for the country. All it really takes, however, are individual differences in the willingness to fight over the same issues. Wholesale reorientations of policy goals (even if we could reliably identify them) are unnecessary for leaders to matter.

23. Integrating these approaches, Colgan and Weeks (2015) show that revolutions are more likely to produce leaders who initiate conflict when those revolutions also produce personalist dictatorships.

24. See also Dube and Harish (2017), who show that European states led by queens in the early modern period were more war-prone than those led by kings—consistent with enduring gender stereotypes.

25. Koch and Fulton (2011) find that female executives are associated with increased conflict behavior as measured in event data (as distinct from the militarized conflict with which I am concerned), though the strength of the relationship is moderated by female representation in the legislature.

26. In related work Mattes and Weeks (2019) present survey evidence consistent with Schultz's (2005) equilibrium in which hawkish leaders can more easily sell rapprochement to the public.

27. Though they do not focus on leader change itself, Owsiak, Diehl, and Goertz (chapter 10, this volume) find that changes in domestic support coalitions (Leeds et al. 2016) can move states between categories in their peace continuum.

28. Licht and Allen (2018) use reputation incentives to explain why new leaders also engage in domestic repression soon after taking office.

29. This is easy to show in an ultimatum game where an uninformed A proposes the division of a unit-sized prize to the informed B, who can accept (implementing the deal for consumption) or reject (which means war). Notation follows Fearon's (1995), but for the fact that A's costs of war are $a > 0$ and B's are $c(1-b)$, with $c \sim U(0, \overline{c})$ unknown to A and $b \sim (0, 1)$ representing a common element of B's cost sensitivity. B accepts when $1 - x \geq 1 - p - c(1-b)$, or when $c \geq \frac{x-p}{1-b} = \hat{c}$, and A's proposal x^* maximizes $\Pr(c < \hat{c})(p-a) + \Pr(c \geq \hat{c})(x)$. The probability of war is $\Pr\left(c\langle\hat{c}\middle|x^*\right) = \frac{1}{2} + (1 - a / (\overline{c}(1-b))$, which decreases in both b and a. Therefore, the probability of war increases in B's war payoffs and decreases in A's, independent of who might have initiated the dispute.

30. I'm indebted to Steve Miller for this point.

31. Schub (2020) takes a different approach, measuring uncertainty over a future leader's preferences and showing that it is associated with a reduced risk of war when peace is costly and the risk of turnover is high.

Chapter Fifteen

War Financing and Foreign Debt

Rosella Cappella Zielinski and Paul Poast

INTRODUCTION

War requires resources. Many of these resources, in turn, require money. Leaders must decide where to secure the funds. Do they raise money domestically or abroad? If sources are domestic, who should pay and how should resources be extracted? If abroad, where should leaders look? These are the questions every leader or governing regime must grapple with when entering a military conflict.

Scholars studying war finance seek to understand how leaders make these decisions. In addition to understanding the range of options available to leaders, they look to understand their preferences. Under what conditions do leaders prefer one option over the other? What do citizens prefer? Scholars also look to understand under what conditions leaders are able to enact their preferences. Thus they explore the ways in which domestic and international institutions enable or constrain leaders.

This chapter provides a review of these questions and then explores a contemporary phenomenon, the rise in foreign debt, notably from other sovereigns. Unlike wars financed by taxation or even domestic debt, foreign debt, especially when acquired from other states, allows the state's war-fighting effort to continue beyond the financial confines of the state. In addition to providing resources beyond the state's capacity, it insulates the public from the financial costs of the war effort. In other words, foreign borrowing can enable a state to avoid the "guns versus butter" trade-off that lies at the heart of arms acquisition (Poast 2019a). Foreign war finance, however, is not without cost. In addition to interest paid on the debt, holders of said debt, particularly when these holders are other sovereigns, are able to extract political concessions from the debtor states. Simply stated, foreign debt provides a window of opportunity for the creditor state to engage in quid pro quo behavior.

The study of war finance has implications for the immediate war effort and beyond. How leaders and governing regimes finance their wars has implications beyond the immediate financing of the war effort. In addition to the critical component of immediate mobilization to sustain military participation in conflict, the manner in which a state pays for conflict can affect the war's duration and settlement (Slantchev 2012; Shea 2014), the fate of the leaders themselves (Carter and Nordstrom 2017), the total cost of the conflict (Poast 2006; Stiglitz and Bilmes 2008), economic stability during and after the conflict (DiGiuseppe, Barry, and Frank 2012), and the long-lasting economic redistributive effects of the conflict (Scheve and Stasavage 2010, 2012; Piketty 2017; Cappella Zielinski 2018).

WHY DEBT?

Scholars have long noted the importance of war debt to the rise of military powers (Rasler and Thompson 1983, 1989; Kennedy 1987; Norrlof and Wohlforth 2019). Indeed, borrowing is the foremost war finance method.[1] From 1823 to 2003, 88 percent of states fighting long wars (i.e., wars over six months) have engaged in at least some form of borrowing, either domestic or from abroad, to confront the cost of war. In regards to domestic debt, 63 percent of states financed at least some of their war efforts via domestic debt. But domestic debt is on the decline. Prior to 1950, 68 percent of states incorporated domestic debt into their strategy, whereas in the postwar era only 40 percent have (Cappella Zielinski 2016, 109). Conversely, foreign borrowing appears to be on the rise; 57 percent of states engaged in some form of foreign debt. Prior to 1950, 52 percent of belligerents engaged in foreign debt, whereas 72 percent have in the postwar era (109).

The United States' war finance since its inception reflects this trend. As Figure 15.1 suggests, borrowing has been the primary component of US war finance since the War of 1812, with the exception of the Korean War, which was paid for by taxation.[2] Beginning with the Gulf War in 1991, US wars have been increasingly funded by sources outside US borders. Indeed, the wars associated with the global war on terror have been coined the "Credit Card Wars" with about 40 percent of the debt held abroad.

What makes debt, particularly foreign debt, so attractive during wartime and why would governments ever choose another option? We can begin to answer this question by considering two factors: *Leader and Citizen Preferences* and *State Capacity*.

Leader and Citizen Preferences

"How do states finance wars?" is the grounding question in the study of war finance. States can finance their wars via forced savings plans, direct and indirect taxation, domestic debt, bond campaigns, printing, austerity measures, existing coffers, foreign debt, foreign grants, and plunder. Which methods are utilized and why?

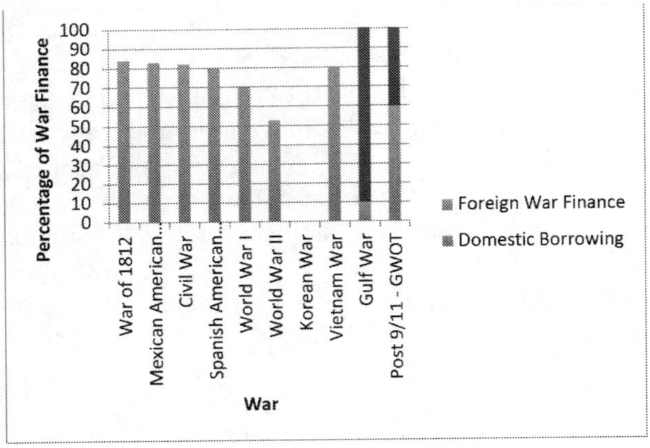

Figure 15.1. Types of War Financing.

To answer this question war finance scholars make two fundamental assumptions: (1) Citizens prefer not to surrender their income, and (2) leaders prefer to stay in power, which often means avoiding the ire of the public. These yoked assumptions are often traced back to liberal economic thought and are pervasive in contemporary war finance scholarship (e.g., Lamborn 1983; Levi 1988; Mastanduno, Lake, and Ikenberry 1989; Shea 2014; Cappella Zielinski 2016; Carter and Palmer 2016; DiGiuseppe and Shea 2016; Kreps 2018). Since the assumptions focus on citizen preferences and leader survival, they are largely oriented toward explaining how democracies can finance via debt.[3] It is worth exploring the merit of these assumptions.

Do citizens prefer to hold their income during wartime? Works at the intersection of war finance and public opinion suggest this is mostly the case, but not always. Under certain conditions, specifically when the war in question is popular or deemed legitimate, citizens will part with their income with little complaint (Berinsky 2007, 2009; Cappella Zielinski 2016; Geys 2010; Geys and Konrad 2016; Kreps 2018). For example, during the Korean War citizens supported war taxation. In a July 1950 Gallup poll 44 percent of people answered that Americans had not been asked to make enough sacrifices in support of the war. Between July 1950 and February 1951 this survey was administered seven times. The average response was that 38 percent felt they had not been asked to sacrifice enough (Cappella Zielinski 2016, 43).[4] A related literature on exploring the relationship between the means of war finance and support for the war itself finds mixed conclusions. For example, while Flores-Macias and Kreps (2017) find that public support for war declines when the war is financed by taxation, Kriner, Lachase, and Cappella Zielinski (2018) find that a decline in public support is mediated by the type of taxes implemented. Kreps (2018) finds party identification as a predictor of propensity for fiscal sacrifice and attitudes about the war itself.

Regardless of the actual effect of war finance on citizen support for a particular war, scholars studying war finance often begin from the premise that leaders have internalized these costs and act accordingly. Leaders, knowing that they will eventually have to face the cost of financing war via domestic resource extraction (which means incurring the political consequences of disgruntled citizens), attempt to finance war in a manner that shields citizens from the true costs of war. During the Vietnam War, for example, the Johnson administration attempted to avoid public scrutiny by not raising taxes. Secretary of State Dean Rusk explained this logic at a staff conference on October 5, 1967. "The Administration made a deliberate decision not to create a war psychology in the United States. There have been no war bond campaigns, etc. The decision was made because it is too dangerous for this country really to get worked up. Maybe this was a mistake; maybe it would have been better to take steps to build up a sense of a nation at war. The course we have taken has meant expecting a great deal of our men in Vietnam, against the background of a home front going about business as usual" (quoted in Cappella Zielinski 2016, 53).

Cappella Zielinski (2016) presents a war finance continuum of more to less visible forms of war finance. On one end of the continuum lies direct resource extraction—encompassing forced labor, forced savings plans, and direct taxation—which refers to resources paid to the state directly by citizens within society. Because this method of extraction is imposed on individuals and one cannot opt out of the policy or transfer the costs of said policy to others, citizens are highly aware of this policy. In the middle of the continuum lies indirect resource extraction, such as indirect taxation, printing money, and austerity measures. Here the government procures resources without directly asking for citizen contributions.[5] On the far end of the spectrum lies external resource extraction. External funding is a broad category that includes securities floated on foreign markets, sovereign-to-sovereign loans, grants, plunder, and diaspora remittances. These forms of finance remove the citizen from interacting with the state's war finance policy. Foreign grants or plunder, for example, do not need to be paid back, resulting in no financial interaction with society. Hence citizens may or may not be aware of these war finance policies.

Kreps (2018), following Cappella Zielinski (2016), provides a war finance typology of leader choice. She argues that leaders either aim to shield the policy from the public (what she calls "Hide-and-Seek" financing) or aim to encourage shared effort by making the war financing means explicit and direct. Prior to World War II, according to Kreps, the American public was willing to incur taxes or directly invest in government bonds to fund the war effort, whether those efforts were limited—meaning wars aimed at well-defined objectives that do not demand the utmost military effort—or absolute—meaning wars that seemingly observe no limits and require massive mobilization of the population and economy. Limited wars included the Quasi-War of 1798, the War of 1812, the Mexican-American War, and the Spanish-American War. The absolute wars were the American Civil War and World War I. Indeed, it was only during times of war that pre–World War II Americans felt the pinch of income taxation. But since World War II the US government has largely shied away from wartime taxation. For Kreps, the post–World War II American public is now willing to accept a permanently higher level of peacetime taxes for the sake of welfare provision and a large, permanent military. Neither of these existed prior to World War II.[6] However, the public is less tolerant of new taxes for executing individual wars. This has led the government to "hide" the cost of wars within the overall budget, rather than calling for new funding. Such hiding is more easily accomplished via debt than imposing new taxes.[7]

State Capacity

The ability of leaders to act on their war finance preferences is constrained by state capacity to mobilize and extract revenue. By state capacity, we are specifically referring to a state's bureaucratic institutions, which determines the state's ability, regardless of leader and citizen preferences, to extract taxes from its populace (Hood 2003). Some forms of tax revenue, such as income or corporate taxes, require extensive bureaucracies, while others, such as tariffs or value added taxes, do not (Cappella Zielinski 2016, 24–25).[8]

From a purely financial and economic efficiency standpoint, taxes provide an optimal source of revenue. Raising war funds via taxation requires drawing from the existing tax structure (e.g., income tax brackets, tax rates, etc.) and imposing supplementary war taxes (e.g., new taxes, including surtaxes, or increased rates on existing taxes for the explicit purpose of financing the war). Unlike most other war finance options, taxation does not require the secured funds to be paid back. Also, because the state is extracting from available resources (i.e., the tax base), it can help control widespread inflation. This is because taxes can serve as a means of controlling civilian spending and do not create an incentive to print money in order to meet debt obligations (Poast 2006).

While optimal, complications with taxation arise once one accounts for bureaucratic capacity. For instance, scholars have noted that states characterized by a low bureaucratic capacity to extract revenue will be confined to indirect or external revenue extraction (e.g., Cappella Zielinski 2016; Centeno 2002; Queralt 2019). If so inclined, these states may also engage in state building during the war effort, allowing higher levels of direct revenue extraction over time than would be otherwise available. In contrast, states with high bureaucratic capacity will be able to implement taxes such as corporate and income taxes that are more elastic to policy change and yield higher revenue. In this case revenue from direct extraction may comprise a larger percentage of a high-capacity state's war finance strategy.

Of course bureaucratic tax capacity can be endogenous to the war effort itself. The war effort may induce development of bureaucratic capacity (Tilly 1990). At the same time the war can place downward pressure on a state's tax capacity and sources of tax revenue. For example, bombings and invasion can force administrative offices to flee, hindering tax collection efforts. China during the Third Sino-Japanese War (1937–1945) provides an example. Chinese leaders favored a tax-based war finance strategy, attempting to raise taxes immediately before and during the war to pay for it. However, the course of the war affected state capacity to raise revenue. Indiscriminate Japanese bombing behind the lines added confusion to tax collection. Moreover, when the government was driven from Nanking to Hankow and then to Chungking, many experienced administrators could not follow (Young 1965). Thus the ability to collect taxes decreased greatly. War can also reduce tax revenue sources. Blockades can hinder tariffs and invaded territory can reduce the size of the tax base.

But low bureaucratic tax capacity need not result in defeat. States can turn to other forms of revenue such as printing, plunder, or, most notably, debt. Indeed, scholars studying debt will often look to its enabling nature. While debt must be paid back with interest and does not share taxation's anti-inflationary properties, it enables leaders to mobilize resources quickly (Shea 2014) and access more resources than the tax base may allow (Schultz and Weingast 2003).

Of particular attraction to all states, not just those with low capacity, is the ability to borrow from abroad. This is because acquiring funds from outside the country can obscure the cost of war (Cappella Zielinski 2016; Kreps 2018). This is why, for many countries, a key source of the money borrowed is *external* debt, primarily in the hands of private financiers (Flaudreau and Flores 2012; Poast 2015). While taxation relies on bureaucratic tax capacity, external borrowing requires attracting funders.[9] Unlike with the domestic populace, where a state can rely on its coercive capacity to compel lending (or alternatively rely on patriotic fervor to induce citizen lending), the state must take steps to encourage foreign loans. This can prove challenging. Foreign lenders are always wary of borrower default and wars can make investors especially nervous due to possible macroeconomic disruptions (Kirshner 2007). This is why financial markets will react to wartime events both on and off the battlefield (Willard, Guinnane, and Rosen 1996; Choudhry 2010; Guidolin and Ferrara 2010), and capital flight and exchange rate pressures can occur during war (Collier 1999; Vu Le and Zak 2006).[10] Given their concerns foreign lenders will look for indicators of quality,[11] be it regime type (Schultz and Weingast 2003), central bank presence and operation (Broz 1998; Poast 2015), adherence to the gold standard or currency status (see Strange 1994)

EXTERNAL BORROWING FROM SOVEREIGNS: THE PREFERRED MEANS OF WAR FINANCE?

All of the "signs of good financial housekeeping" listed earlier in this chapter are vital for attracting *private* finance. But over time, especially starting in the mid-twentieth century, the importance of private finance as a primary means of financing wartime sovereign debt has lessened.[12] In its place has emerged direct sovereign-to-sovereign lending. This is the direct extension of public money with an expectation of repayment (with interest).[13] Private lending is characterized by structured repayment terms, interest rates that reflect the market and probability of repayment, risk premiums, and collateral. But these features are largely absent from sovereign-to-sovereign loans. Moreover, sovereign-to-sovereign loans may not involve the direct transfer of money, as these loans can commonly take the form of material (i.e., equipment and other war inputs). And since the material loans can take place in theater on shared supply lines, parsing out the exact amount of material loaned in the moment is difficult.[14] Hence repayment schedules of sovereign-to-sovereign loans are often negotiated after resources have been extended.

One could go as far as claiming that the rise of sovereign-to-sovereign wartime lending is central to understanding the modern global monetary system. It was US government loans to European belligerents during World War I that substantially contributed to the US dollar supplanting the British pound as the world's major reserve currency.[15] Moreover, the debt crises that marred the global economy of the 1980s and 1990s were in no small part due to sovereign-to-sovereign military loans offered by the United States and Soviet Union to developing countries during the Cold War.

Sovereign-to-sovereign loans have advantages to both the borrowing government and the lending government. The advantage to the borrower is that these loans can overturn the notion of an "Arms versus Allies" trade-off (Morrow 1993). Instead of a trade-off, sovereign-to-sovereign loans show how "allies" can be vital to how many states acquire "arms."[16] This is because sovereign-to-sovereign loans enable a state to sidestep the "guns versus butter" trade-off inherent to arming decisions (Poast 2019). For example, the cost of fighting World War I was overwhelming for Britain.[17] The massive and sudden war expenditures coupled with the need for dollars to purchase war inputs from the United States could not be covered by taxation or domestic borrowing. To compensate, the British sought funds from the only major economy not involved in the fight, the United States. Unable to procure a direct loan from the American government due to neutrality, the British initially purchased private loans, sold their American assets, and shipped gold from Canada.[18] By 1917 the orders were increasing and private lenders were nervous and less willing to continue the extension of credit. The British were nearly dollar bankrupt.[19] But when the United States entered the war in April 1917, immediate financial relief was offered.[20] Upon congressional approval, America extended loans to its newfound allies (or, more accurately, "associate powers").[21] The American government loans had two notable features. First, they were under market rates. British loans floated on Wall Street were at 5 percent and 5.5 percent interest,[22] but the first issue of US Liberty Loans, via the War Finance Bill, was for $2,000,000,000 at 3 percent. Second, the repayment period was open-ended. The War Finance Act stated no exact dates for repayment by the Allies, but stipulated that the loans should bear the same rate of interest as the American bond issued to raise the necessary funds.[23] In total, from April 1917 to November 1920 the United States extended $4,277 million to Great Britain.[24]

The advantages of offering sovereign-to-sovereign loans might seem apparent to the lending state, but they are worth making explicit. First, loans—unlike a grant of aid—hold out the possibility of repayment. This can make the provision of loans more attractive than grants of foreign aid, which are, in one sense, "free"—they are a transfer of wealth from domestic taxpayers and voters to foreigners who are not represented in the creditor state.[25] Hence loans may be easier to sell to the public or elites when they are hesitant to sacrifice for the war effort and likely unaware of the prospects for repayment.[26]

Second, loans can create postwar dependency. While grants can have a quid pro quo element, their ability to ensure policy concessions is limited to withholding the funds. Threatening withholding is not appealing, as the patron state is often providing aid to an ally that it wants to win a war (indeed, at times the patron state is a cobelligerent). But a sovereign-to-sovereign loan carries at least the expectation of an ongoing relationship after the war, as the creditor expects repayment.[27] Because loan repayment is commonly a multiyear process, this gives a patron the means to justify continued monitoring well into the postwar period. Moreover, the possibility of debt forgiveness accords the patron an additional source of postwar leverage. This was surely the view of the United States regarding its sovereign-to-sovereign loans to the British during World War I. Though the British might have preferred a grant that did not entail postwar repayment, key members of the Wilson administration, including Wilson himself, appear to have favored loans. The loans would accord the United States influence over the European states *following* the war. On July 21, 1917, Wilson, in a correspondence with Colonel House, confirmed that the idea of offering loans gave the United States the ability to exercise leverage over the European powers in crafting the peace after the war: "England and France have not the same views with regard to peace that we have by any means. *When the war is over we can force them to our way of thinking, because by that time they will, among other things, be financially in our hands.*"[28]

This second advantage is why the political dynamics of sovereign-to-sovereign lending echo the broader phenomenon of "issue linkage" within an asymmetric alliance relationship (Morrow 1991; Poast 2012, 2013). Alliances, like any codified relationship, are the product of a diplomatic process and vigorous negotiations: the terms of the alliance must be set (Poast 2019). But while initial agreement might be secured via a recognition of shared security interests—perhaps the countering of a specific threat or the recognition of stabilizing a given region—ensuring the alliance's continuance often requires that all parties view and continue to view the alliance as beneficial. When a negotiation is between a minor power and a major power, this identification of mutual benefit can be straightforward: the minor power requires the protection of the major power and, in order to achieve it, is willing to grant concessions to the major power (Morrow 1991). These concessions in turn bolster the willingness of the major power to remain committed to the security of the minor power, even when the minor power is in a difficult security environment (Poast 2013).

While the ability to acquire concessions appears to be a major advantage to the lending state, there is a downside: such loans can create an enormous "moral hazard" problem for the patron.[29] As Krugman recently described it, "[Moral hazard is] any situation in which one person makes the decision about how much risk to take, while someone else bears the cost if things go badly."[30] Even if things go well, the borrower, once the exigencies of war have passed, may have an incentive to default on its debt or use the threat of default to induce the lending of additional funds.[31] If the patron becomes overly leveraged in its lending to the recipient, the fear of default will turn the tables by giving the recipient the ability to influence the patron.[32] This situation is described well by Siegel regarding British and French loans to Russia during World War I: "Imperial Russia's pre-war and wartime obligations to its entente partners were so extensive that the mere threat of national collapse and potential default allowed it to extract not just further investment from its principal creditors, but often compliance with its policy objectives as well."[33] Indeed, some have claimed this moral hazard dynamic compelled the United States to provide Marshall Plan assistance within a few years of World War II's cessation. In 1952 Robert Marjolin, head of the French postwar reconstruction efforts and the first secretary-general of the Organization for European Economic Cooperation (OEEC) (established to implement the Marshall Plan), noted: "Although American aid has been a necessary remedy over a period, and will continue to be for a time, one is bound to acknowledge that in the long run it has had dangerous psychological and political effects … The idea that it is always possible to call on American aid, that here is the ever-present cure for external payment deficits, is a factor destructive of willpower. It is difficult to hope that, while this recourse continues to exist, the nations of Western Europe will apply, for a sufficient length of time, the courageous economic and financial policy that will enable them to meet their needs from their own resources without the contribution of external aid."[34]

While the dangers of moral hazard are real, sovereign-to-sovereign loans remain popular. Recent examples include Ukraine obtaining a $1 billion loan guarantee from the United States government in March 2014 (and another $1 billion in May 2015) to counter Russian-backed military aggression,[35] and Iraqi prime minister Haider al-Abadi coming to Washington, DC, in April 2015 looking to acquire US government loans to cover the costs of fighting the Islamic State in Iraq and Syria.[36] The provision of wartime sovereign-to-sovereign loans—more precisely, the withholding of more than $1 billion of loan guarantees to Ukraine—was at the heart of the 2019 impeachment of US president Donald Trump.[37] Even China has begun leveraging this instrument, as Chinese loans to the Pakistani government enabled it to acquire $6 billion worth of big-ticket military hardware, primarily submarines.[38] In short we see no reason for the use of sovereign-to-sovereign wartime loans to diminish.

CONCLUSION

The study of war finance is rich in ideas and insight regarding the relationship between leaders and their citizenry, between the citizenry and the war effort, and the capacity of the state to raise funds. In addition to reviewing those works that address leader and citizen preferences regarding resource extraction and the war effort, we explore a contemporary phenomenon, sovereign-to-sovereign loans. We suggest scholars should explore the conditions under which states seek to secure a sovereign-to-sovereign loan as well as when creditor states choose to extend one to belligerents. As the trend of relying on external creditors increases due to financial globalization, austerity measures, the reluctance of leaders to demand fiscal sacrifice from their citizens, or the inability to extract resources, sovereign-to-sovereign loans will proliferate. We suggest such wartime loans are an instrument of international power: states become creditors to gain influence over the economic and military policies of another state in order to shape the postwar order. In shaping the postwar order, creditor states shape the distribution of power in their favor.

NOTES

1. Taxation has consistently comprised a small percentage of a state's war finance strategy and appears to be further decreasing in the postwar era. Only half of all belligerents fighting major wars between 1823 to 2003 enacted new taxes to confront the costs of war. Of those states, only nine belligerents have financed 25 percent or more of their wars using tax revenues: Turkey during the Russo-Turkish War of 1828–1829, Denmark in the First Schleswig-Holstein War, France during the Franco-Turkish War, Britain during the Crimean War and World War II, the United States during World War I, World War II, and the Korean War, and Russia during World War II (Cappella Zielinski 2016, 108).

2. Cappella Zielinski 2014.

3. The ability of democracies to finance wars through debt is a key focus of the war finance literature (Shultz and Weingast 2003; Scheve and Stasavage 2010; Shea 2014; Carter and Palmer 2016).

4. For historical polling of American support for war taxes, see Kreps (2018, Appendix A).

5. McDonald (2009, 2015) argues that these politically "free" revenue structures help leaders insulate themselves from making concessions to domestic groups, thus providing leaders greater autonomy and allowing leaders to remain in office longer.

6. Some argue, perhaps mistakenly, that the large permanent military is itself a "welfare" policy of job creation (Garrett-Peltier 2017).

7. In addition to visibility, others argue that debt reduces political costs. Rooted in the tax-smoothing literature, such scholars argue that when a government faces an exogenous expenditure, increasing taxes distorts macroeconomic incentives, with negative consequences for public support (e.g., Barro 1979; Lucas and Stokey 1983; Ohanian 1997, 1998). States with fiscal flexibility, those who can borrow cheaply and easily due to favorable credit terms, have resources to both finance war and respond to citizen demands (Clay and DiGiuseppe 2017). Leaders thus prefer to borrow to pay for war in order to mitigate said costs (Shea 2014; DiGiuseppe 2015).

8. Scholars studying state formation and state-society relations (Ardant 1975; Tilly 1975, 1990; Barnett 1992; Centeno 1997) and international relations scholars studying the projection of military power (Knorr 1975; Organski and Kugler 1980; Kugler and Domke 1986) have long noted how bureaucratic tax capacity is critical to the mobilization of economic resources for war.

9. Tax capacity and debt capacity are associated as potential lenders base lending decisions on ability and willingness to repay; a critical factor of said ability is a robust tax capacity (Tomz 2007b; McDonald 2009; Scheve and Stasavage 2012)

10. As well as attacks to financial infrastructure (Olson 2004).

11. Or what Shea and Poast (2018) refer to as indicators of "economic wherewithal."

12. Though such loans were not exclusively a twentieth-century phenomenon as British loans funded many Latin American wars during the mid-to-late 1800s (see Smith 1919). France played a small role as a wartime financier during the early 1900s—namely to Russia and to Estonia and Latvia during their respective wars of independence.

13. Sovereign-to-sovereign loans differ from cases in which a government supports the extension of a loan underwritten by a private creditor to a belligerent state. Loans extended to Russia during the Russo-Japanese War provide an example. Russia was dependent on the Paris financial market (e.g., Credit Lyonnais) for funds to prosecute the war and due to the size of the loan, needed the French government's backing (see Long 1974).

14. During the Korean War the British Commonwealth of Forces borrowed from the United States. The unexpected increase in intensity and duration due to the entry of the Chinese into the war meant that it was more important to get allied troops in the theater than to negotiate the details of the loan. As Grey writes,

> With forces arriving in a piece-meal fashion, and with one 'expending bandages while … another is expending Centurion tanks and quantities of expensive ammunition', Roberson advised his government that 'the financial interests of each country should be assessed on a man/day basis at the end of the campaign and not each time a major change in strength is deemed to have taken place.' An added bonus was that such a scheme would release combat units and 'heavily pressed depots' from the extensive paperwork necessary in any pool account arrangement" … Moreover, logistic support flowed into a common [American] supply line, making it difficult to determine the amount of support received by any individual unit … it was not until the summer of 1951 that a system of control and accounting was actually fully working. (1988, 173–76).

British repayment of Korean War debt to the American government was not negotiated until 1958 (Grey 1988; Farrar-Hockley 1995).

15. This is in addition to the dollar's centrality as the unit of account in the global oil market. See Croteau and Poast (forthcoming).

16. Some work, such as Horowitz, Poast, and Stam (2017), actually view arms and allies as complements, though not in the manner that allies are important for acquiring arms.

17. According to contemporary estimates, if the war had *not* taken place, British government expenditure in the years 1915 through 1921 would have amounted to a total of £1.5 billion. In contrast, the *actual* expenditures were £12.5 billion. Even if one removes British loans to allies (since these might be repaid), the total difference was still £9.8 billion. Figures from Kirkaldy (1921, 216).

18. Brown 1940, 64–65; Morgan 1952, 324–33; Chernow 2010.

19. The adverse trade balance with the United States in 1917 was £316.2 million and rose to £487.6 million by 1918 (Morgan 1952, 309).

20. Indeed, this was the first time the United States offered sovereign-to-sovereign loans. As reported by Kenwood and Lougheed (1999, 34), prior to 1870, US foreign loans were largely private, limited to mining and manufacturing ventures, and confined to the Western Hemisphere.

21. The Senate vote was 82 to 6 with eight abstentions. In the House, the vote was 373 to 50 with nine abstentions.

22. Kirkaldy (1921, 180). For a discussion of how the various interest rates were agreed upon and the ultimate failure of the loans, see Burk (1985, 70–75, 80–86, 93) and Cooper (1976, 213–14).

23. Kirkaldy 1921, 181.

24. Fisk 1924, 190.

25. Milner and Tingley (2010). But in another sense, grants are not free. The larger the grant, the higher the expectation on the part of the patron of a concession by the client. See Bueno de Mesquita and Smith (2009, 321). However, our focus is on the incentives of the patron.

26. In addition, loans with interest count as an asset rather than a liability in the national budget and thus reduce the appearance of a budget deficit.

27. The borrower will accrue interest of size $i*L*$t, where t is the number of years of the loan and i is the annual interest rate. Hence, upon maturity, the lender will receive a total payment of $L + i*L*$t, unless the borrower defaults. The risk of default is built into the interest rate. The interest rate is $i = \pi^e + RP$, where π^e is the expected rate of inflation and RP is the risk premium, meaning the amount of interest a borrower must provide to compensate for the risk of default (the risk premium is an additional payment over what is deemed by the lender to be the most secure asset).

28. Quoted in Burk (1979, 243). Emphasis added.

29. Kletzer 1984; Innes 1990; Arnott and Stiglitz 1991; Atkeson 1991; Bolton and Dewatripont 2005, 162–68.

30. Krugman 2009, 63.

31. Moral hazard can also be present in alliance relationships (Benson 2012) and the onset of intrastate conflict (Kuperman 2008), but these pertain to decision-making *prior* to the initiation of a conflict.

32. This is reflected in the adage "Owe the bank five thousand dollars, the bank owns you. Owe the bank five million dollars, you own the bank."

33. Siegel 2014.

34. Quoted in Marjolin 1989, 241. See also Ellerman 2007, 572.

35. Tiron and Wallbank 2014; USAID Press Office 2015.

36. Northam 2015.

37. Demirjian, Dawsey, Nakashima, and Leonnig 2019.

38. Ansari 2015. See also BBC News 2015.

Chapter Sixteen

Trends in Interstate Conflict

Bear F. Braumoeller

INTRODUCTION

The question of whether, why, and to what extent the world has witnessed a decline in interstate warfare has not received extensive treatment in the conflict studies literature to date. For that reason the answer to the question "What do we know?" is, sadly, not a lot. Indeed, the fact that the first two editions of this volume contained no chapter on the subject is indicative of the paucity of relevant research and the extent to which the question has been neglected. That neglect is deeply troubling given the incredible destructive potential of modern warfare. In fact, of all the major threats to human well-being that exhibit the kind of explosive "snowballing" behavior that can snuff out lives on a global scale—pandemics, for example, or climate change—the escalation of warfare may be the least well understood.

Fortunately there has recently been a proliferation of research on the question of whether war is in decline. Much of this research was inspired by Steven Pinker (2011) best seller, *The Better Angels of Our Nature*, which argued that warfare in one form or another has been in decline for decades or centuries. Together with other scholars who have made narrower or more qualified versions of the same argument (Väyrynen 2006; Mueller 2007; Goldstein 2011), Pinker focused the attention of a broader group of scholars, inside political science and out, on the question of systemic trends in warfare.

Pinker's decline-of-war thesis has undeniable appeal. He argues that lethal, system-wide great power war has been far less common recently than it was in the past, which is undeniably true. He also argues that every dimension of war save its lethality has been in decline for centuries across the entirety of the international system, and that even lethality has declined since World War II. Here, however, there is considerable room for debate. War is rare, it is true, but war has always been rare: most countries at most times do not fight one another. The question is not whether war is rare, but whether it has grown *more* rare. Lethality poses an even more extreme version of the same problem: really lethal wars are so incredibly rare that it is difficult to assess the claim that they are in decline. As a result we could be, as Nassim Taleb (2005) put it, "fooled by randomness" into believing that the world has become a more peaceful place. Jack S. Levy and William R. Thompson (2013, 413) make this point forcefully when they ask us to imagine a scholarly discussion of the decline-of-war thesis taking place, not in 2012, but in 1912:

> In terms of quantitative trends in war … our counterparts in 1912 had even more grounds for optimism about the prospects for peace than we do today. There was a more sustained decline in great power war and a longer period without a general war. The great power wars that had occurred in the last century were shorter in duration and involved fewer great powers. We all know what happened 2 years later.

Because the rarity and escalatory potential of war confound human intuition, formal statistical tests are needed to evaluate trends in warfare. Testing the decline-of-war thesis, however, turns out to be no trivial matter. It is not at all clear, first of all, what exactly should be measured. Scholarship has converged around two broad areas—the onset or existence of conflict on one hand and its deadliness on the other—but within those areas different scholars have argued for different measures. To compound the problem, measures of the number of fatalities that result from even very recent wars are controversial, and medical science has made such remarkable advances over the past two centuries that the changing wounded-to-killed ratio on the battlefield may produce the illusion of a decline in violence where none exists (Fazal 2014). Finally, the statistical methods required to evaluate the decline-of-war thesis are not typically taught in political science methods courses. The profoundly skewed distributions of many of these measures raise substantial methodological issues that have only recently been recognized within this literature.

As a result this new decline-of-war literature is one of the more interesting, if challenging, growth areas in the study of war. It is also one of the most urgent. While scholars differ on the question of whether the escalatory potential of international warfare has diminished in recent decades, even the best possible interpretation of the data leads us to conclude that it has not declined much—certainly not enough to render it much less of a threat to humanity.

HISTORY OF THE DEBATE

While discussions of trends in warfare, which often involved little more than tallies of wars and battles over time, have a long history, one of the first systematic data-gathering efforts on all aspects of war was Quincy Wright's monumental *A Study of War* (1942). While Wright's contributions were largely empirical, English mathematician and meteorologist Lewis Fry Richardson soon brought real statistical acumen to bear on Wright's data. Richardson (1944) famously demonstrated that war onsets in the period 1820–1929 follow a Poisson distribution, a distribution of events per time period that can be expected to arise when events occur continuously, with very low probability, and their timing is random.[1]

Richardson also demonstrated in subsequent work (Richardson 1948) the close correspondence between war fatalities and a *power-law distribution*—a function in which one number varies as a power of another. In the case of war Richardson demonstrated that the frequency of war and its lethality are so related: while most wars are relatively small, a few are unbelievably huge.

Richardson's focus on the behavior of aggregates naturally led him to examine the behavior of the international system as a whole. In so doing, he occasionally explored the question of whether the system was becoming more or less warlike. He found little evidence to support either conclusion. In his study of the frequency of war onset he noted that the fit of Wright's data to a Poisson distribution reflected "an absence of any drift toward more or fewer wars" (Richardson 1944, 246). In his most famous work on war, *Statistics of Deadly Quarrels* (Richardson 1960b), which unfortunately was published only after his death, he noted: "There is a suggestion, but not a conclusive proof, that mankind has become less warlike since A.D. 1820. The best available observations show a slight decrease in the number of wars with time … but the distinction is not great enough to show plainly among chance variations" (167).

The work of Wright and Richardson inspired historian Melvin Small and political scientist J. David Singer to co-create the Correlates of War project at the University of Michigan in 1963. By 1970 Small and Singer had published their first descriptive analysis of patterns in international war, based on the first version of the Correlates of War project's canonical data set of post-Napoleonic international wars, in the *Annals of the American Academy of Political and Social Science* (Small and Singer 1970). Just over a decade later they published the much more comprehensive *Resort to Arms* (Small and Singer 1982). In the latter work they used regression analysis to look for trends over time in the number of wars, their lethality, the total number of nation-months of war, and the last two quantities adjusted for the prewar populations of the combatants. For all measures, they concluded that "there is no significant trend upward or down over the past 165 years" (1982, 141).

Two additional data sets that were compiled around this time brought more information to the table. Jack Levy's pioneering research on great power war extended the temporal scope of systematic war data back to 1495 and provided the basis for his conclusion (Levy 1982, 1983; Levy and Morgan 1984; Levy and Thompson 2011) that, over the course of centuries, the frequency of major power war had decreased while its lethality had increased. At about the same time Charles Gochman and Zeev Maoz introduced the Militarized Interstate Dispute data (Gochman and Maoz 1984), which catalogued conflicts well below the threshold of war established by the Correlates of War project—those comprising the threat, display, or use of force. While Gochman and Maoz found that the frequency of militarized interstate disputes (MIDs) had varied over time, they also found that that variation could mostly be accounted for by changes in the number of states in the interstate system.

Despite the proliferation of quantitative studies of international conflict in the next two decades, very few studies were published on trends in warfare. One reason for this odd lacuna may lie in the fact that nearly all of the works listed earlier in this chapter described and explored new data on war. In that context the question of whether systemic trends occur in warfare was seen as an interesting, even obligatory, question to answer. After it had been answered, often in a preliminary and cursory way, however, it fell by the wayside, as study after study focused on explaining international conflict at the dyadic level. With very rare exception, an entire profession seemed profoundly disinterested in devoting its time to studying system-wide trends in international warfare.

There were, of course, a few unicorns (Gaddis 1987; Mueller 1989; Kegley 1991). Hensel (2002) explored trends in the frequencies of different kinds of war since 1815, finding no obvious upward or downward trend in interstate wars over time but a general increase in the frequency of MIDs. Wallensteen and Sollenberg (1995) called attention to the fact that, contrary to popular belief, there had been a drop in conflict in the first six years after the end of the Cold War, measured as the number of active armed conflicts per year. In a series of publications Lars-Erik Cederman demonstrated that his agent-based model of territorial conquest and state formation was a plausible explanation for the puzzling fact that the lethality of war conformed to a power law distribution (1994, 1997, 2003). Cederman also found, with T. Camber Warren and Didier Sornette (2011), that the lethality of war had increased significantly following the Napoleonic Wars, a fact that they attributed to the impact of the *levée en masse* on the number of soldiers on the battlefield.

Two more data innovations in the early 2000s prompted a brief renewal of interest in the question of whether war is in decline. Gleditsch et al. (2002) introduced the newly expanded Uppsala-PRIO Armed Conflict data set, which allowed a more fine-grained look at conflict from 1945 to the present by lowering the conflict threshold to twenty-five battle deaths. The observation of a decline in conflict in the new data set prompted both a critique of the use of mostly non-year-specific data to gauge trends in annual fatalities (Gohdes and Price 2012) and a thoughtful rejoinder (Lacina and Gleditsch 2012).

At about the same time Sarkees, Wayman, and Singer (2003) introduced an update to the Correlates of War's data sets on international (interstate), civil (intrastate), and extra-systemic (extra-state) warfare; taking all three forms of warfare into account, they concluded that no clear upward or downward trend in the lethality of war (normalized by the population of system members) could be observed. That conclusion prompted a rejoinder from Lacina, Gleditsch, and Russett (2006), who used an extended version of the Armed Conflict data set to demonstrate both postwar and post–Cold War declines in the risk of death in battle worldwide.

Up to this point the group of scholars who had written on trends in warfare was very small by academic standards—few enough that they probably would not have filled the shuttle bus at an academic conference—and they had published their work entirely in the pages of academic monographs and journals. With the publication of Steven Pinker (2011), the decline-of-war debate made headlines worldwide. While Pinker's arguments about violence, moral evolution, and the role of the state were wide-ranging and sometimes frustratingly vague, he summarized his four main propositions about long-term trends in interstate warfare as follows: "No cycles. A big dose of randomness. An escalation, recently reversed, in the destructiveness of war. Declines in every other dimension of war, and thus in interstate war as a whole" (Pinker 2011, 192).

While Pinker's thesis garnered widespread acclaim—Bill Gates called it "the most inspiring book I've ever read"—the author's treatment of the data on war fatalities provoked a forceful reaction from Nassim Nicholas Taleb, author of *The Black Swan,* a popular science best seller about the conceptual and statistical challenges of thick-tailed data (Taleb 2007, 2012). The book's bold claims about historical trends in warfare also prompted a trio of more detailed reactions from scholars, including Taleb, who specialize to varying degrees in conflict studies and statistical methodology (Cirillo and Taleb 2016; Clauset 2018; Braumoeller 2019). While the methodologies used in the three studies differed, their conclusions were remarkably similar: The data were not consistent with a downward trend in warfare, either long term or after World War II, regardless of the measure used. Braumoeller's study even found a steady *increase* in the peacetime rate of conflict initiation from the relatively peaceful Concert of Europe period through the highly conflictual Cold War; seen from that perspective, the substantial decrease in the rate of conflict initiation at the end of the Cold War seems less like evidence of long-term decline and more like a return to normalcy.

At this writing, two recent articles have sought to qualify or overturn the conclusion that no decline in war could be seen in the data. Cunen, Hjort, and Nygård (2020) use an innovative change-point methodology and an unconventional distributional assumption (the inverse Burr) to uncover decreases in the lethality of war starting in 1950. Spagat and Van Weezel (2020) similarly find that including intrastate as well as interstate wars and measuring the lethality of war as a percentage of world population also produces a decline in the deadliness of war around 1950.

In sum, then, seventy-five years have passed since Richardson first attempted a scientific answer to the question of whether we could observe trends in warfare, and it is not unfair to say that very little consensus has emerged in that time. This is a humbling outcome, especially given the substantial number of scholars engaged in the scientific study of international conflict. Of course very few of those scholars have actually *tried* to answer the question. But as this survey of the work of those who have demonstrates, the question is considerably more challenging than it might seem.

In the next sections I turn to a discussion of some of the underlying issues that make it difficult to arrive at a scholarly consensus on the question of trends in warfare.

MEASURES OF WARFARE

As this review suggests, no single measure of war exists that can usefully assess the thesis that international warfare is in decline. Pinker's "arson analogy" from *Better Angels of Our Nature* does a good job of capturing the nature of the problem:

> There is no single answer, because "warlike" can refer to two different things. It can refer to how likely people are to go to war, or it can refer to how many people are killed when they do. Imagine two rural counties with the same size population. One of them has a hundred teenage arsonists who delight in setting forest fires. But the forests are in isolated patches, so each fire dies out before doing much damage. The other county has just two arsonists, but its forests are connected, so that a small blaze is likely to spread, as they say, like wildfire. Which county has the worse forest fire problem? One could argue it either way. As far as the amount of reckless depravity is concerned, the first county is worse; as far as the risk of serious damage is concerned, the second is. (Pinker 2011, 210)

The analogy focuses our attention usefully on two questions that we can use to begin to triangulate on the question of whether war has declined:

- Has the rate of international conflict initiation changed over time?
- Has the lethality of war changed over time?

These are very specific questions, and it's worth dwelling for a moment on what they do and do not ask.

THE RATE OF CONFLICT INITIATION

First, the term *international conflict* merits a bit of unpacking. The literature on the decline of international war is unanimous in considering full-scale wars, as defined by the Correlates of War project—"sustained combat, involving organized armed forces, resulting in a minimum of 1,000 battle-related fatalities" (Sarkees n.d., 1; see also Small and Singer 1982, chapter 2). There is less consensus regarding international conflict short of war. The Uppsala-PRIO Armed Conflict data (Gleditsch et al. 2002) focus on conflicts with at least twenty-five battle deaths; the Militarized Interstate Dispute data (Jones, Bremer, and Singer 1996) include threats, displays, or uses of force; the International Conflict Behavior (ICB) project focuses on crises, understood to be a perceived change in the state's internal or external environment that generates a threat to basic values, a high probability of military conflict, and the need for a short-term response (Brecher 1977; Brecher and Wilkenfeld 1997).[2]

The main advantage to using lower-lever conflicts as proxies for war is that full-scale international wars are relatively rare—so rare that trends are very difficult to tease out of the data. To the extent that lower-level conflicts either reflect the concept of interest (the initiation of force or violence, e.g., rather than the initiation of violence *and* its escalation past the one thousand battle deaths threshold) or represent potential wars that failed to escalate, their inclusion can be informative. If they do not, however—as in the case of many mundane Coast Guard seizures and detentions, or the sorts of attacks on vessels flagged by third parties that took place during the "Tanker War" (Gibler, Miller, and Little 2016)—their inclusion will at best add noise to the data and at worst produce illusory trends or hide real ones (Gibler and Little 2017).

While the overwhelming majority of political scientists studying conflict focus either on interstate or intrastate conflict, more recent work from outside the field (Fagan et al. 2019; Spagat and Van Weezel 2020) pool data on the two types of warfare prior to looking for trends. There are two compelling reasons not to do so. The first is that substantial evidence already exists to suggest that interstate and intrastate conflicts cannot be explained by a single theory (Levy and Thompson 2013). The second is that, because the power-law distributions that describe the fatality of intrastate and interstate conflicts have different slope coefficients (Spagat et al. 2019), a change in the slope coefficient of the aggregated data could be an artifact of a shift in the ratio of subnational to international conflicts. Indeed, the change point that Spagat and Van Weezel (2020) find in the aggregate data coincides with a significant upsurge in intrastate war initiations relative to interstate ones.[3]

A separate but equally important question has to do with how the *rate* of international conflict initiation should be calculated. A rate is a normalized frequency; it takes into account changes in the number of opportunities for conflict initiation. Because the number of states that comprise the international system has changed dramatically over the past few centuries, the raw frequency of conflict initiation would increase even if the rate remained constant. If we are interested in changes in the probability that states will fight, we have to account for the number of states (or, technically, pairs of states) that can fight—i.e., the number of "politically relevant dyads" (Maoz and Russett 1993).

The fact that data exist on politically relevant dyads should not inspire complacency, however. As Bennett (2006) and others have pointed out, some 15–25 percent of all MIDs occur between states that Maoz and Russett's original measure of political relevance would code as irrelevant. In part to address this concern, Braumoeller (2019, chapter 4) uses a continuous measure of political relevance, which allows the relationship between the covariates and the latent measure to be estimated rather than assumed (Braumoeller and Carson 2011).[4] Braumoeller uses the area under a receiver-operator curve (ROC) to gauge the measures' ability to distinguish between MIDs and non-MIDs and finds that the continuous measure does a better job than the original.

THE LETHALITY OF WAR

Just as there are different answers to the questions of what should count as conflict and how politically relevant dyads should be measured, the *lethality* of war is a contested concept. The scholars who pioneered the original Correlates of War data envisioned three distinct but useful dimensions along which the size of war might be captured: *magnitude* or nation-months of war, *severity* or raw battle deaths, and *intensity*, conceived of as either magnitude or severity divided by the pooled prewar populations of the combatants (Small and Singer 1982, chapter 3). For the most part the literature has paid less attention to magnitude as a measure of war, focusing on either severity or intensity (in the sense of severity normalized by pooled population). More recently a third measure of lethality—battle deaths divided by world population or *prevalence* (Pinker 2011; Braumoeller 2019)—has become a focus of study. To make matters worse, scholars differ on whether lethality should be measured on a war-to-war or a year-to-year basis, with different operationalizations producing different conclusions.

While at times the disagreement over which measure to use takes on some elements of a religious debate, there are nevertheless some useful principles from which to proceed when selecting a measure. If the goal is to understand war as a public health problem, in comparison with other prominent causes of death like cancer or heart disease, a per capita metric can provide a common metric for comparison. Such comparisons must be made with care, of course, as the lethality of thick-tailed phenomena like wars and epidemic disease will be deceptively low much of the time. It's true that war has killed only a tiny fraction of humanity in the past ten years. The same was true of the early to mid-1930s, right before World War II caused the deaths of about 3 percent of the global population. Short-term snapshots comparing war to other public health problems generally fail to account for its extreme volatility.

If, on the other hand, the goal is to understand the lethality of war from a social-scientific perspective—that is, as the result of human behavior and decisions—a good case can be made that the most appropriate measure to use is the one that actually influenced those decisions (Uslaner 1976, 132). This principle suggests severity as a measure in tests of behavioral theories: Within the literature on war termination and public opinion, for example, scholars are all but unanimous in arguing that raw battle deaths influence leaders' decisions (see, e.g., Mueller 1985; Gelpi, Feaver, and Reifler 2005; Kriner and Shen 2014; Sirin and Koch 2015). The case for intensity is that, in so doing, they implicitly normalize for overall population size (Braumoeller 2019, 43).

Regardless of the measure chosen, challenges to inference remain. Overall fatality numbers are so notoriously uncertain, even for recent conflicts, that scholars have chosen to use battle deaths as a proxy, but even battle deaths are often difficult to estimate with certainty (Gohdes and Price 2012; Lacina and Gleditsch 2012). While scholars have explored alternative methodologies for measuring fatality rates with greater precision—Obermeyer, Murray, and Gakidou (2008), for example, use survey data to do so and arrived at significantly higher estimates of lethality than those contained in the Uppsala-PRIO Armed Conflic data—such approaches are often applicable only to the most recent wars.

To complicate matters further, as Fazal (2014) and Cirillo (2008) have pointed out, improvements in military medicine have greatly reduced the number of battle fatalities attributable to disease, a welcome development that biases the data toward accepting the decline-of-war hypothesis. Indeed, the magnitude of the effects of battlefield medicine is so substantial—a twofold to threefold increase in the ratio of wounded to killed, according to Fazal's data—that it could very easily mask an *increase* in the frequency of soldiers' attempts to kill one another. Such an outcome would of course be deeply problematic for theories that posit that humanitarian norms and values are driving a decline in warfare.

Changes in military technology too could be expected to have an impact on the lethality of war. As mentioned earlier in this chapter, Cederman et al. (2011) argue that changes in military doctrine during the Napoleonic Wars had a lasting impact on the lethality of war. Historian Michael Howard (1976, chapter 6), looking at the same period, argues that the advent of rail transport in the middle of the 1800s, which dramatically extended states' ability to move and supply troops, is an equally if not more plausible driver of lethality. Today's weapons of war are becoming more lethal and more capable of precision targeting, a trend that could point in either direction. More confounding still, because the distribution of lethal war technology throughout the system remains highly unequal, its effects will be spread out over time and across space.

Studies of the lethality of war, moreover, are typically agnostic regarding one of the biggest processes by which wars become lethal—conflict diffusion or contagion (Kadera 1998; Beardsley 2011), which can be enhanced or suppressed by military alliances (McDonald and Rosecrance 1985; Siverson and Starr 1989; Christensen and Snyder 1990; Leeds 2003; Colaresi and Thompson 2005; Fang, Johnson, and Leeds 2014). The most *severe* wars, unambiguously, are the two world wars, which achieved high levels of lethality in absolute terms by virtue of having spread to cover much of the globe. The most *intense* wars, on the other hand, are the Paraguayan War of the 1860s, the Chaco War of the 1930s, and the Iran-Iraq War of the 1980s, all of which took breathtaking tolls on their populations but remained relatively contained. One measure, therefore, very much privileges wars that spread, while the other does not.

Measurement, in short, is a major challenge facing the growing literature on systemic trends in warfare. While many thousands of person-hours have gone into collecting, cleaning, and disseminating the data that form the core of these studies, there are very real limits to the accuracy and precision of what we can know about the events they describe. Pushing those limits and characterizing the uncertainty that remains should be high priorities for research in the short term, as much else depends on them.

Moreover, the lack of consensus on which measures best reflect quantities of interest produces a disturbingly high number of "researcher degrees of freedom" that both invite *p*-hacking and complicate attempts to reach consensus. Exemplary in this regard is Spagat and Van Weezel (2020), which varies five elements of the research design to produce hundreds of p-tests grouped into sixteen distinct scenarios, only two of which yield a statistically significant change point.[5] The authors nevertheless conclude that the findings "shift the debate in favour of the decline-of-war thesis" (140).

METHODOLOGICAL ISSUES

This review also reveals a diversity of methodological approaches to the study of trends in warfare. Leaving aside studies that eschew statistical inference entirely (e.g., Pinker 2011; see also Pinker 2012; Taleb 2012), two methodological issues come to the fore: the challenges of dealing with the thick-tailed data that are associated with lethality, and the problem of finding significant change points in time-series data.

The most substantial methodological issue with early studies of war severity and intensity has to do with the statistical properties of extremely thick-tailed data. While most social science statistics courses note that the central limit theorem works best with normally distributed data, there is little appreciation for how poorly it works with extremely skewed data or for what its failure implies.

Figure 16.1 makes this point graphically by charting the sample means of data drawn from normal and power-law distributions as the number of observations goes from one to five thousand. Each panel contains twenty lines that represent separate runs with different random seeds. In the top panel we can see that draws from a normal distribution quickly converge to the population mean (the dashed horizontal line). The bottom panel, by contrast, is an inferential nightmare. While most lines trend somewhere well below the population mean, some do not, and even fairly large samples may have means that are twice the population mean.

A standard approach to dealing with this problem has been to log the fatalities measure to induce normality and then include it in a standard generalized linear model of some sort (see, e.g., Weisiger 2013). As Clauset, Shalizi, and Newman (2009, 690–92) and Packard (2009) point out, however, this approach is seriously flawed. Understanding their solution requires a brief discussion of the nature of power-law distributions.

A power-law relationship is one in which one variable varies as a function of the power of another. In the simple and common case most often discussed in the conflict literature, the probability that a war will exceed a certain level of lethality is a function of that lethality raised to some power: $\Pr(X>x) \sim x^{-\alpha}$. The exponent, α, is referred to as the *slope coefficient* because it captures the slope of the characteristic power-law line in log-log space, as illustrated in Figure 16.2.

While power-law calculations are initially a challenge to intuition, they yield a straightforward interpretation: A war that is x times the lethality of another war is x^{α} times as rare (see Table 16.1). So for $\alpha = 2$, a war that is twice as lethal as another is four times as rare, and a war that is three times as lethal is nine times as rare; but if $\alpha = 3$, those wars are eight times and twenty-seven times as rare, respectively. Larger values of α indicate that large wars are even more rare, while smaller values indicate that they are more common.

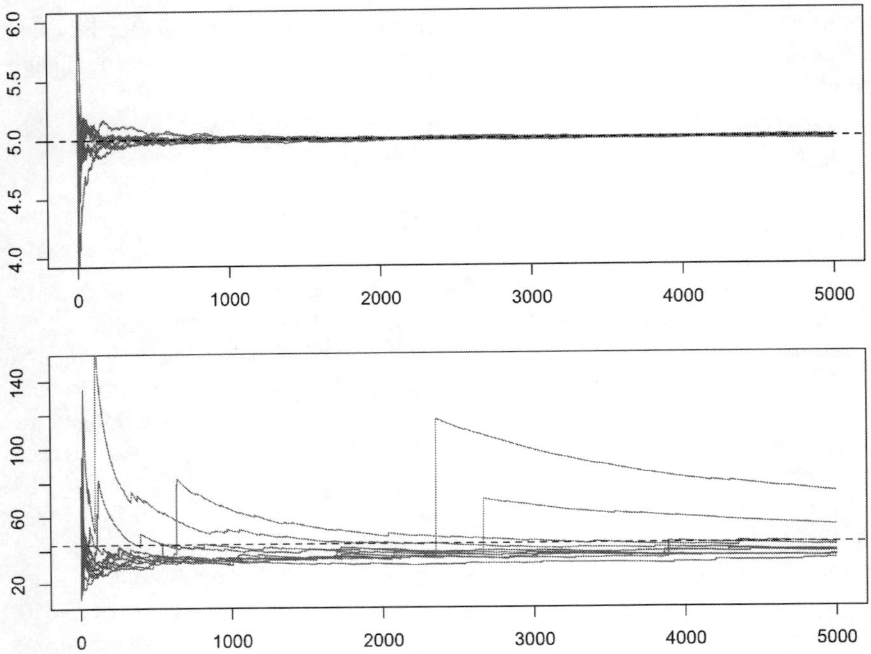

Figure 16.1. The failure of the central limit theorem when data are extremely skewed. Running means of draws from a normal distribution quickly converge to the population mean, while running means of a power-law distribution fail to do so even in large samples.

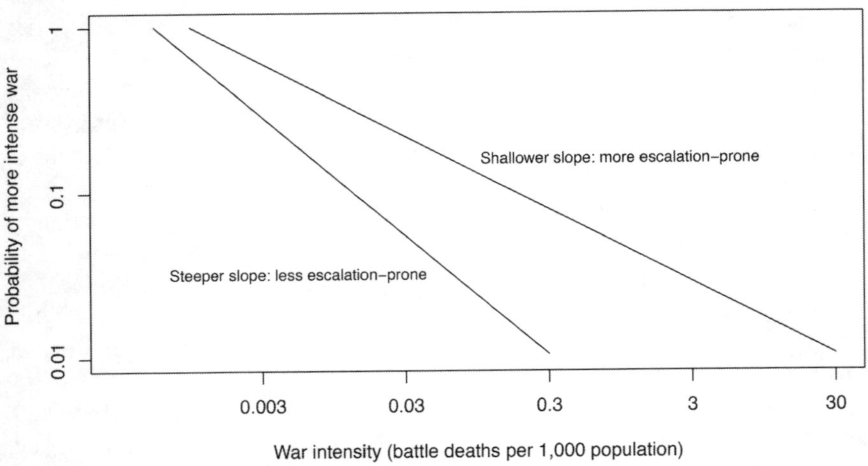

Figure 16.2. Two lines describing power-law relationships. The steeper slope, which reflects a larger slope coefficient (α), indicates a set of wars that are less escalation-prone, while the shallower slope indicates a set of wars that are more escalation-prone.

Table 16.1. The power-law distribution of war fatalities implies that, while most wars will kill few people, there is a nontrivial probability of war becoming incredibly lethal. Rounded fatality calculations are from Braumoeller's (2019) estimates of post–World War II power-law parameters, based on the methodology of Clauset et al. (2009).

With this probability...	90%	70%	50%	30%	10%	1%	0.1%
... a new war will kill at least this many people	1,300	2,100	4,000	11,000	92,000	8,000,000	700,000,000

Aside from its substantive interpretation, the slope coefficient is of considerable methodological importance: devising a statistical test based on it is the key to getting around the challenges that power law-distributed data represent for the central limit theorem. Simply put, the slope coefficient is better behaved in small samples than the mean, and the mean is a function of the slope coefficient. By comparing slope coefficients we can draw inferences about the relative lethality of subsets of wars even with relatively small samples.

Clauset et al. (2009) propose what has become a widely accepted methodology for evaluating the fit of data to a power-law distribution, easily executed in the R programming language (Gillespie 2015). While that methodology is not designed to test for change points in power-law data, Cederman et al. (2011) advocate a two-step approach—a standard Kolmogorov-Smirnov test to look for candidate change points, followed by an F-test of the ratio of the slope coefficients to confirm them. Braumoeller (2019, chapter 5, 238–42) modifies and extends this procedure by adding an Anderson-Darling test, a modification of the Kolmogorov-Smirnov test that weights data in the tails more heavily, and using the bootstrap to test for equality of slope coefficients, as estimated via Clauset et al.'s method.

Different inferential strategies are of course possible. Clauset (2018), for example, looks for a post-1945 decline in the severity of war by simulating post-1945 wars based on the parameters of a pre-1945 model and showing that actual post-1945 outcomes are not at all inconsistent with the prewar data-generating process.

To date these methods have been used to assess the impact of a single binary variable—typically, time period—on lethality. More recently Cunen et al. (2020) have introduced a change-point detection algorithm for thick-tailed data, first developed in Cunen, Hermansen, and Hjort (2018), and use its results to argue in favor of a decline of the lethality of war after 1950. The technique differs from those that precede it in that it utilizes the full range of the data; because power-law behavior typically occurs only in the thick tail of the data, the earlier studies listed all discard some range of less lethal wars when estimating the slope coefficient.

Informed opinion differs on the question of whether this is the best way to look for changes in war's lethality. The argument in favor of doing so is that all observations, not just the largest, are relevant to the question of whether two samples differ significantly. The argument against is that the curve of best fit through the bulk of the data does not always provide a good fit to the tail, and the slope of the power-law tail is the quantity of interest when comparing the escalatory behavior of wars. Indeed, the post-1950 curve (Cunen et al. 2020, Figure 4B) fits the data for the largest wars quite poorly. The question of whether to believe these findings or the power-law results, which the authors also conclude show no change, cannot be resolved statistically: both findings are correct given their assumptions about the distribution of war lethality. We need more insights into the *theoretical* mechanisms that produce war lethality before we can make an informed judgment about which statistical distribution is appropriate.

Compared to the distribution of fatalities, the distribution of conflict onsets is not nearly as skewed, though its degree of skew should inspire at least some caution among scholars inclined to use standard generalized linear models to look for change points. Fortunately recent developments in change-point detection algorithms have made powerful and very general techniques easily available (e.g., James and Matteson 2014; Matteson and James 2014; see also the application in Braumoeller 2019, chapter 4).

The overall picture on the methodological front is thus relatively bright, though challenges remain. The past decade has brought methodological advances that have revolutionized the study of trends in interstate conflict and pushed the debate forward in interesting and useful ways. Further developments and debates can be anticipated of course, and to the best of the author's knowledge nothing as neat and tidy as an R package that will permit easy multivariate inference about the covariates that drive changes in power-law distributions currently exists. But these are soluble problems, and new work continues to expand what we can know about trends in the lethality of war.

THEORETICAL MECHANISMS

Perhaps the greatest opportunity and the most urgent priority in this literature lies in the area of theory. Surprisingly little is known about why wars escalate so dramatically. While quite a bit more is known about conflict initiation, most extant theory is situated at the dyadic level, and generalization from countries or dyads to the level of the international system is not necessarily intuitive.

The Lethality of War

Empirical studies have repeatedly shown that war fatalities are plausibly power-law distributed and, further, that their escalatory potential is alarming: as Cirillo and Taleb (2019, 2) put it, war is "the mother of fat tails, far worse than the popular 80/20 rule: there are few phenomena such as fluid turbulence or thermal spikes on the surface of the sun that can rival the fat-tailedness of violence." When wars do snowball, in other words, they can easily become incredibly, spectacularly lethal.[6]

What we don't know is *why* wars escalate with such ferocity. Given their remarkable potential to kill large numbers of people, understanding the process by which they do so should be a high priority both for social science and for humanity. Yet at present we know shockingly little.

That lack of knowledge is not due to a lack of ideas. Dozens of theoretical mechanisms produce power law-distributed data (Andriani and McKelvey 2009), and more than a few have been proposed to explain the lethality of war. In no particular order, they include:

- *Self-organized criticality* (Bak, Tang, and Wiesenfeld 1987), both in its most general form (Cioffi-Revilla and Midlarsky 2013) and with forest fire dynamics (Roberts and Turcotte 1998), natural disasters (Chatterjee and Chakrabarti 2017), and state-building (Cederman 1997, 2003) as specific models
- A two-sided *gamblers' ruin*, a model of the process by which two actors gambling against one another eventually produce a loser (Braumoeller 2019)
- The Matthew effect (Merton 1968), a process by which initial advantages accumulate, named after the Parable of the Talents in the biblical book of Matthew, the source of the common observation that "the rich get richer" (Pinker 2011)
- The Red Queen effect (Van Valen 1973), an evolutionary hypothesis that explains increases in relative fitness as a function of competitive selection pressures (Johnson et al. 2011)
- A unified theory of human violence across levels of aggregation, first suggested by Richardson (1948) and expanded by Pinker (2011) and Spagat et al. (2019); see also Levy and Thompson (2013) for a critique

At the same time we have no shortage of empirical studies that explore the relationship of various factors, such as arms races, balances of capabilities, and domestic political audience costs, to escalation—conventionally understood as escalation *to* war rather than escalation *within* war (Wallace 1982; Diehl 1983; Huth, Gelpi, and Bennett 1993; Fearon 1994; Partell 1997; Sample 1997; Smith 1999; Reed 2000). Extensions of these insights to within-war escalation has been relatively rare, however, and at present none has been shown to explain the power-law distribution of war fatalities.

The problem, then, is not an absence of ideas about why wars escalate, but rather an inability to distinguish among them. All of the candidate explanations point to an essentially stochastic process of escalation; they differ only in the mechanism that gives rise to that process. The similarity of their empirical implications may make it exceptionally difficult to distinguish among them solely based on data on escalation.

Oddly enough, although we know very little about the reasons for which war intensity or severity follows a thick-tailed distribution, we *do* know something about why wars get more or less deadly—though we don't know much. In a book published thirty-five years after the inception of the Correlates of War project, Professors Daniel Geller and J. David Singer reviewed more than five hundred studies that had utilized Correlates of War data (Geller and Singer 1998). While they listed sixteen factors that had consistently been shown to correlate with the onset of conflict, they noted only two that consistently correlate with duration or severity: the involvement of one or more major power, and the existence of a highly polarized alliance system. While these correlations are suggestive, it's not clear exactly what to make of them. Both the involvement of major powers in war and the existence of highly polarized alliance systems prior to war could produce bigger and deadlier wars, or they could be indicative that the conflict that prompted war was more serious to begin with.[7]

More recently, as noted earlier in this chapter, Cederman et al. (2011) have argued that military doctrine can have a dramatic impact on the severity of international war. They examine the case of the French *levée en masse*, the system of mass conscription implemented by the Convention nationale in 1793. Until that point European wars had largely been fought by small, professional militaries led by aristocratic officers. By pressing all able-bodied young men into service and putting the rest of society to work in the service of the state, the French threatened to overwhelm their opponents, who were forced to adopt similar measures in their own armies. Although European rulers managed to put the Napoleonic genie back in the bottle temporarily by re-forming their postwar armies along eighteenth-century lines, the advent of rail transport in the mid-1800s soon solved two major problems—moving and supplying troops—that had generally limited the size of armies to eighty thousand men. By 1870 Germany was able to deploy 1,200,000 men against France—twice the number that Napoleon had led into Russia (Howard 1976, chapter 6). As Cederman et al. demonstrate, this fundamental change in military doctrine corresponded to a significant increase in the lethality of the wars that countries fight.[8]

As the example of the railroad suggests, technology may be one key to understanding changes in the parameters that govern escalation. Railroads and the *levée en masse* had one obvious effect: putting more human beings on the battlefield to try to kill other human beings. This straightforward increment in the lethality of each step on the escalation ladder could very plausibly reduce the overall lethality of the war, even absent changes in the stochastic process that drives states from one rung on the escalation ladder to the next. It is not hard to imagine other technological advances driving escalation: Cirillo and Taleb (2015) have argued, for example, that the lethality of military weapons could theoretically produce a shift toward deadlier conflicts.

At the same time technological advances can reduce the lethality of warfare, as Fazal (2014) has shown in the case of military medicine. Moreover, the advent of drone warfare may well have the opposite effect of the *levée en masse* by significantly reducing the number of deaths required to achieve battlefield goals. One caveat is in order: technologies that make wars less costly may have the perverse effect of making them more common (Jervis 1984).

More candidate explanations for drivers of the parameters that govern escalation are needed, and given the lethality of war they should be seen as a high priority. While it is not essential that these explanations themselves be capable of generating power-law distributions of lethality, there should at minimum be a bridge to the processes that do. So, for example, while the profitability of peace and the rise of the state have been proposed as determinants of conflict lethality (e.g., Gat 2008, 2013), their connection to processes of escalation remain unclear.

One promising candidate is described by Diehl, Owsiak, and Goertz (chapter 10 in the present volume), who argue that declining territorial conflict has helped account for the growth of peace.

The Rate of Conflict Initiation

As far as theoretical explanations for trends in conflict initiation are concerned, the picture is much different. As the rest of this book documents, the conflict studies community has developed and tested a staggering array of explanations of conflict at the dyadic level. Occasionally, when the prevalence of a cause changes over time, these dyadic effects have been generalized to the level of the international system. The most obvious example of a candidate cause of systemic peace is the spread of democracy (Gleditsch and Hegre 1997; Mitchell, Gates, and Hegre 1999; Gartzke and Weisiger 2013; Crescenzi and Kadera 2016). Other candidates, explored elsewhere in this book, include changes in the polarity of the international system, trends in the issues that give rise to war, and trends in institutionalization, conflict management, security communities, the spread of nuclear weapons, and so on.

There is in principle no reason that the sorts of monadic or dyadic phenomena that we typically study as sources of conflict behavior could not explain systemic outcomes. There is also no reason to believe that they aggregate unproblematically and in a direct, linear fashion to outcomes at the level of the international system. The entire field of complex systems is founded on the insight that outcomes at higher levels of aggregation cannot easily be inferred from behavior at lower levels; in Thomas Schelling's memorable formulation, "micromotives" do not automatically translate into "macrobehavior" (Schelling 1978).

Even in an apparently simple case such as the spread of democracy, which by most accounts pacifies dyads unconditionally, the systemic implications are not obvious. It makes little sense, for example, to expect the democratization of Bolivia and Botswana to increase the prospects for peace between them, as they likely could not fight in the first place. Once we remove such politically irrelevant dyads from the sample, however, we find that the topography of the network of international connections heavily conditions the impact of democratization on systemic peace. If democratization begins with the major powers, which are most central to the network, and proceeds outward in order of centrality, the impact of democratization on systemic conflict will be roughly linear. If, by contrast, democratization begins in the periphery, 80–90 percent of the states in the system would have to democratize before even a modest impact on systemic peace would be felt (see Figure 16.3). The reason is straightforward: the states most likely to democratize in the latter scenario are also the states least able to reach one another.

Topography is of course far from the only factor that complicates the connections between monadic processes and systemic outcomes. Because aggregation across levels so often produces emergent behavior, nothing short of actually taking a complex-systems approach to modeling it will suffice to bridge the gap. In short, as theorists from a wide range of international relations traditions who agree on little else (Waltz 1979; Jervis 1997; Wendt 1999; Braumoeller 2013) have repeatedly pointed out, systemic insights require systemic theory.

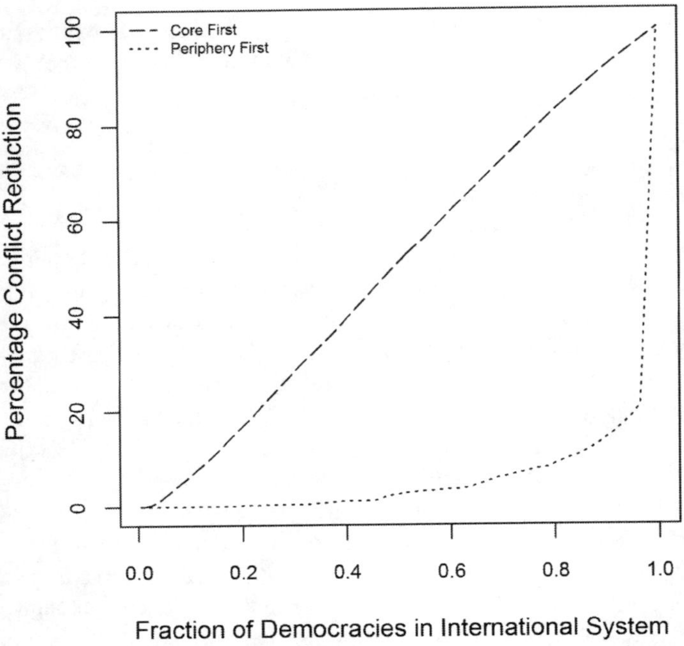

Figure 16.3. The relationship between monadic democracy and systemic conflict, conditional on order of democratization, assuming for the sake of illustration that the probability that democratic dyads will fight is zero.

CONCLUSION

Given the history of marked inattention to the question of systemic trends in interstate conflict and the recent surge in interest on the issue, it is perhaps not surprising that this chapter has focused as much on what we don't know about war as on what we do. Quite a bit remains to be done in this literature, from improving measurements, to developing and popularizing new methods, to understanding conflict escalation and recasting processes of conflict initiation in systemic perspective.

At the same time vastly more progress has been made on understanding systemic trends in interstate conflict in the past ten years than had been made in the sixty years prior. Where the empirical study of trends and outcomes had previously been limited to a small number of isolated efforts, a growing community of scholars has now turned to the task with focus and energy. While their early efforts were clearly inspired by the gauntlet thrown down by Pinker the debate over whether Pinker was right has gradually taken a back seat to a more diverse and (to academics, anyway) more interesting question of what drives changes in warfare large enough to be felt across the international system. Such research is welcome, and at the present moment, very much needed.

NOTES

1. The most memorable application of the Poisson distribution to data was Bortkiewicz's (1898) demonstration that the number of Prussian soldiers kicked to death by horses conformed closely to a Poisson distribution.

2. Following Pinker (2011), a growing number of studies use data from Brecke's "Conflict Catalog" (Brecke 1999) to measure trends in even very low-level violent conflict going back to AD 1400 (see, e.g., Iyigun, Nunn, and Qian 2017; Martelloni, Patti, and Bardi 2019). While the availability of this data set, recently extended to AD 900 (Lee et al. 2019), makes long-range analysis tempting, the scholar who compiled the data set is candid about the fact that it is unfinished and contains substantial errors, especially in some regions and especially in earlier periods (Brecke 2012). While it could hardly be otherwise, of course, few who use the data seem aware of its severe limitations. Bavel et al.'s (2019, 7) critique of the data should strike fear into the heart of even the most adventuresome.

3. A *change point* is a point at which the probability distribution of data changes significantly. In the decline-of-war literature a change point is typically taken to mean a point at which either the lethality of war or the rate of conflict initiation changes to a greater degree than we would expect it to by chance.

4. Specifically, political relevance = $\Lambda(4.801 + 4.50 \times \text{contiguity} - 1.051 \times \log(\text{distance}) + 2.901 \times \text{major power})$. Here, $\Lambda()$ denotes the standard logistic function, $f(x)=1/(1+e^{-x})$. Discrimination of fit was evaluated using the area under a receiver-operator curve (ROC).

5. The first four elements are the data set used, the measure used (severity vs. prevalence), the proposed temporal cut-point, and the type of war included (only interstate vs. pooled interstate and intrastate). Those binary elements yield sixteen scenarios, only eight of which are presented; the authors' admission that "adjusting for world population levels is essential to get anything resembling the results in this chapter" (138) indicates that the other eight do not yield significant results. Within each scenario the fifth element, the lower bound of the power-law distribution, is varied continuously, yielding something like fifty distinct *p*-tests per scenario, for a grand total in the neighborhood of eight hundred *p*-values, which are not adjusted to take into account multiple inference.

6. It is worth noting that, because power-law distributions generally have a lower limit, not all wars can be expected to snowball: only those that reach a certain size can be expected to exhibit shocking escalatory behavior. This mathematical point could have useful connections to the argument made by Vasquez and Valeriano (2010) that major systemic wars are a different category of wars with distinct causes.

7. Some isolated studies have taken on the issue of war intensity and severity: Goldstein (1988), for example, argues that war intensity is related to long, cyclical variations in the global economy, and Hopf (1991) demonstrates that war intensity is unrelated to polarity, or the number of great powers in the international system at a given time.

8. Cederman et al. (2011, 623–24) put the timing of this change between 1780 and 1790, which is puzzling given that Howard (1976) points out that European armies continued to be formed along eighteenth-century lines until the 1860s. The most reasonable interpretation, I think, is that the nationalist wars of the early 1800s were small enough and sufficiently localized not to run up against the technological limitations on the movement and supply of troops.

Part IV

CONCLUSION

Chapter Seventeen

Some Brief Observations on the Contemporary Study of War

William R. Thompson

A half century ago I was supposed to write a critique of quantitative work on war for a directed readings course.[1] I never completed the course because I was unsure where to begin and had other topics that needed attention.[2] Five decades later I have been a sometime participant in the study of war yet I remain skeptical about some aspects of our collective undertaking. By no means does my skepticism extend to the entirety of the study of war. We have better arguments, better data, and richer findings than ever before. Readers are invited to go back to the literature of the late 1960s and early 1970s and see for themselves just how thin our initial understanding of war behavior was back then in comparison to the work summarized in the three editions of *What Do We Know about War?* The study of war has definitely progressed just as it still possesses problem areas that need more or different kinds of attention. These areas fall into two clusters: theory and cumulation/conceptualization/operationalization. Rather than praise the progress, I prefer to curmudgeonly harp on how we might improve our craft further.

THEORY

A few years ago Mearsheimer and Walt (2013) made a bit of a splash with a paper criticizing international relations (IR) work as addicted to simplistic hypothesis testing. While they explicitly disavowed their paper as a *cri de coeur* by two grumpy realists who were opposed to quantitative analysis, they did publish the paper in a journal that is most unlikely to ever publish quantitative analysis, they have never done quantitative analysis themselves, and their evident target was in fact quantitative analysis in IR. The folk wisdom that if it sounds and walks like a duck, it probably is a duck seems to pertain. All that notwithstanding, they have a point but not exactly for the reasons they advance. Moreover, the point applies directly to the study of war, which has been a prominent area for quantitative applications within IR scholarship.

Their argument begins with the uncontestable assertion that theory creation and hypothesis testing are both central to IR scholarship. Of the two, they privilege theory creation over hypothesis testing as the more important of the two types of activity. That may be true in the sense that without theory creation hypothesis testing is difficult to justify, but one can also ask what is the point of theory creation if the outcome goes untested or remains untestable? It is better to think of them as complementary activities as opposed to one being inherently superior to the other.

Mearsheimer and Walt also go astray or, one might say, reveal their true colors by the way in which they justify the privileging of theory over hypothesis testing. They note that the most famous and prestigious scholars in the IR field according to surveys are people known for their theories more than for their empirical work. Most IR books they consider classics are theoretical in nature. Third, grand theories, the "isms," are prominent in graduate course syllabi and the beginning chapters of undergraduate texts. While they acknowledge that middle range and formal theories have roles to play, it is fairly clear that, for them, theory is primarily synonymous with the isms (e.g., realism, liberalism, constructivism). The most prominent scholars are central progenitors of the isms. Classic readings are classic because they advance "ismatic" arguments. Yet Mearsheimer and Walt lament that job postings in the field are more likely to ask for hypothesis testing skills than theory—read ism—proficiency.

The problem here is that the isms have not been all that productive for the creation of empirical theory (see, for example, Lake 2011). Isms themselves are clusters of assumptions about how the world works that have attitudinal (pessimistic, optimistic) and ideological (conservative, liberal) connotations (Rathbun 2012). Whatever else they may be, they tend to be pre-theoretical. That generalization does not mean that they cannot be used to generate empirical theory. Mearsheimer and Walt have both demonstrated that realism can be used to develop testable theories. Others have used liberal and constructivist arguments to develop testable notions about the democratic peace and the impact of ideas. Yet, by and large, most empirical theory in IR and war studies are not directly linked to the isms and, even if they were, probably should not be.[3] I come back to the second part of this assertion later in this chapter.

Mearsheimer and Walt (2013, 431) are back on the right track when they describe theories as simplifications of reality that single out selected factors to provide a causal story.

> A theory says how [key conceptualizations and variables] are defined, which involves making assumptions about the key actors. Theories also identify how independent, intervening, and dependent variables fit together which enables us to infer testable hypotheses. … Most importantly, a theory explains why a particular hypothesis should be true, by identifying the causal mechanism that produces the expected outcome(s). Those mechanisms—that are often unobservable—are supposed to reflect what is actually happening in the real world.

Isms per se do not perform these functions but theories should, and the better they do these things, the stronger, presumably, is the theory. What Mearsheimer and Walt dislike, "simplistic hypothesis testing," does not perform these same functions very well. As they characterize the object of their scorn, the recipe for simplistic hypothesis testing involves:

a. selecting a dependent variable,
b. then selecting one or more independent variables that might account for some variation in the dependent variable,
c. finding a data set(s) that contains appropriate measures of the variables in question,
d. testing hypothesized linkages between the independent and dependent variables,
e. anointing the relationships with statistical significance as the main assessment criterion,
f. advancing the research outcomes as contributions to our knowledge of international behavior.

Put differently, Mearsheimer and Walt's main complaint is that many (most? all?) quantitative analyses do a better job at testing hypotheses than they do drawing the hypotheses from explicit theories or constructing appropriate theories when necessary. I think it is fair to say that many of us in the study of war, as well as in the study of IR and exempting formal theorists, are guilty to varying extents of putting more effort into hypothesis testing than empirical theory construction and hypothesis derivation.[4] One very real motivation for doing so is that journal reviewers often put more stress on statistical analysis than on the theoretical value of the findings. This observation is not a call for less emphasis on methodological questions. Neither is it meant to suggest that reviewers rarely say anything about theory. But there is a tendency to either like or dislike what theories say as opposed to assessing theoretical coherence or the net theoretical value added. At the same time, executing the statistical analyses appropriately is rarely a simple or easy matter. Care must be exercised to get things as right as one can.[5]

But the problem remains that too many analysts of war are more comfortable with hypothesis testing than they are with theory construction and evaluation. Part of the problem goes back to the culture of IR scholarship in respect to folkways that encourage embedding theoretical arguments in discursive verbiage. It is not always easy to extract exactly what the theory is in such presentations. Agreement on what a less-than-explicit theory says, as a consequence, is also less likely. The solution is to make our theoretical arguments more explicit and spend more time delineating the causal mechanisms at work. That has not been our custom, but it is a trait that could be turned around to considerable benefit.[6]

Still, making theory construction more explicit is not enough. We also need more complicated theories that integrate multiple levels of analysis and multiple independent variables.[7] The modal tendency in the study of war has been to choose one level of analysis and one topical independent variable to explain war. Thus we have systemic, national, and individual theories and analyses. We also have arguments that highlight hegemony, territory, capabilities, or misperceptions as the key to war behavior. What are needed are more complicated arguments linking all four of these (levels and variables) or other levels and variables to better account for the complexities of war behavior. This is hardly a new complaint—just one that, unfortunately, remains germane despite considerable explanatory progress. Neither does the complaint imply that we do not have attempts at doing just that—that is, theories linking multiple levels and independent variables.[8] We have some but they have been long in coming, remain few in number, and need rival interpretations of how the levels and variables go together. Another way of putting it is that we need more arguments about war that mix and match the presumed complexities of warfare—as opposed to singling out another dimension of the warfare elephant for close scrutiny. We have a lot of the latter; we need more of the former if we are to continue making progress on explaining the onset of war.

One desideratum that falls in the theory section and overlaps with the section to come on conceptualization/operationalization is that we need to reinvigorate the notion of levels of analysis. Yet I hasten to add that we do not need to restore levels of analysis discourse as existed before. When the levels of analysis discussion emerged with three, four, or five layers (take your pick), we acknowledged their existence and then went our separate ways. Each analyst more or less picked the level with which they were most comfortable and largely ignored what went on in the others. Then came dyadic analyses in the 1990s. Their success, partially enabled by very large N sizes, blew the other levels out of the water for a period of time. Now other levels are starting to reemerge thanks to changing geopolitics and perhaps the growing realization that everything of interest in IR is not a bilateral, state-to-state interaction. Hostility among major powers has returned, giving some restored credence to systemic-level analysis. The much overlooked regional level of analysis is getting a second look because regional differences seem more pronounced—or maybe they always were and we were not paying sufficient attention. The individual level of analysis is also on the uptake thanks to new arguments and data and perhaps even more pronounced decision maker quirks than ever before.

Greater complexity in theory construction demands that we combine levels of analysis—as opposed to treating them as autonomous theaters of operation. No doubt that will not be easy. But it seems to reflect reality a little better than proceeding as if only individuals matter or that regional and systemic contexts matter little.

One more reason for more complex theories is that less complex theories whittle down reality parsimoniously. That is something theories are supposed to do. But the generation of too many simple theories on the same or similar topics leads to a lot of theoretical fratricide. Here I have in mind missile fratricide that occurs when too many missiles are fired simultaneously in a narrow space. Some of these missiles will destroy some portion of the other missiles fired at the same time. I do not really know whether too many narrow theories are apt to destroy one another. But when we have arguments that say a + b leads to z, b + c leads to z, and c + d leads to z as well, would it not be obvious to see whether an all-encompassing theory that combines a, b, c, and d could be developed to supplant the three more narrow theories on z?[9] Unfortunately that does not seem to be part of our research culture. We are more likely to generate more narrow theories introducing e, f, and g to the explanation of z. Lest one think this complaint is an exaggeration, we do have at least seventeen theories/models of rivalry termination (Thompson 2018b). How many interpretations of the democratic peace are out there? For that matter, how many different ways are there to account theoretically for peace?[10] Rather than only continuing to generate new explanations, some consolidation efforts might be useful.

I echo others as well in predicting that hierarchy will prove to be a more critical variable in most complex theories than will our peculiar hang-up with anarchy (Lake 2009; McDonald 2015; Bially Mattern and Zarakol 2016; Zarakol 2017; Volgy et al. 2018; Geller and Travlos 2019).[11] Anarchy allegedly differentiates domestic politics from international politics. Domestic politics has central government and international politics does not. That may be a useful distinction for some purposes, but it is treating the distinction as a constant that gets us into trouble. To simplify greatly, the more hierarchy that exists at various levels, the more governance is possible and probable. Governance in international politics sometimes looks like domestic-style legislature products, but it often comes in rules laid down by stronger states. For that matter, governance in domestic politics sometimes looks like rules established by stronger actors too. But if we take levels of analysis seriously, then what is interesting is how global, regional, and national hierarchies interact—not just whether there are hierarchical arrangements. More to the point perhaps, does hierarchy restrain conflict behavior? These questions will not be easy problems to resolve either.

CUMULATION, CONCEPTUALIZATION, AND OPERATIONALIZATION

One of the more frustrating aspects of war etiology studies is the lingering path dependencies pertaining to rather basic conceptualization and operationalization questions. As a consequence, we have less cumulation than we should and we still have to spend time debating basic issues. For example, we still treat the Composite Indicator of National Capability (CINC) as a sort of reflexive icon of operationalization even though its substantive implications are so obviously wrong (Kadera and Sorokin 2004; Rausch 2017). Any indicator that portrays both the USSR and pre-twenty-first-century China as transiting past the United States has dubious face validity as a measure of capability. Its mixture of quantity and quality indicators may be attractive in the abstract until one thinks about how those indicators are distributed in the annals of IR. By and large, states lead on some indicators and not on others. What happens then is that once we assume that all of the six indicators are of equal significance, states near the top of the scale tend to be lumped together as roughly equal, powerful states. For great or major powers, this tendency corresponds to a curious conceptualization that all members of this elite group are more or less alike. That has never been the case. Some major powers privilege sea power over land power while others reverse the bias.

Some elite states are content to focus on their relative standing in the home region while others have focused on expanding market and political control far away from the home region. Some states are large with agrarian-based economies. Other states are relatively small but are dynamos in trade and innovation. These distinctions have made and continue to make for behavioral differences in major power behavior. If nothing else, it means that some major powers can project power over long distances while others cannot. Why this elementary observation is not more fundamentally enshrined in theoretical arguments is a puzzle that has endured since the discussion of great powers began in the early to mid-nineteenth century. As long as the issue is largely ignored or suppressed, one can expect less explanatory progress than might otherwise be the case.[12]

Another variable that has been around with some prominence for a considerable period of time is polarity. We spend a fair amount of time talking about the implications of unipolarity, bipolarity, and multipolarity but very rarely is there any effort to operationalize what these distributions of power mean other than counting the number of powers.[13] Yet since we do not usually have minimal qualifications for designating states as major powers, the process tends to be circular, like a dog chasing its tail. Moreover, since we do not operationalize these terms we also tend not to appreciate some of their auxiliary characteristics. For instance, bipolarity, I would argue, tends to be asymmetrical (or at least it has in the past hundred years) and unipolarity is not a constant but is highly subject to entropy if nothing else. It may very well prove to be that the dynamics of polarity are more important than the categorizations. If bipolarity tends to be asymmetrical, is it whether the second-ranking state is catching up or someone believes that they are catching up that is more important than whether there are two states that are more powerful than all others? If unipolarity tends toward erosion, do very strong unipolarity situations work the same way as very weak unipolarity structures? What if unipolarity is based exclusively on military capabilities and not on economic capabilities? Should not that make some difference?

Alternatively there is some tendency to treat polarity as a condition that is immune to interacting with other variables (see my earlier comment on the need for more complex arguments). What if polarity does interact with the other variables to make some difference? Tunsjø (2018), for instance, argues that analysts talk about bipolarity as if it influences behavior universally and without consideration for temporal differences.[14] If one contemplates comparing bipolar situations, however, geographical configurations might make some difference. A US-China bipolar confrontation centered on East Asia might work much differently than a US-Soviet confrontation focused on Europe. Why? The European region has few obstacles to territorial expansion and occupation by a large land army. To deter an attack, one has to threaten extreme nuclear consequences in order to make the effort too costly. East Asia, in contrast, is part continental and part maritime. Occupation by the largest land army would not be as easy and therefore the reliance on threatening extreme nuclear consequences may be less necessary to maintaining a stalemate between an aspiring continental hegemon and a defending maritime coalition. Thus, assuming bipolarity characterized the Cold War, the way the second half of the twentieth century played out may not be the best guide to the way the first half of the twenty-first century may behave—again, assuming the Sino-American rivalry takes on a bipolar structure in East Asia.[15]

Not surprisingly, I think rivalry recidivism in conflict behavior should matter a great deal and occupy a more salient starting position than it does now. I will not trot out differences of opinion on how best to measure rivalry. More important is the idea that most actors in international politics do not engage in much conflict. Most conflict is about antagonists repeatedly beating up on each other. We should make better use of that fact both in theorizing and in developing research designs to test theories. The way it tends to work now is that analysts either work on rivalries per se or they work on some other topic within war studies. What I am advocating is that the other topics within war studies need more rivalry in their theoretical arguments. Processes of rivalry are too important to leave to scholars who choose to study rivalries.

Rivalry topics can be encompassed by four categories: origins, fluctuations in hostility, terminations, and consequences. We know a fair amount about origins (primarily spatial, positional, and ideological in nature) and, as previously noted, have a number of ideas about how they end. We do not have a good fix yet on why rivalry relationships go hot and cold.[16] I think it is fair to say that we know very little about what difference the existence of rivalries has on the countries that participate in them.

The study of arms races remains incomplete as far as I am concerned. Some might even argue that their early study helped spawn quantitative IR, courtesy of an unusually concerned meteorologist and former ambulance driver. Yet Richardson focused on military expenditures and so have most subsequent analyses. I would be willing to consider rising military expenditures as one species of arms race but there is always the interpretation problem that they reflect general preparations for war. As a measure of the perception of increasing tensions and heightened threat expectations they could have some interest. People preparing intensely for war may be more likely to participate in war along the lines of thought that paranoids sometimes have reasons to be ultra-suspicious. But another species of arms race—as reflected in the racheting up of the numbers and lethality of ships, planes, and missiles—remains underdeveloped.[17]

The basic idea is that competitive arming leads to irrational outcomes such as war because an environment in which adversaries are improving their military standing induces fear. Presumably, the faster the arming, the more fear is generated. If enough fear is generated, the probability of war is greater than it might otherwise be. All of this hinges on actions and reactions—something we do not capture very well in our equations. Yet the fundamental question remains whether arms races really do make war more likely. I do not know the answer to this question. Neither can I nominate any wars in which arms races (of the second type), or the fear of falling behind an adversary's success in creating more lethal weaponry, were clearly a major cause of war onset.[18] Given my unsatisfactory answers to these questions my suspicion is that arms races are probably not a major cause of warfare.[19] They may even contribute to suppressing the probability of warfare as much as they encourage it. But it would be good to have something more concrete than a hunch to go on.

We are beginning to spend more time on peace processes now and that is to the good. We should have been doing it all along. War and peace processes are not independent. Instead of zeroing in exclusively on the ends of the war-peace continuum, we should make some progress by modeling the movements along the full continuum from warring rivals to security communities. For too long we gave too much credit to one possible source of more pacific relations—democratization. It seems likely that we will find that more peaceful behavior is as complicated as warring behavior (and that our understanding of one should inform how we think about the other).

Finally, among all of the possible topics on the causes of war, we have not done justice to the consequences of war. Yes, we have looked at war consequences (see Quackenbush, chapter 7 in this volume), but not as comprehensively as we might have. Clearly there is a difference between causes and consequences—one precedes and the other follows. But not always. Sometimes war consequences become war causes down the road. This observation relates back to the earlier comment on recidivism. Some wars come in obvious sets—think Arab-Israeli, Indo-Pakistani, or the multiple Indochinese wars. I am sure I must be overlooking someone's work (and, if so, I apologize in advance), but I do not recall any empirical examples of taking a set of interdependent wars as something to be explained as opposed to implicitly assuming that they are independent events to be analyzed en masse.[20] There are of course statistical assumptions and reservations pertaining to independent versus interdependent events, but I am thinking more of the possible substantive payoffs of examining clusters of warfare as interdependent sets. There might be a problem with small N sizes but fortunately that is often the case with war examinations.

That said, the generalization also applies to the consequences of war as consequences as well. Even if all consequences do not become causes in some future iteration, we should know more about them, just as more of us need to be working on peace processes as complementary (or otherwise) to studies of war. Whether Heraclitus is right about war being the father of all things, the positioning of warfare in the general scheme of behavior has been considerably central. Maybe it will not be in the future but we should have a more accurate sense of its centrality (or not) in the past.

In sum, a large number of scholars have devoted their entire careers to the study of warfare. That is something that could not be said prior to the 1970s. As a consequence we are much better off today than we were a generation or two ago in the sense that we know more now than we once did about the phenomenon of war. But as always is the case, there is more to do. We need to build on what has been done; we need to consolidate what we have learned. Then all we have to do is improve further on what has been accomplished to date. More attention to empirical theory construction should aid in that endeavor.

NOTES

1. Thanks to John Vasquez for his comments on an earlier draft.

2. I did a dissertation on military coups and earned tenure and promotion to associate and full professor largely on the basis of papers on what was then comparative politics and derived from the dissertation. Interestingly, in the past two decades the topics of civil wars and military coups have been co-opted to an impressive extent by IR scholars in part

because interstate warfare has waned so markedly. Granted, it may be difficult to demonstrate statistically that interstate warfare has waned (see Braumoeller, chapter 16, this volume and Mitchell and Vasquez, chapter 19, in this volume for instance) but conventional interstate wars in this century have definitely been scarce so far. It is true that civil wars have been more prominent in recent decades and it may be that nominally domestic warfare is masking an increase in regional proxy warfare (Castellano da Silva 2019; Twagiranungu et al. 2019). Nonetheless, classical warfare between major powers has been even more scarce since 1945 or 1953, regardless of how one counts the number of great powers. That may be the best place to look for a decline in warfare. Ironically, one of the reasons I quit studying coups back in the early 1980s was that coup proofing was working too well in too many of the most familiar countries to make it sufficiently interesting to study. We students of conflict get restless when our subjects seem to become more pacific.

3. The first part of this assertion is supported by an examination of the four volumes of empirical IR theory published as part of the Oxford Research Encyclopedia on Politics project (Thompson 2018a). Ism arguments are certainly present but scarce. Surveyed researchers have elsewhere indicated that their work is increasingly ism unrelated. Parenthetically, it should be noted that one-third of the four volumes' contents address war and peace phenomena directly. More general approaches such as bargaining theory and the steps to war are discussed in Choucri and Agarwal (2018), and Owsiak (2018), and Gartzke and Poast (2018). Variations on approaches to peace processes are found in Bayer (2018), Bernhard, Örsün, and Bayer (2017), Chang and Kastner (2018), Diehl and Goertz (2018), Gibler (2018), Goldsmith (2018), Hutchison and Starr (2018), Mares (2018), Mousseau and Cao (2018), Reiter (2018), Schneider (2018), and Weede (2018). Regional processes are examined in Goldsmith (2018), Mares (2018), B. Miller (2018), Rhamey and Volgy (2018), Sakuwa (2018), and Volgy et al. (2018). Other topics include power transitions (DiCiccio 2018; Souva 2018; Tammen, Kugler, and Lemke 2018), rivalry (Dryer 2018; Pardesi 2018; Thompson 2018b), arms races (Rider 2018; Stoll 2018), deterrence (Carcelli and Gartzke 2018; Morgan 2018; Quakenbush 2018; Zagare 2018), military intervention (Pickering and Mitchell 2018; Shirkey 2018), peacekeeping (DiSalavatore and Ruggeri 2018; Gizelis, Dorussen, and Petrova 2018), and territory (Hensel 2018; Wiegand 2018). Related processes are found in Barbieri (2018), Carlson and Dacey (2018), Colaresi (2018), Fordham (2018), Geller (2018), Haas (2018), Lutmar and Terris (2018), Machain (2018), Masterson and Weeks (2018), Murray (2018), and Rasler and Thompson (2018). Another three of the four volumes are devoted to work on civil war processes.

4. Part of the problem is how we train students in methods. It is not surprising that these applied courses feature little if any attention to deriving hypotheses from existing empirical theory. If one is attempting to learn how to correlate two variables, what difference does it make whether the variables make any theoretical sense? Then again, the instructors may know something about one of the more than six subfields of political science but little about the others. But there is also something of a tendency to make up empirical theory as we go along, à la MacGyver, that is particularly endemic in the international politics subfield, although probably not something on which we possess a monopoly in political science.

5. At the same time, as Colgan (2016, 487) notes, no one has actually systematically assessed a sample of quantitative IR to see whether Mearsheimer and Walt's complaints are accurate, including, of course, Mearsheimer and Walt.

6. I think journal practices probably encourage hypothesis testing over theory development in the sense that space devoted to theory development is not really encouraged as much as it is made available to hypothesis testing as a matter of course. I am pretty sure as well that a paper devoted entirely to theory development, other things being equal, would fare more poorly in the review process than one devoted primarily to hypothesis testing.

7. A stronger grounding in historical changes would not hurt either.

8. The most obvious example is the steps-to-war argument (Senese and Vasquez 2008).

9. Truth in advertising dictates that I confess that I am coauthoring a book (Antony and Thompson, under review) that argues (and demonstrates) that the democratic peace, the capitalist peace, and the territorial peace are all interdependent parts of an ensemble or nest of variables historically generated by industrialization. All explanatory problems will not work this way, but there is ample potential for theory integration in our subfield.

10. Granted, explanations of peace may deserve a temporary pass since it is only slowly becoming a more popular research focus. But it is not difficult to come up with fifteen explanations of what drives peace processes.

11. Fortunately anarchy rarely makes it past the assumptive preamble of theories in IR, and some approaches to war such as power transition and leadership long cycle have always been focused on the implications of hierarchy (Ikenberry and Nexon 2019, 408–9).

12. Naturally I would say this because my own favorite approach, leadership long cycles, privileges global reach and technological innovation over bulk or aggregate indicators of military and economic capability when it comes time to distinguish between global and regional IR (more recently, see Lee and Thompson 2018; Thompson and Zakhirova 2018, 2019; Thompson 2020, 2021).

13. I am putting aside the old reluctance to distinguish between polarity and polarization that once plagued the discipline and that seemingly has been resolved, but it was not a quick process.

14. Assumptions about universality and the absence of contextuality are other problems in the study of war.

15. It also ignores the fact that the Soviet Union's claim to bipolarity was largely a function of military capabilities and not economic capabilities while the Sino-American bipolarity will be more economic in nature initially, albeit more about economic size than technological innovation—although that is changing and may or may not become more symmetrical in terms of military capabilities over time.

16. One recent effort in this vein is Mansour and Thompson (2020).

17. See, for instance, Lambelet (1975, 1976) and Stoll (1992) as examples of the second species of arms races. Susan Sample replied quite vigorously at the University of Iowa meeting to discuss the chapters in this book that Gibler, Rider, and Hutchinson (2005) had taken care of this problem by examining which military buildups actually encompassed competitive arms races. The problem is that the Gibler et al. study is based on military expenditures in the context of strategic rivalry, which is a step in the right direction but falls short of actually examining competitive interactions concerning specific weapons systems.

18. Some scholars argue that Germany preferred war with Russia in World War I early as opposed to later because the Russians would be a formidable opponent at a later point, but this strikes me as more a matter of preemptive tactics than arms race fodder. We need to distinguish between adversaries that have already decided that war is probable or desirable and ones that are genuinely reacting to fears about the implications of an opponent's apparent arming.

19. We have been cursed with a number of bad ideas handed down as World War I legacies.

20. Perhaps Organski and Kugler's (1977) Phoenix factor argument would qualify. Not exactly an empirical social scientist, Herodotus, sometimes called the "Father of History," made his reputation looking at a long sequence of Persian wars on the presumption that they were linked.

Chapter Eighteen

War and the Orient Express

Andrew P. Owsiak and Douglas B. Atkinson

Mr. Ratchett died mysteriously one early morning aboard the Orient Express.[1] To investigate the murder, the train company pressed Inspector Poirot into service. His thorough investigation uncovered two possible solutions to the murder puzzle. First, a lone individual boarded the train, murdered Mr. Ratchett, and escaped. Although theoretically possible, this solution failed to account for some key evidence. Inspector Poirot therefore proposed a second, rather unusual solution: that twelve suspects collaborated to murder Mr. Ratchett—carefully coordinating their actions in a sequence of events that left each partially and directly responsible for the murder (i.e., each of the twelve stabbed Mr. Ratchett), while also covering their individual tracks to prevent indictment (i.e., providing alibis for one another). This second solution, though complex, fit the facts of the case best (Christie 1988).

Researchers wear many hats, one being that of detective (Abelson 1995; Collier 2011). A puzzle presents itself and the researcher moves back and forth between theory and evidence—to propose a solution and determine whether the evidence generally supports that solution. More often than not the proposed solution considers individual suspects and circumstantial evidence. Was the top suspect near the scene of the crime when it happened?[2] Or, in the study of war, were the two involved states (not) both democratic when war broke out (Russett and Oneal 2001)? Did they contest territory (Gibler 2012)? What was their relative power (Fearon 1995)? Each answer highlights a single suspect, pushing others into the background (i.e., as control variables, for which one need not account in one's own theory). Moreover, each essentially asks: was the study's top suspect in the vicinity when the crime of war occurred? This approach narrows the list of possible suspects, but leaves the dilemma we now face. It unearths evidence consistent with many detectives' preferred suspect.

Numerous chapters in this volume lament the lack of theory and assert that we need more of it. These general sentiments appear often, but how *exactly* do we produce better theorizing? In this commentary we propose a path forward, imploring future researchers to think more like Poirot aboard the Orient Express to break the deadlock we face. Three fronts, in particular, need further advancement—each related to theoretical integration. First, we must entertain the idea that more than one suspect committed the crime. This involves not only identifying the suspects (e.g., suspect x *and* z bear responsibility) but also theorizing about how and why they (would) work together. Poirot, for example, learns that all twelve suspects, though of different backgrounds, share a common connection to a previous event that motivates their collaborative search for justice. Theories of war likewise need to explain whether and how suspects in the causes of war relate to one another (e.g., conceptual overlap)—and work either together, independently, or at cross-purposes. Second, we need to theorize further about the process through which the crime of war occurs. This brings dynamics back to the forefront. Cause and effect—the sine qua non of any theoretical argument—imply a simple temporal ordering. We must go farther, though. What is the sequence of events? Are events contingent upon one another (i.e., interdependent), and in what ways? Do effects change over time? And so on. Finally, we need to uncover better evidence consistent with any processes we propose. Rather than say that our suspect *could* have committed the crime, or that the evidence is consistent with the fact that s/he might have, can we know whether s/he actually did? That requires a deeper, more creative investigation—one that triangulates evidence to confirm we got it right.[3]

CONCEPTUAL OVERLAP

Any investigation must identify the main suspect(s). What factors, then, cause war? As this volume attests, scholars provide no shortage of answers. Yet, as their list of correlates expands (e.g., territorial disputes, alliances, arms races, contract-intensive economies, sovereign debt lending, systemic factors), one wonders how the items on that list relate to one another. It is possible that each item exerts a completely independent effect on war. In that case one could read the probability of war similar to a barometer.[4] The addition of each discrete item builds greater pressure for war (i.e., increases the probability that war occurs). If the items are truly independent, then the sum contribution of any two given factors ($x + z$) will be predictable, stable, and positive (i.e., $\beta x > 0$ and $\beta z > 0$).[5] Moreover, studying each factor on its own presents little danger; once we understand the individual factors, we merely sum their distinct effects.

Theory and empirics suggest, however, that the onset of war does not work this way (for an early indication, see Bremer 1992). The steps-to-war thesis, for example, combines territorial disputes, rivalry (or dispute recurrence), alliances, and arms races. In advancing it, Senese and Vasquez (2008, 23) not only argue that these factors are "mutually reinforcing" (i.e., each step increases the likelihood that states take another) but also find corresponding, significant evidence that interactions among the steps exist (see also Sample 2018a). Arms race research points in a similar direction. Rider (2013) proposes that arms race competitions (a correlate of war; see chapter 4) arise most often when rivals compete over territorial issues. His analysis uncovers no independent effect of these two factors on the probability that an arms race develops; only the interaction matters (see also Lektzian, Prins, and Souva 2010). Even the now well-worn bargaining model of war implicitly mixes causal factors. Fearon's (1995) formulation of it assumes an issue (i.e., a territorial dispute) and starts from the premise that rational states would reach a prewar bargain that mirrors the expected postwar settlement—if only they understood their relative power and the costs of war. A state's resolve, alliances (and their reliability), intelligence, technology, strategy, funding, and public opinion (e.g., regime type, as a crude measure), as well as the salience of the disputed issue, all determine what war ultimately costs, what costs each side will find acceptable to bear, and therefore the operation of the model.

Despite knowing that multiple factors—working independently or interdependently—influence the path to war, many studies highlight one factor at the expense of others, especially in quantitative work. Owsiak (2012) reflects the typical approach.[6] One variable—in this case, settled borders—takes center stage, while all remaining variables fade into the background, controlling for alternative explanations that do not fit neatly into the theory at hand. Such an approach makes intuitive sense, especially in the early stages of research. Before one can build a complex explanation, one must understand the constituent parts. That explains why Poirot initially interviews each suspect independently aboard the Orient Express. Nevertheless, integration must proceed at some point, particularly after we understand the constituent parts better and know integration makes sense.

If the individual items on scholars' correlates list fit together into larger causal stories, then developing these causal stories next requires theorizing about the combinations of variables that cause war. That theorizing, however, will not produce results identical to the barometer example discussed earlier in this chapter. Independent events by definition do not work together. Imagine two (or more) suspects who try to murder a victim, unaware of one another's efforts. The suspects operate independently, and the victim's danger (or probability of death) likely grows incrementally as the number of murderers pursuing him increases (see barometer example).[7] In contrast, two (or more) murderers might work *together*—either in collaboration, or at cross-purposes. The net effect of their collective efforts need not be the sum of their individual contributions. To study their effects, then, we need to follow Poirot's lead. Which suspects are working together, what connects them (if anything), and how?

Three routes supply answers to these questions, thereby promising to integrate and advance our causal stories about war. The first considers conceptual overlap (Goertz 2012). Do our suspects relate to one another (conceptually), and if so, how? No greater research program demonstrates the need for such thinking better than the democratic peace (for an overview of the democratic peace research program, see chapter 10 of this volume; Chan 2010; Mitchell 2012; see also chapter 17 of this volume). After discovering the empirical regularity that democracies do not fight wars against one another, research blossomed. It first emphasized democratic characteristics—or the qualities that differentiate democracies from non-democracies. More recently, however, a series of challengers suggest that those qualities misrepresent the relationship between democracy and "not-war."[8] Various scholars argue that capitalism (Gartzke 2007), contract-intensive economies (chapter 10 of this volume; Mousseau 2009), dispensing with territorial (border) issues (chapter 9 of this volume; Gibler 2012), or systemic hegemons (McDonald 2015) could instead explain the same cases—even rendering the democracy/not-war relationship spurious.

Without wading into those debates too deeply, we offer a simple observation. Even as scholars present their alternative explanations as competing theories, many of the concepts underlying those theories overlap significantly. How do we think about theoretical integration when this occurs? To illustrate the challenge, consider the relationship between two "competing" theories of "not-war"—the democratic peace (see Russett and Oneal 2001) and contract-intensive economies (Mousseau 2013). In a contract-intensive economy:

> Making contracts with strangers promotes loyalty not to patrons but to a state that enforces these contracts with impartiality and equal application of the rule of law. … [V]oters in contract-intensive societies are more likely to support candidates for office who stress individual freedoms, at home and abroad, and who advocate government transparency and equal enforcement of the law. (Mousseau 2009, 58–59)

This passage suggests significant overlap between the two competing theories' main concepts (i.e., democracy and contract-intensive economy). Both contain many of the same characteristics, including the rule of law, equality (particularly before the law), voting (and therefore elections), individual freedoms, and government transparency.[9] The data bear out this overlap. When a state has a contract-intensive economy in a given year, it is also democratic 97 percent of the time.[10]

What do we do in cases like these, when one theory's concept appears to be a subset of another's? Quantitative researchers would advise that we check the correlation coefficient (i.e., collinearity). Between democratic and conflict-intensive-economy state-years, it is 0.1991—an acceptably low level that suggests no major issue. A set theoretic analysis, in contract, suggests a problem. Set theory "treats concepts as sets or categories in which cases (or observations) can have membership," and then conceives of relationships between sets "using ideas of necessity and/or sufficiency—or, equivalently, superset/subset relationships" (Goertz and Mahoney 2012, 18).

Figure 18.1 visually depicts the sets "democratic state" (lighter circle) and "contract-intensive economy state" (darker circle).[11] Neither set lies *completely* within the other, but they overlap significantly (i.e., 97 percent of the time). The "contract-intensive economy" set is nearly a subset of the "democracy" set. Being democratic is a (near) necessary condition for a state being a contract-intensive

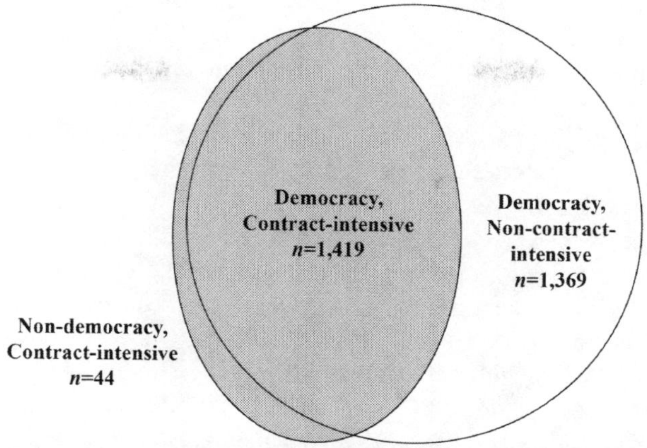

Figure 18.1. Democracy and Contract-Intensive Economies within State-Years, 1920–2010.

economy in a given year, or, conversely, being a contract-intensive economy is a (near) sufficient condition for a state being democratic in a given year (i.e., almost all members of the "contract-intensive economy" set fall within the "democracy" set). That should give us pause, for it suggests that in a statistical model, the forty-four nondemocratic, contract-intensive-economy states drive any results associated with the contract-intensive variable. Of these latter state-year observations, thirty-eight (or 86.36 percent) result from only three states: Taiwan, Malaysia, and Singapore.[12]

One cannot deny the statistical finding that contract-intensive economies go to war less often than their counterparts. How do we make sense of that in light of Figure 18.1? How do we integrate the related concepts? As a first cut, we could examine temporal ordering. Owsiak and Vasquez (2019) adopt such a strategy to explore the relationship between democratic dyads and territorial disputes. Democratic peace scholars argue that democratic dyads handle their disputes more peacefully than nondemocratic dyads, regardless of the issues involved (e.g., see Dixon 1994; Russett and Oneal 2001). Territorial peace scholars, in contrast, propose that democratic dyads behave peacefully because they settle the most salient, conflict-prone, interstate issues—namely, the delimitation of interstate borders—before both states become democratic (e.g., see Gibler 2012; see also Vasquez 2009). This gives democratic dyads a qualitatively different set of issues to manage, ones more easily handled via less violent means.

To adjudicate these viewpoints, Owsiak and Vasquez (2019) consider the temporal ordering of two events within each dyad: its border settlement date and the date on which joint democracy emerges. They find that, in the majority of cases, border settlement precedes joint democracy within both contiguous (i.e., neighboring) and noncontiguous dyads. Given such an ordering, democracy could not possibly contribute to the management of territorial issues within most dyads. The democratic suspect was not in the vicinity when border settlement occurred. Moreover, Gibler and Owsiak (2018) examine the few cases where it was (i.e., border settlement followed joint democracy) and conclude that democratic status did not grant the involved states any bargaining advantage during the border settlement process. Democratic dyads may handle *some* issues more successfully than nondemocratic dyads, but not those involving the placement of mutual borders.

A temporal analysis may assist in the democracy/contract-intensive economy case too. Do states become democracies *before* they become contract-intensive economies? The answer is almost always yes.[13] Chile, for example, qualifies as a non-contract-intensive economy from 1920 (the beginning of that data set) to 1980, has no classification between 1981 and 1995, and obtains membership in the contract-intensive economy set from 1996 to 2010 (the end of that data set). During the blackout period (1981–1995) it becomes a democratic state (specifically, in 1989; Marshall and Jaggers 2019). The United Kingdom similarly lacks classification from 1920 to 1930, but then qualifies as a contract-intensive economy during the period 1931–2010. Its membership in the democracy set long predates contract-intensive classification. Other states display a similar trend (e.g., Israel or Botswana). This indicates that, if there is a causal relationship between the two concepts—or if they sit together within a larger, dynamic process model—democracy likely leads to contract-intensive economies, although the mechanism through which and the conditions under which it does so remain unspecified here.

Temporal ordering need not imply that a later factor exerts no causal import, even if part of a process. How, then, do we assess the relative merits of democracy and contract-intensive economies when explaining not-war—as we attempt to integrate them? Set theoretic logic assists here too (Mahoney and Vanderpoel 2015). Figure 18.2 illustrates how each theory's main variable covers the "not-war" outcome of interest. We construct this figure from an analysis of dyad-year observations during the period 1920–2010, and it reiterates many points from our earlier discussion.[14] Joint democracy and joint contract-intensive economies are subsets of the not-war category; they are (individually) each sufficient for observing not-war.[15] Moreover, joint democracy is a near necessary condition for a joint contract-intensive economy; 93 percent of dyad-year observations that contain a contract-intensive economy lie within the joint democracy set.

What concerns us most here, however, is coverage—that is, how many "not-war" cases each variable of interest can potentially explain, or visually, how much of the "not-war" space the democracy or contract-intensive economy spaces cover. There are 772,104 dyad-year observations in which war does not occur between 1920 and 2010. Of these, joint contract-intensive economies can, at best, explain 15,347 (or 1.99 percent).[16] Joint democracy, in contrast, accounts for a maximum of 218,215 of these cases (or 28.26 percent). Neither performs fantastically. The vast majority of not-war dyads (71–98 percent) lack an explanation under either theory. Nevertheless, from a set-theoretic perspective, joint democracy is more "important" than contract-intensive economies. As Goertz and Mahoney (2012, 34) note, "[t]he sufficient condition *X* will become more important as its coverage of *Y* increases; that is, the subset *X* becomes more important as it approaches perfect overlap with *Y*. … [W]hen multiple sufficient

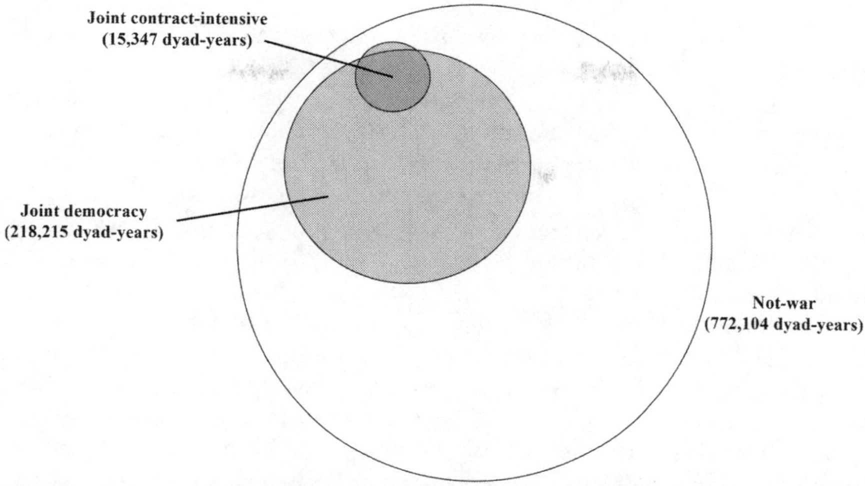

Figure 18.2. Joint Democracy, Joint Contract-Intensive Economies, and "Not-War" Dyad-Years, 1920-2010.

conditions are identified for a given kind of outcome, the more frequently present ones are the more important ones." Both theories surveyed here propose a sufficient condition, and in Figure 18.2 democracy "covers" more of the "not-war" set, thereby outperforming contract-intensive economies. Omitting the democratic peace in favor of contract-intensive economies therefore gives us less purchase power over explaining the "not-war" cases.

Lest we be misunderstood, we do not intend the foregoing discussion to denigrate the contract-intensive-economy research program, but rather to illustrate the obstacles we face.[17] The field continues to debate which single factor matters most in the causes of (not-) war. Indeed, some critics of the democratic peace adopt such a position (e.g., Gibler 2012; Mousseau 2013; McDonald 2015). It is, however, unlikely to be a fruitful path. When concepts overlap significantly, alternative explanations blend together empirically. Moreover, scholars can obtain strong statistical relationships (e.g., contract-intensive economies) even when controlling for conceptual overlap, and these quantitative relationships lose import on closer inspection (see also Narang and Nelson 2009). To untangle the relationships, we need to step back and (re)conceptualize. What essential characteristics grant a case membership in a given set? How does that set relate to the sets we already have available? What is the "usefulness" of the new set?

Past breakthroughs originated from similar moments of reconceptualization. As an illustration, the idea of rivalry predates modern research on war, yet its systematic conceptualization transformed the field of conflict studies (see chapter 5 in this volume; Diehl and Goertz 2000; Colaresi et al. 2007). Focus shifted from explaining individual interstate conflicts as isolated events to (i) the observation that most of these conflicts cluster in a small number of dyads, and then (ii) understanding what makes these dyadic *relationships* unique. More modern research now broadens the rivalry perspective further to conceptualize all dyadic relationships—a move that holds promise for explaining interstate peace (see chapter 11 in this volume; Goertz, Diehl, and Balas 2016).

At times it may be difficult to untangle related concepts, and theoretical relationships may not be immediately obvious. Model building then stalls. To jump-start it again, one could turn independent variables into dependent variables (and vice versa). Although research on territory as covered in this volume (chapter 1), alliances (chapter 3), arms races (chapter 4), and rivalries (chapter 5) has already done this, other research areas have not (e.g., power; see chapter 2). To see its value, assume that research on contract-intensive economies has identified the subset of democracies most responsible for peace.[18] We would then want to know why and how some democracies develop contract-intensive economies, and what other factors play a role in that process. Similarly, as Morey and Kadera (chapter 2) note, war might be part of a larger process in which it (as an independent variable) produces other outcomes of interest (i.e., dependent variables). All of this suggests we should build models that place various factors within a larger process—one in which individual causes of war, as well as war itself, might influence both each other and the likelihood that war occurs.[19]

PROCESSES THAT LEAD TO WAR

A second path to theoretical integration investigates the processes that lead to war. Overlapping concepts (and theories) aside, explanations that isolate one variable at the expense of others seem dissatisfying because war results from a complex process. World War I supplies a good example (see Levy and Goertz 2007). As Thompson (2007, 120) notes:

> One of the more frustrating aspects of World War I analyses is that practically every explanation for conflict seems to find some resonance in the events leading to war in 1914. … Authors can construct plausible explanations of what happened without seeking to be fully comprehensive in circumstances in which a good number of the explanatory foci in international relations seemed to be at work.

The list of causal factors that theoretically affect the origins of World War I indeed seems daunting: a long-standing territorial disagreement between Austria and Serbia in the Balkans (e.g., Austria annexed Bosnia), repeated international crises in which Germany felt slighted (e.g., the First and Second Moroccan Crises), repeated wars in the Balkans, the formation and tightening of alliance structures (i.e., the Triple Alliance and Triple Entente), the drift in governments toward more hard-line policies (e.g., see Austria), the decline of the Austro-Hungarian Empire, numerous intersecting rivalries, and an Anglo-German naval arms race. In June–July 1914 still further causes appeared, which set the war machinery more immediately in motion. The archduke's assassination created a wake in which Germany offered support for Austria to behave aggressively (i.e., issue an ultimatum and pursue a limited war), key allies obfuscated their intentions (perhaps unintentionally; see the United Kingdom), officials deliberately misled their leaders on key points (e.g., Russian officials told the czar that partial mobilization was possible when it was not), leaders felt constrained (e.g., see the Nicky-Willy telegrams), and so on.[20] In short, a long list of—sometimes overlapping—causal factors overdetermine the war (i.e., they all predict war). How do we distinguish among their effects?

The problem mirrors that of overlapping concepts (Goertz 2017, chapter 3). In response, quantitative scholars typically propose controlling for the alternatives and addressing collinearity, yet the problem cannot be solved via a large-*n* strategy like this. Better answers lie with theoretical model building. A model is a simplified representation of reality that helps us understand a phenomenon of interest (Lave and March 1993; Clarke and Primo 2012, chapter 3). How much a model simplifies reality depends on the researcher as well as the purpose for which the model will be used. Nevertheless, we believe the field needs *less* simple—or less parsimonious—theoretical models.[21] If multiple factors operate simultaneously to cause war, then models that simplify to one factor each may be *too* simple. Adding complexity, including the relationship(s) between factors, promises advancement (e.g., see Bremer 1992).[22]

How we organize variables within process models requires frameworks. One such framework relies on underlying and proximate causes. "Underlying causes are fundamental causes [or factors] that set off a chain [or sequence] of [more immediate] events (the proximate causes) that end in war" (Vasquez 2009, 7). Within this framework some causal variables (e.g., territorial disputes) become background conditions (or necessary conditions) against which others appear (e.g., power politics; Vasquez 2009). A second framework uses two alternative dimensions: opportunity and willingness (Most and Starr 2015). For any given outcome to occur, the involved agents must possess both the opportunity (i.e., the capability, defined materially and structurally) and willingness (i.e., the desire) to pursue that outcome.[23] Variables—from the system to the state, substate group, and individual (e.g., leader or key decision maker) levels—all influence these two dimensions, giving it wide appeal.

A third framework unpacks necessary and sufficient conditions. We admit a slight preference for this framework. The first two frameworks supply a way to organize the variables that belong in any process-oriented model, but do not specify the relationship between these variables. Necessary and sufficient conditions, in contrast, specify such relationships and imply a temporal ordering between two or more variables as well. Both features prove essential in a process-oriented model.

A process-oriented model recognizes that one event (X_1) leads to another event (X_2) in a sequence of (X_n) events that ultimately cause an outcome of interest (Y). All causal arguments advance a process-oriented model, even if the accompanying empirical analyses do not examine each link in the process's chain (see next section). Our models of war, however, typically abstract away from that process, leaving us with simple, uni-causal processes (Goertz 2017). Does democracy (X_1)

reduce the likelihood of war (Y)? Do unsettled borders (X_1) increase the probability of war (Y)? Do arms races (X_1) increase the likelihood of war (Y)? And so on.

With a solid understanding of the individual ($X_{1...}$ X_n) factors involved (e.g., studying some as dependent variables), however, their integration into larger theoretical arguments that consider sequences (i.e., event order) and dynamics (i.e., time) can proceed. This integration offers a second solution to the overlap among "competing" theories. If, for example, settled borders (X_1) precede—perhaps even cause—democracy (X_2), and some democracies become contract-intensive economies (X_3), then a sequence of events ($X_1 \rightarrow X_2 \rightarrow X_3$) may exist through which some "not-war" (Y) occurs, one that combines issue-based research with that on the liberal peace. Other sequences may also emerge. "Not-war" occurs in many nondemocratic dyads, but these dyads never experience X_2 or X_3. What sequence of events yields "not-war" in these dyads?[24]

To step away from the democracy-"not-war" nexus, relative power equality (X_1) often serves as a precondition for rivalry (X_2). The rivalry relationship in turn constrains the policy options available to leaders—in some cases, promoting arms races (X_3). This pathway ($X_1 \rightarrow X_2 \rightarrow X_3$) might also lead to war. And once again, other sequences are theoretically possible. If rivals pursue arms races—a long-term strategy that unfolds over many years—to enhance their security, what do they do in the short run? Does alliance building serve such a function? If so, a possible sequence might be: $X_1 \rightarrow X_2 \rightarrow$ alliance building $\rightarrow X_3 \rightarrow$ war. Through such thinking, various "pathways" to (not-)war emerge. If we can identify those pathways theoretically, we can then examine their prevalence, as well as how the mechanisms connecting adjacent events work.

CAUSAL MECHANISMS AND THEIR EVIDENCE

As we build more complex (i.e., more than one suspect), dynamic (i.e., process-oriented) theoretical models of war, two additional considerations will be paramount. The first concerns a more explicit elaboration (and general acceptance) of scope conditions, which limit the applicability of a given model (or argument, mechanism, or empirical relationship; Goertz 2017; see also chapter 2 in this volume).[25] Models vary in purpose—from foundational to organizational to exploratory to predictive (Clarke and Primo 2014). They also differ in their generalizability along a continuum—from applying to all cases to (perhaps) a single case. As one pushes the model to cover more content (or more cases) satisfactorily, it often gains greater complexity.[26] Scope conditions allow the researcher to manage that complexity.

This can be a valuable trait, particularly when using set-theoretic logic. Quantitative research typically allows for many empirical anomalies, so long as most of the evidence coincides with the argument's thrust. Were one to complain that a handful of cases do not fit the models' prediction, a quantitative researcher would respond with one of two arguments: (i) most cases fit the pattern, or relatedly (ii) the model, being probabilistic, will get a few predictions wrong.[27] The frustration the complainant experiences mirrors that of arguing with a weather forecaster. Sure, the forecaster promised an 80 percent chance of rain today; when the sun comes out instead, however, the forecaster merely replies, "that was the other 20% of my prediction."

Models that rely on set-theoretic logic, in contrast, leave less room for error, particularly those that elaborate sufficient conditions.[28] Following the logic "if *X*, then always *Y*," a truly sufficient condition does not have anomalies.[29] Schenoni et al. (2020, 59) illustrate the challenge. They propose a two-level model in which attention, altered preferences, and third-party assistance are individually necessary and jointly sufficient for the settlement of resistant territorial disputes. The argument can be falsified in two ways—namely, by finding *any single case* in which *either* (i) settlement occurs without one of the three necessary conditions, *or* (ii) settlement fails to occur when all three necessary conditions are present simultaneously. A single anomaly threatens the theoretical argument.

The Schenoni et al. model operates under three scope conditions. First, a "resistant" territorial dispute must exist. Second, that dispute must occur in the post–World War II period. Because the norms for territorial settlement shift around this time, the authors argue that a different model needs to explain similar settlements before 1945. Finally, the general version of the model applies widely, but the particular version applies only to Latin America. Attention, altered preferences, and third-party assistance manifest in Latin America after 1945 as militarization, democratization, and mediation, respectively. In other contexts (e.g., regions) the same general factors might adopt different expressions. Under these scope conditions, Schenoni et al. (2020) find no anomalies. Resistant territorial disputes in post-1945 Latin America always settle when the three factors are present simultaneously, but never settle when any one factor is absent.

Reviewers often react negatively to scope conditions. We suspect that reaction results from suspicion—namely, that an author has raised a red flag by supplying a model that does not generalize to all places and times. But is that a red flag? Could one not advance a valid model of US foreign policy that applied to the Obama administration, but not the Trump administration? Could one likewise not advance a valid model of how resistant territorial disputes settle in Latin America after 1945—or more broadly, of minor state wars, wars of rivalry, territorial wars, and so on? Of course one could.[30]

The larger issue is why we as a discipline often find scope conditions suspicious. This suspicion encourages scholars not to state their scope conditions explicitly. More surprisingly, reviewers then ask for them, usually indirectly. Present a quantitative model of militarized interstate dispute (MID) onset, for example, and the questions roll in. What about the distinction between dyads that started versus joined the MID? What about fatal MIDs? What about the MIDs that escalate? What about the "fishing boat" MIDs?[31] What about the pre-1945 MIDs? And so on.[32] Some interpret this as probing the "robustness" of the results. That seems an odd interpretation, though; even if the results do not hold across all contexts, publication proceeds—with the appropriate caveats (e.g., "except in the post–Cold War period"). We therefore conclude that reviewers *want* authors to specify scope conditions, even as doing so makes them jittery.

We concur with the importance of explicitly theorizing and stating scope conditions. The greater transparency not only produces better science but also allows us to see how various models could fit together—or to identify different paths to the same outcome (i.e., war). Consider the debate between the territorial and democratic peace. The territorial peace argument requires land-contiguous states; the democratic peace argument does not. Could these theories be integrated? Owsiak (2019) argues that it is possible, largely because the scope conditions differ. Many noncontiguous dyads never (or rarely) experience war. Because the territorial peace cannot possibly explain any of these not-war dyads, other models *must* do so. The democratic peace offers one such potential model. More broadly, complex phenomena—like war—will generate many models simply because many causal pathways lead to it. As a result no single model will explain all wars (see Vasquez and Valeriano 2010). Explicit specification of scope conditions permits us to understand where each model operates and where it does not, pointing the way toward future research or opportunities for integration.

In addition to elaborating scope conditions more explicitly, a second consideration concerns our search for evidence. As noted at the outset, scholars—particularly in the quantitative tradition—propose a theory, replete with a proposed causal mechanism. They often, however, do not test whether the causal mechanism operates as they claim. Figure 18.3 illustrates the point, deriving a sample causal mechanism from Clay and Owsiak (2016). The authors argue that border settlements diffuse because a state (A) uses a settlement with one neighbor (B) to signal to another neighbor (C) both its willingness to settle and the possible terms it will find acceptable. Through this process, a settlement in Dyad A-B diffuses to Dyad A-C.

We stress four features of the figure. First, a process need not imply that all factors reinforce one another (or overlap). A willingness to settle and the terms of settlement to select are somewhat independent. One easily exists without the

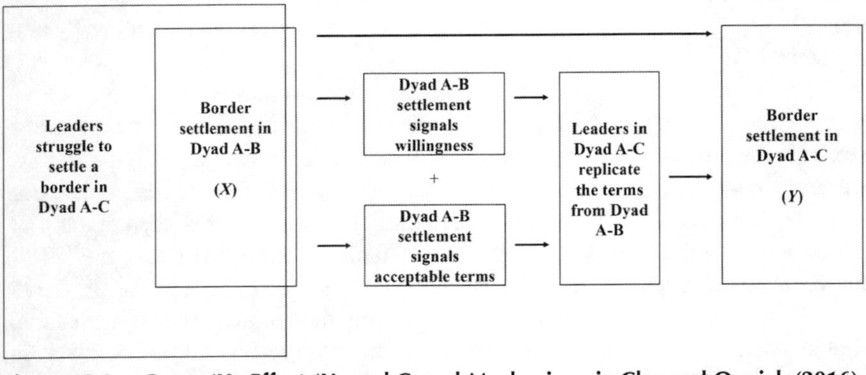

Figure 18.3. Cause (*X*), Effect (*Y*), and Causal Mechanisms in Clay and Owsiak (2016).

other. Moreover, they each can exert an independent effect on diffusion, even if their joint effect is stronger, and be situated within a causal process. Second, a single factor might lead to a given outcome through multiple causal mechanisms (e.g., third-party assistance, as opposed to the signaling mechanism described in Figure 18.3). Third, much quantitative research sidesteps an evaluation of the intermediate causal process. Clay and Owsiak (2016), for example, test the relationship between dyadic border settlement in Dyad A-B (*X*) and border settlement in Dyad A-C (*Y*). This follows the topmost arrow in the diagram and means that statistical analyses "could provide convincing evidence linking *X* to *Y*, but *could not explain how* X *produces or causes* Y" (Goertz 2017, 31). Finally, and relatedly, the authors could delve into the mechanism to look for evidence that it operates as they claim, by pursuing evidence that would exist if that were the case (e.g., causal-process observations; see Collier 2011; Seawright 2016). To do that, they would need more than statistics.

More specifically, Goertz (2017) proposes a research triad that consists of causal mechanisms (e.g., game theory), within-case inference (e.g., counterfactuals or process tracing), and cross-case inference (e.g., experiments or statistics). Valid causal inference—and true multi-method research—involves moving from a single point along the triangle (e.g., cross-case statistical analysis alone) to embrace all its points. That embrace requires more than the examples, plausibility probes, or vignettes that permeate quantitative work. In Clay and Owsiak (2016), for example, a statistical analysis backs the theoretical argument, but they also supply a plausibility probe in which Chinese leaders claim to do exactly as the mechanism proposes.

The inclusion of a plausibility probe supplies better evidence than statistics alone, but does not fully embrace a within-case inference strategy (e.g., process tracing or counterfactuals). For the latter, the authors would need to dive into the cases to show exactly how and why leaders negotiate border settlements as they do. Doing so offers tremendous value, for:

> To put it bluntly: process tracing is almost always the best (if not only) way to examine decision makers' strategic thinking. We can often overhear a decision maker considering her opponent's calculus to derive her best strategic play. Moreover, historical case studies can make it possible to match the explanatory theoretical variables to the real world, often more accurately than with a large data set with many cases. (Goemans and Spaniel 2016, 28)

Researchers increasingly recognize the value of multi-method research, even as many graduate programs fail to train students sufficiently in them.[33] Fazal (2018), for example, uses a multi-method approach to investigate the laws of war, while Poast (2019a) does the same for negotiations over alliance commitments. Nevertheless, multi-method research remains less common in the study of war. If we want to take the evaluation of causal mechanisms seriously, then that needs to change.

CONCLUSION

Were we now aboard the Orient Express, we would likely accept Poirot's first solution, blame one of the twelve suspects, or debate among ourselves which single suspect bears most responsibility—or so the state of current research on war leads us to believe. That would be a mistake. As this volume demonstrates, many suspects bear responsibility in the outbreak of war, and we have increasing evidence that these suspects are working together. Moving forward therefore requires three large steps. First, we need to identify the particular suspects working together and theorize about how they relate to one another. (Re)conceptualization and interchanging independent and dependent variables can assist in this endeavor. We then must return to the idea that war results from a causal process of events. Rather than building more uni-causal models—or ones that stress a single variable at the expense of the others—we instead need to construct multi-causal models that integrate causal factors and elaborate on the sequences of events through which they produce war. Thinking in set-theoretic logic assists here, for it forces us to think about the relationship between variables, including their temporal ordering. Finally, the models we develop need to state their scope conditions explicitly and use multi-method strategies to uncover direct evidence that the causal mechanisms we put forward operate as we say. These steps will not be easy to take, but solving crime never is.

NOTES

1. We thank Paul Diehl, Gary Goertz, Chad Clay, and the editors for comments on earlier versions of this chapter. All remaining errors are our own.

2. Police now use cell phone records to answer this question. Even if the suspect does not confess, they can obtain evidence that the suspect was in the vicinity of the crime when it occurred—a fact the suspect must explain away.

3. Any commentary necessarily reflects the perspective of those writing it. Acknowledging our own biases, we recognize that other avenues for research both exist and may be fruitful.

4. On a risk barometer for war, see Senese and Vasquez (2008).

5. Before critics hastily disagree with this statement, we offer two observations. First, were combinations not stable, temporal dynamics would come into play. Most research on war, however, uses static indicators in the year(s) prior to a war, and does not generally discuss how an indicator's effect on the probability of war varies over time. Second, although items on the list can decrease the probability of war, most quantitative research on war assumes causal symmetry (i.e., the presence of an item and its absence are two sides of the same coin; see Goertz and Mahoney 2012, chapter 5) and uses dichotomous variables to measure diverse concepts (e.g., issues, alliances, arms races). We think it is difficult to sustain the assumption of causal symmetry (see chapter 11 of this volume). Nevertheless, the quantitative research designs that result from it typically mean that negating a variable that dampens the probability of war (i.e., using *non-democracy* instead of democracy, or switching the coded 0s and 1s) will increase the probability of war.

6. Out of fairness, we recognize that our own work deserves the same critique we broadly give to others.

7. As long as the independent would-be murderers are somewhat competent, their individual chance of successfully committing the crime—however small—will exceed 0.

8. We use "not-war" instead of "peace" because (i) the two are not equivalent (see chapter 11 of this volume) and (ii) not-war captures the empirical prediction most accurately (i.e., war does not occur).

9. In a slightly later passage we learn that contract-intensive economies overlap with capitalist ones too (Mousseau 2009, 61). It also just so happens that the contract-intensive economies have been the world's hegemons—at least during the time the data cover (1920–2010)—and often settle their borders long before the contract-intensive economy emerges.

10. $\chi^2 = 3{,}400$, $p < 0.000$; $\gamma = 0.9857$, $p < 0.002$.

11. Data on contract-intensive economies only cover the period 1920–2010 (CIEb variable; Mousseau 2019a). A state is a democracy if, in a given year, it scores +6 or higher on the Polity IV Project's autocracy-democracy index (Polity2 variable; Marshall and Jaggers 2019).

12. For a similar exercise and conclusion in research on whether democratization increases the likelihood of war, see Narang and Nelson (2009).

13. Given Figure 18.1 the only other (likely) logical possibility would be that contract-intensive economies emerge simultaneously with democracy within some states.

14. Data on contract-intensive economies only cover this temporal range, making it the appropriate boundary to any analysis. The Militarized Interstate Dispute (MID) 4 data set identifies wars (Palmer et al. 2015). A dyad is a joint democracy if, in a given year, both members of the dyad score +6 or higher on the Polity IV Project's autocracy-democracy index (Polity2 variable; Marshall and Jaggers 2019). It is a joint contract-intensive economy if, in a given year, both members have a contract-intensive economy (CIEb variable; Mousseau 2019a).

15. The MID data identify some democratic dyads that fight wars during this period ($n = 16$ dyad-years, covering ten distinct dyads). Scholars generally agree, however, that democratic dyads do not fight wars and then argue over anomalies. We avoid that debate and omit those observations here. It does not change the discussion that follows.

16. We say "at best" purposely here. If other, related variables (better) explain some of these cases instead, then contract-intensive economies account for fewer cases than the maximum listed here.

17. For alternative examples, consider the relationship between human rights violations and democracy, or rivalry and arms races.

18. If it has, we can explain far fewer not-war dyad-years than we could with democracy. That underscores the importance of thinking about both causal (a)symmetry (i.e., are "not-war" and "peace" the same; see chapter 11 of this volume) and how our various models might integrate to explain a greater number of "not-war" cases.

19. Achen (2002) makes a similar point. His "rule of three" derives from better understanding collinearity among independent variables. That in turn requires better specified scope conditions (Achen 2002, 446). We thank Sara Mitchell for noting this connection.

20. See Williamson and Van Wyk (2003); Goertz and Levy (2006); Levy and Vasquez (2014); Williamson (2014).

21. The theoretical model's complexity may increase, even as the empirical model's complexity decreases—for example, if one limited the number of control variables and theorized carefully about the relationships between any included variables (Achen 2002).

22. Note, however, that we do not advocate the return of the -isms, which differentiates our recommendation from analytic eclecticism (Sil and Katzenstein 2010). We also are agnostic about using multiple levels of analysis (see chapter 17 of this volume). Doing so offers one way to build model complexity, but not the only one.

23. This is true unless the outcome "accidentally" occurs. Purely accidental wars, however, do not occur. One actor may *prefer* not to fight a given war or feel constrained to choose war from among its available policy options, but that does not make the war accidental.

24. See Metzger and Jones (2016).

25. Relatedly, we need to think about where models do *not* apply, even though we might observe the outcome the model predicts. I thank Paul Diehl for noting this issue.

26. To cover more content, a theorist could also reduce complexity. The resulting model, however, unsatisfactorily covers the content—at least from a causal mechanism standpoint (e.g., see Waltz 1979).

27. As Clarke and Primo (2014) note, most models we see in political science are predictive—in sample (a backward-looking prediction) or out of sample (a forward-looking prediction).

28. Sufficient and necessary conditions relate to one another, though. If X is necessary for Y, then $\sim X$ is sufficient for $\sim Y$.

29. This may explain why critics of the democratic peace's sufficient conditions face quick rejoinders. See, for example, Doyle (2005), Kinsella (2005), and Slantchev, Alexandrova, and Gartzke (2005) in response to Rosato (2005).

30. Interaction terms can also capture nuance within a theoretical argument. Too many interaction terms in the same model, however, could indicate a need to think more systematically about scope conditions.

31. These are MIDs in which one state seizes (and later, usually releases) a vessel in its waters.

32. Such questions explain why a lengthy appendix now accompanies many quantitative research articles. Qualitative authors are not immune from the trend, though; in fact, they may need similar appendices to provide convincing case histories, details, and analyses. See Schenoni et al. (2020).

33. In the United States many graduate programs in political science include a compulsory, multi-course sequence in quantitative methods. Qualitative methods, in contrast, are often not compulsory, and if offered at all, frequently appear as a single "Qualitative Methods" course. For more on training in qualitative methods, see the Institute for Qualitative and Multi-Method Research (IQMR) housed at Syracuse University.

Chapter Nineteen

What Do We Know about War?

Sara McLaughlin Mitchell and John A. Vasquez

It has been more than twenty years since the publication of the first edition of this book (Vasquez 2000). At the time a number of us believed that the scientific study of peace and war had accumulated enough findings that an attempt should be made to bring those findings together for a larger audience as well as students. Although the scientific study of peace and war could be dated from the work of Lewis F. Richardson beginning in 1919, most advancement did not occur until the collection of replicable data by J. David Singer and Melvin Small in the Correlates of War (COW) project (see Singer and Small 1966b, 1972). Contributions to this volume focus on what we know about war based on quantitative analyses of data compiled since the pathbreaking work of the COW project began.

It was not until Singer and Small (1966b) that the field compiled a list of interstate wars using scientific criteria. These data permitted us to know how much war there was in the system and whether it varies. Trends in warfare have been a subject of debate in terms of whether it is declining. As Braumoeller (chapter 16) demonstrates, much depends on the benchmark. If one starts from the two world wars, then war is declining, but if one starts from 1816 with the end of the Napoleonic wars, then the amount of peace now is no greater than during the Concert of Europe period (1816–1848), as portrayed by Sarkees and Wayman (2010, 189, Figure 3.2).

This chapter is divided into two main parts followed by a conclusion. We look at what we know about (1) factors that increase the probability of war—with a focus on territorial disputes, alliances, rivalry, and arms races and (2) factors related to peace with a focus on the democratic peace and the territorial peace. We do not attempt to summarize the previous chapters; however, we will highlight the main findings on each of these topics, as well as going into more depth on our individual research programs in each of these areas.

FACTORS THAT BRING ABOUT WAR

Progress often involves the creation of new data sets that permit researchers to address different questions than they have before. The COW project was initiated to provide data that could address questions that had been primarily discussed theoretically with evidence consisting of historical anecdotes. Singer and Small (1972) sought to go beyond that by a systematic analysis of all historical cases through the collection of replicable data on interstate war, formal alliances, and capability going back to 1816. Over the years this was extended to include data on militarized interstate disputes (MIDs) (Jones, Bremer, and Singer 1996) and civil wars (Small and Singer 1982). These efforts were also advanced by those of Michael Brecher (1977, 1984) in the creation of data on international crises from 1919 in the International Crisis Behavior (ICB) project (see Brecher and Wilkenfeld 1997).[1]

By the year 2000 the MID data had greatly expanded the kind of knowledge that could be generated because it made it possible to shift from systemic characteristics as correlates of war to the characteristics of those MIDs associated with escalation to war. This shift was accompanied by a change in research design away from correlational analysis to one that examined dyads and their probability of going to war. Stuart Bremer (1992) was responsible for this transformation, and his analysis of "dangerous dyads" became the primer for much of the research in peace science during the twenty-first century. His important chapter in the first edition, "Who Fights Whom, When, Where and Why," serves as a benchmark for seeing what we have learned in the past twenty years and how we have moved beyond his summary of findings.

Territorial Disputes

Bremer (2000) listed four key factors that we knew then that we did not know before: states that are contiguous, that are major states, and that have a history of fighting are most likely to fight each other, while states that are jointly democratic are not. Of these four Bremer was most confident about contiguity as a key factor promoting escalation to war. There is still much that is true about this relationship in that states that are far away are not likely to fight each other (see Bennett and Stam 2004; Halvard and Gleditsch 2006) because they cannot reach each other.

The problem with this finding, however, is that it seems obvious and without much theoretical significance. Bremer addressed this by coming up with an ingenious general statistical explanation often used to explain the decrease in traffic accidents the further away from home. Most traffic accidents occur not far from our work and home because that is where we do most of our driving. Underlying this probabilistic explanation, however, is a realist assumption that war is pervasive and that states would always fight each other if they could (Morgenthau 1960).

One of the early problems with the contiguity hypothesis is that a large number of contiguous states do not fight each other. These cases do not affect the statistical analysis because they are overwhelmed by the large number of noncontiguous states that do not fight each other. Vasquez (1995) argues that neighbors often do not fight, and when they do, it is not because they are proximate or contiguous but because they have real grievances. For Vasquez (1995; see also Vasquez 1993, chapter 4), what separates neighbors that go to war from those that do not is the presence of border grievances.

Work by Hensel (1996a) and Senese (1996) provide some empirical evidence on this claim in that both scholars find that territorial MIDs are statistically more likely to escalate to war than non-territorial MIDs. By 2001 Vasquez and Henehan, using a variety of data samples and breaking down the categorization of MIDs into territory, regime, and policy, demonstrate that territorial MIDs are more likely to escalate to war than regime MIDs, while policy disputes are statistically not likely to escalate.

This differentiation, however, is among disputes that already employ some sort of armed force. What about territorial issues before they give rise to armed force? Are they more likely to give rise to MIDs than other types of issues? For the longest time there was no empirical answer to this question because there were no data on diplomatic conflicts (or issue claims) before they became militarized. The Issue Correlates of War (ICOW) project, codirected by Paul Hensel and Sara Mitchell, was founded to collect data on such issues (Hensel and Mitchell 2017b).[2] The ICOW project collects data on territorial claims for the entire globe from 1816 to 2001 (Frederick, Hensel, and Macaulay 2017), and maritime and river claims from 1900 to 2001 (Hensel et al. 2008). A comparison of these three issues shows that territorial issues are much more likely to give rise to conflict than maritime or river issues, which are more apt not to produce violent conflict. These studies along with a host of others over the past twenty years (see Miller et al. 2020) make us conclude that one of the main factors increasing the probability of MIDs escalating to war is the presence of territorial disputes.

With regard to contiguity and territorial disputes, both Hensel (2000, 2012) and Vasquez (2001) show that while contiguous states are more likely to have MIDs that escalate to war, the presence of territorial disputes make escalation to war more likely than the presence of contiguity. Thus noncontiguous dyads are not only just as likely to goto war as contiguous states, but as Vasquez's study (2001, 161, Table 4) shows, they have the highest conditional probability of going to war when territorial disputes are present. Selection issues are a concern in this kind of research and are typically addressed with Heckman models. Senese (2005) finds that contiguity increases the likelihood of having some sort of a MID, but the presence of a territorial MID is what increases the likelihood of escalation of war, not contiguity.[3]

Why do territorial MIDs have a greater propensity to escalate to war? There are two answers to this question. The first argues that territorial MIDs are prone to war because of certain characteristics these issues have, and the second is that they are prone to war because of the way they are handled. Whereas Bremer (1992) is able to ask what separates MIDs that escalate to war from those that do not, the collection of territorial issues before they became militarized permits researchers to ask the much broader question of what separates territorial issues that give rise to the use of force (i.e., a MID) from those that do not.

Huth (1996) demonstrates that different types of territorial issues have different propensities to escalating to war and to the use of force. Ethnic territory is the most conflict prone, followed by strategic territory, with territory involving economic disputes actually prone to peaceful settlement.[4]

The main limitation of Huth's (1996) and Huth and Allee's (2002) data is that they do not go back to 1816. This limitation is overcome by the ICOW project. The ICOW project did much more than extend Huth's data back to 1816, however. In many ways it transforms the kinds of data available to the field because it collects data on three new issue areas that were never empirically investigated—maritime claims, river claims, and identity claims. In addition, this project collected a host of variables on the salience characteristics of these types of claims that might help distinguish which were more prone to MIDs and which might be more likely to be peacefully resolved. Early on they find that the tangibility of territorial claims is a key factor in the conflict process. Hensel and Mitchell (2005) find that territorial issues that consist of tangible stakes that can be divided are more likely to be resolved, whereas issues that consist of intangible stakes like honor or shared identity are more prone to the use of force and war (see also Hensel et al. 2008, see Powell and Wiegand, chapter 11, for more detailed findings on intangiblity). Tangible stakes are considered divisible and therefore more easily open to negotiated resolution. As with Huth, Hensel et al. (2008) also find that ethnic issues often give rise to a MID and typically involve intangible characteristics.[5]

The ICOW project used the findings on tangibility to develop a twelve-point salience index that could measure the importance of these issues to decision makers. For example, territorial claims are more salient if they involve populated areas with more resources and if states claim historical ownership over the areas, while maritime claims are more salient if the contested areas involve oil, migratory fishing stocks, or strategic choke points. They find that salience predicts not only the conflict-proneness of issue claims but also the use of peaceful attempts to resolve the issue.[6]

In fact, one of the more subtle findings of Hensel et al. (2008) is their uncovering of how these two seeming contradictory actions are intimately connected. They find that when repeated MIDs fail to resolve an issue, actors switch to peaceful attempts and when peaceful attempts fail, actors resort to the use of MIDs. This sort of dynamic behavior suggests that contention over issue claims is not only a function of their intrinsic characteristics but is also affected by ongoing strategic interactions. For example, as Hensel and Goemans (chapter 1) point out, Huth, Croco, and Appel (2011) and Prorok and Huth (2015) find that involvement in legal proceedings can affect the behavior of states depending on previous outcomes and the extent to which they have a better legal claim.

The ICOW project is also important because it demonstrates that despite the statistical tendency of territorial MIDs to escalate to war, most territorial claims do not. Most territorial (59 percent), maritime (73 percent), and river (89 percent) issues do not involve a single use of force (MIDs) (Hensel 2001, 100–3, Tables 3–5; Hensel et al. 2008, 135; Hensel 2012, Table 1.7; Hensel and Mitchell

2017). Coding issues from their diplomatic origin thus demonstrate that even the most salient issues connected to wars in history are more typically resolved by states through peaceful negotiations.

In recent years much of the research on the peaceful resolution of territorial disputes has investigated "forum shopping" in terms of which international organizations are most apt to provide a better outcome (Wiegand and Powell 2011; see also Powell 2020 on Islamic states). The use of the courts and arbitration has been the focus of much of this work (Powell and Wiegand, chapter 11). A good deal of research on border negotiations has looked at the impact of territorial settlement on trade (Simmons 1999; Schultz 2014). Some of this work has conceptualized borders as "institutions" that can create new cooperation opportunities (Simmons 2005; Schultz 2015). Carter and Goemans (2018) show how trade actually increases depending on the type of border that is drawn by negotiations.

While the salience index is very predictive, what is more important, especially for future research and the precision of knowledge, is which components of the index are most associated with conflict or with peaceful attempts at resolution. Hensel and Mitchell (2005) and Hensel et al.(2008) find that claims that involve homeland territory (as opposed to colonial territory) that have been previously possessed or that have had a long history of sovereign rule are more likely to have MIDs and escalate to war. Statistical research on these component factors coupled with the judicious use of case studies might improve our understanding of the underlying causal processes associated with the escalation to conflict, especially the escalation to war. An examination of the components and how they might be related temporally would also help explain the sequence of dyadic interactions and whether there are certain paths to war (Vasquez 2011b; Levy 2012).

The main limitation of much of the ICOW research discussed is that almost all of it has been confined to a sample of data limited to certain regions, typically the Western Hemisphere, Western Europe, and the Middle East. Not until 2013 were territorial data for the entire globe released, and then not for all the variables in the data set (http://www.paulhensel.org/icowterr.html). One of the pressing tasks for ICOW and the field as a whole is to replicate all their findings to see if they hold globally. Already Frederick (2012) in his dissertation uncovers heterogeneity in the data with findings differing across regions. Why some regions might differ and what this can tell us about territorial conflict is an important area for future inquiry. One factor that might be at play is that there may be a neighborhood effect that shapes behavior. Hensel (1998) pointed out early on that certain territorial disputes, like those in the Middle East or South Asia, can poison relations not only in the dyad but also in an entire region. Senese and Vasquez (2008, 278–79) at the end of their study call for an examination of "neighborhood effects" on conflict.

Although a great deal of the research on why territorial issues have a high probability of going to war has centered on the characteristic of these issues, another strand of research has focused on how these issues have been handled. In terms of process, early on it was found that territorial MIDs tend to be reciprocated more than non-territorial MIDs (Hensel and Diehl 1994; Hensel 1996a, 2000; Mitchell and Prins 1999, 174n13). In addition, they tend to recur (Hensel 1996), and recurring MIDs are more likely to escalate to war (Senese and Vasquez 2008, chapters 5–6).[7] Presumably the salience of these issues makes actors unwilling to give in and increase the use of force as the issue fails to be resolved. Leng (1983) finds this to be a general bargaining pattern among crises that repeat among major states.

Senese and Vasquez (2008) find that as states resort to power politics (e.g., alliances, arms buildups) to handle territorial disputes, there is a stepwise increase in the probability of escalation to war as different power politics practices are adopted. These findings, however, hold primarily for the 1816–1945 period. In the nuclear Cold War period alliances and arms races become nonsignificant.[8]

Finding that specific practices, like alliance making and arms races, make territorial disputes more likely to escalate to war suggests that these same practices might make non-territorial issues prone to war if handled in this manner. Senese and Vasquez (2008, 134–35, 149, 152, 156, 159) find that to be the case, although these issues (namely, policy and regime disputes) experience lower probabilities for war. The way issues are handled thus has a general effect on the probability of war, while it has an enhanced effect on territorial issues.

In recent years there has been more work on the biological basis of war. In international relations Johnson and Toft (2014) explored the biological basis of human territoriality as a source for territorial conflict. This is part of a general trend within political science to examine the genetic, evolutionary, and hormonal basis of political behavior, including violence (see Hatemi and McDermott 2011). Experimentation is one way of getting at this question within political science. Powers (2018), using an experimental design, has shown that individuals facing territorial conflicts of interest (as opposed to non-territorial conflicts of interest) are more apt to react aggressively both in interpersonal and in foreign policy contexts. These findings hold not only for subject pool samples, but for a random sample public opinion survey conducted in Kentucky.[9] The next step in experimental design is to move away from survey responses to the measuring of electrodermal activity and cortisol levels as indicators of physiological reactions and eventually to the use of brain scans, as in neuroscience, to see if territorial issues differ from non-territorial issues.

Last, outside of political science, Mitani, Watts, and Amsler (2010) while doing fieldwork in Uganda found that chimps not only fight but also conquer territory, providing evidence relevant to the territorial explanation of war. The biological foundation of political behavior presents a new area of research within political science that attempts to use experimental designs to probe these relationships; this is an exciting area of new research that will provide a more complete understanding of our genetic and evolutionary inheritance with regard to territoriality and violence and will go a long way in filling in the micro-foundations of the causes of war.

Two of the questions raised in the last edition (Vasquez 2012, 307) as pressing areas for further investigation are (1) the development of a typology of territorial disputes that would clearly distinguish those that go to war from those that do not and (2) the impact of territorial MIDs on domestic politics. Recent work has added answers on both questions. In terms of a typology, Gibler (2017b) constructs an empirical-based typology to see what types of territorial MIDs are most likely to escalate. Based on a review of MID data that eliminated certain cases for lack of documentation and corrected errors, he finds that certain territorial MIDs account for more of the conflict in the data.[10]

Gibler (2017b, 211–13) finds that the most conflict-escalatory MIDs involve (a) disputed ownership of territory (60 percent), (b) conflict over border delimitation (13.33 percent), (c) new states entering the system (17 percent), and (d) opportunity conflict (e.g., revanche, predation) (6 percent). Because Gibler's (2017b) typology uncovers what kinds of territorial MIDs produce fatalities and wars it adds to our substantive and descriptive knowledge. Concerning future research, it provides food for thought and theoretical reflection in terms of coming up with a more detailed theoretical topology that would explain what it is about these various types that "causes" actors to go to war. Here an attempt to combine Gibler's descriptive knowledge with a breakdown of the components of the Hensel and Mitchell (2005) salience index might be worth pursuing.

A number of studies have extended our knowledge of the impact of territorial disputes on domestic politics. This work focuses on how the presence (or absence) of territorial MIDs affects levels of democracy at the domestic level. Much of this follows from Tilly's argument that war leads to centralization (and hence the absence of democracy).[11] Gibler (2010) shows that an external territorial threat can be used to remove veto players and thereby consolidate power. Gibler and Miller (2014) argue that external territorial threat will not only increase centralization but also reduce the likelihood of civil war and unrest, although as Goemans (2000) notes, based on the case of the First World War, losses in a war can lead to revolution.

Several studies find that an external territorial threat leads individuals to more strongly identify with and support the state. Gibler, Hutchison, and Miller (2012) in an analysis of African domestic politics find that external territorial threat helps unify the populace by making individuals identify with the state rather than with their subgroup. S. Miller in several studies (2013, 2017, 2018) finds that territorial threat leads individuals to support strong government and even autocratic rule if its purpose is to defend territory.

The more novel studies on domestic politics are those that show the relationship of the absence of territorial threat and the emergence of democratic traits. Hutchison and Gibler (2007) replicate a well-known study (Peffley and Rohrschneider 2003) on tolerance to show that the presence of a targeted territorial MID reduces the likelihood of internal tolerance even when other variables are present that would normally predict it. The study shows that the absence of external territorial threat is an important variable in the evolution of tolerance, normally thought to be solely a function of domestic variables. Hutchison (2011a, 2011b) also explores the effect of territorial threat on trust in government. These studies provide an area for future research, particularly to see what other democratic attitudes are reduced by an external territorial threat. Also, Gibler (2018,

27–28) notes that states that experience a "democracy reversal" are likely to have an outside territorial MID. However, an independent judiciary can inoculate such states so long as the territorial threat does not entail a long-term rivalry (Gibler and Randazzo 2011; Randazzo, Gibler, and Reid 2016). In a larger context, they support the more general claim of Gourevitch (1978) that external relations can shape domestic politics (for more details on these findings see Gibler and Miller, chapter 9).

Last, Vasquez and Valeriano (2010, 300) show that territorial wars are only one type of war; policy wars make up 24 percent of the wars and regime wars constitute 11 percent. The modal war in the system is a two-party war between neighbors over territory—54.4 percent (forty-three of seventy-nine) of the dyadic wars. This is an important foundation for advancing knowledge in that a great many wars of the past and the future may be explained on the basis of the modal war. Some of the dyadic conflicts associated with the breakup of the Soviet Union, like the wars between Russia and Georgia, Armenia and Azerbaijan, or the conflict between Russia and Ukraine, can all be explained by the territorial explanation of war. There are also important policy implications for this finding in that if these disputes can be resolved before they escalate, then war might be avoided. With better knowledge on how to handle such border disputes, diplomacy might be able to eliminate an entire category (indeed the modal category) of war. In terms of purely *peace* research, more emphasis needs to be placed on how to prevent territorial disputes and how to actively resolve or settle them (see Powell and Wiegand, chapter 11). The decline of interstate warfare over time reflects the resolution of many historical border disputes (see Hensel and Goemans, chapter 1; Braumoeller, chapter 16).

Given these findings what do we know about territory and war?

- States involved in territorial disputes have a much higher probability of going to war than states involved in other types of MIDs.
- States that have territorial disagreements and competing claims are more apt to have MIDs than states with non-territorial claims such as river disputes.
- Neighbors are more apt to have MIDs and go to war not because they are contiguous, but because they have territorial disputes.
- Some kinds of territorial issues are more apt to escalate to the use of armed force and to war than others. Specifically territorial issues involving ethnic claims are more apt to escalate, followed by disagreements over strategic territory. Disputes over territory involving economic questions are more apt to be settled peacefully.
- Territory involving intangible stakes are much more apt to escalate to conflict and to war than those involving tangible stakes.
- Territorial MIDs go to war because they are more apt to result in fatalities, they are more likely to recur, and recurring MIDs are more apt to escalate to war.
- Issues that are handled by the practices of power politics are more apt to result in war even if they are not territorial disputes, although territorial disputes have a higher probability of escalating with the use of power politics.

- The presence of territorial MIDs has a significant impact on domestic politics. Generally, since they pose an external threat, they lead to centralization of power. The presence of a territorial MID makes citizens less tolerant, seems to have an effect on trust, and can make for democracy reversal. More importantly, the absence of territorial threat can lead to democratization.

There is still much that we do not know. Here are some of the key inquiries that will further advance our knowledge:

- Most territorial issues do not give rise to MIDs or wars. Why? What is it specifically about these irenic territorial issues *or the context* in which they occur that makes them less prone to war?
- What is the relationship between human territoriality and state use of violence to handle territorial disputes, and how has it changed over time?
- What more can we learn about the characteristics of issues and the way they are handled that will make clear why some territorial claims give rise to the use of force and war, while most do not? Can this new knowledge be used to construct a theoretical typology of territorial issues?

Alliances

The role of alliances in peace and conflict has been a topic of international relations (IR) theory going back to Thucydides, mostly because of realist concerns about power and balancing (Morgenthau 1948). The early work of Singer and Small (1966a, 1966b), which was at the systemic level, finds some relationship between the number of alliances in the system and war. Levy (1981) collects data on major state alliances going back to 1495 and finds with the exception of the nineteenth century that alliances involving at least one major state tend to be followed by a war within five years. These findings are contrary to the classic realist notion that alliances balance power and thereby prevent war. Whether wars occur because the alliances in question fail to balance or whether, having balance, war occurs anyway cannot be determined without further research. At the same time Levy also finds that most wars are not preceded by alliances, which indicates that alliances are not a necessary condition for war. However, Vasquez and Rundlett (2016) find that alliances are in fact a necessary condition for multiparty wars, as in the two world wars.

Levy's (1981) overall findings are consistent with Ostrom and Hoole (1978), who use COW data for the 1816–1965 period and find that alliances are likely to be followed by war in the first three years after their formation, are negatively associated with war from four to eleven years after their formation, and have no relationship with war twelve years after they are made. All of these findings suggest that some alliances are followed by war, but others are not. This implies that a typology of alliances might provide some clarification about the "dangerous alliances." Gibler (2000) developed such a typology and finds that alliances consisting of major states, that have been successful in their most recent war, and that are dissatisfied with the status quo are more apt to be followed by war than alliances lacking these characteristics (see also Gibler and Vasquez 1998). Maoz (2000) also finds that alliances consisting of major states have different effects than those consisting of minor states and that alliances made by democratic states are less likely to be followed by war. Last, Gibler (1996) finds that one type of alliance—territorial settlement treaties that create an alliance to seal the bargain—is rarely followed by war.

Further evidence that alliances are associated with war is provided by the steps-to-war analysis of Senese and Vasquez (2008). They find that the presence of outside alliances increases the likelihood of a MID if both sides have an outside ally to support them. This alliance variable is not wiped out by rivalry or ongoing arms races. However, these findings hold only for the 1816–1945 period. During the Cold War (1946–1989) period both sides having an ally actually makes escalation to war *less* likely.

Why this is the case needs further research. Presumably one reason for the Cold War patterns is nuclear weapons, but it might also involve something about the peculiar nature of East-West alliance polarization (Vasquez and Kang 2013). Alliances formed in the Cold War also involved larger multilateral agreements with many democratic partners. Senese and Vasquez (2008, chapters 5–6) also argue that alliance formation is connected with recurrent disputes and with arms racing. The extent to which this is correct needs to be empirically investigated with a sensitivity to the problems of endogeneity.

The COW data dominated most of the alliance research to the end of the twentieth century. Gibler and Sarkees (2004) updated those data, and Gibler (2009) published a handbook of all the treaties that established the formal alliances. Starting in 2000 the Alliance Treaty Obligations and Provision (ATOP) project released a new data set on alliances (see Leeds et al. 2002). The main innovation is the inclusion of the specific treaty obligations that committed allies to each other. Using these data, Leeds, Long, and Mitchell (2000) criticize Sabrosky's (1980) argument that allies are not reliable. He finds that only slightly over 25 percent of allies entered a war when their ally was attacked. Leeds et al. show that it is important to take into account whether an alliance treaty's provisions are relevant to a war (e.g., does the alliance apply to this specific geographical location or adversary?). When obligations are considered more carefully, the authors find that in fact 75 percent of the time, alliance partners are reliable in that they enter the war, contrary to Sabrosky's findings.

Later work in ATOP examines the factors that may reduce the reliability of alliance. Leeds and Savun (2007) and Leeds, Mattes, and Vogel (2009) find that domestic leadership changes can lead to an increased risk of unreliability, as can changes in capability of individual allies (Johnson and Joiner 2019). Reliability of course also affects future reputation. Crescenzi et al. (2012) find that states that uphold their alliances are more likely to have future alliances with other states. Conversely, states that do not have a record of reliability are likely to face tougher demands from future alliance partners (Mattes 2012, 2015).

Kang (2012) provides some evidence as to the process by which some alliances promote war. He finds that when minor states have an alliance with a major state, they may be emboldened to resist demands and even go to war. The Crimean War is a classic historical example of this with the Ottoman Empire being emboldened by anticipated support from Britain and France. This is part of the general problem of alliances creating a moral hazard, a question Benson (2012) analyzes in detail. In other analyses, Kang (2012) finds for the 1816–1945 period that states with defense pacts with major states are more likely to have MIDs initiated against them, further evidence that findings will vary by types of alliances (Leeds 2003).

Since the last edition there has been some investigation of the internal economic impact of alliance making. Kang (forthcoming) finds that making an alliance permits a state to shift resources away from defense spending and thereby increase domestic economic growth. He also finds that alliances encourage trade amongst members.

The use of the ATOP data that has attracted the most attention has been their findings on deterrence. Leeds (2003) argues that defense pacts, because they have written commitments to defend, should have deterrent effects. She finds for the 1815–1944 period that states with an outside defense pact have fewer MIDs initiated toward them than those without such alliances. She also finds that defensive alliances reduce the likelihood that a potential target will be attacked, while offensive and neutrality alliances increase the risks for conflict. Thus, Leeds (2003) finds that if the challenger has an offensive alliance and calls upon that ally to join the dispute, then a MID is more likely (see also Johnson and Leeds 2011). Likewise, Leeds (2003) finds that the same pattern holds for neutrality pacts (see Quackenbush's (2018, 693) discussion of these articles).

These studies show that different alliances can have contradictory effects. Leeds's (2003) differences in findings with the steps-to-war analysis stem in part from the different empirical strategies used. Leeds connects alliances to MIDs in the same directed dyad year, while Senese and Vasquez (2008) consider the onset of MIDs in the five-year period following a MID.[12] Differences may also be a result of the coding of an alliance, especially whether it is a defensive pact. Benson (2011, 2012) questions Leeds's (2003) coding and uses Smith (1995) to construct more precise categories based on whether the defense obligation is conditional or unconditional and whether the aim is compellence or deterrence. He finds that compellent alliances and active deterrent alliances are conflict prone; only passive deterrent alliances reduce MIDs, and these are not a large percentage of all alliances.

Kenwick, Vasquez, and Powers (2015) also address the deterrence question. They look at both MID onset and war onset. They eschew a dyad-year research design and examine only individual dyads that have made a defensive alliance and then seek to draw inferences about deterrence by whether the number of MIDs and wars after an alliance is significantly lower than the number of MIDs and wars the dyad had before the alliance. For MID onset, they find for the 1816–1945 period that there is actually an increase in the number of MIDs within five years after the formation of an alliance compared to the five years before. They find that forty-seven cases have an increase in MIDs while twenty-seven have a decrease, which is a statistically significant difference (Kenwick et al. 2015).

No doubt such alliances are formed because actors anticipate conflict. A key difference with Leeds's (2003) research design is that they look at alliance formation and she does not. Alliance formation may reflect a reaction to threat perception. At any rate, these findings are consistent with those of Morgan and Palmer (2003), who find that joining an alliance increases the probability of MID initiation. The fact that alliance formation may be endogenous to conflict escalation does not undercut the fact that they fail to deter MIDs. However, for the nuclear Cold War era, Kenwick et al. (2015) find no significant relationship.

Kenwick et al.'s (2015: 952; see also Table 2.2 online appendix) findings on war onset are similar. They find that thirty-one cases have more wars after an alliance formation while fourteen cases have fewer, which is a statistically significant difference. For the nuclear Cold War era, they also find weak evidence that wars increase (Kenwick et al. 2015, 952).

The findings on war are not inconsistent with Leeds (2005), when she finds that *if* a dispute escalates to war and defensive allies are present, then allies live up to their commitments and are likely to intervene. Thus if alliances fail to deter the onset of a MID, they may be associated with larger and more violent wars if the MID escalates.

Given that Kenwick et al. do not find the same pattern for the Cold War period, Morrow's (2017) claim that future research should look not at whether deterrence works, but when it does is a useful tack to take (see Kenwick and McManus, chapter 3). One way to start this process is to examine the cases in Kenwick et al. (2015) in more detail. What is it about the twenty-seven MID divergent cases and the fourteen war onset cases that makes them seemingly cases of deterrence success? Morrow (2017, Figure 2) finds for 1816–1945 that defensive alliances fail to deter when he does not add controls, but when he adds controls, then he finds that they do deter, which is consistent with Leeds (2003). The control variables are his key for determining when alliances deter, but most of these control variables—like joint democracy, lack of contiguity, and the presence of offensive alliance or neutrality pacts—are all well-known predictors of peace or conflict, so including them as controls that wipe out the relationship obscures rather than clarifies the true relationship. For example, we would not expect opposing alliances consisting of joint democracies to be followed by more MIDs; neither would alliances consisting of states that were not contiguous. We need to have a better empirical strategy for determining the relationship between alliances and other correlates of war.[13]

Last, despite disagreements and the complexity of statistical studies, one thing is clear—the onset of the First World War and the Second World War in 1914 and 1939 must be seen as cases of massive deterrence failure. Further case work as to why deterrence failed remains an interesting avenue for exploration (for a start, see Vasquez 2018, 353–56).

What have we learned about alliances and war to date?

- Except for the nineteenth century, from 1495 on, alliance formation that includes major states is more likely to be followed by war than by peace within five years.[14]

- The *presence* of an outside alliance is positively associated with the escalation of a MID to war in the 1816–1945 period if a five-year window is examined, but not during the Cold War 1946–1989 period.
- The most dangerous alliances regardless of period seem to be those consisting of major states, those that are dissatisfied, or those that have been successful in their most recent war. Alliances that are part of a territorial settlement treaty are almost always followed by peace.
- Most wars involving major states have not been preceded by an alliance. However, alliances seem to be a necessary condition for multiparty interstate wars.
- Leeds (2003) finds that the *presence* of defensive alliances has a muting (deterrent) effect on MIDs onset in the 1815–1944 period. Kenwick et al. (2015) find that *formation* of alliances in the 1816–1945 period increases the probability of MID onset and war onset.
- The First World War and the Second World War are cases of massive deterrence failure in the presence of alliances.

These and other findings in the literature suggest a number of unanswered questions:

- Since findings with regard to alliance vary by type and temporal domain, one way of advancing research is to find out *when* they result in war and *when* they lead to successful deterrence.
- Given that initial analyses suggest that the 1990–2001 period is like 1816–1945 with regard to alliances and war, will the future be more like it or the Cold War, where alliances are negatively related to war?
- When deterrence fails, why does it fail?
- What factors occur within five years after the formation of an alliance that increase the probability of war?
- What effects do war-prone alliances (i.e., those consisting of major states, dissatisfaction, and success in the most recent war) have on increased armament levels, repeated MIDs, and the making of counter-alliances?
- What effect does alliance making have on the domestic political environment of the signatories and possible targets, particularly the balance between hardliners and accommodationists?
- Does the causal sequence of wars that break out without a preceding alliance differ in a theoretically significant manner from the causal sequence of wars that are preceded by an alliance?

Rivalry

In Bremer's (2000: 25) analysis of who fights whom, the second of four key things he thought we knew about war is that war emerges from recurring disputes or dyads that "have a history of fighting one another." He gives two reasons why disputes might recur. First, an issue is rarely settled by one fight and so long as the issue remains unresolved, it will continue to periodically give rise to fights. Second, "the mere act of fighting produces feelings of grievances and hostility, even hatred" (25). This emphasis on repeated disputes became a focus of the rivalry literature.

Leng (1983) was among the first to show that as crises repeat, contenders engage in more escalatory acts, and that by the third crisis war is very likely. He argues that to treat each crisis as independent is to fail to understand that war emerges out of a process of the interaction of states over time with previous crises having an impact on subsequent ones (see Akcinaroglu and Radziszewski, chapter 5).

Leng's analysis set the stage for distinguishing dyads in terms of whether they have recurring disputes. Goertz and Diehl (1992a) and Diehl and Goertz (2000) reconceptualize recurring disputes as rivalry and identify rivalry as a source of war. Wayman and Jones (1991) examine how previous disputes have a negative impact on subsequent disputes. The crucial and original insight of these scholars is that war emerges from an underlying *relationship* between states. War cannot be understood simply by the characteristics of the issues under contention or the states contending over them, but by the relationship that emerges as conflict repeats. It thus makes sense theoretically to study relations between states over time rather than to statistically analyze and compare individual MIDs without controlling for the particular states involved and MID history.

Once Diehl and Goetz (2000) advocated studying individual dyads as opposed to MIDs, they found that dyads that are enduring rivals (with six or more MIDs) have a pronounced proclivity to go to war. A related finding is also produced in ICB data (Brecher and Wilkenfeld 1997, 832–33) for what they call states in protracted conflict. Indeed, Diehl and Goertz (2000) find that enduring rivalries account for almost 50 percent of wars from 1816 to 1992. Colaresi, Rasler, and Thompson (2007) find a similar pattern by identifying rivals through historical statements, showing that 75 percent of all wars are linked to what they identify as principal (or strategic) rivalries.

Senese and Vasquez (2008, chapters 5–6) also find that the presence of rivals increases the probability of a MID escalating to war for all periods. They show that rivalry is one of a series of steps that progressively increases the probability of war—the others being alliance making and arms races. Examining how rivalry is related to these other two steps is a fruitful avenue for future research. Some of these findings on the steps to war are independently confirmed by Colaresi and Thompson (2005) and Colaresi et al. (2007) using ICB data.

Since 2000 several scholars have looked at how states become rivals. Wayman (2000) distinguishes between rivals that are born feuding and go to war on their first MID, like India and Pakistan, and those that go to war after several MIDs, like Germany and Britain.[15] This suggests there may be different types of rivals in terms of when and how they go to war. Valeriano (2013), building on Leng (1983), finds that rivals that increase their level of hostility as they go from their early MIDs to later ones have a greater propensity for war. He labels these "escalating rivalries," thereby helping us move toward a typology of rivalries (see Akcinaroglu and Radziszewski, chapter 5, especially Tables 5.1 and 5.2 for a review of key factors related to escalation).

Rivalry has also been viewed as a context for war, increasing the chances for escalation in conjunction with other risk factors such as arms races (Gibler and Hutchison 2005), high inflation (Mitchell and Prins 2004), territorial disputes (Rasler and Thompson 2006; Lektzian, Prins, and Souva 2010), ongoing war (Vasquez et al. 2011; Vasquez 2018), or power transitions (Geller 2000; Kadera 2001).

Bennett (1996, 1997) has led the way in the study of rivalry termination. By defining a rivalry as ended when the underlying issue is fully resolved, he established more historically informed end dates for rivalry data. The ending of the Cold War provides an important case study on this with the most important volume in terms of ending rivalry being written by Kupchan (2010). Cox (2010) sees rivalry termination as linked to domestic realignment, especially if it is coupled with a foreign policy failure. Owsiak and Rider (2013) examine the role of borders in rivalry termination and find that states that successfully settle their borders are likely to see their rivalry come to end (Owsiak, Diehl, and Goertz 2017; Rider and Owsiak forthcoming).

What have we learned about rivalry and war?

- Most enduring rivalries go to war and more than half of interstate wars are associated with rivalries.
- Rivalry has a separate and independent effect on the probability of MIDs escalating to war when controlling for territorial MIDs, recurring disputes, outside alliances, and arms races for both the 1816–1945 and the Cold War 1946–1989 periods.
- States are likely to become rivals if they have territorial disputes and this is a function of the fact that territorial MIDs recur frequently.
- Other risk factors for war are enhanced in the presence of rivalry.
- Conflict resolution techniques have limited success in ending rivalry; however, if states settle their borders, that is apt to lead to the termination of their rivalry. Political shocks often contribute to rivalry termination.

A number of theoretical questions remain unanswered:

- What interactions make rivalry occur and then escalate to war? As disputes recur they seem to have an impact on each other. Exactly what that impact is needs further research, but it is thought to produce increased hostility, threat perception, and some form of escalation.
- Is there a difference between rivalries that are born feuding and those that evolve over time? Likewise, is there a difference between major state rivalries and minor state rivalries?
- How do rivalries produce an expansion of war (and MIDs) through contagion?
- What is the role of domestic politics in the origin and persistence rivalry? What is the domestic impact of recurring disputes within each state (e.g., on the balance between hard-liners and accommodationists)?

- How are rivalries connected through networks of rivalry ties? If states end one rivalry to focus on another rivalry, how do they make those choices?
- Why do conflict dynamics vary across rivalries? We see punctuated equilibrium, evolutionary, and other dynamic patterns in rivalry MID data, but we do not have theoretical explanations to predict these varying patterns.

Arms Races

The final factor that we know plays some significant role in making states war prone is arms races. Some of the earliest work in peace science was devoted to measuring and modeling arms races (Richardson 1960a; Choucri and North 1975; Zinnes 1976). The major advance in research came with Wallace (1979), who rather than looking at whether arms races "cause" war, focuses on MIDs involving major powers, showing that ongoing arms races increase the likelihood that MIDs escalate to war. This research generated a heated criticism by Weede (1980) and Diehl (1983) as to the accuracy of the research design (see Sample 2012 and chapter 4 for a review).

Sample (1997) resolved much of the debate by showing that mutual military buildups that do not go to war produce a MID that goes to war within five years. The five-year window is important because typically the first MID does not go to war. Thus for Germany and Britain, there were several MIDs in the 1930s before the 1939 MID went to war. With this control Sample finds a statistically significant relationship between arms racing and escalation to war for the 1816–1945 era, but not for the nuclear era. In later work Sample (1998) also finds that the hostile spiral model is more accurate than the deterrence model (see also Wallace 1982). Interestingly, she finds that capability shifts do not enhance the probability of war in the presence of arms racing (Sample, chapter 4), consistent with the mixed findings on capability discussed by Morey and Kadera (chapter 2). Subsequently she collects new data for minor states (Sample 2002) and finds that the same relationship holds for minor-minor dyads, but not for major-minor dyads. She also greatly improves upon Wallace's data in that she just does not collect data for MIDs, but for every year there is a dyad, recording their military expenditures for that year. Senese and Vasquez (2008), using Susan Sample's data, also find that arms races have a separate and independent effect from alliances, rivalry and territory. Although arms races are extremely dangerous, they are fairly rare; only a few wars are preceded by arms races (see Diehl 1983; Vasquez and Valeriano 2010).

Sample has addressed the question of the relationship of arms races to rivalry, responding to criticism by Diehl and Crescenzi (1998). Rider, Findley, and Diehl (2011) provide some evidence for their claim, that rivalry is key but they do not use a five-year window or control for the nuclear era. When Sample (2012) restores these controls, the significance is of arms races restored. She also finds that arms races increase the probability of war in cases *without* an ongoing rivalry.

To further disentangle how the steps to war are related to both arms races and rivalry, Sample (2014, 2018a) traces the temporal sequence of each step in the history of forty-two cases involving Huth and Allee's (2002) territorial claims and war from 1919 to 1995. She finds that the order varies with rivalry sometimes preceding arms races and at other times arms races preceding rivalry. In twelve cases where all the steps occur, she concludes that rivalry occurs early enough that she thinks it might be producing the steps rather than the other way around (Sample 2018a, 497–98, but also see 2014, 282). In other cases, especially those involving territorial disputes, however, she finds that arms races increase risks for war as anticipated by the steps-to-war model (Sample 2018a, 508). Vasquez (2011b) has also identified the various paths to war from 1816 to 2001 and has found, supporting Levy (2000), that there are multiple paths to war with some combination of steps, especially those involving arms races, having a higher probability of resulting in war than others (probability ranges from 0.17 to 0.78, Vasquez 2011b, 142).[16]

What conclusions can be made about arms races?

- There is a statistically significant relationship between the presence of a mutual military buildup and the escalation of a MID to war for the 1816–1945 period, if a five-year window is examined.
- The probability of war breaking out in the presence of a mutual military buildup greatly increases in the presence of other factors—namely, territorial disputes, outside alliances, and rivalry.
- There are multiple paths to war, and arms races only play a role in a limited set of wars.
- Conversely, the presence of nuclear weapons, most likely eliminates the effect of mutual military buildups on war onset by raising the costs of war to unacceptable levels.

A number of things about arms races are still worth exploring:

- Arms races seem to have an independent effect on the probability of rivals going to war. Statistical evidence controlling for both factors supports this conclusion. Also split population studies show that in the absence of rivalry but in the presence of arms races, war is likely. Nonetheless, there are wars where rivalry comes sufficiently early on that it may be driving the relationship. More case studies or sensitive statistical tests are needed to unravel this relationship.
- Why arms races are present in only a limited set of wars needs to be explained—in particular, scholarship needs to investigate whether certain types of conflicts escalate to war before both sides get a chance to increase their arms.
- There is a need to identify and collect data on other indicators of military buildups—such as pursuing new types of weapons, increasing the stockpiles of certain weapons, initiating conscription, extending the time of military service, or increasing the number of reserves. Such indicators may also be useful in identifying arms races in the nineteenth century, which with present measures appears not to have many arms races.
- How are systemic changes in offense and defense technologies related to mutual military buildups and the onset of war?

- Are there any theoretically significant differences between the way wars preceded by arms races break out and the way wars without preceding arms races break out?

The next section examines factors associated with peace. Before doing so, something should be said about what we know about who wins wars and the impact of war (see Quackenbush, chapter 7). Early on Rosen (1972) showed that most interstate wars are won by states that have more economic revenue at their disposal followed by the states that lose a lower percentage of their population. The former is consistent with Rasler and Thompson (1983), who show that major states with greater access to credit are apt to win their wars (see Cappella Zielinski and Poast, chapter 15, on war financing more generally). Nonetheless some well-known "Davids" defeat "Goliaths" and these have been explained by Organski and Kugler (1978). Last, most major states recover from major defeats within twenty years in what Organski and Kugler (1977) call a phoenix factor.

FACTORS PROMOTING PEACE

The Democratic Peace

The most fundamental thing we know about peace is that it actually occurs in the international system; indeed it is much more common than one would expect from realist IR theory. Morgenthau (1960, 38), for instance, states that: "All history shows that nations active in international politics are continuously preparing for, actively involved in, or recovering from organized violence in the form of war." For this, among other reasons, the study of peace has generally been ignored (see Owsiak, Diehl and Goertz, chapter 10).

Nonetheless, data show that war is relatively rare. Geller and Singer (1998, 1) note that in the post-Napoleonic era 150 states have never even experienced one war, and 49 only have had one or two. Indeed, by 2007 there were only 95 wars among legally recognized states (Sarkees and Wayman 2010, 75).

The best evidence that states are not constantly at war with one another is the democratic peace. Ray (2000, 299) stated in the last edition of this book: "One thing that 'we know about war' with unusual, or even unparalleled confidence is that it almost never involves two democratic states." Levy (1988, 662) in an oft-quoted phrase called the democratic peace one of the few laws we have in political science. The landmark studies are Rummel (1983), Doyle (1986), Maoz and Russett (1993), Ray (1995), and Russett and Oneal (2001). Babst (1964) was the first to publish findings on this phenomenon, but he was largely ignored at the time.

Bremer (2000) includes the democratic peace as one of his top four factors that we know about war—namely, that democratic states do not fight each other. Other studies have shown that joint democracies are also less likely to have MIDs (Maoz and Russett 1993; Russett and Oneal 2001) and are more likely to use peaceful conflict resolution techniques to handle their disputes (Dixon 1994; Raymond 1994). Weart (1998) also argues that democratic states in the ancient world were less likely to fight each other.

There are two explanations of the democratic peace—the structural (or institutional) and the normative. The structural has attracted more attention mostly because of the formal analyses associated with it (see Bueno de Mesquita et al. 2003). In its original form the rationale of the institutional explanation follows Kant and maintains that domestic constituencies and institutions restrain democratic leaders from going to war because the people do not see war as in their interest. The normative explanation maintains that democratic states share norms about how to deal with conflict. Generally it is found that jointly democratic states are more apt to negotiate than to resort to force, even over territorial issues (Huth and Allee 2002). Mitchell (2002) goes so far as to argue (and finds) that these norms are so powerful that they can create a Kantian system or culture that will even influence nondemocratic states to adopt these norms as the percentage of democratic states in the system increases. This is a very important study using early ICOW data and should be replicated on data for the entire world.

An important distinction in the democratic peace is between the dyadic and monadic democratic peace. The former argues that democracy has its pacific effect primarily for joint democracies, whereas the latter argues, following Woodrow Wilson, that even at the monadic level democracies are intrinsically peaceful. Most statistical studies are based on the dyadic level, but Ray (2000) finds some evidence for the monadic democratic peace (see also work by Rummel). Quackenbush and Rudy (2009) find little statistical evidence for the monadic peace and Vasquez (2020), examining the largest wars of the three most liberal states in the nineteenth century—France, England, and the United States—finds that the executive tries to restrain the legislature and public gripped by war fever rather than the other way around.

Russett and Oneal (2001) have expanded the democratic peace into a liberal peace consisting of three pillars—joint democracy, economic interdependence, and shared membership in an international organization—all of which reinforce each other (see Mousseau, chapter 8). Empirical evidence is strongest for the democratic peace, mixed for the trade peace, and weaker for the international governmental organization (IGO) peace (Anderson, Mitchell, and Schilling 2016).

THE TERRITORIAL PEACE

Gibler (2007) was among the first to challenge the democratic peace by looking at the role settling territorial borders plays in peace. He argues that democratic states do not fight each other because they have stable borders and do not have territorial disputes. His data analysis of the relationship between stable borders and peace wipes out the statistical significance of joint democracy.[17] He then goes on in an original analysis to speak of a territorial peace (Gibler 2012) in contradiction to a democratic peace. He argues that just as war can lead to state centralization, settling borders can lead to peace and eventually to a democratization of the state. For Gibler, border settlement precedes democratization. His study makes two lasting contributions to our understanding of peace. First, he shows that without a preceding territorial peace there is no democratic peace. Second, a neighborhood with stable borders creates a benign environment for the emergence of democracy.

Gibler's (2007, 2012) analysis is consistent with early evidence that democratic states have few if any territorial disputes (Mitchell and Prins 1999; Huth and Allee 2002, 267). Likewise, several studies demonstrate the pacific effect of settling borders. Kocs (1995) finds that states that legally settle their borders are considerably less likely to have a war. The problem with Kocs's study, however, is that it is limited to 1945–1987. Owsiak (2012) expands the study, adding more precise measures of settled borders for the 1816–2001 period, and finds that once two neighbors legally settle all of their contested borders, they are less likely to have a war or a MID. Similarly, Hensel (2006) shows that settling a territorial claim significantly reduces the likelihood of any future MID of any type, territorial or non-territorial.[18] Owsiak and Rider (2013) find that settling borders leads to the termination of rivalries (see also Tir 2006a and Gibler and Tir 2010). In addition to these dyadic studies, Henehan and Vasquez (2006) show that certain historical periods that lack wars among major states are periods with few or no territorial MIDs. These periods include the Concert of Europe (1816–1848) and the Bismarckian Peace (1871–1895). All of this evidence on border settlement is consistent with Vasquez's (1993, 146) early proposition that:

> If the territorial divisions among neighbors are not challenged but accepted as legitimate, peaceful relations can govern. Most borders once satisfactorily settled remain so for long periods of time.

This prediction was made before any systematic research had shown that territorial disputes were highly war prone. What Vasquez had in mind was that if territorial disputes were one of the main reasons for neighbors fighting, then eliminating that cause should reduce the probability of war. It would also increase the prospects for peace because it reduces the hostile relationship associated with severe territorial contention.

Vasquez's proposition on neighbors settling their borders is a much broader conception of the territorial peace than Gibler's (2012), who focuses on democratization. Vasquez conceives of the territorial peace as including any set of dyads that have settled their border, which includes many nondemocratic and noncontiguous states. Taking this broader conception of the territorial peace, Owsiak and Vasquez (2021) examine all dyads that have never had a war (for certain long stretches of time). Looking at a sample of politically relevant dyads, they find for the Cold War period (1946–1989) that of the 981 dyads that have never had a war, 78 (7.95 percent) have never had a territorial MID. Conversely, of these 981 peaceful dyads only 57 (5.81 percent) are jointly democratic. This may not seem like much of a difference until they show that 56 (98.25 percent) of the 57 joint democracies never have had a territorial MID (Owsiak and Vasquez 2019, Tables 3–4). This finding suggests that the territorial peace can subsume the democratic peace.

To investigate the statistical significance of these findings, they conduct a firth logit and find that the absence of MIDs statistically wipes out the significance of joint democracy when the dependent variable is "never having a war."[19] When a dyad has one or more territorial MIDs in a given period, it is significantly less likely to be classified as peaceful during that period (the logistic regression coefficient for any territorial MIDs is −3.36 [0.53] [significant at < 0.01] while joint democracy is 0.51 [1.45] and nonsignificant at the 0.05 level) (Owsiak and Vasquez 2021, Table 2).

One criticism of such findings is that the reasons joint democracies do not have many territorial MIDs is that they handle territorial claims peacefully (see Park and James 2015). The counter-response of the advocates of the territorial peace is that joint democracies do not have territorial disputes because they settled their borders before they become democracies. Owsiak and Vasquez (2019) find this to be the case. Gibler and Owsiak (2018) also show that border settlement typically precedes democratization (see also Gibler and Owsiak 2018 and Thompson 1996).[20] In other words, border settlement leads to democracy and not the other way around, but of course this only holds between the borders of contiguous states.[21]

Related to the work on peaceful dyads that never go to war is the work of Vasquez and Barrett (2015) on dyads that never have a MID in the first place. They argue that MID-free dyads are apt to be those that do not have territorial claims rather than those that are joint democracies. They find that of the 1,068 dyads in the Cold War 1946–1989 period that never had a MID, 94 percent (1,004) never had a territorial claim.[22] In other words, the absence of territorial disagreement is a big component in the absence of armed force. The findings on joint democracy are equally interesting. Vasquez and Barrett (2015, 17–19) find that of the 1,068 MID-free dyads, only 74 (about 7 percent) are joint democracies.[23] Further they find that of these 74 joint democracies, 65 (87.84 percent) never had a territorial claim.[24] As with Owsiak and Vasquez (2021), these findings strongly suggest that the territorial peace can subsume the democratic peace in that the territorial peace can explain why contiguous joint democracies are at peace. Vasquez and Barrett (2015, 20) go on to examine whether the territorial peace can also account for large numbers of nondemocratic dyads at peace. They find that of the 994 nondemocratic dyads that are MID-free, 94.47 percent (939) do not have a territorial claim, using Huth's (1996) data. The Vasquez and Barrett (2015) study was based on Huth and Allee's data, which ended in 1995. It would be interesting to see if it could be replicated on the new ICOW data set.

While the research on the territorial peace is recent, several things are known:

- Dyads that settle their borders tend to have a lower probability of going to war or even having a MID.
- Dyads that have never been at war for long periods of time are those that have no territorial claims or MIDs.

- Joint democracies rarely have territorial MIDs; this is because they settle their borders before they become joint democracies. This implies that the territorial peace can subsume the democratic peace, at least for contiguous dyads.
- Most dyads that have never had a MID are those that have never had a territorial claim. Likewise, about 85 percent of the joint democracies that have never had a MID have never had a territorial claim.

LEWIS RICHARDSON AND THE FUTURE OF PEACE SCIENCE

Slightly over one hundred years ago Lewis Richardson began to apply scientific techniques to the study of war in hopes that by identifying the causes of war they might be eliminated. For a long time he worked alone. Today more scholars are using his general approach to working on this set of related research questions than at any other time in history, and it is not unreasonable to think that someday his hope may be fulfilled. Understanding the causes of war is a conceptual, theoretical, and empirical problem, and peace science to date has made progress on all three fronts, but there is still a long way to go. Nonetheless, as the chapters in this volume have demonstrated, some things have been learned and some avenues for future progress have been outlined.

We have gone beyond Bremer's (2000) four things we know about who fights whom. We still know that contiguity or proximity, which he identified as a key factor, is important. But now we know that the reason this is often the case is that neighbors have territorial grievances over their borders, and such wars are the modal war in the system since 1815. This is one cause of war that can be mitigated, if not eliminated, and in doing so an entire class of war might be greatly reduced. How and why territorial disputes can be peacefully resolved has received attention in research and should receive more attention in diplomacy.

A second factor that helps determine who fights whom is recurring disputes. The repeated (and failing) attempt to resolve disputes through the use of force, Bremer saw as a key factor leading states to fight each other. Today we have more clearly conceptualized this process as a function of an underlying relationship—namely, that it is *rivalry* that leads to increased hostility and escalation across repeated crises.

What Bremer did not discuss and what seems to be a factor, especially before the nuclear era, is the presence of arms races, a factor that concerned Richardson (1960a; see also Zinnes 1976). Arms races when coupled with other factors can lead to a high probability of war (Sample 1997, chapter 4). However, only a few wars, albeit very intense wars, are associated with arms races.

The absence of war among the two major states in the Cold War raises the question of whether nuclear deterrence can prevent war (Fuhrmann, chapter 6). Conceptually a key insight is that nuclear weapons raise the "provocation threshold" (Lebow 1981, 277), but whether such deterrence can deter nuclear war for all time is another question. Theoretically the logic of nuclear mutually assured destruction (MAD) is more compelling than the logic of conventional deterrence where mutual and inexorable defeat of both sides is rarely evident before the fact. This has led some to claim that deterrence did not work in the pre-nuclear era with regard to war onset or even MID onset (Kenwick et al. 2015; but see also Leeds 2003 and Leeds and Johnson 2017). Part of the debate must be seen in the larger context of the role of capability in bringing about or preventing war in general (see Morey and Kadera, chapter 2). Nonetheless, it is known that alliances can sometimes embolden minor states (Kang 2012) and act as a moral hazard (Benson 2012). Clearly the alliances blocs before the First World War and the Second World War resulted in major deterrence failures. Nonetheless, the role of alliances in the onset of war is still an area where more needs to be known, especially since the Cold War era exhibits a different pattern from the previous 1816–1945 era.

The last factor Bremer discussed was that democracies rarely fight each other. Since he wrote this, the research agenda has been deeply enriched with numerous findings and expanded into a broader theoretical liberal peace (Russett and Oneal 2001, see Mousseau, chapter 8). Nevertheless, since Bremer wrote, we have seen democracies close to the brink of war, such as the Kargil War between India and Pakistan. Various criticisms of the democratic peace have identified alternate explanations of why these countries do not fight each other (e.g., Mousseau's notion of contract culture).

As a field, Bremer's (1992) notion of dangerous dyads coupled with J. David Singer's (1972, 1979) commitment to investigating the sources of war through the collection and statistical analysis of replicable data has made for progress in understanding the causes of interstate war. Our volume shows many promising avenues of research for identifying additional causes of war such as cyber threats (Valeriano, Maness, and Jensen, chapter 12), changes in war financing (Cappella Zielinski and Poast, chapter 15), leadership incentives and changes (Akcinaroglu and Radziszewski, chapter 5; Wolford, chapter 14), and climate change and natural disasters (Mitchell and Yang, chapter 13). Richardson's belief that the application of scientific thinking and techniques can solve intractable problems has not been in vain.

NOTES

1. These data have now been expanded to include "near crises" (Iakhnis and James forthcoming). For recent findings based on the ICB not covered, see James (2019).

2. Paul Huth (1996) was the first scholar in the quantitative conflict community to compile a data set of territorial claims—collecting data from 1950 to 1990 (later expanded to 1919–1995 in Huth and Allee 2002). ICOW data build upon this information.

3. In the two-stage model the relationship between the presence of a territorial MID and war is still significant while controlling for contiguity in the first stage.

4. These findings are reconfirmed by Goemans and Schultz (2017).

5. This latter finding gave rise to the ICOW project collecting data on identity claims between states (Hensel and Mitchell 2017).

6. Vasquez (2009, 349) interprets this finding to mean that the more "salient" an issue, the more likely decision makers will feel "pressured to do something" about it, whether that be a peaceful attempt at resolution or the use of force.

7. Senese and Vasquez (2008, 146–48, 162) find that recurring disputes have a separate and independent effect on an MID escalating to war from rivalry in the 1816–1945 period, but not the Cold War period, where rivalry wipes out the significance of recurring disputes. In their short post–Cold War period from 1990 to 2001 they find both factors are significant, however.

8. There is some evidence that they hold for the post–Cold War period from 1990 to 2001. Why this is the case may have something to do with nuclear weapons raising the "provocation threshold" for war (see Lebow 1981, 277), but also see Vasquez and Kang (2013) and Sense and Vasquez (2008, 157–58), who argue that the singular East-West alliance structure may have stripped alliances of their belligerent effect.

9. See Tanaka (2016) for an experiment in Japan on territory and conflict.

10. For instance, he found that 6.32 percent should be dropped and another 4.39 percent should be merged—Gibler (2017b, 211–12). For criticisms of his changes, see Palmer et al. (2020) and response by Gibler, Little, and Miller (2020).

11. For more details on territorial threat and centralization, see Gibler and Miller, chapter 9.

12. A similar observation applies to differences with Kenwick et al. (2015).

13. My thanks to Mike Kenwick for comments on this section.

14. This holds for NATO if one counts the Korean War.

15. See also Maoz (1989b).

16. Vasquez (2011b) lists the cases that fit each path. For example, he finds that the 1859 War of Italian Unification has a path consisting of territory, rivalry, and alliances present, while World War II has these three plus arms races.

17. For a criticism of Gibler's (2007) data analysis, see Park and Colaresi (2014), as well as Gibler's (2014) reply. Much of the criticism related to the lack of replicable data, however, was eliminated by the publication of Gibler (2012).

18. This particular study, which was based on a limited regional domain and early data, is an important area for further investigation and replication.

19. Firth logit deals with the problem of separation in logistic regression.

20. See also Gibler and Braithwaite (2013), who show that regional stability is related to the emergence of a territorial peace.

21. Russett and Oneal's results are in part driven by noncontiguous relationships that democratic hegemonic states foster, and these are an important part of the democratic peace results.

22. This analysis is based on Huth and Allee's (2002) data and not ICOW since it was not complete for the entire globe. For the post–Cold War 1990–1995 period, 91 percent have no claims. Note at the time of writing the Huth and Allee (2002) data ended at 1995.

23. For the 1990–1995 period, the number of joint democracies without an MID jumps to 377; this accounts for about 22 percent of the peaceful cases.

24. Of the 377 joint democracies in the 1990–1995 period without a MID, 318 (84.35 percent) do not have a territorial claim.

Chapter Twenty

Afterword

The Russia-Ukraine War—What Can Quantitative Conflict Research Tell Us?

Sara McLaughlin Mitchell and John A. Vasquez

The Russia-Ukraine War broke out on February 24, 2022, with the Russian invasion. What can the quantitative conflict research summarized in this book tell us about the onset of this war? This afterword addresses this question and shows that many findings of empirical research conducted in the past few decades can elucidate the causes of the war and provide useful insights about how the war might be managed to limit its expansion.[1]

The most important analytical thing to keep in mind about the war is that *it is typical of the modal war fought in the international system* since 1816. Vasquez and Valeriano (2010, 300) find that the modal war in the period from 1816 to 1997 is a two-party (dyadic) war between neighbors over territory. These constitute 54.4 percent (forty-three) of the seventy-nine wars in this period. Identifying the Russia-Ukraine War as a modal war means that it fits a pattern common within international history. This is of immense help in utilizing peace science research to identify the likely causes of such wars.

The chapter sees quantitative research on interstate conflict contributing three things. First, its findings can help explain the causes of the war by placing it within the larger pattern of such wars. It does this by identifying what *type* of war it is and connecting its outbreak to what we know about the causes of war, which helps to uncover why this war occurred. Second, it can use the considerable amount of empirical research reviewed in this edited volume to analyze whether the two sides had a high probability of war and whether the war might have even been forecasted by some risk barometer of war (Senese and Vasquez 2008, 270–76; Maness and Valeriano 2012). Third, its findings about war expansion can suggest how major states, like the United States and China, can manage their relations to avoid being dragged into the war. Put another way, what do past findings suggest about how the local two-party war between Russia and Ukraine might expand to become a war between major states? We begin, however, with an overview of how the war began.

OVERVIEW OF THE WAR

The seeds of the war originated with Russia's annexation of Crimea in March 2014. Shortly thereafter in May 2014, Peter Poroshenko was elected president of Ukraine, replacing the pro-Russian Viktor Yanukovych, who was removed by Parliament in February and fled to Moscow. Within a few months of his election, Poroshenko launched an offensive against the separatists in Eastern Ukraine, which failed, but the conflict in the Donbas continued. At this time, he returned to a policy of greater integration with the West, and he expressed a desire to join the North Atlantic Treaty Organization (NATO). In 2019, he was replaced by Volodymyr Zelenskyy, who ran on an anti-corruption platform and a pledge to bring the Donbas back into Ukraine. Tensions with Russia and Putin increased after President Biden took office in the United States, with Russia amassing troops on the border in March/April 2021 and warning the West that if they threatened Russia's security interests, they would respond militarily. In June, NATO renewed its pledge of 2008 that Ukraine could become a member and then conducted joint military exercises with the country in the Black Sea that Russia protested. In September 2021, Belarus and Russia conducted joint military exercises. On December 21, Russia accused Ukraine of massing forces in the east to attack separatists. Two days later, American intelligence openly warned that Russia was planning an invasion. On February 22, 2022, Russia invaded.

EXPLANATIONS FOR THE RUSSIA-UKRAINE WAR

Many individuals in academic, military, policy, and media circles expressed skepticism that Russia would invade Ukraine in 2022. Most of these predictions focused on the large costs that Russia would incur in such a conflict (Driedger and Polianskii 2023), something we have observed to date.[2] On the other hand, a survey of 352 international relations experts by the TRIP Project (December 2021 to January 2022) showed that 56 percent of respondents predicted the Russian invasion, compared with 20 percent of respondents predicting no war, a three-to-one margin for the correct prediction by international relations scholars (Entringer et al. 2022). The outbreak of the war defied the purported trend in the decline of interstate war (Braumoeller, chapter 16 in this volume). Some saw the Russian threats of weapons of mass destruction usage to be non-credible,[3] although the case illustrated a repeated pattern found in empirical research of a nuclear state initiating a conflict against a non-nuclear state (Fuhrmann, chapter 6 in this volume; Bell and Miller 2015).

Some scholars focused on the security dilemma generated by NATO and EU expansion, especially in Eastern European countries near Russia's borders to explain the war. Mearsheimer (2014), who had previously argued for Ukraine keeping its nuclear weapons (1993), saw the expansion and Western foreign policy on Ukraine as provocative toward Russia, and blamed US foreign policy in the region for contributing to the war's onset (Mearsheimer 2022). Maness and Valeriano (2012) agreed that NATO expansion raised the risk of conflict based on statistical findings that the formation of new alliances increases threat and the probability of war (see also Kenwick, Vasquez, and Powers 2015; Kenwick and McManus, chapter 3 in this volume). Their implication was that more distance, if not even a spheres of influence policy, should be maintained as a way of keeping the post–Cold War peace. Putin, of course, was not above exploiting these views as a way of justifying his Ukrainian policy. For example, on February 28, 2022, the Russian Ministry of Foreign Affairs posted Mearsheimer's (2014) article on its Twitter account.[4]

There is ample evidence that quantitative conflict research could have predicted the war. Maness and Valeriano (2012), for example, did so about a decade before Russia invaded. Using the steps to war theory and risk barometer of Senese and Vasquez (2008), they reviewed the conflicts in Estonia, Georgia, and Ukraine and saw the latter as having a risk of war. The risk barometer gives one point for every step to war—presence of a territorial dispute, rivalry, making of an alliance, and ongoing arms race. For Russia-Ukraine relations in 2012, they concluded: "War is a very real possibility because of . . . a territorial dispute (the Crimean) [and] the recent anti-Russian government's attempts at allying with the West," along with the election of a more pro-Western leader (Maness and Valeriano 2012, 142).

As we show in this chapter, much of the quantitative literature on interstate war provides insights about why the war occurred despite some pundits and academics predicting that it was unlikely.[5] The territorial conflict that emerged after Russia took the Crimea in 2014, combined with identity conflicts involving the treatment of Russians within Ukraine and Russian support for insurgents in the Donbas, increased the salience of the contested areas and promoted more aggressive foreign policies. In addition to unstable borders fueling rivalry generally, Russia and Ukraine were also issue rivals, with contestation over other diplomatic issues as well (e.g., delimitation of the Sea of Azov, identity issues involving Russians in Ukraine). All these increased the chances for the escalation to war. A more powerful state like Russia initiating a war against a weaker adversary like Ukraine also fits a pattern observed in directed dyadic analysis of conflict onset (Morey and Kadera, chapter 2 in this volume). The presence of many outside alliances for both countries combined with arms buildups by Ukraine may have increased conflict risks but were not immediate factors. As a mixed regime dyad, the likelihood for conflict escalation was higher, although the two sides' very advanced economies should have reduced war risks. Putin had diversionary incentives for war as well, including rally effects that the actions against Crimea had generated (but waned), and his efforts to secure a historical legacy for his leadership. In short, peace science researchers are not surprised by the outbreak of this modern war given all the conditions in place before the war began. We detail these arguments in the following sections and review other conflicts in the region that portended trouble.

TERRITORIAL CONFLICT, THE ISSUE APPROACH, AND RIVALRY

What are the causes of such territorial wars and rivalries between neighbors, and do they fit the Russia-Ukraine War? How do the issues at stake in the conflict, such as Crimean sovereignty, maritime delimitation, the use of the Sevastopol naval port, and the treatment of Russian speakers within Ukraine relate to the outbreak of war? The territorial explanation of peace and the issue and rivalry approaches provide some answers.

Quite some time ago, Vasquez (1993, chapter 4; see also Vasquez 1995) argued that the well documented finding of Bremer (1992) that contiguous states are more apt to fight each other than distant states was true primarily because contiguous states have territorial (i.e., border) disputes and distant states do not. Subsequent research by Hensel (2000), Vasquez (2001), and Senese (2005) that controlled for the two variables (contiguity and territory) showed that this was empirically the case (see Hensel and Goemans, chapter 1 in this volume).[6] Vasquez (1993) argued that states go to war because they have serious disagreements over issues that are not resolved by negotiation or other diplomatic efforts. The most dangerous issues over which they disagree are territorial, and as these territorial disputes repeat because they are not resolved, this creates a spiral of interactions that makes war more probable. Territorial disputes have a higher probability of war than other issues.[7] Much of the research supporting this claim has been summarized in the previous chapter by Mitchell and Vasquez (chapter 19 in this volume).

Two of the more important findings are that states that have repeated territorial disputes are more likely to become rivals and rivals are more likely to go to war (Vasquez and Leskiw 2001, 305–10). Unstable borders fuel rivalry (Owsiak, Diehl, and Goertz, chapter 10 in this volume), especially when one side is dissatisfied with the outcome of the previous conflict (Quackenbush, chapter 7 in this volume), thus it makes sense that Russia's taking the Crimea and the resumption of a territorial dispute between the two sides shifted the dyad to a more severe rivalry. Furthermore, the two countries were engaged in interstate disputes over multiple issues, such as territorial, maritime, and identity issues, which increased the risks for conflict escalation (Mitchell and Thies 2011; Powell and Wiegand, chapter 11 in this volume; Mitchell and Yang, chapter 13 in this volume).[8]

Vasquez (1993, chapter 4) argues that *territoriality* makes human collectivities, like states, want to establish borders where they have exclusive control—in modern terms, states seek to protect their territorial integrity and political sovereignty

(Morgenthau 1952, 973). This territoriality axiom, as he called it, implies that states are unusually sensitive to any threats to their borders. If this axiom is true, Vasquez (1993, 141, 143, 145) said we could expect certain things within international relations, such as the division of the world into territorial units and that equal neighbors not separated by natural barriers were apt to contest their borders at some point in their history. This means that aggressive displays and force, including violence, are often used to establish boundaries.

The same is true of new states that enter the system. They must make a place for themselves in a neighborhood. The independence of India/Pakistan and of Israel and the subsequent India-Pakistan (1947–1949) and Arab-Israeli (1948–1949) wars exemplify the process. Maoz (1989a) shows that attempts by new states to join the club of nations typically results in militarized disputes, but that those that entered the system through violent or revolutionary means are more likely to have wars than states that entered the system through evolutionary means. In many ways from the territorial theoretical perspective, the breakup of the former Soviet Union, although evolutionary and not violent, set the stage for the possibility that the new states replacing the USSR might have problems with their "new" borders under certain conditions. Russian leaders agreed to cede sovereignty of the Crimea to Ukraine in negotiations following the breakup of the Soviet Union in exchange for naval basing rights at Sevastopol and other concessions, a move that seemed to settle the land border issue. Yet the loss of the Crimea was never fully accepted by Russia due to the large presence of ethnic Russians in the territory, demands for Russian language rights in Crimea, and the strategic importance of the area.[9] Such territorial disagreements put countries on a potential path to war (see Gibler and Miller, chapter 9 in this volume).

Broader Border Problems

The breakup of the Soviet Union generated border problems that resulted in four wars, one of which is the Russia-Ukraine War. The two conditions that raised border problems after the breakup of the Soviet Union in 1990 were nationalism and strategic concerns that made certain territory valuable for security reasons. The first of three wars over boundaries involved Nagorno-Karabakh with several armed clashes in 1991–1993, and 2020 between Armenia and Azerbaijan, including a full-scale war (Sarkees and Wayman 2010, 178–80, 457–58). The first war goes back to 1988 and predates the breakup of the Soviet Union, but the independence of the two states aggravated the situation in that now an international border was at issue; the conflict continues to date.[10] Underlying the border question is a nationalist issue. Nagorno-Karabakh is mostly inhabited by ethnic Armenians, but geographically within Azerbaijan. The dispute has been characterized by forced expulsions by both sides and the creation of the breakaway Republic of Artsakh, which has de facto control of part of Nagorno-Karabakh. What is important about this case for our purposes is that this case is similar to the Russia-Ukraine War, but Putin is not involved, meaning that there are larger underlying causes in these sorts of wars than just Putin.

The second case is the Russian-Georgia War of 2008, where Russia supported the autonomy and then independence of two Russian-speaking areas of Georgia within the old borders of that federated state—Abkhazia and South Ossetia. Both were separatist states, and in March 2008 sought recognition from Russia after Western states recognized the separatist state of Kosovo. By August, Russia and Georgia would go to war, which ended with a Russian military victory and recognition of the independence of the two states as it withdrew its troops from Georgia. Russia then signed a bilateral military agreement with Abkhazia and South Ossetia and left troops there.

Nationalism was a key issue in this case, but security issues, including the strategic location of Abkhazia on the Black Sea, played a role for Russia. President Mikheil Saakashvilli, who replaced President Eduard Shevardnadze[11] in 2004 during the "Rose Revolution" in Georgia, wanted to join NATO. The situation was aggravated in April 2008 when President George W. Bush advocated for admission of Georgia and Ukraine to NATO. Then NATO considered a path to membership for Georgia, overcoming objections by France and Germany. Georgia's admission to NATO was anathema to Putin, who warned in the same month that admission to NATO would be taken as a direct threat to Russia's security. As the crisis came closer to armed conflict, Saakashvilli mistakenly thought that the United States would come to his assistance if war broke out, and this seemed to embolden him. The prospect of Georgia's admission in this sense may have contributed to war. Nevertheless, a key underlying cause was that the immediate change in borders resulting from the breakup of the Soviet federation created a new border for Georgia that also raised nationalist concerns for two internal states and for Russia itself.

The third case is the Crimea Seizure of 2014. Here strategic concerns that a hostile Ukraine (and one that was seeking admission to NATO and the European Union) would jeopardize the major and historic Russian naval base in the Black Sea were paramount. The fact that most inhabitants were Russian speakers and that the economy was mostly associated with the base and Russian tourism added to the legitimacy that this territory was historically and currently "Russian" and had only become part of the Ukraine because of what was later seen as an ill-advised 1954 agreement made by Khrushchev. While the land border had been settled in 1997, Russia and Ukraine did not agree on delimitation of the Sea of Azov. These tensions created pressures for territorial revisionism as Ukraine increased rental costs for the Sevastopol naval base in response to Russia increasing natural gas prices. Russia's construction of a land bridge in the Kerch Strait also contributed to clashes at sea between the two countries.[12]

Putin was willing to let the matter rest so long as he could control the domestic situation within Ukraine by keeping a pro-Russian government in power, but once President Viktor Yanukovych was pushed out by the Rada (parliament), and eventually replaced by anti-Russian leaders, Putin became concerned with Crimea and in due course turned to its occupation and absorption as a permanent solution. Each of the previously discussed three wars involve new states contesting the borders they inherited with the breakup of the federated USSR because of ethnic-nationalist and/or strategic territorial concerns, with the latter sometimes related to security concerns that one side might join the alliance of an adversary and former rival.

Past as Prologue?

The current Russia-Ukraine War follows the pattern of the previous three. Keeping aside the question of Crimea, it involves the issue of nationalism with the concentration of Russian speakers in the Donbas region (Donetsk and Luhansk) that rebelled in March 2014 at the beginning of Putin's Crimean Seizure. Putin initially supported the rebels covertly, and they eventually separated, which was followed by a Ukrainian counterattack.

In terms of explaining why the Russia-Ukraine War occurred, it is important to see that this war fits the pattern of wars between neighbors over borders. This is true of the immediate three wars that follow the breakup of the Soviet Union and a larger pattern of two-party dyadic wars since 1815. Two factors led to disputes over the new borders. The first was the norm of nationalism and self-determination that supported Russian speakers who found themselves in a new "non-Russian" state that was sometimes seen as hostile or outright oppressive. The second had to do with the strategic value of the territory. For example, with Azerbaijan, the secession of Nagorno-Karabakh and/or its absorption by Armenia—its enemy and rival—would create a hole within its very state. For Russia, Crimea was a key naval base of current and historical value that could not be lost to a hostile power potentially aligned with a former adversary and threatening coalition of a powerful bloc.

The pattern of new states entering the system and having border problems is also illustrated by two conflicts that did not escalate to war—the conflict with Estonia and the secession of Transnistria from Moldova. The Estonia conflict began soon after it declared its independence on August 20, 1991. The government claimed the old borders before Stalin invaded just prior to the Second World War. This meant that parts of Russia were now being claimed, although it only involved 2000 square kilometers (Berg and Saima 2000, cited in Maness and Valeriano 2012, 136). Later and more serious is how Russia reacted when the Estonian government asserted that ethnic Russians were not true Estonians and that they should not be considered citizens nor have the right to vote, as they had in the former Soviet Estonia. In the 1992 national elections, for instance, only Russians who were citizens before 1940 were permitted to vote, which meant that about five hundred thousand Russian residents were not allowed to vote (Grofman, Mikkel, and Taagepera 1999). These ethnic Russians objected, as did Russia itself, to these kinds of restrictions.

Both these conflicts, especially the latter, which was seen as potentially much more serious, were resolved by the insistence of the European Union that to become a member, Estonia had to resolve its existing border disputes, a requirement of all nations seeking membership, and second that Estonia had to have a certain level of democracy which it would not meet if it stripped categories of people of the right to vote. In 1999, it declared it accepted its current borders and complied with the voting requirements (Maness and Valeriano 2012, 136). Estonia was admitted to both the European Union and NATO in 2004. In this way, an international organization helped defuse the crisis with the new Russian Federation. Ironically, the European Union helped keep the peace even though Russia regarded Estonia joining it and NATO as provocative.

A similar example is the conflict between Moldova and Transnistria, who seceded (see Rumer 2017). With the breakup of the Soviet Union, Moldova became its own state, but there was some talk of unifying with Romania because of the close ethnic ties between the two peoples. This sort of discussion led a number of ethnic Russians and Russian speakers who were concentrated near the Dniester to become concerned about their status and rights. Although only about four hundred thousand to five hundred thousand by 1992, their identity claims encouraged them to secede. Russian troops who were still there from the Cold War intervened to protect them, and a cease-fire was arranged in July 1992. The situation has been frozen since, with Russian troops still there and Transnistria a de facto state. Again, the emergence of a new state, in this case Moldova with its Russian minority, led to a conflict with an intervention (of previous Cold War–deployed) Russian troops playing a part in establishing a state within a state. This was not unlike Abkhazia and South Ossetia, except in this case, Moldova, in part because of its miliary weakness, has not sought a war to restore the separatist state.[13]

The Russia-Ukraine War not only follows the pattern of the immediate three wars before it in the region, but also a longer pattern of dyadic wars since 1815. It is representative of the modal territorial war fought in the international system. Competitive territorial claims to Crimea, Russia's identity claims against Ukraine, and escalating tensions in the maritime arena all contributed to war risks. While territorial issues, other contentious issues, and rivalry give us a great deal of leverage for understanding why war occurred, other causes of war discussed in the book were also present in this case, to which we now turn our attention.

POWER PREPONDERANCE

As discussed earlier, many analysts thought war was unlikely because of the potential costs involved, with Russia and Ukraine being large states with advanced economies and extensive military capabilities. Considering data on national capabilities from the Correlates of War Project[14] (from 2016, the most recent data), Russia had 3.6 percent of global military, economic, and demographic capabilities versus Ukraine's 0.7 percent, or a five times larger share. As Bremer (1992) showed, interstate war is more likely in dyads with no large power differences (e.g., parity), consistent with theories connecting power preponderance and peace (e.g., power transition theory, hegemonic stability theory, bargaining model of war). Morey and Kadera (chapter 2 in this volume) note, however, that the relationship depends on the unit of analysis, with Bremer's finding holding for dyads but not directed dyads. They show that preponderance is linked to stronger states initiating conflicts against weaker opponents for directed dyads (e.g., Russia initiator versus Ukraine target). In this regard, we can understand why Russia, as a potential challenger that is five times larger than Ukraine in terms of capabilities, initiated the conflict, although we recognize that disparate findings in the literature make any conclusions drawn more difficult.

ALLIANCES

Ukraine's potential membership in NATO caused much consternation for Russia, as noted earlier. Past research suggests that membership in a defensive alliance can deter attacks (Leeds 2003) and if Ukraine had joined NATO, this might have deterred a Russian invasion.

Even though Ukraine never joined NATO, we can consider how the other outside alliances that both Russia and Ukraine had in place before the start of the war might have had an impact on their decision making. Using data from the Alliance Treaty Obligations and Provisions (ATOP) dataset (Leeds et al. 2002),[15] we can count the number of alliances active in the last year of the data set (2018) for each country. Russia had fifty-nine alliances that were active that include defense, neutrality, nonaggression, and consultation obligations. Ukraine had 9 active alliances that included neutrality, nonaggression, and consultation obligations. Most of Ukraine's security commitments were bilateral alliance treaties signed shortly after independence (1991–1992) with several states (e.g., Bulgaria, Estonia, France, Hungary, India, Poland, Turkey, and Uzbekistan). Other than efforts to join NATO, Ukraine was not actively seeking out other alliance partners.

Russia's primary defense pact, the Collective Security Treaty Organization, had already been weakened prior to the conflict due to Russia's unwillingness to come to Armenia's aid in the Nagorno-Karabakh conflict. Yet the presence of so many outside alliance ties for both sides increased risks for conflict escalation, especially in the post–Cold War era, largely because such outside support emboldens allies to make stronger demands against their adversaries (Senese and Vasquez 2008, 167). New alliances are also more dangerous if they alter the security environment (Kenwick and McManus, chapter 3 in this volume), and Russia signed five new alliances in the five years before the 2022 war. On the other hand, these were mostly nonaggression bilateral pacts signed with countries not in the region (Belize, Dominica, Guyana, Honduras, and Turkmenistan). The lack of defensive allies for Ukraine and a situation where both sides had outside alliances increased the potential for war, but it was not these alliances that were the main factor in Russia's calculation for the invasion; rather, it was the prospect of Ukraine joining NATO.

ARMS RACES AND NUCLEAR WEAPONS

Arms races typically increase the risks for war by enhancing security dilemmas in dyadic interactions (Sample, chapter 4 in this volume). Scholars consider mutual military buildups to occur when both sides are increasing military expenditures at a rate of 8 percent or higher for several years in a row. Using data from the Stockholm International Peace Research Institute, we can assess how arms spending contributed to the outbreak of war.[16] Not surprisingly, Russia's taking of the Crimea in 2014 resulted in increased military expenditures by Ukraine, with four years of spending increases between 9 and 30 percent between 2015 and 2021 and an average growth in arms expenditures by 11.3 percent. Russia's spending, on the other hand, declined on average by 3.9 percent, although some years saw increased spending (2016, 2019, 2021). Thus, it is not clear that the two sides were engaged in a mutual military buildup that could precipitate war.

Furthermore, Sample (2002) shows that mutual military buildups do not make war more likely in major power–minor power dyads, which is the situation in the Russia versus Ukraine dyad. As a nuclear power, Russia could potentially use its larger capabilities to extract better demands in bargaining with Ukraine; yet the inability of the two sides to reach peaceful agreements on most of their underlying territorial, maritime, and identity issues suggests otherwise. Russia as a nuclear state initiating a conflict against Ukraine as a non-nuclear state, however, is an empirical pattern observed consistently in history (Fuhrmann, chapter 6 in this volume; Bell and Miller 2015). Nonetheless, as described earlier, the threat to use nuclear weapons was not very credible by Russia in this situation.

LIBERAL PEACE

The liberal peace literature (Mousseau, chapter 8 in this volume) shows that mixed regime and autocratic dyads have higher conflict risks among all dyadic regime pairings (Gleditsch and Hegre 1997), and this case fits that with that general pattern. We can treat this either as an autocratic dyad or a mixed dyad, depending on how Ukraine's regime is scored. In 2023, Freedom House scores Russia as 16/100 on their scale of political rights and civil liberties (i.e. not free), while Ukraine scores partly free with 50/100.[17] The Varieties of Democracy project similarly scores Russia in the bottom 10 to 20 percent using their 2022 liberal democracy index, while Ukraine is just below the threshold that marks the fiftieth percentile. Ukraine's score declined between 2022 and 2021, which makes sense given that countries often restrict freedoms in wartime. In general, though, Ukraine was on a path of liberalization and integration with the West, generating regime differences between itself and Russia, and increasing conflict chances. In fact, Putin regarded this specific integration as a hostile act and a threat.

On the other hand, both states are economically advanced, which Bremer (1992) found to decrease the risks for war. He considers a country to be economically advanced if its economic Composite Indicator of National Capability share exceeds its military share, something that is true for both Russia (5 percent economic versus 3 percent military) and Ukraine (1 percent economic versus 0.1 percent military) using data from the COW Project in 2016. Yet Mousseau (chapter 8 in this volume) points out that many dynamics of liberal peace are driven by market-oriented economic development, such as contractual compliance and protection of private property rights. While both Russia and Ukraine have large economies, albeit with some degree of corruption, Ukraine's organization as a more capitalist system and its moves to enter the European Union created economic competition between the states. While both sides had much to lose economically from a costly war, their lack of economic similarities resulted in those opportunity costs failing to restrain escalation, as we typically see in interactions between advanced economies.

LEADERS' MOTIVES FOR WAR

Although we do not have precise evidence on Putin's motives, it is possible that he may have had personal interests in the war to rally the Russian public around his leadership and to cement his historical legacy. Following the return of Crimea to Russia in 2014, Putin's approval jumped from 65 to 89 percent.[18] Those rally effects have waned in the intervening years, giving Putin a potential diversionary incentive for war, especially if he thought the invasion would result in a quick taking of the Ukrainian capital in a matter of days. Research on diversionary war shows that autocratic leaders are more likely to engage in diversionary conflicts than democratic leaders when faced with higher inflation rates (Mitchell and Prins 2004); Russia's inflation had increased several points in the two-year period leading up to the invasion.[19] The invasion of Ukraine produced a similar rally effect for Putin as the retaking of Crimea, increasing his approval from 71 to 83 percent between February 2022 and May 2023. Wolford (chapter 14 in this volume) notes that leaders who see foreign threats as stemming from other states' politics are more willing to use force (Saunders 2011), a pattern that fits with how Putin framed his motivations for the conflict.

Another possibility is that Putin's attempts to unify Russians, Ukrainians, and Belarussians in the "ancient Rus" may have been driven both by diversionary motives and his desire to create a historical legacy for himself as leader. He is known to ask historians how they will be writing about him in a hundred years, showing his concern with his own legacy.[20] In the summer before the invasion, Putin authored an essay articulating his irredentist motives to bring Ukrainian territory back into the Russia fold and undo the separation that past leaders like Lenin had created.[21] He saw Kyiv as a territory central to this unification of the Ancient Rus and justified his irredentist motives based on mistreatment of Russians in Ukraine.[22] The combination of domestic conditions and historical legacy motives provides leverage for understanding why Putin may have engaged in this potentially costly war, especially if he predicted a quick victory based on his country's military superiority.

As for the United States and NATO, the publicly stated motives have been an outrage at Putin's and Russia's "aggression" coupled with a hostility to his authoritarian regime. The war has intensified a return to the West's rivalry with Russia (on the latter see Maness and Valeriano 2015). These sentiments, along with the onslaught of war refugees into the bordering NATO countries, provided the immediate motive for active intervention in terms of aid. As the Ukrainians survived the initial attack, especially on Kyiv, a second complementary but not publicly acknowledged motive, at least in the United States, seems to have taken root. This motive stems from the belief that the ongoing military action is de-grading Russian capability in a significant manner, especially in terms of tangible material assets, like ammunition and conventional weaponry.[23] As the war exacts its toll on Russia and the latter turns to outsiders, like Iran and even North Korea, to take up the slack, there is a feeling that Putin and Russia are being so de-graded that it will be some time before they will ever be able to mount a threat to their neighbors. This outcome has provided a motive to continue aid and the war effort. So long as the Ukrainians are willing to fight on, this secondary motive gives the United States an additional incentive to stay involved, if the costs remain within bounds. However, when military intelligence reveals that a de-grading of Russian capability has plateaued, then this motive may make for a new incentive for negotiation.[24]

WILL THE WAR EXPAND TO BECOME A MAJOR STATE WAR?

This is the main policy concern of the ongoing war at this writing since drawing in the United States and other NATO states would also raise the possibility that it could escalate to a nuclear confrontation. Even if it did not, a conventional World War III on the European continent could be horrific and greatly exceed the damage already done in the local war.[25] What can quantitative conflict theory and research tell us about the danger of such a major state war occurring?

Peace science scholars have conducted some research on war contagion and the factors associated with the expansion of local wars into major state wars. A review of these factors can aid us in providing insights about what factors present in the current war could play a role in expanding the war to including major states on each side. In addition, an analysis of why war was avoided between the United States and the Soviet Union during the Cold War might also provide some useful guidelines about how to manage the current war.

We begin by reviewing who are the major states in the current post-1945 period. According to the Correlates of War project, the major states are the United States, Russia, Britain, France, China after 1949, and Germany and Japan after 1990 (Sarkees and Wayman 2010, 34–35). This listing is based on a consensus of scholars regarding the capability and status of states. Using this designation, the last major state war fought within the international system was the Korean War, which expanded to include one major state on each side when China intervened to attack US and UN forces. The local Korean War can be said to have expanded into a major war with the massive counterattack by Chinese forces on November 25, 1950, against MacArthur's newly launched offensive on November 24 (George and Smoke 1974, 210–11; Whiting 1991, 105; see also Langer 1980, 1348).[26] Theoretically, one would want to ask why there has been no major state war for over seventy years. There must be a reason, and identifying the factors for the absence of such a war can help us identify the obstacles that have prevented such wars in the post-1945 era. At the same time, understanding why China entered the Korean War can give us some clue of how local wars might expand in this era.

One possible reason is the presence of nuclear weapons. Although the US nuclear superiority did not deter China, Gaddis (1987) has suggested that nuclear weapons, especially after the Soviet Union tested an atomic bomb in 1949 and later its acquisition of a hydrogen bomb in 1953, made for mutual deterrence in the Cold War. The emergence of a second nuclear power made war less likely because nuclear war was seen as catastrophic and not winnable (Brodie 1946). Both the Soviet Union and the United States sought to avoid war with each other during the Cold War and sought to manage confrontations that would increase the risk of escalation to nuclear war (see Kahn's 1965 warnings on the latter). While it is widely agreed that nuclear weapons raised the provocation threshold (Lebow 1981, 277) in that provocations that would have resulted in war in the past did not do so now because of the costs of nuclear war, it is not agreed by all that nuclear deterrence would work to always prevent war (see Vasquez 1991; Kang and Kugler 2023).

Mueller (1988) argued that the experience of total war in the two World Wars played a separate and more important role in making both superpowers averse to fighting World War III, and one can see this as especially likely in the early crises in the Cold War in 1948 in Berlin and 1950 in Korea. Despite later nuclear threats by the United States, the Korean War did not seem that influenced by nuclear weapons and did not escalate because Stalin and Truman did not want to fight a larger war given the recent Second World War. Nevertheless, no major state war has been fought since then, and here we would have to say that the prospect of nuclear catastrophe has played a role. Yet, in the end, despite Gaddis's (1987) emphasis, it must be said that the destructiveness of nuclear weapons worked in conjunction with several factors that ameliorated the likelihood of war during the Cold War. These included the lack of direct territorial disputes between the United States and USSR, tolerance of the status quo, the experience of the two World Wars, creation of rules of the game to govern interactions, crisis management, and arms control (see Vasquez 1991, 220).

The creation of rules of the game and norms played a key role in preventing war between the two superpowers. These rules regulated behavior to implicitly govern superpower interactions and regulate both what they could and could not do. After Korea, the main unwritten rule about intervention into ongoing wars was that whoever intervened first with a large number of troops would have this war as their own and that the other could do whatever it wanted to aid the other side except send troops in large numbers. Even during Korea, Stalin abided by this unwritten rule in large part. The same was true for Russia in the Vietnam War. For the United States, this was true in Russia's invasion of Afghanistan. The United States armed the Mujahadin and bled the Soviets as much as it could without direct intervention. In 2001 when the United States went to war against Afghanistan, the tables turned and Russia did what it could to aid the war effort against the United States, but did not intervene with troops.

Such a norm and its long history is the key for determining whether the current Russia-Ukraine War will remain a local war or expand into a major state war. Given this unwritten rule, so long as the United States and NATO do not actually send troops, which seems unlikely, the war should not expand. Even though this rule has worked in the past, the question is whether certain actions by the West will spiral out of control and lead Russia to attack the West directly. One scenario is to ask under what conditions would Russia attack NATO. To date there are those in the United States who want to use the war to degrade Russian military capability and feel the effort is remarkably successful from this point of view. This is within the unwritten rules, but how might it spin out of control?

One possibility is that direct attacks on Russia proper, especially ones that succeeded, might cross a threshold. The sinking of Russian warships is a case in point; certainly, continued actions of this sort might lead to Russian retaliation toward the West if they were having a large impact. Russia certainly would not permit Western-supported efforts to damage the entire Black Sea fleet. Such efforts, if successful, might lead Putin to attack those providing the aid that made such attacks possible. Likewise, any significant Ukrainian sustained attack on Russian territory proper, which at the beginning of the war had been scrupulously avoided, but now has become part of an effort to bring the war to Russia, might provoke retaliation. This is especially true if it turns out that NATO and the United States may not just be training the Ukrainians but providing strategic advice and planning of how to fight the war, about which the Russians have openly complained.[27] In addition, since the United States thinks that success against Russia might help deter China from attacking Taiwan in the future, this adds to the complexity of the situation and to the perceived benefit of continuing the war (and even the willingness to take risks).

The other scenario to examine to see how the war might expand is to ask under what conditions NATO might attack Russia. This is less likely, but the obvious case would be in response to a military action involving one of their members. The 2023 incident involving the Russian air force damaging a US drone is an example. Instead of just protesting the downing, the United States or NATO could have retaliated against the base in Russia from which the planes left. This could then easily escalate out of control. Of course, the greatest danger would be if Putin used tactical nuclear weapons against the Ukrainians and NATO responded to that use, especially toward bases in Russia or ships that launched them.

Work on conflict contagion offers some general models on how wars might expand. Vasquez (2018) examining the expansion of the First World War from the local Austro Hungarian-Serbian War sees major states entering that war if they have alliances, are rivals, or have ongoing arms races with one side; are contiguous to a belligerent; or have a belligerent that is economically dependent on it. Almost all these conditions are present at least at the latent level in the current Russia-Ukraine war, which does not bode well. Although Putin invaded in part to prevent the Ukraine from becoming a member of NATO, the war has made the West its de facto ally with it supplying a large and increasing amount of lethal aid. It has also created a renewed intense rivalry between the United States and Russia, ending the post–Cold War era of lesser conflict. Ukraine is contiguous to several NATO allies, which facilitates contagion, but is not contiguous to the United States. Most importantly, Ukraine's war effort is now entirely economically dependent on the West, which means that at some point, Russia—like Germany in 1917—might seek to end the war by attacking those who are supplying its opponent. The only contagion factor not present is an ongoing arms race between Russia and the United States, which ended with the Cold War and could now be resurrected. Nevertheless, these contagion factors, which are pushing toward war expansion, must be seen as being opposed by the obstacles to war expansion presented by norms and the rules of the game, as well as the deterrence effects of nuclear weapons.

Another factor that might encourage negotiation is the realization that aspects of each side's strategy have already backfired. For Putin and Russia, the decision to invade means that for at least a generation or more, it will not be possible for any pro-Russian leader to be elected in Ukraine, and it will take a long time for any public sentiment to become sympathetic to Russia. What was once a Ukraine that could swing either way is now a Ukraine that will remain hostile for a long time.[28] Additionally, a key Russian objective in the war was to keep Ukraine out of NATO, but because of the invasion, Finland and Sweden, long-time neutrals during the Cold War, have sought membership and are now members of NATO. Furthermore, a much larger number of NATO members are meeting or close to meeting the expected 2 percent defense contribution, a dynamic driven in large part by Russia's actions toward Ukraine.[29]

For the United States and the West, economic sanctions and especially blocking Russia from the use of SWIFT codes in the global banking system have backfired in that it has led to attacks on the US dollar as the world's major currency. The BRICS nations—Brazil, Russia, India, China, and South Africa—have considered challenging the dollar as the world's reserve currency.[30] The dollar is not likely to be overthrown, but it may be partially replaced as the petro dollar, at least while sanctions are in place.

At the same time, Western aligning with Ukraine has led Russia and China to come closer and also led each of them to reach out to states regarded as "pariahs" by the West; namely Iran and North Korea, reducing the isolation of the latter two.

While such unanticipated negative consequences of the war will not make either side seek negotiations, they do make each side think twice about its previous decision making. Some of the older Cold War norms are more relevant and suggest how the present conflict might be managed and what the end game might look like that avoids an expansion of the war. Negotiation, if it is to occur, would be premised on some sort of military stalemate that both sides would want to end. The bargaining space might involve the following: Russia would likely concede not pursuing overthrowing the current regime and permitting its territory to be intact except for Crimea, the Donbas region, and perhaps a land bridge to Crimea. Ukraine might, depending on the military situation, be willing to accept the former two conditions in return for an end to the bombing and some recognition of the current regime, but contest the land bridge. Admission to NATO and the European Union would be subject to negotiation, but Russia might accept an application to NATO, especially if it thought one or more members of NATO would veto admission. In the end, Ukraine's allies (rather than Ukraine itself) would be in a position to negotiate this aspect of the end game. A key factor for the United States and Europe would be to not let a weaker ally draw it into a major state war,[31] since they might be more willing to compromise on these issues than Zelenskyy. Ultimately, even if there is no formal end to the war, a cease-fire could establish a de facto end to the war and a settlement on the ground based on a line of control. Of course, whether any of these scenarios would play out in negotiations could rapidly shift, depending on battlefield conditions.

CONCLUSION

In this chapter, we consider how quantitative conflict research reviewed in this book provides leverage for understanding the onset of the Russia-Ukraine War. We show that many theories and empirical findings are relevant to explaining the conflict, including the presence of a territorial dispute, multiple issues in contention, rivalry, regime differences, and domestic incentives for war. Other factors, such as the presence of outside alliances or arms buildups by at least one side, may have been less relevant for understanding the origins of this war, but the prospect of joining NATO was an expressed concern of Putin's. We also discussed work on the contagion of conflict and concluded that expansion of this conflict to a larger major state war is unlikely, although the initial warring parties in World War I and II did not foresee such a possibility either. Hopefully, a peaceful resolution of the conflict is found soon.

NOTES

1. In this chapter, we take an empirical approach and try to understand the war in terms of its causes. We are not seeking to justify one side or the other or engaging in a normative analysis. Plenty of that is being done elsewhere by others. We are generally opposed to war and this one specifically. The war has clearly reminded us of war's cruelty and injustice no matter what the cause. That, however, is a separate analysis and one that is fundamentally different in its logic and method than what we attempt here.

2. See Reuters. September 9, 2023. "Russia Ramps Up Artillery Production, but Still Falling Short, Western Officials Say." https://www.reuters.com/world/europe/russia-ramps-up-artillery-production-still-falling-short-western-official-says-2023-09-09/.

3. Weitz, Richard. July 27, 2022. "Russia's War in Ukraine: WMD Issues." Hudson Institute. https://www.hudson.org/national-security-defense/russia-s-war-in-ukraine-wmd-issues.

4. https://x.com/mfa_russia/status/1498336076229976076?s=20.

5. Paul Poast, one of the contributing authors in this volume, reviewed many scholarly contributions whose conclusions would have predicted war in this case in a May 31, 2022, Twitter thread: https://x.com/ProfPaulPoast/status/1531605978420064258?s=20.

6. This is true even though many states (but not major states) do not fight because they are not capable of reaching each other (Buhaug and Gleditsch 2006).

7. Other issues can give rise to war, but they are not as likely to do so. This means that territorial issues, although they increase the probability of war, are not necessary conditions of war.

8. See also the review of how crisis bargaining and rivalry are related in chapter 5 of this volume (Akcinaroglu and Radziszewski). Rivals are more likely to escalate disputes to war, so the rivalry that reignited after 2014 also increased overall risks for war.

9. Hensel, Paul, Sara Mitchell, Andy Owsiak, and Krista Wiegand. March 4, 2022. "Russia's Land Grabs in Ukraine Could Break the International Order." *Washington Post*. https://www.washingtonpost.com/politics/2022/03/04/putin-sovereignty-ukraine-irredentism.

10. *New York Times*. September 24, 2023. "Refugees Flee to Armenia as Breakaway Enclave Comes under Azerbaijan's Control." https://www.nytimes.com/2023/09/24/world/europe/armenians-nagorno-karabakh-azerbaijan.html.

11. Shevardnadze was the former foreign minister under Gorbachev.

12. Mitchell, Sara McLaughlin. December 5, 2018. "Could the New Fighting between Russia and Ukraine Escalate into All-Out War?" *Washington Post*. https://www.washingtonpost.com/news/monkey-cage/wp/2018/12/05/russia-and-ukraine-are-clashing-at-sea-what-do-we-know-about-the-danger-of-escalation/.

13. Russia seems to be playing a similar card in Moldova recently to its initial taking of the Crimea, though, which could escalate the conflict beyond Ukraine. Yet the lack of territorial gains in the war with Ukraine may limit those plans. See *CNN*. February 26, 2023. "Why Moldova Fears It Could Be Next for Putin." https://www.cnn.com/2023/02/26/europe/moldova-transnistria-russia-tensions-explainer-intl/index.html.

14. Correlates of War. "National Material Capabilities." https://correlatesofwar.org/data-sets/national-material-capabilities.

15. The Alliance Treaty Obligation and Provision. http://www.atopdata.org/.

16. Stockholm International Peace Research Institute. "SIPRI Military Expenditure Database." https://www.sipri.org/databases/milex.

17. Freedom House. "Global Freedom Status." https://freedomhouse.org/explore-the-map?type=fiw&year=2023.

18. Statista. "Do You Approve of the Activities of Vladimir Putin as the President (Prime Minister) of Russia?" https://www.statista.com/statistics/896181/putin-approval-rating-russia/.

19. Trading Economics. "Russia Inflation Rate." https://tradingeconomics.com/russia/inflation-cpi.

20. Matthews, Dylan. February 6, 2022. "'It's Not About Russia, It's About Putin.' An Expert Explains Putin's End Game in Ukraine." *Vox*. https://www.vox.com/22917832/vladimir-putin-ukraine-military-invasion.

21. *New York Times*. February 21, 2022. "Putin Calls Ukrainian Statehood a Fiction. History Suggests Otherwise." https://www.nytimes.com/2022/02/21/world/europe/putin-ukraine.html.

22. http://en.kremlin.ru/events/president/news/66181.

23. See, for instance, the information provided in *Oryz*. February 24, 2022. "Attack on Europe: Documenting Russian Equipment Losses During the Russian Invasion of Ukraine." https://www.oryxspioenkop.com/2022/02/attack-on-europe-documenting-equipment.html. On the de-grading motive, see Senator Tim Scott's comments in the third Republican presidential primary debate on November 8, 2023: ". . . what is [in] America's national vital interest in Ukraine [is that] . . . It is actually . . . degrading the Russian military. https://www.rev.com/blog/transcripts/rnc-third-republican-presidential-primary-debate-transcript.

24. Republican opposition for providing military aid to Ukraine also increased in the United States, especially many Congressmen that demanded these budget cuts as part of the 2023 budget negotiations.

25. As of August 2023, the two countries had suffered five hundred thousand casualties, including more than two hundred thousand battle deaths and the destruction of thousands of tanks and artillery vehicles. See *New York Times*. August 18, 2023. "Troop Deaths and Injuries in Ukraine War Near 500,0000, U.S. Officials Say." https://www.nytimes.com/2023/08/18/us/politics/ukraine-russia-war-casualties.html#:~:text=Russia's%20military%20casualties%2C%20the%20officials,and%20100%2C000%20to%20120%2C000%20wounded.

26. Chinese volunteers initially crossed the Yalu secretly in small numbers in October (Whiting 1960, vii).

27. *Reuters*. May 4, 2023. "Russia Accuses U.S. of Orchestrating Kremlin Drone Attack." https://www.reuters.com/world/russia-accuses-us-being-behind-alleged-kremlin-drone-attack-2023-05-04/.

28. On the role of friendship-hostility as an attitudinal variable and how it is produced by other aspects of cooperation and conflict, see Mansbach and Vasquez (1981, 234–54).

29. https://www.nato.int/cps/en/natohq/topics_49198.htm.

30. *Reuters*. August 23, 2023. "What Is a BRICS Currency and Is the U.S. Dollar in Trouble?" https://www.reuters.com/markets/currencies/what-is-brics-currency-could-one-be-adopted-2023-08-23/.

31. One of Morgenthau's (1960, 565–66) major rules of diplomacy is "Never allow a weak ally to make decisions for you."

References

Abbott, Andrew. 1995. "Things of Boundaries." *Social Research* 62 (4): 857–82.

Abbott, Kenneth W., and Duncan Snidal. 1998. "Why States Act through Formal International Organizations." *Journal of Conflict Resolution* 42 (1): 3–32.

Abelson, Robert P. 1995. *Statistics As Principled Argument.* Hillsdale, NJ: Erlbaum.

Abramson, Scott F., and David B. Carter. 2016. "The Historical Origins of Territorial Disputes." *American Political Science Review* 110 (4): 675–98.

Acemoglu, Daron, and James A. Robinson. 2006. *Economic Origins of Dictatorship and Democracy.* New York: Cambridge University Press.

Acharya, Avidit, and Alexander Lee. 2019. "Path Dependence in European Development: Medieval Politics, Conflict and State Building." *Comparative Political Studies* 52 (13–14): 2171–206.

Achen, Christopher H. 2002. "Toward a New Political Methodology: Microfoundations and ART." *Annual Review of Political Science* 5: 423–50.

Achen, Christopher H., and Duncan Snidal. 1989. "Rational Deterrence Theory and Comparative Case Studies." *World Politics* 41 (2): 143–69.

Akcinaroglu, Seden, Jonathan DiCicco, and Elizabeth Radziszewski. 2011. "Avalanches and Olive Branches: A Multimethod Analysis of Disasters and Peacemaking in Interstate Rivalries." *Political Research Quarterly* 64 (2): 260–75.

Akcinaroglu, Seden, and Elizabeth Radziszewski. 2017. "Web of Links: Rival Connections and Strategic Accommodation in Response to Threats." *Journal of Global Security Studies* 2 (3): 237–52.

Akcinaroglu, Seden, Elizabeth Radziszewski, and Paul F. Diehl. 2014. "The Effects of Rivalry on Rivalry: Accommodation and the Management of Threats." *Foreign Policy Analysis* 10 (1): 81–100.

Akerman, James R. 1995. "The Structuring of Political Territory in Printed Atlases." *Imago Mundi* 47: 138–54.

Allee, Todd, and Paul K. Huth. 2006a. "Legitimizing Dispute Settlement: International Adjudication As Domestic Political Cover." *American Political Science Review* 100 (2): 219–34.

——.2006b. "The Pursuit of Legal Settlements to Territorial Disputes." *Conflict Management and Peace Science* 23 (4): 285–307.

Allen, Susan H., and Carla Martinez Machain. 2018. "Choosing Air Strikes." *Journal of Global Security Studies* 3 (2): 150–62.

Allison, Graham T. 1971. *Essence of Decision: Explaining the Cuban Missile Crisis*. Boston: Little, Brown.

Allison, Graham T., and Frederic A. Morris. 1975. "Armaments and Arms Control: Exploring the Determinants of Military Weapons." *Arms, Defense Policy, and Arms Control* 104 (3): 99–129.

Allison, Michael E., Paul R. Hensel, and Ahmed Khanani. 2009. "Territorial Integrity Treaties and Armed Conflict over Territory." *Conflict Management and Peace Science* 26 (2): 120–43.

Alsharabati, Carole, and Jacek Kugler. 2008. "War Initiation in a Changing World." *International Interactions* 34 (4): 358–81.

Altfeld, Michael. 1983. "Arms Races?—And Escalation? A Comment on Wallace." *International Studies Quarterly* 27 (2): 225–31.

Altman, Dan. 2017. "By Fait Accompli, Not Coercion: How States Wrest Territory from Their Adversaries." *International Studies Quarterly* 61 (4): 881–91.

——. 2020. "The Evolution of Territorial Conquest after 1945 and the Limits of the Territorial Integrity Norm." *International Organization* 74 (3): 490–522.

Anderson, Christopher C., Sara McLaughlin Mitchell, and Emily U. Schilling. 2016. "Kantian Dynamics Revisited: Time-Varying Analyses of Dyadic IGO-Conflict Relationships." *International Interactions* 42 (4): 644–76.

Anderton, Charles H., and John R. Carter. 2001. "The Impact of War on Trade: An Interrupted Times-Series Study." *Journal of Peace Research* 38 (4): 445–57.

Andriani, Pierpaolo, and Bill McKelvey. 2009. "From Gaussian to Paretian Thinking: Causes and Implications of Power Laws in Organizations." *Organization Science* 20 (6): 1053–71.

Angell, Norman. 1910. *The Great Illusion: A Study of the Relation of Military Power in Nations to Their Economic and Social Advantage*. W. Heinemann.

Ansari, Usman. 2015. "Pakistan Boosts Defense Budget." *DefenseNews* (June 5). Available at http://www.defensenews.com/story/defense/2015/06/05/pakistan-boosts-defense-budget/28565379.

Antony, Alexander K. and William R. Thompson. Under Review. *Piecing Together the Peaces: The Industrialization Transition and the Rise of Zones of Peace*.

Ardant, Gabriel. 1975. "Financial Policy and Economic Infrastructure of Modern States and Nations." In *Formation of the National State in Western Europe*, edited by Charles Tilly, 164–242. Princeton, NJ: Princeton University Pres.

Arena, Philip, and Daehee Bak. 2015. "Diversionary Incentives, Rally Effects, and Crisis Bargaining." *Foreign Policy Analysis* 11 (2): 233–50.

Arendt, Hannah. 1958. *The Human Condition*. Chicago: University of Chicago Press.

Arky, Aaron. 2010. "Trading Nets for Guns: The Impact of Illegal Fishing on Piracy in Somalia." Thesis and dissertation, Naval Post Graduate School.

Arnott, Richard J., and Joseph E. Stiglitz. 1991. "Moral Hazard and Nonmarket Institutions: Dysfunctional Crowding Out of Peer Monitoring?" *American Economic Review* 81: 179–90.

Arquilla, John, and David Ronfeldt. 1996. *The Advent of Netwar*. Santa Monica, CA: RAND.

Arreguin-Toft, Ivan. 2001. "How the Weak Win Wars: A Theory of Asymmetric Conflict." *International Security* 26 (1): 93–128.

——. 2005. *How the Weak Win Wars*. Cambridge: Cambridge University Press.

Asgeirsdottir, Aslaug. 2007. "Oceans of Trouble: Domestic Influence on International Fisheries Cooperation in the North Atlantic and Barents Sea." *Global Environmental Politics* 7 (1): 120–44.

Atkeson, Andrew. 1991. "International Lending with Moral Hazard and Risk of Repudiation." *Econometrica: Journal of the Econometric Society* 59 (4): 1069–1108.

Atzili, Boaz. 2012. *Good Fences, Bad Neighbors: Border Fixity and International Conflict*. Chicago: University of Chicago Press.

Atzili, Boaz, and Burak Kadercan, eds. 2017. *Territorial Designs and International Politics: Inside-out and Outside-in*. London: Routledge.

Avey, Paul. 2019. *Tempting Fate: Why Nonnuclear States Confront Nuclear Opponents*. Ithaca, NY: Cornell University Press.

Axelrod, Robert. 1984. *The Evolution of Cooperation*. New York: Basic Books.

Aytac, Erdem, Michael Mousseau, and Ömer Örsün. 2016. "Why Some Countries Are Immune from the Resource Curse." *Democratization* 23 (1): 71–92.

Babst, Dean V. 1964. "Elective Government—A Force for Peace." *Wisconsin Sociologist* 3 (1): 9–14.

Baezner, Marie. 2019. *Iranian Cyber-activities in the Context of Regional Rivalries and International Tensions*. Zurich: ETH Zurich.

Bail, Christopher A., Brian Guay, Emily Maloney, Aidan Combs, D. Sunshine Hillygus, Friedolin Merhout, Deen Freelon, and Alexander Volfovsky. 2020. "Assessing the Russian Internet Research Agency's Impact on the Political Attitudes and Behaviors of American Twitter Users in Late 2017." *Proceedings of the National Academy of Sciences* 117 (1): 243–50.

Bailey, Jennifer L. 1996. "Hot Fish and Bargaining Chips." *Journal of Peace Research* 33 (3): 257–62.

Bak, Daehee. 2018. "Alliance Proximity and the Effectiveness of Extended Deterrence." *International Interactions* 44 (1): 107–31.

Bak, Daehee, Kerry Chavez, and Toby J. Rider. 2020. "Domestic Political Consequences of International Rivalry." *Journal of Conflict Resolution* 64 (4):703–28.

Bak, Daehee, and Glenn Palmer. 2010. "Testing the Biden Hypothesis: Leader Tenure, Age, and International Conflict." *Foreign Policy Analysis* 6 (3): 257–73.

Bak, Per, Chao Tang, and Kurt Wiesenfeld. 1987. "Self-Organized Criticality: An Explanation of the 1/f Noise." *Physical Review Letters* 59 (4): 381–84.

Barbieri, Katherine. 1996. "Economic Interdependence: A Path to Peace or a Source of Interstate Conflict?" *Journal of Peace Research* 33 (1): 29–49.

——. 2018. "Economic Incentives As Weapons of War." In *The Oxford Encyclopedia of Empirical International Relations Theory*, edited by William R. Thompson, Vol. 1, 609–28. New York: Oxford University Press.

Barbieri, Katherine, and Jack S. Levy. 1999. "Sleeping with the Enemy: The Impact of War on Trade." *Journal of Peace Research* 36 (4): 463–79.

Barceló, Joan. 2020. "Are Western-Educated Leaders Less Prone to Initiate Militarized Disputes?" *British Journal of Political Science* 50 (2): 535–66.

Barnett, Jon, and W. Neil Adger. 2007. "Climate Change, Human Security, and Violent Conflict." *Political Geography* 26 (6): 639–55.

Barnett, Jon, and Michael Webber. 2009. "Accommodating Migration to Promote Adaptation to Climate Change." *Commission on Climate Change and Development*, https://www.preventionweb.net/publications/view/11872.

Barnett, Michael N. 1992. *Confronting the Costs of War: Military Power, State, and Society in Egypt and Israel*. Princeton, NJ: Princeton University Press.

Barnett, Michael N., and Jack S. Levy. 1991. "Domestic Sources of Alliances and Alignments: The Case of Egypt, 1962–73." *International Organization* 45 (3): 369–95.

Barzashka, Ivanka. 2013. "Are Cyber-weapons Effective? Assessing Stuxnet's Impact on the Iranian Enrichment Programme." *RUSI Journal* 158 (2): 48–56.

Bas, Muhamet, and Andrew Coe. 2016. "A Dynamic Theory of Nuclear Proliferation and Preventive War." *International Organization* 70 (4): 655–85.

Bayer, Reşat. 2010. "Peaceful Transitions and Democracy." *Journal of Peace Research* 47 (5): 535–46.

——. 2018. "Empirics of Stable Peace." In *The Oxford Encyclopedia of Empirical International Relations Theory*, edited by William R. Thompson, Vol. 1, 768–83. New York: Oxford University Press.

BBC News. 2013. "Obama: North Korea's Crisis-for-Concession Days Over." (May 7). https://www.bbc.com/news/world-us-canada-22443305

BBC News. 2015. "Is China-Pakistan 'Silk Road' a Game-changer?" (April 22). Available at http://www.bbc.com/news/world-asia–32400091.

Bearce, David H., and Stacy Bondanella. 2007. "Intergovernmental Organizations, Socialization, and Member-State Interest Convergence." *International Organization* 61 (4): 703–33.

Beardsley, Kyle. 2011. "Peacekeeping and the Contagion of Armed Conflict." *Journal of Politics* 73 (4): 1051–64.

Beardsley, Kyle, and Victor Asal. 2009a. "Nuclear Weapons As Shields." *Conflict Management and Peace Science* 26 (3): 235–55.

——. 2009b. "Winning with the Bomb." *Journal of Conflict Resolution* 53 (2): 278–301.

Beardsley, Kyle C., David M. Quinn, Bidisha Biswas, and Jonathan Wilkenfeld. 2006. "Mediation Style and Crisis Outcome." *Journal of Conflict Resolution* 50 (1): 58–86.

Beck, Andrea. 2014. "Drought, Dams, and Survival: Linking Water to Conflict and Cooperation in Syria's Civil War." *International Affairs Forum* 5: 11–22.

Bell, Mark, and Nicholas Miller. 2015. "Questioning the Effect of Nuclear Weapons on Conflict." *Journal of Conflict Resolution* 59 (1): 74–92.

Bell, Sam R., and Jesse C. Johnson. 2015. "Shifting Power, Commitment Problems, and Preventive War." *International Studies Quarterly* 59 (1): 124–32.

Bennett, D. Scott. 1996. "Security, Bargaining, and the End of Interstate Rivalry." *International Studies Quarterly* 40 (2): 157–83.

——. 1997. "Measuring Rivalry Termination." *Journal of Conflict Resolution* 41 (April): 277–54.

——. 2006. "Exploring Operationalizations of Political Relevance." *Conflict Management and Peace Science* 23 (3): 245–61.

Bennett, D. Scott, and Allan C. Stam III. 1996. "The Duration of Interstate Wars, 1816–1985." *American Political Science Review* 90 (2): 239–57.

——. 1998. "The Declining Advantages of Democracy." *Journal of Conflict Resolution* 42 (3): 344–66.

——. 2004. *The Behavioral Origins of War*. Ann Arbor: University of Michigan Press.

——. 2006. "Predicting the Length of the 2003 US-Iraq War." *Foreign Policy Analysis* 2 (2): 101–16.

Benson, Brett V. 2011. "Unpacking Alliances: Deterrent and Compellent Alliances and Their Relationship with Conflict, 1816–2000." *Journal of Politics* 73 (4): 1111–27.

——. 2012. *Constructing International Security: Alliances, Deterrence, and Moral Hazard*. New York: Cambridge University Press.

Benson, Brett V., Patrick R. Bentley, and James Lee Ray. 2013. "Ally Provocateur: Why Allies Do Not Always Behave." *Journal of Peace Research* 50 (1): 47–58.

Benson, Brett V., Adam Meirowitz, and Kristopher W. Ramsay. 2014. "Inducing Deterrence through Moral Hazard in Alliance Contracts." *Journal of Conflict Resolution* 58 (2): 307–35.

Benziman, Uzi. 2019. "Israel-Egypt Treaty, 40 Years Later: When Israel Made Peace a Priority." Haaretz, March 30.

Bercovitch, Jacob, and Gerald Schneider. 2000. "Who Mediates? The Political Economy of International Conflict Management." *Journal of Peace Research* 37 (2): 145–65.

Berg, Eiki, and Oras Saima. 2000. "Writing Post-Soviet Estonia on to the World Map." *Political Geography* 19 (5): 601–25.

Berinsky, Adam J. 2007. "Assuming the Costs of War: Events, Elites, and American Public Support for Military Conflict." *Journal of Politics* 69 (4): 975–97.

——. 2009. *Understanding American Public Opinion from World War II to Iraq*. Chicago: University of Chicago Press.

Berkemeier, Molly, and Matthew Fuhrmann. 2017. "Nuclear Weapons in Foreign Policy." In *Oxford Encyclopedia of Foreign Policy Analysis*, edited by Cameron G. Thies. New York: Oxford University Press. https://doi.org/10.1093/acrefore/9780190228637.013.457.

——. 2018. "Reassessing the Fulfillment of Alliance Commitments in War." *Research and Politics* 5 (2): 1–5.

Berman, Eli, Jacob N. Shapiro, and Joseph H. Felter. 2011. "Can Hearts and Minds Be Bought? The Economics of Counterinsurgency in Iraq." *Journal of Political Economy* 119 (4): 766–819.

Bernama News Agency. 2002. Malaysia to Focus on Island Dispute with Singapore, Says Deputy Premier. December 18.

Bernauer, Thomas, and Tobias Siegfried. 2012. "Climate Change and International Water Conflict in Central Asia." *Journal of Peace Research* 49 (1): 227–39.

Bernhard, Michael, Ömer Faruk Örsün, and Reşat Bayer. 2017. "Democratization in Conflict Research: How Conceptualization Affects Operationalization and Testing Outcomes." *International Interactions* 43 (6): 941–66.

Bertoli, Andrew, Allan Dafoe, and Robert F. Trager. 2019. "Is There a War Party? Party Change, the Left-Right Divide, and International Conflict." *Journal of Conflict Resolution* 63 (4): 950–75.

Besley, Timothy, and Torsten Persson. 2009. "The Origins of State Capacity: Property Rights, Taxation, and Politics." *American Economic Review* 99 (4): 1218–44.

——. 2010. "State Capacity, Conflict, and Development." *Econometrica* 78 (1): 1–34.

Betts, Richard K. 1987. *Nuclear Blackmail and Nuclear Balance*. Washington, DC: Brookings.

Bially Mattern, Janice, and Ayşe Zarakol. 2016. "Hierarchies in World Politics." *International Organization* 70 (3): 623–54.

Biddle, Stephen. 2004. *Military Power: Explaining Victory and Defeat in Modern Battle*. Princeton, NJ: Princeton University Press.

——. 2007. "Strategy in War." *PS: Political Science and Politics* 40 (3): 461–66.

Bienen, Henry, and Nicholas van de Walle. 1991. *Of Time and Power: Leadership Duration in the Modern World*. Stanford, CA: Stanford University Press.

Bilmanis, Alfred. 1947. *Lativia as an Independent State.* Washington, DC: Latvian Legation.

Bitzinger, Richard A. 2010. "A New Arms Race? Explaining Recent Southeast Asian Military Acquisitions." *Contemporary Southeast Asia* 32 (1): 50–69.

Black, Samuel. 2010. *The Changing Political Utility of Nuclear Weapons: Nuclear Threats from 1970 to 2010*. Washington, DC: Stimson Center. https://www.stimson.org/sites/default/files/file-attachments/Nuclear_Final_1.pdf.

Blagden, David. 2019. "Do Democracies Possess the Wisdom of Crowds? Decision Group Size, Regime Type, and Strategic Effectiveness." *International Studies Quarterly* 63 (4):1192–95.

Blainey, Geoffrey. 1988. *The Causes of War*, 3rd edition. London: Macmillan.

Blimes, Randall. 2009. "*Great Expectations: Domestic Expectations, Conflict Outcomes, and the Duration of Leadership Tenure.*" PhD thesis. Boulder: University of Colorado.

Blum, Gabriella. 2007. *Islands of Agreement: Managing Enduring Armed Rivalries*. Cambridge, MA: Harvard University Press.

Boehmer, Charles, and Mark Daube. 2013. "The Curvilinear Effects of Economic Development on Domestic Terrorism." *Peace Economics, Peace Science, and Public Policy* 19 (3): 359–68.

Boehmer, Charles, Erik Gartzke, and Timothy Nordstrom. 2004. "Do Intergovernmental Organizations Promote Peace?" *World Politics* 57 (1): 1–38.

Boix, Carles. 2003. *Democracy and Redistribution.* New York: Cambridge University Press.

Bolks, Sean, and Richard J. Stoll. 2000. "The Arms Acquisition Process: The Effect of Internal and External Constraints on Arms Race Dynamics." *Journal of Conflict Resolution* 44 (5): 580–603.

Bolton, Patrick, and Mathias Dewatripont. 2005. *Contract Theory*. Cambridge, MA: MIT Press.

Borghard, Erica D., and Shawn W. Lonergan. 2019. "Cyber Operations As Imperfect Tools of Escalation." *Strategic Studies Quarterly* 13 (3): 122–45.

Boschee, Elizabeth, Jennifer Lautenschlager, Sean O'Brien, Steve Shellman, and James Starz. 2018. "*ICEWS Automated Daily Event Data.*" https://doi.org/10.7910/DVN/QI2T9A, Harvard Dataverse, V260.

Boulding, Kenneth E. 1962. *Conflict and Defense*. New York: Harper and Row.

Bracken, Paul. 1983. *The Command and Control of Nuclear Forces*. New Haven, CT: Yale University Press.

Braithwaite, Alex. 2006. "The Geographic Spread of Militarized Disputes." *Journal of Peace Research* 43 (5): 507–22.

——. 2010. "MIDLOC: Introducing the Militarized Interstate Dispute Location Dataset." *Journal of Peace Research* 47 (1): 91–98.

Braithwaite, Alex, and Douglas Lemke. 2011. "Unpacking Escalation." *Conflict Management and Peace Science* 28 (2): 111–23.

Branch, Jordan. 2014. *The Cartographic State: Maps, Territory, and the Origins of Sovereignty*. Cambridge: Cambridge University Press.

Brantingham, P. Jeffrey, George E. Tita, Martin B. Short, and Shannon E. Reid. 2012. "The Ecology of Gang Territorial Boundaries." *Criminology* 40 (3): 851–85.

Brantly, Aaron F. 2018. "The Cyber Deterrence Problem." In *2018 10th International Conference on Cyber Conflict (CyCon)*, 31–54. IEEE. https://ieeexplore.ieee.org/abstract/document/8405009?casa_token=EXPpO0Xk0GIAAAAA:w-5mPjfS1CeJx-vFMe2SYECZjdhf8EYEOLMI-gQPoIW8HgfPYAvyiucL5Be6HFDUJV9qTxub-wck

Braumoeller, Bear F. 2013. *The Great Powers and the International System: Systemic Theory in Empirical Perspective*. Cambridge: Cambridge University Press.

——. 2019. *Only the Dead: The Persistence of War in the Modern Age*. Oxford: Oxford University Press.

Braumoeller, Bear F., and Austin Carson. 2011. "Political Irrelevance, Democracy, and the Limits of Militarized Conflict." *Journal of Conflict Resolution* 55 (2): 292–320.

Braut-Hegghammer, Malfrid. 2011. "Revisiting Osirak: Preventive Attacks and Nuclear Proliferation Risks." *International Security* 36 (1): 101–32.

Brecher, Michael. 1977. "Toward a Theory of International Crisis Behavior: A Preliminary Report." *International Studies Quarterly* 21 (1): 39–74.

——. 1984. "International Crises, Protracted Conflicts." *International Interactions* 11 (3/4): 237–98.

——. 1993. *Crises in World Politics: Theory and Reality*. Oxford: Pergamon.

Brecher, Michael, and Jonathan Wilkenfeld. 1989. *Crisis, Conflict and Instability*. Oxford: Pergamon Press.

——. 1997. *A Study of Crisis*. Ann Arbor: University of Michigan Press.

Brecke, Peter. 1999. "Violent Conflicts 1400 A.D. to the Present in Different Regions of the World. http://web.archive.org/web/20121023154335/http://www.inta.gatech.edu/peter/PSS99_paper.html.

——. 2012. "Notes about Conflict Catalog." http://www.cgeh.nl/sites/default/files/Notes about Conflict Catalog.pdf

Bremer, Stuart A. 1992. "Dangerous Dyads: Conditions Affecting the Likelihood of Interstate War, 1816–1965." *Journal of Conflict Resolution* 36 (2): 309–41.

——. 1993. "Advancing the Scientific Study of War." *International Interactions* 19 (1–2): 1–26.

——. 2000. "Who Fights Whom, When, Where, and Why?" In *What Do We Know about War?* edited by John A. Vasquez, 23–36. Lanham, MD: Rowman and Littlefield.

Brewer, John. 1989. The Sinews of Power: *War, Money, and the English State,* 1688-1783. New York: Knop.

Brochmann, Marit. 2012. Basin Asymmetries and the Risk of Conflict in International River Basins. In "*Bridge over Troubled Water: Interaction in International River Basins.*" Ph D dissertation, University of Oslo.

Brochmann, Marit, and Nils Petter Gleditsch. 2012. "Shared Rivers and Conflict: A Reconsideration." *Political Geography* 31 (8): 519–27.

Brochmann, Marit, and Paul R. Hensel. 2009. "Peaceful Management of International River Claims." *International Negotiation* 14 (2): 391–416.

——. 2011. "The Effectiveness of Negotiations over International River Claims." *International Studies Quarterly* 55 (3): 859–82.

Brodie, Bernard. 1946. *The Absolute Weapon: Atomic Power and World Order*. New York: Harcourt, Brace.

Brooks, Stephen G. 1999. "The Globalization of the Production and the Changing Benefits of Conquests." *Journal of Conflict Resolution* 43 (5): 646–70.

Brown, William Adam, Jr. 1940. *The International Gold Standard Reinterpreted, 1914–1934*. New York: National Bureau of Economic Research.

Broz, J. Lawrence. 1998. "The Origins of Central Banking: Solutions to the Free Rider Problem." *International Organization* 52 (2): 231–68.

Buchanan, Ben. 2016. *The Cybersecurity Dilemma: Hacking, Trust, and Fear between Nations*. Oxford: Oxford University Press.

Bueno de Mesquita, Bruce. 1978. "Systemic Polarization and the Occurrence and Duration of War." *Journal of Conflict Resolution* 22 (2): 241–67.

——. 1981. *The War Trap*. New Haven, CT: Yale University Press.

——. 1990. "Pride of Place: The Origins of German Hegemony." *World Politics* 43 (1): 28–52.

Bueno de Mesquita, Bruce, James D. Morrow, Randolph M. Siverson, and Alastair Smith. 1999. "An Institutional Explanation for the Democratic Peace." *American Political Science Review* 93 (4): 791–807.

——. 2003. *The Logic of Political Survival*. Cambridge, MA: MIT Press.

——. 2004. "Testing Novel Implications from the Selectorate Theory of War." *World Politics* 56 (3): 363–88.

Bueno de Mesquita, Bruce, and Randolph M. Siverson. 1995. "War and the Survival of Political Leaders: A Comparative Study of Regime Types and Political Accountability." *American Political Science Review* 89 (4): 841–55.

——. 1997. "Nasty or Nice? Political Systems, Endogenous Norms, and the Treatment of Adversaries." *Journal of Conflict Resolution* 41 (1): 75–199.

Bueno de Mesquita, Bruce, Randolph M. Siverson, and Gary Woller. 1992. "War and the Fate of Regimes: A Comparative Analysis." *American Political Science Review* 86 (3): 638–46.

Bueno de Mesquita, Bruce, and Alastair Smith. 2009. "A Political Economy of Aid." *International Organization* 63 (2): 309–40.

Buhaug, Halvard. 2005. "Dangerous Democracies Revisited: Democracies May Not Be That Peaceful after All." *Conflict Management and Peace Science* 22 (2): 95–111.

——. 2010. "Climate Not to Blame for African Civil Wars." *Proceedings of the National Academy of Sciences in the United States of America* 107 (88): 16477–82.

Buhaug, Halvard, and Nils Petter Gleditsch. 2006. "The Death of Distance? The Globalization of Armed Conflict." In *Territoriality and Conflict in the Era of*

Globalization of Armed Conflict, edited by Miles Kahler and Barbara Walter, 187–216. Cambridge: Cambridge University Press.

Burk, Kathleen. 1979. "Great Britain in the United States, 1917–1918: The Turning Point." *International History Review* 1 (2): 228–45.

——. 1985. *Britain, America and the Sinews of War, 1914–1918*. London: Allen and Unwin.

Burke, Marshall B., Solomon M. Hsiang, and Edward Miguel. 2015. "Climate and Conflict." *Annual Review of Economics* 7: 577–617.

Burke, Marshall B., Edward Miguel, Shanker Satyanath, John A. Dykema, and David B. Lobell. 2009. "Warming Increases the Risk of Civil War in Africa." *Proceedings of the National Academy of Sciences* 106 (49): 20670–74.

Burkhart, Ross E., and Michael S. Lewis-Beck. 1994. "Comparative Democracy: The Economic Development Thesis." *American Political Science Review* 88 (4): 111–31.

Busby, Joshua W. 2008. "Who Cares about the Weather? Climate Change and U.S. National Security." *Security Studies* 17 (3): 468–504.

Bussmann, Margit, and John R. Oneal. 2007. "Do Hegemons Distribute Private Goods? A Test of Power-Transition Theory." *Journal of Conflict Resolution* 51 (1): 88–111.

Butterworth, Robert L. 1978. "Do Conflict Managers Matter? An Empirical Assessment of Interstate Security Disputes and Resolution Efforts, 1945–1974." *International Studies Quarterly* 22 (2): 195–214.

Cappella Zielinski, Rosella. 2014. Confronting the Costs of War Dataset. http://sites.bu.edu/cappella/confronting-the-cost-of-war-data/

——. 2016. *How Do States Pay for War?* Ithaca, NY: Cornell University Press.

——. 2018. "U.S. Wars Abroad Increase Inequality at Home: Who Foots the Bill for American Hegemony?" *Foreign Affairs*. https://www.foreignaffairs.com/articles/2018–10-05/us-wars-abroad-increase-inequality-home

Carcelli, Shannon, and Eric Gartzke. 2018. "The Diversification of Deterrence: New Data and Novel Realities." In *The Oxford Encyclopedia of Empirical International Relations Theory*, edited by William R. Thompson, Vol. 1, 535–54. New York: Oxford University Press.

Carlson, Lisa A., and Raymond Dacey. 2018. "A Unified Analysis of the Diversionary and Constraint Accounts of Crisis Initiation." In *The Oxford Encyclopedia of Empirical International Relations Theory*, edited by William R. Thompson, Vol. 4, 459–76. New York: Oxford University Press.

Carter, David B. 2010. "The Strategy of Territorial Conflict." *American Journal of Political Science* 54 (4): 969–87.

Carter, David B., and Hein E. Goemans. 2011. "The Making of the Territorial Order: New Borders and the Emergence of Interstate Conflict." *International Organization* 65 (2): 275–310.

——. 2018. "International Trade and Coordination: Tracing Border Effects." *World Politics* 70 (1): 1–52.

Carter, David B., and Paul Poast. 2015. "Why Do States Build Walls? Political Economy, Security, and Border Stability." *Journal of Conflict Resolution* 61 (2): 239–70.

Carter, Jeff. 2017. "The Political Cost of War Mobilization in Democracies and Dictatorships." *Journal of Conflict Resolution* 61 (8): 1768–94.

Carter, Jeff, and Giacomo Chiozza. 2018. "State Leaders and Foreign Policy." In *The Oxford Encyclopedia of Foreign Policy Analysis*, edited by Cameron G. Thies. New York: Oxford University Press. https://doi.org/10.1093/acrefore/9780190228637.013.487

Carter, Jeff, and Timothy Nordstrom. 2017. "Term Limits, Leader Preferences, and Interstate Conflict." *International Studies Quarterly* 61 (3): 721–35.

Carter, Jeff, and Glenn Palmer. 2016. "Regime Type and Interstate War Finance." *Foreign Policy Analysis* 12 (4): 695–719.

Carter, Jeff, and Charles E. Smith Jr. 2020. "A Framework for Measuring Leaders' Willingness to Use Force." *American Political Science Review* 114 (4): 1352–58.

Castellano da Silva. 2019. "Explaining Co-operation and Conflict in Southern Africa: State-Building, Foreign Policy and Regional Order." *Contexto International* 38 (2). https://doi.org/10.1590/S0102-8529.2016380200003

Cederman, Lars-Erik. 1994. "Emergent Polarity: Analyzing State-Formation and Power Politics." *International Studies Quarterly* 38 (4): 501–33.

——. 1997. *Emergent Actors in World Politics*. Princeton, NJ: Princeton University Press.

——. 2003. "Modeling the Size of Wars: From Billiard Balls to Sandpiles." *American Political Science Review* 97 (1): 135–50.

Cederman, Lars-Erik, Camber Warren, and Didier Sornette. 2011. "Testing Clausewitz: Nationalism, Mass Mobilization, and the Severity of War." *International Organization* 65 (4): 605–38.

Centeno, Miguel Angel. 1997. "Blood and Debt: War and Taxation in Nineteenth-Century Latin America." *American Journal of Sociology* 102 (6): 1565–605.

——. 2002. *Blood and Debt: War and the Nation-State in Latin America*. University Park: Pennsylvania State University Press.

Cerny, Philip G. 1995. "Globalization and the Changing Logic of Collective Action." *International Organization* 49 (4): 595–625.

Chan, Steve. 1984. "Mirror, Mirror on the Wall … Are the Freer Countries More Pacific?" *Journal of Conflict Resolution* 28 (4): 617–48.

——. 2010. "Progress in the Democratic Peace Research Agenda." In *International Studies Encyclopedia*, edited by Robert Denemark, 5924–42. London: Wiley-Blackwell.

Chang, Hyojoon, and Scott L. Kastner. 2018. "Economic Interdependence and Conflict." In *The Oxford Encyclopedia of Empirical International Relations Theory*, edited by William R. Thompson, Vol. 1, 628–45. New York: Oxford University Press.

Chapman, Terrence L., Patrick J. McDonald, and Scott Moser. 2015. "The Domestic Politics of Strategic Retrenchment, Power Shifts, and Preventive War." *International Studies Quarterly* 59 (1): 133–44.

Chari, P. R. 2012. "*Nuclear Crisis, Escalation, Control and Deterrence in South Asia*." L. L. Stimson Working Paper. Henry L. Stimson Center. Washington, DC.

Chatterjee, Arnab, and Bikas K. Chakrabarti. 2017. "Fat Tailed Distributions for Deaths in Conflicts and Disasters." *Reports in Advances of Physical Sciences* 1 (1): 1740007.

Chaudhuri, Rudra. 2016. "Arms and Assistance in South Asia 1953–1965: Why Racing Alone Explains Little." In *Arms Races in International Politics*, edited

by Thomas Mahnken, Joseph Maiolo, and David Stevenson, 225–45. New York: Oxford University Press.
Chernow, Ron. 2010. *The House of Morgan: An American Banking Dynasty and the Rise of Modern Finance*. New York: Grove Press.
Chiozza, Giacomo, and Hein E. Goemans. 2003. "Peace through Insecurity: Tenure and International Conflict." *Journal of Conflict Resolution* 47 (4): 443–67.
——. 2004b. "International Conflict and the Tenure of Leaders: Is War Still 'Ex Post' Inefficient?" *American Journal of Political Science* 48 (3): 604–19.
——. 2004. "Avoiding Diversionary Targets." *Journal of Peace Research* 41 (4): 423–43.
——. 2011. *Leaders and International Conflict*. Cambridge: Cambridge University Press.
Choe, Hang-Sung. 2013. "North Korea Threatens to Attack U.S. With 'Lighter and Smaller Nukes.'" *New York Times* (March 5).
Choi, Ajin. 2003. "The Power of Democratic Cooperation." *International Security* 28 (1): 142–53.
——. 2004. "Democratic Synergy and Victory in War, 1816–1992." *International Studies Quarterly* 48 (3): 663–82.
Choi, Ajin, and Giacomo Chiozza. 2003. "Guess Who Did What: Political Leaders and the Management of Territorial Disputes." *Journal of Conflict Resolution* 47 (3): 251–78.
Choi, Seung-Whan, Patrick James, and Johann Park. 2006. "Democracy and Conflict Management: Territorial Claims in the Western Hemisphere Revisited." *International Studies Quarterly* 50 (4): 803–17.
Choucri, Nazli, and Gaurav Agarwal. 2018. "The Theory of Lateral Pressure: Highlights of Quantification and Empirical Analysis." In *The Oxford Encyclopedia of Empirical International Relations Theory*, edited by William R. Thompson, Vol. 4, 294–322. New York: Oxford University Press.
Choucri, Nazli, and Robert C. North. 1975. *Nations in Conflict: National Growth and International Violence*. San Francisco: W. H. Freeman.
Choudhry, Taufiq. 2010. "World War II Events and the Dow Jones Industrial Index." *Journal of Banking and Finance* 34 (5): 1022–131.
Christensen, Thomas, and Jack Snyder. 1990. "Chain Gangs and Passed Bucks: Predicting Alliance Patterns in Multipolarity." *International Organization* 44 (Spring): 137–68.
Christie, Agatha. 1988. *Murder on the Orient Express*. London: Harper Collins.
Cioffi-Revilla, Claudio, and Manus A. Midlarsky. 2013. "Power Laws, Scaling, and Fractals in the Most Lethal International and Civil Wars." *SSRN Electronic Journal*. http://www.ssrn.com/abstract=2291166.
Cirillo, Pasquale, and Nassim Nicholas Taleb. 2015. "On the Tail Risk of Violent Conflict and Its Underestimation." eprint arXiv:1505.04722. http://arxiv.org/abs/1505.04722v1.
———. 2016. "On the Statistical Properties and Tail Risk of Violent Conflicts." *Physica A: Statistical Mechanics and its Applications* 452: 29–45.
———. 2019. "The Decline of Violent Conflicts: What Do the Data Really Say?" In *The Causes of Peace: What We Now Know*, edited by Asle Toje and Bård Nikolas Vik Steen, 57–86. Bergen, Norway: Nobel Symposium Proceedings.

Cirillo, Vincent J. 2008. "Two Faces of Death: Fatalities from Disease and Combat in America's Principal Wars, 1775 to Present." *Perspectives in Biology and Medicine* 51 (1): 121–33.

Clague, Christopher, Philip Keefer, Stephen Knack, and Mancur Olson. 1999. "Contract-Intensive Money: Contract Enforcement, Property Rights, and Economic Performance." *Journal of Economic Growth* 4 (2): 185–211.

Clare, Joe, and Vesna Danilovic. 2010. "Multiple Audiences and Reputation Building in International Conflicts." *Journal of Conflict Resolution* 54 (6): 860–82.

Clark, David, and Timothy Nordstrom. 2005. "Democratic Variants and Democratic Variance: How Domestic Constraints Shape Interstate Conflict." *Journal of Politics* 67 (1): 250–70.

Clarke, Kevin A., and David M. Primo. 2012. *A Model Discipline: Political Science and the Logic of Representations*. New York: Oxford University Press.

Clarke, Richard A., and Robert K. Knake. 2014. *Cyber War*. Old Saybrook: Tantor Media.

Clauset, Aaron. 2018. "Trends and Fluctuations in the Severity of Interstate Wars." *Science Advances* 4 (2): eaao3580.

Clauset, Aaron, Cosma Rohilla Shalizi, and M. E. J. Newman. 2009. "Power-Law Distributions in Empirical Data." *SIAM Review* 51 (4): 661–703.

Clay, K. Chad, and Matthew R. DiGiuseppe. 2017. "The Physical Consequences of Fiscal Flexibility: Sovereign Credit and Physical Integrity Rights." *British Journal of Political Science* 47 (4): 783–807.

Clay, K. Chad and Andrew P. Owsiak. 2016. "The Diffusion of International Border Agreements." *Journal of Politics* 78 (2): 427–42.

Cockburn, Andrew. 1983. *The Threat: Inside the Soviet Military Machine*. New York: Random House.

Cohen, Dara Kay. 2013. "Explaining Rape during Civil War: Cross-National Evidence (1980–2009)." *American Political Science Review* 107 (3): 461–77.

Cohn, Carol. 1987. "Sex and Death in the Rational World of Defense Intellectuals." *Signs* 12 (4): 687–718.

Colaresi, Michael P. 2001. "Shocks to the System: Great Power Rivalry and the Leadership Long Cycle." *Journal of Conflict Resolution* 45 (5): 569–93.

——. 2004. "When Doves Cry: International Rivalry, Unreciprocated Cooperation, and Leadership Turnover." *American Journal of Political Science* 48 (3): 555–70.

——. 2005. *Scare Tactics*. Syracuse, NY: Syracuse University Press.

——. 2018. "Fighting Abroad, Fighting at Home (and Vice Versa): Identifying the Relationship between Civil and Interstate Conflict with Fewer Assumptions." In *The Oxford Encyclopedia of Empirical International Relations Theory*, edited by William R. Thompson, Vol. 2, 1–19. New York: Oxford University Press.

Colaresi, Michael P., Karen A. Rasler, and William R. Thompson. 2007. *Strategic Rivalries in World Politics: Position, Space and Conflict Escalation*. Cambridge: Cambridge University Press.

Colaresi, Michael P., and William R. Thompson. 2002. "Strategic Rivalries, Protracted Conflict, and Crisis Escalation." *Journal of Peace Research* 39 (3): 263–87.

Colgan, Jeff D. 2013. "Domestic Revolutionary Leaders and International Conflict." *World Politics* 65 (4): 656–90.

———. 2016. "Where Is International Relations Going? Evidence from Graduate Training." *International Studies Quarterly* 60 (3): 486–98.

Colgan, Jeff D., and Jessica L. P. Weeks. 2015. "Revolution, Personalist Dictatorships, and International Conflict." *International Organization* 69 (1): 163–94.

Collier, David. 2011. "Understanding Process Tracing." *PS: Political Science and Politics* 44 (4): 823–30.

Collier, John G., and Vaughan Lowe. 1999. *The Settlement of Disputes in International Law Institutions and Procedures*. Oxford and New York: Oxford University Press.

Collier, Paul. 1999. "On the Economic Consequences of Civil War." *Oxford Economic Papers* 51 (1): 168–83.

Conrad, Courtenay R., and Emily Hencken Ritter. 2013. "Tenure, Treaties, and Torture: The Conflicting Domestic Effects of International Law." *Journal of Politics* 75 (2): 397–409.

Cooper, John M., Jr. 1976. "The Command of Gold Reversed: American Loans to Britain, 1915–1917." *Pacific Historical Review* 45 (2): 209–30.

Cornwell, Derekh, and Michael P. Colaresi. 2002. "Holy Trinities, Rivalry Termination, and Conflict." *International Interactions* 28 (4): 325–53.

Coser, Lewis A. 1956. *The Functions of Social Conflict*. Glencoe, IL: Free Press.

Cox, Eric W. 2010. *Why Enduring Rivalries Do—or Don't—End*. Boulder, CO: First Forum Press.

Craig, Anthony, and Brandon Valeriano. 2016. "Conceptualising Cyber Arms Races." In 8th International Conference on Cyber Conflict (Cy Con), edited by N. Pissanidis, H. Roigas, and M. Veenendaal, 141-58, Estonia: IEEE.

Cranmer, Skyler J., Bruce A. Desmarais, and Elizabeth J. Menninga. 2012. "Complex Dependencies in the Alliance Network." *Conflict Management and Peace Science* 29 (3): 279–313.

Crescenzi, Mark J. C. 2018. *Of Friends and Foes: Reputation and Learning in International Politics*. New York: Oxford University Press.

Crescenzi, Mark J. C., and Kelly M. Kadera. 2016. "Built to Last: Understanding the Link between Democracy and Conflict in the International System." *International Studies Quarterly* 60 (3): 565–72.

Crocker, Chester, Fen Osler Hampson, and Pamel Aall. 2004. *Taming Intractable Conflicts: Mediation in the Hardest Cases*. Washington, DC: United States Institute of Peace.

Croco, Sarah E. 2011. "The Decider's Dilemma: Leader Culpability, Domestic Politics, and War Termination." *American Political Science Review* 105 (3): 457–77.

Croco, Sarah E., and Jessica L. P. Weeks. 2016. "War Outcomes and Leader Tenure." *World Politics* 68 (4): 577–607.

Croteau, Sabreena, and Paul Poast. 2020. "Dollars for Oil" in Benjamin J. Cohen, Sabreena Croteau, Aashna Khanna, Daniel McDowell, Hongying Wang, and W. Kindred Winecoff. Symposium on Global Monetary Order and the Liberal Order Debate. *International Studies Perspectives* 21 (2): 109–53.

Cunen, Céline, Gudmund Hermansen, and Nils Lid Hjort. 2018. "Confidence Distributions for Change-Points and Regime Shifts." *Journal of Statistical Planning and Inference* 195 (May): 14–34.

Cunen, Céline, Nils Lid Hjort, and Håvard Mokleiv Nygård. 2020. "Statistical Sightings of Better Angels: Analysing the Distribution of Battle-Deaths in Interstate Conflict over Time." *Journal of Peace Research* 57 (2): 221–34.

Cunningham, Fiona S., and M. Taylor Fravel. 2015. "Assuring Assured Retaliation: China's Nuclear Posture and U.S.-China Strategic Stability." *International Security* 40 (2): 7–50.

Cyberspace Solarium Commission Report. 2020. Available at https://www.solarium.gov/

Dafoe, Allan. 2012. "*Resolve, Reputation, and War: Cultures of Honor and Leaders' Time-in- Office*." PhD thesis, University of California, Berkeley.

Dafoe, Allan, and Devin Caughey. 2016. "Honor and War." *World Politics* 68 (2): 341–81.

Dahl, Robert A. 1957. "The Concept of Power." *Behavioral Science* 2 (3): 201–15.

——. 1997. "Development and Democratic Culture." In *Consolidating the Third Wave of Democracies, Themes and Perspectives*, edited by Larry Diamond, Marc F. Plattner, Yun-han Chu, and Hung-mao Tien, 34–39. Baltimore, MD: Johns Hopkins University Press.

Daniels, Kelly, and Sara McLaughlin Mitchell. 2017. "Bones of Democratic Contention: Maritime Disputes." *International Area Studies Review* 20 (4): 293–310.

Danilovic, Vesna. 2002. *When the Stakes Are High: Deterrence and Conflict among Major Powers*. Ann Arbor: University of Michigan Press.

Danilovic, Vesna, and Joe Clare. 2007. "Global Power Transitions and Regional Interests." *International Interactions* 33 (3): 289–304.

Daoudy, Marwa. 2009. "Asymmetric Power: Negotiating Water in the Euphrates and Tigris." *International Negotiation* 14 (2): 361–91.

Dassel, Kurt, and Eric Reinhart. 1999. "Domestic Strife and the Initiation of Violence at Home and Abroad." *American Journal of Political Science* 43 (1): 56–85.

Davenport, Christian, Erik Melander, and Patrick Regan. 2018. *The Peace Continuum: What It Is and How to Study It*. Oxford: Oxford University Press.

Davis, Christina L. 2012. *Why Adjudicate? Enforcing Trade Rules in WTO*. Princeton, NJ: Princeton University Press.

de Figueiredo, Rui J. P. 2002. "Electoral Competition, Political Uncertainty, and Policy Insulation." *American Political Science Review* 96 (2): 321–33.

Debs, Alexandre, and Hein E. Goemans. 2010. "Regime Type, the Fate of Leaders, and War." *American Political Science Review* 104 (3): 430–45.

Debs, Alexandre, and Nuno N. Monteiro. 2014. "Known Unknowns: Power Shifts, Uncertainty, and War." *International Organization* 68 (1): 1–31.

Demirjian, Karoun, Josh Dawsey, Ellen Nakashima, and Carol D. Leonnig. 2019. "Trump Ordered Hold on Military Aid Days before Calling Ukrainian President, Officials Say." *Washington Post* (September 23). Available at https://www.washingtonpost.com/national-security/trump-ordered-hold-on-military-aid-days-before-calling-ukrainian-president-officials-say/2019/09/23/df93a6ca-de38-11e9-8dc8-498eabc129a0_story.html.

Department of the Army. 2007. *US Army/Marine Corps Counterinsurgency Field Manual*. Chicago: University of Chicago Press.

Desch, Michael. 2003. "Democracy and Victory: Fair Fights or Food Fights?" *International Security* 28 (1): 180–94.

DeSombre, Elizabeth R., and J. Samuel Barkin. 2011. *Fish*. Cambridge: Polity.

de Soysa, Indra, John R. Oneal, and Yong-Hee Park. 1997. "Testing Power-Transition Theory Using Alternative Measures of National Capabilities." *Journal of Conflict Resolution* 41 (4): 509–28.

Deutsch, Karl W., Sidney A. Burrell, and Robert A. Kann. 1957. *Political Community and the North Atlantic Area: International Organization in Light of Historical Experience*. Princeton, NJ: Princeton University Press.

Deutsch, Karl W., and J. David Singer. 1964. "Multipolar Systems and International Stability." *World Politics* 16 (3): 390–406.

DiCicco, Jonathan M. 2018. "Power Transition Theory and the Essence of Revisionism." In *The Oxford Encyclopedia of Empirical International Relations Theory*, edited by William R. Thompson, Vol. 3, 188–214. Oxford: Oxford University Press.

DiCicco, Jonathan M., and Jack Levy. 1999. "Power Shifts and Problem Shifts: The Evolution of the Power Transition Research Design." *Journal of Conflict Resolution* 43 (6): 675–704.

Diehl, Paul F. 1983. "Arms Race and Escalation: A Closer Look." *Journal of Peace Research* 20 (5): 249–59.

——. 1985. "Arms Races to War: Testing Some Empirical Linkages." *Sociology Quarterly* 26 (3): 331–49.

——. 1992. "What Are They Fighting For? The Importance of Issues in International Conflict Research." *Journal of Peace Research* 29 (3): 333–44.

——. ed. 1998. *The Dynamics of Enduring Rivalries*. Champaign: University of Illinois Press.

——. 2016. "Exploring Peace: Looking beyond War and Negative Peace." *International Studies Quarterly* 60 (1): 1–10.

——. 2019. "Peace: A Conceptual Survey." In *The Oxford Research Encyclopedia of International Studies*. Oxford: Oxford University Press. Available at https://doi.org/10.1093/acrefore/9780190846626.013.515.

Diehl, Paul F., and Mark J. C. Crescenzi. 1998. "Reconfiguring the Arms Race-War Debate." *Journal of Peace Research* 35 (1): 111–18.

Diehl, Paul F., and Gary Goertz. 1992, 2002. *Territorial Changes and International Conflict*. New York: Routledge.

——. 2000. *War and Peace in International Rivalry*. Ann Arbor: University of Michigan Press.

——. 2012. "The Rivalry Process: How Rivalries are Sustained and Terminated." *In What Do We Know about War?* 2nd edition, edited by John A. Vasquez, 83–109. Lanham, MD: Rowman and Littlefield.

——. 2018. "Theories of Interstate Peace." In *The Oxford Encyclopedia of Empirical International Relations Theory*, edited by William R. Thompson, Vol. 4, 254-71. New York: Oxford University Press.

——. 2019. "Peace Data: Concept, Measurement, Patterns, and Research Agenda." *Conflict Management and Peace Science*, forthcoming.

Diehl, Paul F., Gary Goertz, and Yahve Gallegos. Forthcoming. "Peace Data: Concept, Measurement, Patterns, and Research Agenda." *Conflict Management and Peace Science*. doi:10.1177/0738894219870288

Diehl, Paul F., Gary Goertz, and Daniel Saeedi. 2005. "Theoretical Specifications of Enduring Rivalries: Applications to the India-Pakistan Case." In *The India-Pakistan Conflict: An Enduring Rivalry*, edited by T. V. Paul, 27–53. Cambridge: Cambridge University Press.

Diehl, Paul F., and Jaroslav Tir. 2002. "Geographic Dimensions of Enduring Rivalries." *Political Geography* 21 (2): 263–86.

DiGiuseppe, Matthew R. 2015. "Guns, Butter and Debt: Sovereign Creditworthiness and Military Spending." *Journal of Peace Research* 52 (2): 680–93.

DiGiuseppe, Matthew R., Colin M. Barry, and Richard W. Frank. 2012. "Good for the Money: International Finance, State Capacity, and Internal Armed Conflict." *Journal of Peace Research* 49 (3): 391–405.

DiGiuseppe, Matthew R., and Patrick Shea. 2016. "Borrowed Time: Sovereign Finance, Regime Type and Leader Survival." *Economics and Politics* 28 (3): 342–67.

Dinar, Shlomi. 2009. "Scarcity and Cooperation along International Rivers." *Global Environmental Politics* 9 (1): 109–35.

Di Salvatore, Jessica and Andrea Ruggeri. 2018. "The Effectiveness of Peacekeeping Operations." In *The Oxford Encyclopedia of Empirical International Relations Theory*, edited by William R. Thompson, Vol. 1, 645–66. New York: Oxford University Press.

Dixon, William J. 1994. "Democracy and the Peaceful Settlement of International Conflict." *American Political Science Review* 88 (1): 14–32.

Dixon, William J., and Paul D. Senese. 2002. "Democracy, Disputes, and Negotiated Settlements." *Journal of Conflict Resolution* 46 (4): 547–71.

Domke, William. 1988. *War and the Changing Global System*. New Haven, CT: Yale University Press.

Dorussen, Han, and Hugh Ward. 2008. "Intergovernmental Organizations and the Kantian Peace: A Network Perspective." *Journal of Conflict Resolution* 52 (2): 189–212.

Downes, Alexander B. 2007. "Draining the Sea by Filling the Graves: Investigating the Effectiveness of Indiscriminate Violence As a Counterinsurgency Strategy." *Civil Wars* 9 (4): 420–44.

——. 2009. "How Smart and Tough Are Democracies? Reassessing Theories of Democratic Victory in War." *International Security* 33 (4): 9–51.

Downes, Alexander B., and Todd S. Sechser. 2012. "The Illusion of Democratic Credibility." *International Organization* 66 (3): 457–89.

Downing, Brian M. 1993. *The Military Revolution and Political Change: Origins of Democracy and Autocracy in Early Modern Europe*. Princeton, NJ: Princeton University Press.

Downs, George W., and David M. Rocke. 1994. "Conflict, Agency, and Gambling for Resurrection: The Principal-Agent Problem Goes to War." *American Journal of Political Science* 38 (2): 362–80.

Doyle, Michael W. 1986. "Liberalism and World Politics." *American Political Science Review* 80 (December): 1151–69.

——. 2005. "Three Pillars of the Liberal Peace." *American Political Science Review* 99 (3): 463–66.

Driedger, Jonas J., and Mikhail Polianskii. 2023. "Utility-based Predictions of Military Escalation: Why Experts Forecasted Russia would not Invade Ukraine." *Contemporary Security Policy* 44 (4): 544–60.

Dreyer, David. 2012. "Issue Intractability and the Persistence of International Rivalry." *Conflict Management and Peace Science* 29 (5): 471–89.

—— 2018. "Foundations of Rivalry Research." In *The Oxford Encyclopedia of Empirical International Relations Theory*, edited by William R. Thompson, Vol. 2, 65–80. New York: Oxford University Press.

Dreyer, David, and William R. Thompson. 2011. *Handbook of International Rivalries*. Washington, DC: CQ Press.

Dube, Oeindrila, and S. P. Harish. 2017. "Queens." NBER Working Paper No. 23337.

Duelfer, Charles, and Stephen Benedict Dyson. 2011. "Chronic Misperception and International Conflict." *International Security* 36 (1): 73–100.

Dykstra, John, Glen A. Larson, Richard A. Colla, Alan J. Levi, Richard Hatch, Dirk Benedict, Lorne Greene, et al. 2003. *Battlestar Galactica*. Universal City, CA: Universal.

Eason P. K., G. A. Cobbs, and K. G. Trinca. 1999. "The Use of Landmarks to Define Territorial Boundaries." *Animal Behavior* 58 (1): 85–91.

Ellerman, David. 2007. "Helping Self-Help: The Fundamental Conundrum of Development Assistance." *Journal of Socio-Economics* 36 (4): 561–77.

Ellis, Cali Mortenson, Michael C. Horowitz, and Allan C. Stam III. 2015. "Introducing the LEAD Data Set." *International Interactions* 41 (4): 718–41.

Ellis, Glynn, Sara McLaughlin Mitchell, and Brandon C. Prins. 2010. "How Democracies Keep the Peace: Contextual Factors That Influence Conflict Management Strategies." *Foreign Policy Analysis* 6 (4): 373–98.

Enia, Jason. 2017. "Do Contracts Save Lives? The Relationship between Contract Intensive Economies and Natural Disaster Fatalities." *Risk, Hazards and Crisis in Public Policy* 9 (1): 60–81.

Enterline, Andrew J., Emily Stull, and Joseph Magagnoli. 2013. "Reversal of Fortune? Strategy Change and Counterinsurgency Success by Foreign Powers in the Twentieth Century." *International Studies Perspectives* 14 (2): 176–98.

Entringer, Irene, Garcia Blanes, Ryan Powers, Susan Peterson, and Michael J. Tierney. 2022. "Poll: Will Russia Invade Ukraine?" *Foreign Policy*, January 31. https://foreignpolicy.com/2022/01/31/poll-russia-ukraine-invasion-crisis-biden-response/.

Fagan, Brennen T., Marina I. Knight, Niall J. MacKay, and A. Jamie Wood. 2019. "Changepoint Analysis of Historical Battle Deaths." arXiv:1909.02626 [stat]. http://arxiv.org/abs/1909.02626.

Fair, C. Christine. 2014. *Fighting to the End: The Pakistan Army's Way of War*. New York: Oxford University Press.

Fang, Songying, Jesse C. Johnson, and Brett Ashley Leeds. 2014. "To Concede or to Resist? The Restraining Effect of Military Alliances." *International Organization* 68 (4): 775–809.

Farber, Henry S., and Joanne Gowa. 1995. "Polities and Peace." *International Security* 20 (2): 123–46.

Farrar, Lancelot L., Jr. 1977. "Cycles of War: Historical Speculations on Future International Violence." *International Interactions* 3 (2): 161–79.

Farrar-Hockley, Anthony. 1995. *The British Part in the Korean War*. London: Stationery Office.

Fazal, Tanisha. 2004. "State Death in the International System." *International Organization*. 58 (2): 311–44.

——. 2014. "Dead Wrong? Battle Deaths, Military Medicine, and Exaggerated Reports of War's Demise." *International Security* 39 (1): 95–125.

——. 2018. *Wars of Law: Unintended Consequences in the Regulation of Armed Conflict*. Ithaca, NY: Cornell University Press.

Fearon, James D. 1994a. "Domestic Political Audiences and the Escalation of International Disputes." *American Political Science Review* 88 (3): 577–92.

——. 1994b. "Signaling versus the Balance and Power and Interests: An Empirical Test of a Crisis Bargaining Model." *Journal of Conflict Resolution* 38 (2): 236–69.

——. 1995. "Rationalist Explanations for War." *International Organization* 49 (3): 379–414.

——. 1997. "Signaling Foreign Policy Interests: Tying Hands versus Sinking Costs." *Journal of Conflict Resolution* 41 (1): 68–90.

Feaver, Peter D. 1992. *Guarding the Guardians: Civilian Control of Nuclear Weapons in the United States*. Ithaca, NY: Cornell University Press.

Feitelson, Eran, Abdelrahman Tamimi, and Gad Rosenthal. 2012. "Climate Change and Security in the Israeli–Palestinian Context." *Journal of Peace Research* 49 (1): 241–57.

Filson, Darren, and Suzanne Werner. 2002. "A Bargaining Model of War and Peace: Anticipating the Onset, Duration, and Outcome of War." *American Journal of Political Science* 46 (4): 819–38.

———. 2004. "Bargaining and Fighting: The Impact of Regime Type on War Onset, Duration, and Outcomes." *American Journal of Political Science* 48 (2): 296–313.

Findley, Michael G., and Joseph K. Young. 2007. "Fighting Fire with Fire? How (Not) to Neutralize an Insurgency." *Civil Wars* 9 (4): 378–401.

Fisk, Harvey E. 1924. *The Inter-ally Debts: An Analysis of War and Post-war Public Finance, 1914–1923*. New York: Bankers Trust Company.

Flandreau, Marc and Juan H. Flores. 2012. "The Peaceful Conspiracy: Bond Markets and International Relations during the Pax Britannica." *International Organization* 66 (2): 211–41.

Florea, Adrian. 2018. "Spatial Rivalry and Coups against Dictators." *Security Studies* 27 (1): 1–26.

Flores-Macias, Gustavo A., and Sarah E. Kreps. 2017. "Borrowing Support for War: The Effect of War Finance on Public Attitudes toward Conflict." *Journal of Conflict Resolution* 61 (5): 997–1020.

Fordham, Benjamin, and Paul Poast. 2016. "All Alliances Are Multilateral: Rethinking Alliance Formation in the International System." *Journal of Conflict Resolution* 60 (5): 840–65.

Forsberg, Erika. 2016. "Transnational Dimensions of Civil Wars: Clustering, Contagion, and Connectedness." In *What Do We Know about Civil Wars?* edited by T. David Mason and Sara McLaughlin Mitchell, 75–92. Lanham, MD: Rowman and Littlefield.

Fortna, Virginia Page. 2003. "Scraps of Paper? Agreements and the Durability of Peace." *International Organization* 57 (2): 337–72.

——. 2004a. *Peace Time: Cease-Fire Agreements and the Durability of Peace*. Princeton, NJ: Princeton University Press.

——. 2004b. "Interstate Peacekeeping: Causal Mechanisms and Empirical Effects." *World Politics* 56 (4): 481–519.

Fravel, M. Taylor. 2008. *Strong Borders, Secure Nation: Cooperation and Conflict in China's Territorial Disputes*. Princeton, NJ: Princeton University Press.

——. 2010. "Explaining Stability in the Senkaku (Diaoyu) Dispute." In *Getting the Triangle Straight: Managing China-Japan-US Relations,* edited by Gerald Curtis, Ryosei Kokubun, and Wang Jisi, 144–64. Washington, DC: Brookings.

Frazier, Derrick V. 2006. "Third Party Characteristics, Territory and the Mediation of Militarized Interstate Disputes." *Conflict Management and Peace Science* 23 (4): 267–84.

Frazier, Derrick V., and William J. Dixon. 2006. "Third-Party Intermediaries and Negotiated Settlements, 1946–2000." *International Interactions* 32 (4): 385–408.

Frederick, Bryan A. 2012. "*The Sources of Territorial Stability*." PhD dissertation, Washington, DC, Johns Hopkins University School of Advanced International Studies, OCLC: 814397500.

Frederick, Bryan A., Paul R. Hensel, and Christopher Macaulay. 2017. "The Issue Correlates of War Territorial Claims Data, 1816–2001." *Journal of Peace Research* 54 (1): 99–108.

Freedman, Lawrence. 1988. *The Evolution of Nuclear Strategy*. New York: St. Martins.

Friedman, Jeffrey A. 2011. "Manpower and Counterinsurgency: Empirical Foundations for Theory and Doctrine." *Security Studies* 20 (4): 556–91.

Fuhrmann, Matthew. 2018. "When Preventive War Threats Work for Nuclear Nonproliferation." *Washington Quarterly* 41 (3): 111–35.

Fuhrmann, Matthew, and Sarah Kreps. 2010. "Targeting Nuclear Programs in War and Peace: A Quantitative Empirical Analysis, 1941–2000." *Journal of Conflict Resolution* 54 (6): 831–59.

Fuhrmann, Matthew, Matthew Kroenig, and Todd S. Sechser. 2014. "The Case for Using Statistics to Study Nuclear Security." *H-Diplo/International Security Studies Forum* (2): 37–54.

Fuhrmann, Matthew, and Todd S. Sechser. 2014. "Signaling Alliance Commitments: Hand-Tying and Sunk Costs in Extended Nuclear Deterrence." *American Journal of Political Science* 58 (4): 919–35.

——. 2019. "Can the U.S. Protect Its Nuclear Weapons in Turkey?" *The Washington Post Monkey Cage*, October 18. https://www.washingtonpost.com/politics/2019/10/18/can-us-protect-its-nuclear-weapons-turkey/

Fuhrmann, Matthew, and Benjamin Tkach. 2015. "Almost Nuclear: Introducing the Nuclear Latency Dataset." *Conflict Management and Peace Science* 32 (4): 443–61.

Furlong, Kathryn, Nils Petter Gleditsch, and Håvard Hegre. 2006. "Geographic Opportunity and Neomalthusian Willingness: Boundaries, Shared Rivers, and Conflict." *International Interactions* 32 (1): 79–108.

Gaddis, John Lewis. 1987. *The Long Peace: Inquiries into the History of the Cold War*. New York: Oxford University Press.

——. 2005. *The Cold War: A New History*. New York: Penguin Books.

Gallagher, Maryann E., and Susan H. Allen. 2014. "Presidential Personality: Not Just a Nuisance." *Foreign Policy Analysis* 10 (1): 1–21.

Galtung, Johan. 1959. "Violence, Peace, and Peace Research." *Journal of Peace Research* 6 (3): 167–91.

——. 1996. *Peace by Peaceful Means: Peace and Conflict, Development and Civilization*. Beverly Hills, CA: Sage.

Galula, David. 1964. *Counterinsurgency Warfare: Theory and Practice*. New York: Praeger.

Gamble, John King Jr., and Dana D. Fischer. 1976. *The International Court of Justice: An Analysis of a Failure*. Lexington: Lexington Books.

Ganguly, Sumit, and Devin T. Hagerty. 2005. *Fearful Symmetry: India-Pakistan Crises in the Shadow of Nuclear Weapons*. New Delhi: Oxford University Press.

Ganguly, Sumit, and William Thompson. 2011. "Two-Level Games in Asian Rivalries." In *Asian Rivalries: Conflict, Escalation, and Limitations on Two-Level Games*, edited by Sumit Ganguly and William Thompson, 195–211. Stanford, CA: Stanford University Press.

Gardner, Kyle. 2020. "The Elusive Watershed of Ladakh: Explaining India and China's Missing Border." Published online at: https://www.orfonline.org/expert-speak/elusive-watersheds-ladakh-explaining-india-china-missing-border/.

Garnham, David. 1986. "War-Proneness, War-Weariness, and Regime Type: 1816–1980." *Journal of Peace Research* 23 (3): 279–89.

Garrett-Peltier, Heidi. 2017. "Job Opportunity Cost of War." Report from the Cost of War Project, Watson Institute.

Gartner, Scott Sigmund, and Jacob Bercovitch. 2006. "Overcoming Obstacles to Peace: The Contribution of Short-Lived Conflict Settlements." *International Studies Quarterly* 50 (4): 819–40.

Gartner, Scott Sigmund, and Randolph M. Siverson. 1996. "War Expansion and War Outcome." *Journal of Conflict Resolution* 40 (1): 4–15.

Gartzke, Erik. 2007. "The Capitalist Peace." *American Journal of Political Science* 51 (1): 166–91.

——. 2012. "Could Climate Change Precipitate Peace?" *Journal of Peace Research* 49 (1): 177–92.

——. 2013. "The Myth of Cyberwar: Bringing War in Cyberspace Back Down to Earth." *International Security* 38 (2): 41–73.

Gartzke, Erik, and Dong-Joon Jo. 2009. "Bargaining, Nuclear Proliferation, and Interstate Disputes." *Journal of Conflict Resolution* 53 (2): 209–33.

Gartzke, Erik, Jeffrey M. Kaplow, and Rupal N. Mehta. 2014. "The Determinants of Nuclear Force Structure." *Journal of Conflict Resolution* 58 (3): 481–508.

Gartzke, Erik, and Matthew Kroenig. 2009. "A Strategic Approach to Nuclear Proliferation." *Journal of Conflict Resolution* 53 (2): 151–60.

Gartzke, Erik, Quan Li, and Charles Boehmer. 2001. "Investing in the Peace: Economic Interdependence and International Conflict." *International Organization* 55 (2): 391–438.

Gartzke, Erik, and Jon R. Lindsay. 2015. "Weaving Tangled Webs: Offense, Defense, and Deception in Cyberspace." *Security Studies* 24 (2): 316–48.

Gartzke, Erik, and Paul Poast. 2018. "Empirically Assessing the Bargaining Theory of War: Potential and Challenges." In *The Oxford Encyclopedia of Empirical*

International Relations Theory, edited by William R. Thompson, Vol. 1, 755–68. New York: Oxford University Press.
Gartzke, Erik, and Alex Weisiger. 2013. "Permanent Friends? Dynamic Difference and the Democratic Peace." *International Studies Quarterly* 57 (1): 171–85.
Gat, Azar. 2008. *War in Human Civilization*. Oxford: Oxford University Press.
——. 2013. "Is War Declining—and Why?" *Journal of Peace Research* 50 (2): 149–57.
Gates, Scott, Torbjorn L. Knutsen, and Jonathon W. Moses. 1996. "Democracy and Peace: A More Skeptical View." *Journal of Peace Research* 33 (1): 1–10.
Geller, Daniel S. 1990. "Nuclear Weapons, Deterrence, and Crisis Escalation." *Journal of Conflict Resolution* 34 (2): 291–310.
——. 2000. "Material Capabilities: Power and International Conflict." In *What Do We Know about War?* edited by John A. Vasquez, 259–77. Lanham, MD: Rowman and Littlefield.
——. 2018. "Nuclear Weapons and International Conflict: Theories and Empirical Evidence." In *The Oxford Encyclopedia of Empirical International Relations Theory*, edited by William R. Thompson, Vol. 2, 667–92. New York: Oxford University Press.
Geller, Daniel S., and J. David Singer. 1998. *Nations at War: A Scientific Study of International Conflict*. Cambridge: Cambridge University Press.
Geller, Daniel, S. and Konstantinos Travlos. 2019. "Integrating Realist and Neoliberal Theories of War." *Peace Economics, Peace Science, and Public Policy* 25 (2): 1–29.
Gelpi, Christopher, Peter D. Feaver, and Jason Reifler. 2005. "Success Matters: Casualty Sensitivity and the War in Iraq." *International Security* 30 (3): 7–46.
Gent, Stephen E., and Megan Shannon. 2010. "The Effectiveness of International Arbitration and Adjudication: Getting into a Bind." *Journal of Politics* 72 (2): 366–80.
——. 2011a. "Bias and the Effectiveness of Third Party Conflict Management Mechanisms." *Conflict Management and Peace Science* 28 (2): 124–44.
——. 2011b. "Decision Control and the Pursuit of Conflict Management: Choosing the Ties That Bind." *Journal of Conflict Resolution* 55 (5): 710–34.
George, Alexander L., and Richard Smoke. 1974. *Deterrence and American Foreign Policy: Theory and Practice*. New York: Columbia University Press.
Gerson, Michael S. 2010. "No First Use: The Next Step for U.S. Nuclear Policy." *International Security* 35 (2): 7–47.
Getmansky, Anna, Guy Grossman, and Austin L. Wright. 2019. "Border Walls and Smuggling Spillovers." *Quarterly Journal of Political Science* 14 (3): 329–47.
Geys, Benny. 2010. "Wars, Presidents, and Popularity: The Political Cost(s) of War Re-examined." *Public Opinion Quarterly* 74 (2): 357–74.
Geys, Benny, and Kai A. Konrad. 2016. "Patriotism and Taxation." Working Paper No. 2016–11 Max Planck Institute for Tax Law and Public Finance, November.
Ghosn, Faten, Glenn Palmer, and Stuart A. Bremer. 2004. "MID3 Data Set, 1993–2001: Procedures, Coding Rules, and Description." *Conflict Management and Peace Science* 15 (1): 75–97.
Gibbons, Rebecca Davis, and Keir Lieber. 2019. "How Durable Is the Nuclear Weapons Taboo?" *Journal of Strategic Studies* 42 (1): 29–54.

Gibler, Douglas M. 1996. "Alliances That Never Balance: The Territorial Settlement Treaty." *Conflict Management and Peace Science* 15 (Spring): 75–97.

——. 2000. "Alliances: Why Some Cause War and Why Others Cause Peace." In *What Do We Know about War?* edited by John A. Vasquez, 145–64. Lanham, MD: Rowman and Littlefield.

——. 2007. "Bordering on Peace: Democracy, Territorial Issues, and Conflict." *International Studies Quarterly* 51 (3): 509–32.

——. 2008. "The Costs of Reneging: Reputation and Alliance Formation." *Journal of Conflict Resolution* 52 (3): 426–54.

——. 2009. *International Military Alliances, 1648–2008*. Washington, DC: CQ Press.

——. 2010. "Outside-in: The Effects of Territorial Threat on State Centralization." *Journal of Conflict Resolution* 54 (4): 519–42.

——. 2012. *The Territorial Peace: Borders, State Development, and International Conflict*. Cambridge: Cambridge University Press.

——. 2014. "Contiguous States, Stable Borders, and the Peace between Democracies." *International Studies Quarterly* 58 (1): 126–29.

——. 2017a. "State Development, Parity, and International Conflict." *American Political Science Review* 111 (1): 21–38.

——. 2017b. "What They Fight For: Specific Territorial Issues in Militarized Interstate Disputes, 1816–2001." *Conflict Management and Peace Science* 34 (2): 194–211.

——. 2018. "The Territorial Peace: A Research Program." In *The Oxford Encyclopedia of Empirical International Relations Theory*, edited by William R. Thompson, Vol. 4, 12–34. New York: Oxford University Press.

Gibler, Douglas M., and Alex Braithwaite. 2013. "Dangerous Neighbours, Regional Territorial Conflict and the Democratic Peace." *British Journal of Political Science* 43 (4): 877–87.

Gibler, Douglas M., and Marc L. Hutchison. 2013. "Territorial Issues, Audience Costs, and the Democratic Peace: The Importance of Issue Salience." *Journal of Politics* 75 (4): 879–93.

Gibler, Douglas M., Marc L. Hutchison, and Steven V. Miller. 2012. "Individual Identity Attachments and International Conflict: The Importance of Territorial Threat." *Comparative Political Studies* 45 (12): 1655–83.

Gibler, Douglas M., and Erin K. Little. 2017. "Heterogeneity in the Militarized Interstate Disputes (MIDs), 1816–2001: What Fatal MIDs Cannot Fix." *Political Science Research and Methods* 5 (1): 189–99.

Gibler, Douglas M., Erin K. Little, and Steven V. Miller. 2020. "The Importance of Correct Measurement: A Response to Palmer, et al." *International Studies Quarterly* 64 (2): 476–79.

Gibler, Douglas M., and Steven V. Miller. 2013. "Quick Victories? Territory, Democracies, and Their Disputes." *Journal of Conflict Resolution* 57 (2): 258–84.

———. 2014. "External Territorial Threat, State Capacity, and Civil War." *Journal of Peace Research* 51 (5): 634–46.

Gibler, Douglas M., Steven V. Miller, and Erin K. Little. 2016. "An Analysis of the Militarized Interstate Dispute (MID) Dataset, 1816–2001." *International Studies Quarterly* 60 (4): 719–30.

Gibler, Douglas M., and Andrew P. Owsiak. 2018. "Democracy and the Settlement of International Borders, 1919–2001." *Journal of Conflict Resolution* 62 (9): 1847–75.

Gibler, Douglas M., and Kirk A. Randazzo. 2011. "Testing the Effects of Independent Judiciaries on the Likelihood of Democratic Backsliding." *American Journal of Political Science*. 55 (3): 696–709.

Gibler, Douglas M., Toby J. Rider, and Mark L. Hutchison. 2005. "Taking Arms against a Sea of Troubles: Conventional Arms Races during Periods of Rivalry." *Journal of Peace Research* 42 (2): 131–47.

Gibler, Douglas M., and Meredith Reid Sarkees. 2004. "Measuring Alliances: The Correlates of War Formal Interstate Alliance Dataset, 1816–2000." *Journal of Peace Research* 41 (2): 211–22.

Gibler, Douglas M., and Jaroslav Tir. 2010. "Settled Borders and Regime Type: Democratic Transitions As Consequences of Peaceful Territorial Transfers." *American Journal of Political Science* 54 (4): 951–68.

——. 2014. "Territorial Peace and Democratic Clustering." *Journal of Politics* 76 (1): 27–40.

Gibler, Douglas M., and John A. Vasquez. 1998. "Uncovering the Dangerous Alliances, 1495–1980." *International Studies Quarterly* 42 (December): 785–807.

Gibler, Douglas M., and Scott Wolford. 2006. "Alliances, then Democracy: An Examination of the Relationship between Regime Type and Alliance Formation." *Journal of Conflict Resolution* 50 (1): 129–53.

Gillespie, Colin S. 2015. "Fitting Heavy Tailed Distributions: The Powerlaw Package." *Journal of Statistical Software* 64 (2). http://www.jstatsoft.org/v64/i02/

Gilman, Nils. 2007. *Mandarins of the Future: Modernization Theory in Cold War America*. Baltimore, MD: Johns Hopkins University Press.

Gilpin, Robert. 1988. "The Theory of Hegemonic War." *Journal of Interdisciplinary History* 18 (4): 591–613.

Giordano, Mark F., Meredith A. Giordano, and Aaron T. Wolf. 2005. "International Resource Conflict and Mitigation." *Journal of Peace Research* 42 (1): 47–65.

Gizelis, Theodora-Ismene, Han Dorussen, and Marina Petrova. 2018. "Research Findings on the Evolution of Peacekeeping." In *The Oxford Encyclopedia of Empirical International Relations Theory*, edited by William R. Thompson, Vol. 3, 348–63. New York: Oxford University Press.

Glaser, Charles. 1990. *Analyzing Strategic Nuclear Policy*. Princeton, NJ: Princeton University Press.

——. 2010. Rational Theory of International Politics: *The Logic of Competition and Cooperation*. Princeton, NJ: Princeton University Press.

Gleditsch, Kristian Skrede. 2002a. *All International Politics Is Local*. Ann Arbor: University of Michigan Press.

——. 2002b. "Expanded Trade and GDP Data." *Journal of Conflict Resolution* 46 (5): 712–24.

Gleditsch, Kristian Skrede, and Michael D. Ward. 1999. "A Revised List of Independent States since the Congress of Vienna." *International Interactions* 25 (4): 393–413.

Gleditsch, Nils Petter. 1998. "Armed Conflict and the Environment: A Critique of the Literature." *Journal of Peace Research* 35 (3): 381–400.

Gleditsch, Nils Petter, Kathryn Furlong, Håvard Hegre, Bethany Lacina, and Taylor Owen. 2006. "Conflicts over Shared Rivers: Resource Scarcity or Fuzzy Boundaries?" *Political Geography* 25 (4): 361–82.

Gleditsch, Nils Petter, and Håvard Hegre. 1997. "Peace and Democracy: Three Levels of Analysis." *Journal of Conflict Resolution* 41 (2): 283–310.

Gleditsch, Nils Petter, Steven Pinker, Bradley A. Thayer, Jack S. Levy, and William R. Thompson. 2013. "The Forum: The Decline of War." *International Studies Review* 15 (3): 396–419.

Gleditsch, Nils Petter, Peter Wallensteen, Mikael Eriksson, Margareta Sollenberg, and Håvard Strand. 2002. "Armed Conflict 1946–2001: A New Dataset." *Journal of Peace Research* 39 (5): 625–37.

Gleick, Peter H. 1993. "Water and Conflict: Fresh Water Resources and International Security." *International Security* 18 (1): 79–112.

Gochman, Charles S., and Zeev Maoz. 1984. "Militarized Interstate Disputes, 1816–1976: Procedures, Patterns, and Insights." *Journal of Conflict Resolution* 28 (4): 585–616.

Goddard, Stacie E. 2009. "When Right Makes Might: How Prussia Overturned the European Balance of Power." *International Security* 33 (3): 110–42.

Goddard, Stacie, E. and Daniel H. Nexon. 2016. "The Dynamics of Global Power Politics: A Framework for Analysis." *Journal of Global Security Studies* 1 (1): 4–18.

Goemans, Hein E. 2000. *War and Punishment: The Causes of War Termination and the First World War*. Princeton, NJ: Princeton University Press.

——. 2006. "Bounded Communities: Territoriality, Territorial Attachment, and Conflict." In *Territoriality and Conflict in an Era of Globalization*, edited by Miles Kahler and Barbara F. Walter, 25–61. New York: Cambridge University Press.

——. 2008. "Which Way Out? The Manner and Consequences of Losing Office." *Journal of Conflict Resolution* 53 (6): 771–94.

Goemans, Hein E., and Mark Fey. 2009. "Risky but Rational: War As an Institutionally Induced Gamble." *Journal of Politics* 71 (1): 35–54.

Goemans, Hein E., Kristian Skrede Gleditsch, and Giacomo Chiozza. 2009. "Introducing Archigos: A Data Set of Political Leaders." *Journal of Peace Research* 46 (2): 269–83. Available at http://mail.rochester.edu/~hgoemans/data.htm.

Goemans, Hein E., and Kenneth A. Schultz. 2017. "The Politics of Territorial Disputes: A Geo-spatial Approach applied to Africa." *International Organization* 71 (1): 31–64.

Goemans, Hein E., and William Spaniel. 2016. "Multimethod Research: A Case for Formal Theory." *Security Studies* 25 (1): 25–33.

Goertz, Gary. 2012. *Social Science Concepts: A User's Guide*. Princeton, NJ: Princeton University Press.

——. 2017. *Multimethod Research, Causal Mechanisms, and Case Studies: An Integrated Approach*. Princeton, NJ: Princeton University Press.

Goertz, Gary, and Paul F. Diehl. 1992a. "The Empirical Importance of Enduring Rivalries." *International Interactions* 18 (2): 151–63.

_____. 1992b. *Territorial Changes and International Conflict*. London: Routledge.

Goertz, Gary, Paul F. Diehl, and Alexandru Balas. 2016. *The Puzzle of Peace: The Evolution of Peace in the International System*. Oxford: Oxford University Press.

Goertz, Gary, and James Mahoney. 2012. *A Tale of Two Cultures: Qualitative and Quantitative Research in the Social Sciences*. Princeton, NJ: Princeton University Press.

Goettlich, Kerry. 2019. "The Rise of Linear Borders in World Politics." *European Journal of International Relations* 25 (1): 203–28.

Goh, Evelyn. 2013. *The Struggle for Order: Hegemony, Hierarchy, and Transition in Post–Cold War East Asia.* Oxford: Oxford University Press.

Gohdes, Anita R. 2015. "Pulling the Plug: Network Disruptions and Violence in Civil Conflict." *Journal of Peace Research* 52 (3): 352–67.

_____. 2020. "Repression Technology: Internet Accessibility and State Violence." *American Journal of Political Science* 64 (3): 488–503.

Gohdes, Anita, and Megan Price. 2012. "First Things First: Assessing Data Quality before Model Quality." *Journal of Conflict Resolution* 57 (6): 1090–108.

Goldsmith, Benjamin E. 2018. "Peace, War, Theory, and Evidence in East Asia." In *The Oxford Encyclopedia of Empirical International Relations Theory*, edited by William R. Thompson, Vol. 3, 17–33. New York: Oxford University Press.

Goldstein, Joshua S. 1988. *Long Cycles: Prosperity and War in the Modern Age*. New Haven, CT: Yale University Press.

——. 2011. *Winning the War on War: The Decline of Armed Conflict Worldwide*. New York: Dutton.

Goldstein, Joshua S., and Jon C. Pevehouse. 1997. "Reciprocity, Bullying, and International Cooperation: Time-Series Analysis of the Bosnia Conflict." *American Political Science Review* 91 (3): 515–29.

Goldstein, Joshua S., Jon C. Pevehouse, Deborah J. Gerner, and Shibley Telhami. 2001. "Reciprocity, Triangularity, and Cooperation in the Middle East, 1979–97." *Journal of Conflict Resolution* 45 (5): 594–620.

Gourevitch, Peter. 1978. "The Second Image Reversed: The International Sources of Domestic Politics." *International Organization* 32 (Autumn): 881–912.

Grauer, Ryan, and Stephen L. Quackenbush. 2020. "Initiative and Military Effectiveness: Evidence from the Yom Kippur War." *Journal of Global Security Studies*, forthcoming.

Gray, Clark and Erika Wise. 2016. "Country-Specific Effects of Climate Variability on Human Migration." *Climatic Change* 135 (3–4): 555–68.

Green, Donald, Soo Yeon Kim, and David Yoon. 2001. "Dirty Pool." *International Organization* 55 (2): 441–68.

Greenberg, Andy. 2019. *Sandworm: A New Era of Cyberwar and the Hunt for the Kremlin's Most Dangerous Hackers*. New York: Doubleday.

Greenhill, Kelly M., and Paul Staniland. 2007. "Ten Ways to Lose at Counterinsurgency." *Civil Wars* 9 (4): 402–19.

Greif, Avner, Paul Milgrom, and Barry R. Weingast. 1994. "Coordination, Commitment, and Enforcement: The Case of the Merchant Guild." *Journal of Political Economy* 102 (4): 745–76.

Greig, Michael. 2001. "Moments of Opportunity: Recognizing Conditions for Ripeness for International Mediation between Enduring Rivals." *Journal of Conflict Resolution* 45 (6): 691–718.

Greig, Michael J., Andrew P. Owsiak, and Paul F. Diehl. 2019. *International Conflict Management*. Medford, MA: Polity.

Grey, Jeffrey. 1988. *The Commonwealth Armies and the Korean War*. Manchester: Manchester University Press.

Grofman, Bernard, Evald Mikkel, and Rein Taagepera. 1999. "Electoral Systems Change in Estonia, 1989-1993." *Journal of Baltic Studies* 30 (3): 227–49.

Gross, Michael L., Daphna Canetti, and Dana R. Vashdi. 2017. "Cyberterrorism: Its Effects on Psychological Well-Being, Public Confidence and Political Attitudes." *Journal of Cybersecurity* 3 (1): 49–58.

Guetzkow, Harold. 1968. "Some Correspondences between Simulation and Realities in International Relations." In *New Approaches to International Relations*, edited by Morton A. Kapan, 202–69. New York: St. Martin's.

Guidolin, Massimo, and Eliana La Ferrara. 2010. "The Economic Effects of Violent Conflict: Evidence from Asset Market Reactions." *Journal of Peace Research* 47 (6): 671–84.

Guo, Rongxing. 2007. *Territorial Disputes and Resource Management: A Global Handbook*. Hauppauge, NY: Nova Science Publishers.

Gurr, Ted Robert. 1988. "War, Revolution, and the Growth of the Coercive State." *Comparative Political Studies* 21(1): 45–65.

Haas, Michael L. 2018. "Population Aging and International Conflict." In *The Oxford Encyclopedia of Empirical International Relations Theory*, edited by William R. Thompson, Vol. 3, 91–105. New York: Oxford University Press.

Hafner-Burton, Emilie M., Miles Kahler, and Alexander H. Montgomery. 2009. "Network Analysis for International Relations." *International Organization* 63 (3): 559–92.

Haftendorn, Helga. 2000. "Water and International Conflict." *Third World Quarterly* 21 (1): 51–68.

Hagerty, Devin T. 1995. "Nuclear Deterrence in South Asia: The 1990 Indo-Pakistani Crisis." *International Security* 20 (3): 79–114.

Halvard, Buhaug, and Nils Petter Gleditsch. 2006. "The Death of Distance? The Globalization of Armed Conflict." In *Territoriality and Conflict in the Era of Globalization*, edited by Miles Kahler and Barbara Walker, 187–216. Cambridge: Cambridge University Press.

Hamid, S. Shahid. 1986. *Disastrous Twilight: A Personal Record of the Partition of India*. London: Leo Cooper.

Hamilton, Earl J. 1977. "The Role of War in Modern Inflation." *Journal of Economic History* 37 (1): 13–19.

Hamner, Jesse H., and Aaron T. Wolf. 1998. "Patterns in International Water Resource Treaties: The Transboundary Freshwater Dispute Database." *Colorado Journal of International Environmental Law and Policy* 9: 157–77.

Hansen, Holley E., Sara McLaughlin Mitchell, and Stephen C. Nemeth. 2008. "IO Mediation of Interstate Conflicts: Moving beyond the Global vs. Regional Dichotomy." *Journal of Conflict Resolution* 52 (2): 295–325.

Hardin, Garrett. 1968. "The Tragedy of the Commons." *Science* 162: 1243–48.

Harrington de Santana, Anne. 2009. "Nuclear Weapons As the Currency of Power." *Nonproliferation Review* 16 (3): 325–45.

Hartcup, Guy. 1988. *The War of Invention: Scientific Developments,* 1914–1918. London: Brassey's.

——. 2000. *The Effect of Science on the Second World War*. New York: St. Martin's.

Hassner, Ron E. 2003. "To Halve and to Hold: Conflict over Sacred Space and the Problem of Indivisibility." *Security Studies* 12 (4): 1–33.

Hassner, Ron E., and Jason Wittenberg. 2015. "Barriers to Entry: Who Builds Fortified Boundaries and Why?" *International Security* 40 (1): 157–90.

Hatemi, Peter K., and Rose McDermott, eds. 2011. *Man Is by Nature a Political Animal: Evolution, Biology, and Politics*. Chicago: University of Chicago Press.

Hathaway, Oona A., and Scott J. Shapiro. 2017. *The Internationalists*. New York: Simon and Schuster.

Haynes, Kyle. 2016. "Diversity and Diversion: How Ethnic Composition Affects Diversionary Conflict." *International Studies Quarterly* 60 (2): 258–71.

Hegre, Håvard. 2008. "Gravitating toward War: Preponderance May Pacify, but Power Kills." *Journal of Conflict Resolution* 52 (4): 566–89.

Hegre, Håvard, Michael Bernhard, and Jan Teorell. 2020. "Civil Society and the Democratic Peace." *Journal of Conflict Resolution* 64 (1): 32–62.

Hegre, Håvard, John R. Oneal, and Bruce Russett. 2010. "Trade Does Promote Peace: New Simultaneous Estimates of the Reciprocal Effects of Trade and Conflict." *Journal of Peace Research* 47 (6): 763–74.

Henderson, Errol A., and Resat Bayer. 2013. "Wallets, Ballots, or Bullets: Does Wealth, Democracy, or Military Capabilities Determine War Outcomes?" *International Studies Quarterly* 57 (2): 303–17.

Hendrix, Cullen S., and Idean Salehyan. 2012. "Climate Change, Rainfall, and Social Conflict in Africa." *Journal of Peace Research* 49 (1): 35–50.

Henehan, Marie. T., and John A. Vasquez. 2006. "The Changing Probability of War, 1816–1992: Identifying Peaceful Eras." In *The Waning of War,* edited by Raimo Väyrynen, 280–99. London: Routledge.

Henrich, Joseph, Robert Boyd, Samuel Bowles, Colin Camerer, Ernst Fehr, Herbert Gintis, Richard McElreath, Michael Alvard, Abigail Barr, Jean Ensminger, Natalie Smith Henrich, Kim Hill, Francisco Gil-White, and Michael Gurven. 2005. "'Economic Man' in Cross-Cultural Perspective: Behavioral Experiments in 15 Small-Scale Societies." *Behavioral and Brain Sciences* 28 (6): 795–815.

Hensel, Paul R. 1994. "One Thing Leads to Another: Recurrent Militarized Disputes in Latin America, 1816–1986." *Journal of Peace Research* 31 (3): 281–97.

——. 1996a. "Charting a Course to Conflict: Territorial Issues and Interstate Conflict." *Conflict Management and Peace Science* 15 (1): 43–73.

——. 1996b. "*The Evolution of Interstate Rivalry.*" PhD dissertation, University of Illinois, Urbana-Champaign.

——. 1998. "Interstate Rivalry and the Study of Militarized Conflict." In *Conflict in World Politics: Advances in the Study of Crisis, War and Peace*, edited by Frank Harvey and Ben D. Mor, 162–204. New York: St. Martin's Press.

——. 1999. "Charting a Course to Conflict: Territorial Issues and Interstate Conflict." In *A Road Map to War*, edited by Paul F. Diehl, 115–46. Nashville, TN: Vanderbilt University Press.

——. 2000. "Territory: Theory and Evidence on Geography and Conflict." In *What Do We Know about War?* edited by John A. Vasquez, 57–84. Lanham, MD: Rowman and Littlefield.

——. 2001. "Contentious Issues and World Politics: The Management of Territorial Claims in The Americas." *International Studies Quarterly* 45 (1): 81–109.

——. 2002. "The More Things Change … : Recognizing and Responding to Trends in Armed Conflict." *Conflict Management and Peace Science* 19 (1): 27–52.

——. 2006. "Territorial Claims and Armed Conflict between Neighbors." Paper presented at the Lineae Terrarum International Borders Conference, University of Texas–El Paso, New Mexico State University, Colegio de la Frontera Norte, and Universidad Autonoma de Ciudad Juarez, March.

——. 2012. "Territory: Geography, Contentious Issues, and World Politics." In *What Do We Know about War?* edited by John A. Vasquez, 3–26. Lanham, MD: Rowman and Littlefield.

——. 2018. "Territory, Contentious Issues, and International Relations." In *The Oxford Encyclopedia of Empirical International Relations Theory*, edited by William R. Thompson, Vol. 4, 49–67. New York: Oxford University Press.

Hensel, Paul R., Michael Allison, and Ahmed Khanani. 2009. "Territorial Integrity Treaties and Armed Conflict over Territory." *Conflict Management and Peace Science* 26 (2): 120–43.

Hensel, Paul R., and Paul F. Diehl. 1994. "It Takes Two to Tango: Non-militarized Response in Interstate Disputes." *Journal of Conflict Resolution* 38 (3): 479–506.

Hensel, Paul R., Gary Goerz, and Paul F. Diehl. 2000. "The Democratic Peace and Rivalries." *Journal of Politics* 62 (4): 1173–88.

Hensel, Paul R., and Christopher Macaulay. 2016. "Territorial Claims and Identity Claims under the Territorial Integrity Norm." Paper presented at the Annual Meeting of the International Studies Association, Atlanta, March 18.

Hensel, Paul R., and Sara McLaughlin Mitchell. 2005. "Issue Indivisibility and Territorial Claims." *GeoJournal* 64 (4): 275–85.

——. 2007a. "International Institutions and Compliance with Agreements." *American Journal of Political Science* 51 (4): 721–37.

——. 2007b. "The Issue Correlates of War (ICOW) Project Issue Data Set: Territorial Claims Data." https://doi.org/10.7910/DVN/E6PSGZ, Harvard Dataverse, V2, UNF:3: AT9Kx7uv+vzYYTZA9Fo35Q==[fileUNF].

——. 2017. "From Territorial Claims to Identity Claims: The Issue Correlates of War (ICOW) Project." *Conflict Management and Peace Science* 34 (2): 126–40.

Hensel, Paul R., Sara McLaughlin Mitchell, and Thomas E. Sowers II. 2006. "Conflict Management of Riparian Disputes: A Regional Comparison of Dispute Resolution." *Political Geography* 25 (4): 383–411.

Hensel, Paul R., Sara McLaughlin Mitchell, Thomas E. Sowers II, and Clayton L. Thyne. 2008. "Bones of Contention: Comparing Territorial, Maritime, and River Issues." *Journal of Conflict Resolution* 52 (1): 117–43.

Herb, Guntram. 1997. *Under the Map of Germany: Nationalism and Propaganda 1918–1945*. London: Routledge.

Herz, John. 1950. "Idealist Internationalism and the Security Dilemma." *World Politics* 2 (2): 157–80.

Hidalgo, F. D., S. Naidu, S. Nichter, and N. Richardson. 2010. "Economic Determinants of Land Invasions." *Review of Economics and Statistics* 92 (3): 505–23.

Hirschman, Albert O. 1992. *Views of Market Society and Other Recent Essays*. Cambridge, MA: Harvard University Press.

Holsti, Kalevi J. 1991. *Peace and War: Armed Conflicts and International Order, 1648–1989*. New York: Cambridge University Press.

Holsti, Ole R., Robert C. North, and Richard A. Brody. 1968. "Perception and Action in the 1914 Crisis." In *Quantitative International Politics*, edited by J. David Singer, 123–58. New York: Free Press.

Homer-Dixon, Thomas. 1994. "Environmental Scarcities and Violent Conflict: Evidence from Cases." *International Security* 19 (1): 5–40.

Hong, Mi Hwa, and Nam Kyu Kim. 2019. "How Do External Territorial Threats Affect Mass Killing?" *Journal of Peace Research* 56 (4): 529–44.

Hood, Christopher. 2003. "The Tax State in the Information Age." In *The Nation State in Question*, edited by T. V. Paul, G. John Ikenberry, and John A. Hall, 224–25. Princeton, NJ: Princeton University Press.

Hopf, Ted. 1991. "Polarity, the Offense-Defense Balance, and War." *American Political Science Review* 85 (2): 475–93.

Horowitz, Michael C. 2009. "The Spread of Nuclear Weapons and International Conflict: Does Experience Matter?" *Journal of Conflict Resolution* 53 (2): 234–57.

Horowitz, Michael C., and Matthew Fuhrmann. 2018. "Studying Leaders and Military Conflict: Conceptual Framework and Research Agenda." *Journal of Conflict Resolution* 62 (10): 2072–86.

Horowitz, Michael C., Rose McDermott, and Allan C. Stam. 2005. "Leader Age, Regime Type, and Violent International Relations." *Journal of Conflict Resolution* 49 (5): 661–85.

Horowitz, Michael C., Paul Poast, and Allan C. Stam. 2017. "Domestic Signaling of Commitment Credibility: Military Recruitment and Alliance Formation." *Journal of Conflict Resolution* 61 (8): 1682–710.

Horowitz, Michael C., Philip B. K. Potter, Todd S. Sechser, and Allan C. Stam III. 2018. "Sizing Up the Adversary: Leader Attributes, Credibility, and Reciprocation in International Conflict." *Journal of Conflict Resolution* 62 (10): 2180–204.

Horowitz, Michael C., Erin M. Simpson, and Allan C. Stam III. 2011. "Domestic Institutions and Wartime Casualties." *International Studies Quarterly* 55 (4): 909–36.

Horowitz, Michael C., and Allan C. Stam III. 2014. "How Prior Military Experience Influences the Future Militarized Behavior of Leaders." *International Organization* 68 (3): 527–59.

Horowitz, Michael C., Allan C. Stam III, and Cali M. Ellis. 2015. *Why Leaders Fight*. Cambridge: Cambridge University Press.

Houweling, Henk, and Jan G. Siccama. 1988. "Power Transitions As a Cause of War." *Journal of Conflict Resolution* 32 (1): 87–102.

Howard, Michael. 1976. *War in European History*. Oxford: Oxford University Press.

——. 1981. *The Franco-Prussian War: The German Invasion of France, 1870–1871*. London: Methuen.

Huntington, Samuel. 1958. "Arms Races Prerequisites and Results." *Public Policy* 8 (1): 41–86.

Hutchison, Marc L. 2011a. "Territorial Threat and the Decline of Political Trust in Africa: A Multilevel Analysis." *Polity* 43 (4): 432–61.

——. 2011b. "Territorial Threats, Mobilization, and Political Participation in Africa." *Conflict Management and Peace Science* 28 (3): 183–208.

Hutchison, Marc L., and Douglas M. Gibler. 2007. "Political Tolerance and Territorial Threat: A Cross-National Study." *Journal of Politics* 69 (1): 128–42.

Hutchison, Marc L., and Daniel Starr. 2018. "The Territorial Peace: Theory, Evidence, and Implications." In *The Oxford Encyclopedia of Empirical International Relations Theory*, edited by William R. Thompson, Vol. 4, 34–49. New York: Oxford University Press.

Huth, Paul, D. Scott Bennett, and Christopher C. Gelpi. 1992. "System Uncertainty, Risk Propensity, and International Conflict among the Great Powers." *Journal of Conflict Resolution* 36 (3): 478–517.

Huth, Paul K. 1988a. "Extended Deterrence and the Outbreak of War." *American Political Science Review* 82 (2): 423–43.

——. 1988b. *Extended Deterrence and the Prevention of War*. New Haven, CT: Yale University Press.

——. 1990. "The Extended Deterrent Value of Nuclear Weapons." *Journal of Conflict Resolution* 34 (2): 270–90.

——. 1996. *Standing Your Ground*. Ann Arbor: University of Michigan Press.

——. 2000. "Territory: Why Are Territorial Disputes between States a Central Cause of International Conflict?" In *What Do We Know about War?* edited by John Vasquez, 85–110. Lanham, MD: Rowman and Littlefield.

Huth, Paul K., and Todd Allee. 2002. *The Democratic Peace and Territorial Conflict in the Twentieth Century*. Cambridge: Cambridge University Press.

Huth, Paul K., Sara E. Croco, and Ben J. Appel. 2011. "Does International Law Promote the Peaceful Settlement of International Disputes? Evidence from the Study of Territorial Conflicts since 1945." *American Political Science Review* 105 (2): 415–36.

——. 2013. "Bringing Law to the Table: Legal Claims, Focal Points, and the Settlement of Territorial Disputes since 1945." *American Journal of Political Science* 57 (1): 90–103.

Huth, Paul K., Christopher Gelpi, and D. Scott Bennett. 1993. "The Escalation of Great Power Militarized Disputes: Testing Rational Deterrence Theory and Structural Realism." *American Political Science Review* 87 (3): 609–23.

Huth, Paul K., and Bruce M. Russett. 1984. "What Makes Deterrence Work? Cases from 1900 to 1980." *World Politics* 36 (4): 496–526.

Iakhnis, Evgeniia, and Patrick James. Forthcoming. "Near Crises in World Politics: A New Data Set and Analysis of Escalation Processes." *Conflict Management and Peace Science* https://doi.org/10.1177/0738894219855610

Ikenberry, G. John. 2001. *After Victory: Institutions, Strategic Restraint, and the Rebuilding of Order after Major Wars*. Princeton, NJ: Princeton University Press.

Ikenberry, G. John, and Daniel H. Nexon. 2019. "Hegemony Studies 3.0: The Dynamics of Hegemonic Orders." *Security Studies* 28 (3): 395–421.

Innes, Robert D. 1990. "Limited Liability and Incentive Contracting with Ex-ante Action Choices." *Journal of Economic Theory* 52 (1): 45–67.

IPCC. https://www.ipcc.ch/site/assets/uploads/2018/03/WGII_TAR_full_report-2.pdf

Iyigun, Murat, Nathan Nunn, and Nancy Qian. 2017. *Winter Is Coming: The Long-Run Effects of Climate Change on Conflict, 1400–1900*. Cambridge, MA: National Bureau of Economic Research. http://www.nber.org/papers/w23033.pdf.

James, Nicholas A., and David S. Matteson. 2014. "Ecp: An R Package for Nonparametric Multiple Change Point Analysis of Multivariate Data." *Journal of Statistical Software* 62 (7): 1–25.

James, Patrick. 2019. "What Do We Know about Crisis Escalation and War? A Visual Assessment of the International Crisis Behavior Project." *Conflict Management and Peace Science* 36 (1): 3–19.

James, Patrick, Johann Park, and Seung-Whan Choi. 2006. "Democracy and Conflict Management: Territorial Claims in the Western Hemisphere Revisited." *International Studies Quarterly* 50 (4): 803–17.

James, Patrick, Eric Solberg, and Murray Wolfson. 1999. "An Identified Systemic Model of the Democracy-Peace Nexus." *Defense and Peace Economics* 10 (1): 1–37.

Jenke, Libby, and Christopher Gelpi. 2017. "Theme and Variations: Historical Contingencies in the Causal Model of Interstate Conflict." *Journal of Conflict Resolution* 61 (10): 2262–84.

Jensen, Benjamin, and David Banks. 2018. *Cyber Operations in Conflict: Lessons from Analytic Wargames*. Berkeley, CA: Center for Long-Term Cybersecurity, UC Berkeley.

Jensen, Benjamin, and Brandon Valeriano. 2019. "Cyber Escalation Dynamics: Results from War Game Experiments." Paper presented at the annual meeting of the International Studies Association, Toronto, March 27.

Jervis, Robert. 1976. *Perception and Misperception in International Politics*. Princeton, NJ, and Oxford: Princeton University Press.

——. 1984. *The Illogic of American Nuclear Strategy*. Ithaca, NY: Cornell University Press.

——. 1985. "Perceiving and Coping with Threat." In *Psychology and Deterrence*, edited by Robert Jervis, Richard Ned Lebow, and Janice G. Stein, 13–33. Baltimore, MD: Johns Hopkins University Press.

——. 1989. *The Meaning of the Nuclear Revolution: Statecraft and the Prospect of Armageddon*. Ithaca, NY: Cornell University Press.

——. 1994. "Leadership, Post–Cold War Politics, and Psychology." *Political Psychology* 15 (4): 769–77.

——. 1997. *System Effects: Complexity in Political and Social Life*. Princeton, NJ: Princeton University Press.

——. 2002. "Signaling and Perception: Drawing Inferences and Projecting Images." In *Political Psychology,* edited by Kristen Renwick Monroe, 293–314. Mahwah, NJ: Erlbaum.

——. 2013. "Do Leaders Matter and How Would We Know?" *Security Studies* 22 (2): 153–79.

Jervis, Robert, Richard Ned Lebow, and Janice Gross Stein. 1989. *Psychology and Deterrence*. Baltimore, MD: Johns Hopkins University Press.

Johns, Robert, and Graeme A. M. Davies. 2012. "Democratic Peace or Clash of Civilizations? Target States and Support for War in Britain and the United States." *Journal of Politics* 74 (4): 1038–52.

Johnson, Dominic, and Monica Duffy Toft. 2014. "Grounds for War: The Evolution of Territorial Conflict." *International Security* 38 (3): 7–38.

Johnson, Jesse C., and Tiffany D. Barnes. 2011. "Responsibility and the Diversionary Use of Force." *Conflict Management and Peace Science* 28 (5): 478–96.

Johnson, Jesse C., and Stephen Joiner. 2019. "Power Changes, Alliance Credibility, and Extended Deterrence." *Conflict Management and Peace Science* (Forthcoming).

Johnson, Jesse C., and Brett Ashley Leeds. 2011. "Defense Pacts: A Prescription for Peace?" *Foreign Policy Analysis* 7 (1): 45–65.

Johnson, Neil, Spencer Carran, Joel Botner, Kyle Fontaine, Nathan Laxague, Philip Nuetzel, Jessica Turnley, and Brian Tivnan. 2011. "Pattern in Escalations in Insurgent and Terrorist Activity." *Science* 333 (6038): 81–84.

Johnston, Alastair Iain. 1995. "Thinking about Strategic Culture." *International Security* 19 (4): 32–64.

Joint Comprehensive Plan of Action, US Treasury. Available at https://www.treasury.gov/resource-center/sanctions/Programs/Pages/jpoa_archive.aspx.

Jones, Benjamin T., Eleonora Mattiacci, and Bear F. Braumoeller. 2017. "Food Scarcity and State Vulnerability: Unpacking the Link between Climate Variability and Violent Unrest." *Journal of Peace Research* 54 (3): 335–50.

Jones, Benjamin T., and Shawna K. Metzger. 2018. "Evaluating Conflict Dynamics: A Novel Empirical Approach to Stage Conceptions." *Journal of Conflict Resolution* 62 (4): 819–47.

Jones, Daniel M., Stuart A. Bremer, and J. David Singer. 1996. "Militarized Interstate Disputes, 1816–1992: Rationale, Coding Rules, and Empirical Patterns." *Conflict Management and Peace Science* 15 (2): 163–213.

Justwan, Florian, and Sarah K. Fisher. 2017. "International Adjudication and Public Opinion in Territorial Disputes: Evidence from a Survey Experiment Using Amazon Mechanical Turk." *Peace Economics, Peace Science and Public Policy* 23 (3): 1–18.

——. 2018. "Territorial Claims, Territorial Attachment, and Political Tolerance: Evidence from India." *Politics, Groups, and Identities*, forthcoming. doi: 10.1080/21565503.2018.1528164

——. 2020. "Social Trust and Public Opinion about Territorial Disputes: Evidence from a Survey in India." *Journal of Global Security Studies* 5 (4): 617–33.

Kadera, Kelly M. 1998. "Transmission, Barriers, and Constraints: A Dynamic Model of the Spread of War." *Journal of Conflict Resolution* 42 (3): 367–87.

——. 1999. "The Power-Conflict Story: A Synopsis." *Conflict Management and Peace Science* 17 (2): 149–74.

——. 2001. *The Power-Conflict Story: A Dynamic Model of Interstate Rivalry*. Ann Arbor: University of Michigan Press.

Kadera, Kelly M., Mark J. C. Crescenzi, and Megan L. Shannon. 2003. "Democratic Survival, Peace, and War in the International System." *American Journal of Political Science* 47 (3): 234–47.

Kadera, Kelly M., Sung Woo Kim, and Jonathan Ring. 2010. *States and Spiders: Size, Information, and Persistence in Jumping Spider Contests and MIDs*. Paper presented at the annual meeting of International Studies Association.

Kadera, Kelly M., and Daniel S. Morey. 2008. "The Trade-Offs of Fighting and Investing: A Model of the Evolution of War and Peace." New Orelans, February 17-20 *Conflict Management and Peace Science* 25 (2): 152–70.

Kadera, Kelly M., and Gerald L. Sorokin. 2004. "Measuring National Power." *International Interactions* 30 (3): 211–30.

Kahn, Herman. 1965. *On Escalation: Metaphors and Scenarios*. New York: Praeger.

Kahneman, Daniel, and Amos Tversky. 1979. "Prospect Theory: An Analysis of Decision under Risk." *Econometrica* 47 (2): 263–91.

Kang, Alice J., and Nam Kyu Kim. 2020. "Territorial Threat and Women's Legislative Representation." *Democratization* 27 (2): 340–58.

Kang, Choong-Nam. 2012. "Alliances: Path to Peace or Path to War?" In *What Do We Know about War?* 2nd edition, edited by John Vasquez, 27–44. Lanham, MD: Rowman and Littlefield.

——. 2020. "Not Just Pooling, but Helping Generate Capabilities: Alliances and Capability Growth." *Journal of Political and Military Sociology*, forthcoming.

——. n.d. "Emboldening or Restraining? Capabilities of Allies and Dispute Initiation." Unpublished Manuscript.

Kang, Kyungkook, and Jacek Kugler. 2023. "Beyond Deterrence: Uncertain Stability in the Nuclear Era." *Conflict Management and Peace Science* 40 (6).

Kant, Immanuel. 1939 [1795]. *Perpetual Peace*. Translated and edited by Nicholas Murray Butler. New York: Columbia University Press.

Kaplan, Fred. 2016. *Dark Territory: The Secret History of Cyber War*. London: Simon and Schuster.

Kapur, S. Paul. 2007. *Dangerous Deterrent: Nuclear Weapons Proliferation and Conflict in South Asia*. Stanford, CA: Stanford University Press.

Karim, Sabrina, and Kyle Beardsley. 2016. "Explaining Sexual Exploitation and Abuse in Peacekeeping Missions." *Journal of Peace Research* 53 (1): 100–15.

Kasperson, Jeanne X., Roger E. Kasperson, B. L. Turner II, Wen Hsieh, and Andrew Schiller. 2014. "Vulnerability to Global Environmental Change." In *Social Contours of Risk,* edited by Jeanne X. Kasperson and Roger E. Kasperson, 245–85. New York: Routledge.

Kegley, Charles W., ed. 1991. *The Long Postwar Peace: Contending Explanations and Projections*. New York: HarperCollins.

Kello, Lucas. 2013. "The Meaning of the Cyber Revolution: Perils to Theory and Statecraft." *International Security* 38 (2): 7–40.

______. 2017. *The Virtual Weapon and International Order*. New Haven, CT: Yale University Press.

Kennan, George. 1948. "The Inauguration of Organized Political Warfare." State Department Policy Planning Staff, National Archives and Records Administration, RG 273.

Kennedy, Paul. 1983. *Strategy and Diplomacy 1870–1945*. London: Allen and Unwin.

——. 1987. *The Rise and Fall of the Great Powers: Economic Change and Military Conflict from 1500 to 2000*. New York: Random House.

Kenwick, Michael R., and John A. Vasquez. 2017. "Defense Pacts and Deterrence: Caveat Emptor." *Journal of Politics* 79 (1): 329–34.

Kenwick, Michael R., John A. Vasquez, and Matthew A. Powers. 2015. "Do Alliances Really Deter?" *Journal of Politics* 77 (4): 943–54.

Kenwood, A. George, and Alan L. Lougheed. 1999. *The Growth of the International Economy 1820–2000: An Introductory Text*. New York: Routledge.

Keohane, Robert O., and Lisa L. Martin. 1995. "The Promise of Institutionalist Theory." *International Security* 20 (1): 39–51.

Kertzer, Joshua D. 2016. *Resolve in International Politics*. Princeton, NJ: Princeton University Press.

Keshk, Oar M. G., Brian M. Pollins, and Rafael Reuveny. 2004. "Trade Still Follows the Flag: The Primacy of Politics in a Simultaneous Model of Interdependence and Armed Conflict." *Journal of Politics* 66 (4): 1155–79.

Keynes, John M. 1920. *The Economic Consequences of the Peace*. New York: Harcourt, Brace, and Howe.

Khrushchev, Sergei N. 2000. *Nikita Khrushchev and the Creation of a Superpower*. University Park: Pennsylvania State University Press.

Kilcullen, David. 2009. *The Accidental Guerrilla: Fighting Small Wars in the Midst of a Big One*. Oxford: Oxford University Press.

——. 2010. *Counterinsurgency*. Oxford: Oxford University Press.

Kim, Hyung Min, and David L. Rousseau. 2005. "The Classical Liberals Were Half Right (or Half Wrong): New Tests of the 'Liberal Peace' 1960–88." *Journal of Peace Research* 42 (5): 523–43.

Kim, Nam Kyu. 2019a. "External Territorial Threats and Military Regimes." *Political Research Quarterly* 72 (4): 863–77.

——. 2020. "Territorial Disputes and Individual Willingness to Fight." *Journal of Peace Research* 57 (3): 406–21.

Kim, Woosang. 1991. "Alliance Transitions and Great Power War." *American Journal of Political Science* 35 (4): 833–50.

——. 2002. "Power Parity, Alliance, Dissatisfaction, and Wars in East Asia, 1860–1993." *Journal of Conflict Resolution* 46 (5): 654–71.

King, Gary, Robert Keohane, and Sidney Verba. 1994. *Designing Social Inquiry*. Princeton, NJ: Princeton University Press.

King, Gary, and Langche Zeng. 2001. "Explaining Rare Events in International Relations." *International Organization* 55 (3): 693–715.

Kinne, Brandon J. 2013. "IGO Membership, Network Convergence, and Credible Signaling in Militarized Disputes." *Journal of Peace Research* 50 (6): 659–76.

——. 2018. "Defense Cooperation Agreements and the Emergence of a Global Security Network." *International Organization* 72 (4): 799–837.

——. 2020. "The Defense Cooperation Agreement Dataset (DCAD)." *Journal of Conflict Resolution* 64 (4): 729–55.

Kinsella, David. 2005."No Rest for the Democratic Peace." *American Political Science Review* 99 (3): 453–57.

Kirkaldy, A. W. 1921. *British Finance during and after the War*. London: Pitman.

Kirshner, Jonathan. 2007. *Appeasing Bankers: Financial Caution on the Road to War*. Princeton, NJ: Princeton University Press.

Klare, Michael T. 2002. *Resource Wars: The New Landscape of Global Conflict*. New York: Henry Holt.

——. 2012. *The Race for What's Left: The Global Scramble for the World's Last Resources*. New York: Picador.

Klein, James P., Gary Goertz, and Paul F. Diehl. 2006. "The New Rivalry Data Set: Procedures and Patterns." *Journal of Peace Research* 43 (3): 331–48.

Klein, Natalie. 2011. *Maritime Security and the Law of the Sea*. Oxford: Oxford University Press.

Kletzer, Kenneth M. 1984. "Asymmetries of Information and LDC Borrowing with Sovereign Risk." *Economic Journal* (1984): 287–307.

Knorr, Klaus. 1975. *The Power of Nations: The Political Economy of International Relations*. New York: Basic Books.

Kober, Avi. 2016. "Arms Races and the Arab-Israeli Conflict." In *Arms Races in International Politics*, edited by Thomas Mahnken, Joseph Maiolo, and David Stevenson, 205–23. New York: Oxford University Press.

Koch, Michael T., and Sarah A. Fulton. 2011. "In the Defense of Women: Gender, Office Holding, and National Security Policy in Established Democracies." *Journal of Politics* 73 (1): 1–16.

Kocher, Matthew Adam, Thomas B. Pepinsky, and Stathis N. Kalyvas. 2011. "Aerial Bombing and Counterinsurgency in the Vietnam War." *American Journal of Political Science* 55 (2): 201–18.

Kocs, Stephen. 1995. "Territorial Disputes and Interstate War." *Journal of Politics* 57 (1): 159–75.

Kostyuk, Nadiya, and Yuri M. Zhukov. 2019. "Invisible Digital Front: Can Cyber Attacks Shape Battlefield Events?" *Journal of Conflict Resolution* 63 (2): 317–47.

Koubi, Vally, 2005. "War and Economic Performance." *Journal of Peace Research* 42 (1): 67–82.

——. 2019. "Climate Change and Conflict." *Annual Review of Political Science* 22: 343–60.

Koubi, Vally, Thomas Bernauer, Anna Kalbhenn, and Gabriele Spilker. 2012. "Climate Variability, Economic Growth, and Civil Conflict." *Journal of Peace Research* 49 (1): 113–27.

Koubi, Vally, Gabriele Spilker, Thomas Böhmelt, and Thomas Bernauer. 2014. "Do Natural Resources Matter for Interstate and Intrastate Armed Conflict?" *Journal of Peace Research* 51 (2): 227–43.

Krause, Volker. 2004. "Hazardous Weapons? Effects of Arms Transfers and Defense Pacts on Militarized Disputes, 1950–1995." *International Interactions* 30 (4): 349–71.

Krcmaric, Daniel, Stephen C. Nelson, and Andrew Roberts. 2020. "Studying Leaders and Elites: The Personal Biography Approach." *Annual Review of Political Science* 23: 133–51.

Krepon, Michael (ed.). 2004. *Nuclear Risk Reduction in South Asia*. New York: Palgrave Macmillan.

Kreps, Sarah E. 2018. *Taxing Wars: The American Way of War Finance and the Decline of Democracy*. Oxford: Oxford University Press.

Kreps, Sarah E., and Matthew Fuhrmann. 2011. "Attacking the Atom: Does Bombing Nuclear Facilities Affect Proliferation?" *Journal of Strategic Studies* 34 (2): 161–87.

Krieger, Tim, and Daniel Meierrieks. 2015. "The Rise of Market-Capitalism and the Roots of Anti-American Terrorism." *Journal of Peace Research* 52 (1): 46–61.

Kriner, Douglas L., Breanna Lachase, and Rosella Cappella Zielinski. 2018. "Self-Interest, Partisanship, and the Conditional Influence of Taxation on Support for War in the USA." *Conflict Management and Peace Science* 35 (1): 43–64.

Kriner, Douglas L., and Francis X. Shen. 2014. "Reassessing American Casualty Sensitivity: The Mediating Influence of Inequality." *Journal of Conflict Resolution* 58 (7): 1174–201.

Kroenig, Matthew. 2013. "Nuclear Superiority and the Balance of Resolve: Explaining Nuclear Crisis Outcomes." *International Organization* 67 (1): 141–71.

——. 2018. "The Case for the US ICBM Force." *Strategic Studies Quarterly* 12 (3): 50–69.

Krugman, Paul. 2009. *The Return of Depression Economics and the Crisis of 2008*. New York: W. W. Norton.

Kugler, Jacek. 1984. "Terror without Deterrence: Reassessing the Role of Nuclear Weapons." *Journal of Conflict Resolution* 28 (3): 470–506.

Kugler, Jacek, and Marina Arbetman. 1989. "Exploring the 'Phoenix Factor' with the Collective Goods Perspective." *Journal of Conflict Resolution* 33 (1): 84–112.

Kugler, Jacek, and William Domke. 1986. "Comparing the Strength of Nations." *Comparative Political Studies* 19 (1): 39–69.

Kugler, Tadeusz, Kyung Kook Kang, Jacek Kugler, Marina Arbetman-Rabinowitz, and John Thomas. 2013. "Demographic and Economic Consequences of Conflict." *International Studies Quarterly* 57 (1): 1–12.

Kupchan, Charles A. 2010. *How Enemies Become Friends: The Sources of Stable Peace*. Princeton, NJ: Princeton University Press.

Kuperman, Alan J. 2008. "The Moral Hazard of Humanitarian Intervention: Lessons from the Balkans." *International Studies Quarterly* 52 (1): 49–80.

Kydd, Andrew. 1997. "Sheep in Sheep's Clothing: Why Security Seekers Do Not Fight Each Other. *Security Studies* 7 (1): 114–55.

——. 2000. "Arms Races and Arms Control: Modeling the Hawk Perspective." *American Journal of Political Science* 44 (2): 222–38.

Lacina, Bethany, and Nils Petter Gleditsch. 2012. "The Waning of War Is Real: A Response to Gohdes and Price." *Journal of Conflict Resolution* 57 (6): 1109–27.

Lacina, Bethany, Nils Petter Gleditsch, and Bruce Russett. 2006. "The Declining Risk of Death in Battle." *International Studies Quarterly* 50 (3): 673–80.

Lake, David A. 1992. "Powerful Pacifists: Democratic States and War." *American Political Science Review* 86 (1): 24–37.

——. 2009. *Hierarchy in International Relations*. Ithaca, NY: Cornell University Press.

——. 2011. "Why 'Isms' Are Evil: Theory, Epistemology, and Academic Sects As Impediments to Understanding and Progress." *International Studies Quarterly* 55 (2): 465–80.

Lambelet, John C. 1975. "A Numerical Model of the Anglo-German Dreadnought Race." *Peace Science Society (International) Papers* 24: 29–48.

——. 1976. "A Complementary Analysis of the Anglo-German Dreadnought Race, 1906–1916." *Peace Science Society (International) Papers* 26: 219–66.

Lamborn, Alan C. 1983. "Power and the Politics of Extraction." *International Studies Quarterly* 27: 125–46.

Landau, Mark J. Sheldon Solomon, Jeff Greenberg et al. 2004. "Deliver Us from Evil: The Effects of Mortality Salience and Reminders of 9/11 on Support for President George W. Bush." *Personality and Social Psychology Bulletin* 30: 1136–50.

Langer, William L. 1980. *An Encyclopedia of World History.* Fifth edition. Revised and enlarged. Boston: Houghton Mifflin.

Langø, Hans-Inge. 2016. "Competing Academic Approaches to Cyber Security." In *Conflict in Cyber Space*, edited by Karsten Friis and Jens Ringsmose, 23–42. London: Routledge.

Lathrop, Colater G. 2014. "Why Litigate a Maritime Boundary? Some Contributing Factors." In *Litigating International Law Disputes: Weighing the Options*, edited by Natalie Klein, 230–59. Cambridge: Cambridge University Press.

Lave, Charles A., and James G. March. 1993. *An Introduction to Models in the Social Sciences*. Lanham: University Press of Maryland.

Layne, Christopher. 1994. "Kant or Cant: The Myth of the Democratic Peace." *International Security* 19 (2): 5–49.

Lebow, Richard Ned. 1981. *Between Peace and War: The Nature of International Crisis*. Baltimore, MD: Johns Hopkins University Press.

Lebow, Richard Ned, and Janice Gross Stein. 1995. "Deterrence and the Cold War." *Political Science Quarterly* 110 (2): 157–81.

Lee, Harry F., David D. Zhang, Peter Brecke, and Qing Pei. 2019. "Climate Change, Population Pressure, and Wars in European History." *Asian Geographer* 36 (1): 29–45.

Lee, Hoon, and Sara McLaughlin Mitchell. 2012. "Foreign Direct Investment and Territorial Disputes." *Journal of Conflict Resolution* 56 (4): 675–703.

Lee, Jong R. 1977. "Rally around the Flag: Foreign Policy Events and Presidential Popularity." *Presidential Studies Quarterly* 7 (4): 252–56.

Lee, Kyungsuk, James Kim, Hwalmin Jin, and Matthew Fuhrmann. 2020. "Nuclear Weapons and Low-Level Military Conflict." Unpublished manuscript.

Lee, Michael, and William R. Thompson. 2018. "Major Powers vs. Global Powers: A New Measure of Global Reach and Power Protection Capacity." In *The Oxford Encyclopedia of Empirical International Relations Theory*, edited by William R. Thompson, Vol. 2, 485–511. New York: Oxford University Press.

Lee, Sojeong, and Sara McLaughlin Mitchell. 2019. "Energy Resources and the Risk of Conflict in Shared River Basins." *Journal of Peace Research* 56 (3): 336–51.

Leeds, Brett Ashley. 2003. "Do Alliances Deter Aggression? The Influence of Military Alliances on the Initiation of Militarized Interstate Disputes." *American Journal of Political Science* 47 (3): 427–39.

——. 2005. "Alliances and the Expansion and Escalation of Militarized Interstate Disputes." In *New Directions in International Relations*, edited by Alex Mintz and Bruce Russett, 117–34. Lanham, MD: Lexington.

Leeds, Brett Ashley, and David R. Davis. 1997. "Domestic Political Vulnerability and International Disputes." *Journal of Conflict Resolution* 41 (6): 814–34.

Leeds, Brett Ashley, and Jesse C. Johnson. 2017. "Theory, Data, and Deterrence: A Response to Kenwick, Vasquez, and Powers." *Journal of Politics* 79 (1): 335–40.

Leeds, Brett Ashley, Andrew G. Long, and Sara McLaughlin Mitchell. 2000. "Reevaluating Alliance Reliability: Specific Threats, Specific Promises." *Journal of Conflict Resolution* 44 (5): 686–99.

Leeds, Brett Ashley, Michaela Mattes, and Naoko Matsumura. 2016. "Measuring Change in Source of Leader Support: The CHISOLS Dataset." *Journal of Peace Research* 53 (2): 259–67.

Leeds, Brett Ashley, Michaela Mattes, and Jeremy S. Vogel. 2009. "Interests, Institutions, and the Reliability of International Commitments." *American Journal of Political Science* 53 (2): 461–76.

Leeds, Brett Ashley, Jeffrey M. Ritter, Sara McLaughlin Mitchell, and Andrew G. Long. 2002. "Alliance Treaty Obligations and Provisions, 1815–1944." *International Interactions* 28 (3): 237–60.

Leeds, Brett Ashley, and Burcu Savun. 2007. "Terminating Alliances: Why Do States Abrogate Agreements?" *Journal of Politics* 69 (4): 1118–32.

Lefler, V. A. 2014. "Strategic Forum Selection and Compliance in Interstate Dispute Resolution." *Conflict Management and Peace Science* 32 (1): 1–23.

Lektzian, David, Brandon C. Prins, and Mark Souva. 2010. "Territory, River, and Maritime Claims in the Western Hemisphere: Regime Type, Rivalry, and MIDs from 1901 to 2000." *International Studies Quarterly* 54 (4): 1073–98.

Lemke, Douglas. 2002. *Regions of War and Peace*. Cambridge: Cambridge University Press.

Lemke, Douglas, and William Reed. 1996. "Regime Types and Status Quo Evaluations: Power Transition Theory and the Democratic Peace." *International Interactions* 22 (2): 143–64.

Lemke, Douglas, and Suzanne Werner. 1996. "Power Parity, Commitment to Change, and War." *International Studies Quarterly* 40 (2): 235–60.

——. 1998. "Power Is Not Satisfaction: A Comment on de Soysa, Oneal, and Park." *Journal of Conflict Resolution* 42 (4): 511–16.

——. 2001. "War and Rivalry among Great Powers." *American Journal of Political Science* 45 (2): 457–69.

Leng, Russell J. 1983. "When Will They Ever Learn? Coercive Bargaining in Recurrent Crises." *Journal of Conflict Resolution* 27 (September): 379–419.

——. 2000. *Bargaining and Learning in Recurring Crisis*. Ann Arbor: University of Michigan Press.

Leng, Russell J., and J. David Singer. 1988. "Militarized Interstate Crises: The BCOW Typology and Its Applications." *International Studies Quarterly* 32 (2): 155–73.

Leng, Russell J., and Hugh G. Wheeler. 1979. "Influence Strategies, Success, and War." *Journal of Conflict Resolution* 23 (4): 655–84.

Leonard, Hans. 2020. "*Three Essays on Economic Interaction and Conflict*." PhD dissertation, Department of Political Science, University of Rochester.

LeVeck, Brad L., and Neil Narang. 2017. "How International Reputation Matters: Revisiting Alliance Violations in Context." *International Interactions* 43 (5): 797–821.

Levi, Margaret. 1988. *Of Rule and Revenue*. Berkeley: University of California Press.

Levy, Jack S. 1981. "Alliance Formation and War Behavior: An Analysis of the Great Powers, 1495–1975." *Journal of Conflict Resolution* 25 (4): 581–613.

——. 1982. "Historical Trends in Great Power War, 1495–1975." *International Studies Quarterly* 26 (2): 278–301.

——. 1983. *War in the Modern Great Power System, 1495–1975.* Lexington: University of Kentucky Press.

——. 1988. "Domestic Politics and War." *Journal of Interdisciplinary History* 18 (4): 653–73.

——. 1989. "The Diversionary Theory of War: A Critique." In *Handbook of War Studies*, edited by Manus I. Midlarsky, 259–88. London: Allen and Unwin.

——. 2000. "The 'Paths to War' Concept." In *What Do We Know about War?* edited by John A. Vasquez, 281–90. Lanham, MD: Rowman and Littlefield.

——. 2004. "What Do Great Powers Balance against and When?" In *Balance of Power Revisited: Theory and Practice in the 21st Century*, edited by T. V. Paul,

James Wirtz, and Michel Fortmann, 29–51. Stanford, CA: Stanford University Press.

——. 2012 “The ‘Paths-to-War’ Concept.” In *What Do We Know about War?* edited by John A. Vasquez, 281–90. Lanham, MD: Rowman and Littlefield.

Levy, Jack S., and Salvatore Ali. 1998. “From Commercial Competition to Strategic Rivalry: The Evolution of the Anglo-Dutch Rivalry, 1609–52.” In *The Dynamics of Enduring Rivalries*, edited by Paul Diehl, 29–63. Champaign: University of Illinois Press.

Levy, Jack S., and Katherine Barbieri. 2004. “Trading with the Enemy during Wartime.” *Security Studies* 13 (Spring): 1–47.

Levy, Jack S., and Gary Goertz: , eds 2007. *Explaining War and Peace: Case Studies and Necessary Condition Counterfactuals*. New York: Routledge.

Levy, Jack S., and T. Clifton Morgan. 1984. “The Frequency and Seriousness of War: An Inverse Relationship?” *Journal of Conflict Resolution* 28 (4): 731–49.

——. 1986. “The War-Weariness Hypothesis: An Empirical Test.” *American Journal of Political Science* 30 (1): 26–49.

Levy, Jack S., and William R. Thompson. 2011. *The Arc of War: Origins, Escalation, and Transformation*. Chicago and London: University of Chicago Press.

——— 2013. “The Decline of War? Multiple Trajectories and Diverging Trends.” *International Studies Review* 15 (3): 411–19.

Levy, Jack S., and John A. Vasquez: , eds 2014. *The Outbreak of the First World War: Structure, Politics, and Decision-Making*. Cambridge: Cambridge University Press.

Li, Quan, and Rafael Reuveny. 2011. “Does Trade Prevent or Promote Interstate Conflict Initiation?” *Journal of Peace Research* 48 (July): 437–53.

Li, Quan, and David Sacko. 2002. “The (Ir)Relevance of Militarized Interstate Disputes for International Trade.” *International Studies Quarterly* 46 (1): 11–43.

Libicki, Martin. 2016. *Cyberspace in Peace and War*. Annapolis, MD: Naval Institute Press.

Licht, Amanda A., and Susan Hannah Allen. 2018. “Repressing for Reputation: Leadership Transitions, Uncertainty, and the Repression of Domestic Populations.” *Journal of Peace Research* 55 (5): 582–95.

Lieber, Keir, and Daryl Press. 2019. *The Myth of the Nuclear Revolution: Power Politics in the Atomic Age*. Ithaca, NY: Cornell University Press.

Lim, Lydia. 2003. “Pedra Branca: Key Step Forward Today; Singapore and Malaysia Will Sign an Accord in KL to Refer the Dispute over the Island to International Court of Justice.” *Straits Times*, February 6, 2003.

Lindsay, Jon R. 2013. “Stuxnet and the Limits of Cyber Warfare.” *Security Studies* 22 (3): 365–404.

Lindsay, Jon R., and Erik Gartzke. 2016. “Coercion through Cyberspace: The Stability-Instability Paradox Revisited.” In *Coercion: The Power to Hurt in International Politics*, edited by Kelly M. Greenhill and Peter J. P. Krause, 176–204. New York: Oxford University Press.

Lindsay, Jon R., and Lucas Kello. 2014. “Correspondence: A Cyber Disagreement.” *International Security* 39 (2): 181–92.

Lo, Nigel, Barry Hashimoto, and Dan Reiter. 2008. “Ensuring Peace: Foreign-Imposed Regime Change and Postwar Peace Duration, 1914–2001.” *International Organization* 62 (4): 717–36.

Long, Andrew G. 2008. "Bilateral Trade in the Shadow of Armed Conflict." *International Studies Quarterly* 52 (1): 81–101.

Long, Austin, and Brendan Rittenhouse Green. 2015. "Stalking the Secure Second Strike: Intelligence, Counterforce, and Nuclear Strategy." *Journal of Strategic Studies* 38 (1–2): 38–73.

Long, James W. 1974. "Franco-Russian Relations during the Russo-Japanese War," *Slavonic and East European Review* 52 (127): 213–33.

Long, Stephen. 2003. "The Present and Time Past: Rivalry and the Duration of Interstate Wars, 1846–1985." *International Interactions* 29 (3): 215–36.

Lowi, Miriam R. 1993. "Bridging the Divide: Transboundary Resource Disputes and the Case of West Bank Water." *International Security* 18 (1): 113–38.

Lowi, Theodore. 1964. "American Business, Public Policy, Case Studies, and Political Theory." *World Politics* 16 (4): 677–715.

Luard, Evan. 1970. *The International Regulation of Frontier Disputes*. Santa Barbara, CA: Praeger.

Lucas, Robert E., and Nancy Stokey. 1983. "Optimal Fiscal and Monetary Policy in an Economy without Capital." *Journal of Monetary Economics* 12 (1): 55–93.

Ludvik, Jan. 2019. "Closing the Window of Vulnerability: Nuclear Proliferation and Conventional Retaliation." *Security Studies* 28 (1): 87–115.

Ludwig, Ralf, Roberto Roson, Christos Zografos, and Giorgios Kallis. 2011. "Towards an Inter-disciplinary Research Agenda on Climate Change, Water, and Security in Southern Europe and Neighboring Countries." *Environmental Science & Policy* 14 (7): 794–803.

Lujala, Päivi, Jan Ketil Rød, and Nadja Thieme. 2007. "Fighting over Oil: Introducing a New Dataset." *Conflict Management and Peace Science* 24 (3): 239–56.

Lupton, Danielle L. 2018a. "Reexamining Reputation for Resolve: Leaders, States, and the Onset of International Crises." *Journal of Global Security Studies* 3 (2): 198–216.

——. 2018b. "Signaling Resolve: Leaders, Reputations, and the Importance of Early Interactions." *International Interactions* 44 (1): 59–87.

——. 2020. *Reputation for Resolve: How Leaders Signal Determination in International Politics.* Ithaca, NY: Cornell University Press.

Lutmar, Carmela, and Lesley Terris. 2018. "War Termination." In *The Oxford Encyclopedia of Empirical International Relations Theory*, edited by William R. Thompson, Vol. 4, 519–33. New York: Oxford University Press.

Lyall, Jason. 2009. "Does Indiscriminate Violence Incite Insurgent Attacks? Evidence from Chechnya." *Journal of Conflict Resolution* 53 (3): 331–62.

——. 2010b. "Do Democracies Make Inferior Counterinsurgents? Reassessing Democracy's Impact on War Outcomes and Duration." *International Organization* 64 (1): 167–92.

——. 2010a. "Are Coethnics More Effective Counterinsurgents? Evidence from the Second Chechen War." *American Political Science Review* 104 (1): 1–20.

Lyall, Jason, and Isaiah Wilson III. 2009. "Rage against the Machines: Explaining Outcomes in Counterinsurgency Wars." *International Organization* 63 (1): 67–106.

Mach, K. J., C. M. Kraan, W. N. Adger, W. N. et al. 2019. "Climate As a Risk Factor for Armed Conflict." *Nature* 571: 193–97.

Machain, Carla Martinez. 2018. "Popular Exceptions, Theory, and Empirical Findings on Air Power." In *The Oxford Encyclopedia of Empirical International Relations Theory*, edited by William R. Thompson, Vol. 3, 81–91. New York: Oxford University Press.

Mack, Andrew. 1975. "Why Big Nations Lose Small Wars." *World Politics* 27 (2): 175–200.

Mahoney, James, and Rachel S. Vanderpoel. 2015. "Set Diagrams and Qualitative Research." *Comparative Political Studies* 48 (1): 65–100.

Maiolo, Joseph. 2016. "Naval Armaments Competition between the Two World Wars." In *Arms Races in International Politics*, edited by Thomas Mahnken, Joseph Maiolo, and David Stevenson, 93–114. New York: Oxford University Press.

Mandelbaum, Michael. 1989. "Ending the Cold War." *Foreign Affairs* 68 (2): 16–36.

Maness, Ryan, and Brandon Valeriano. 2012. "Russia and the Near Abroad: Applying a Risk Barometer for War." *The Journal of Slavic Military Studies* 25 (2): 125–48.

——. 2015. *Russia's Coercive Diplomacy: Energy, Cyber, and Maritime Policy As New Sources of Power*. London: Palgrave Macmillan.

Maness, Ryan C., Brandon Valeriano, and Benjamin Jensen. 2019. "The Dyadic Cyber Incident and Campaign Dataset (DCID), Version 1.5." Available at https://drryanmaness.wixsite.com/cyberconflict/cyber-conflict-dataset.

Mansbach, Richard, and John A. Vasquez. 1981. *In Search of Theory: A New Paradigm for Global Politics*. New York: Columbia University Press.

Mansfield, Edward D. 1992. "The Concentration of Capabilities and the Onset of War." *Journal of Conflict Resolution* 36 (1): 3–24.

Mansour, Imad, and William R. Thompson, eds. 2020. *Shocks and Rivalries in the Middle East and North Africa.* Washington, DC: Georgetown University Press.

Maoz, Zeev. 1989a. "Joining the Club of Nations: Political Development and International Conflict, 1816-1976." *International Studies Quarterly* 33 (June): 199–231.

——. 1989b. "Power, Capabilities, and Paradoxical Conflict Outcomes." *World Politics* 41 (2): 239–66.

——. 1990. *Paradoxes of War: On the Art of National Self-Entrapment*. Boston: Unwin Hyman.

——. 1997. "The Controversy over the Democratic Peace: Rearguard Action or Cracks in the Wall?" *International Security* 22 (1): 162–98.

——. 2000. "Alliances: The Street Gangs of World Politics—Their Origins, Management, and Consequences." In *What Do We Know about War?* edited by John A. Vasquez, 111–44. Lanham, MD: Rowman and Littlefield.

Maoz, Zeev, and Nasrin Abdolali. 1989. "Regime Types and International Conflict, 1816–1976." *Journal of Conflict Resolution* 33 (1): 3–36.

Maoz, Zeev, Ranan D. Kuperman, Lesley Terris, and Ilan Talmud. 2006. "Structural Equivalence and International Conflict: A Social Networks Analysis." *Journal of Conflict Resolution*: 50 (5): 664–89.

Maoz, Zeev, and Ben Mor. 2002. *Bound by Struggle: The Strategic Evolution of Enduring International Rivalries*. Ann Arbor: University of Michigan Press.

Maoz, Zeev, and Bruce Russett. 1992. "Alliance, Contiguity, Wealth, and Political Stability: Is the Lack of Conflict among Democracies a Statistical Artifact?" *International Interactions* 17 (3): 245–67.

——. 1993. "Normative and Structural Causes of Democratic Peace, 1946–1986." *American Political Science Review* 87 (3): 624–38.
Mares, David R. 2018. "The Latin American Puzzle for the Study of International Relations." In *The Oxford Encyclopedia of Empirical International Relations Theory*, edited by William R. Thompson, Vol. 2, 389–407. New York: Oxford University Press.
Marjolin, Robert. 1989. *Architect of European Unity: Memoirs, 1911–86*. London: Weidenfeld & Nicolson.
Marshall, Monty G., Ted R. Gurr, and Keith Jaggers. 2019. *Polity IV Project: Political Regime Characteristics and Transitions, 1800–2018*. College Park, MD: Center for Systemic Peace.
Marshall, Monty G., and Keith Jaggers. 2019. *Polity IV Project: Dataset Users' Manual*. Available at http://www.systemicpeace.org/inscr/p4manualv2018.pdf.
Martelloni, Gianluca, Francesca Di Patti, and Ugo Bardi. 2019. "Pattern Analysis of World Conflicts over the Past 600 Years." arXiv: https://arxiv.org/abs/1812.08071.
Martin, Lisa L. 2000. *Democratic Commitments: Legislatures and International Cooperation*. Princeton, NJ: Princeton University Press.
Marwick, Arthur. 1974. *War and Social Change in the Twentieth Century*. Basingstoke: Palgrave Macmillan.
Mastanduno, Michael, David A. Lake, and G. John Ikenberry. 1989. "Toward a Realist Theory of State Action." *International Studies Quarterly* 33 (4): 457–74.
Masterson, Michael, and Jessica L. P. Weeks. 2018. "Power, Institutions and Issues As Causes of Conflict." In *The Oxford Encyclopedia of Empirical International Relations Theory*, edited by William R. Thompson, Vol. 3, 118–38. New York: Oxford University Press.
Mattes, Michaela. 2012. "Reputation, Symmetry, and Alliance Design." *International Organization* 66 (4): 679–707.
——. 2015. "Democratic Reliability, Precommitment of Successor Governments, and the Choice of Alliance Commitment." *International Organization* 66 (1): 153–72.
Mattes, Michaela, Brett Ashley Leeds, and Naoko Matsumura. 2016. "Measuring Change in Source of Leader Support: The CHISOLS Dataset." *Journal of Peace Research* 53 (2): 259–67.
Mattes, Michaela, and Jessica L. P. Weeks. 2019. "Hawks, Doves, and Peace: An Experimental Approach." *American Journal of Political Science* 63 (1): 53–66.
Matteson, David S., and Nicholas A. James. 2014. "A Nonparametric Approach for Multiple Change Point Analysis of Multivariate Data." *Journal of the American Statistical Association* 109 (505): 334–45.
Maurer, John H. 1992. "The Anglo-German Naval Rivalry and Informal Arms Control, 1912–1914." *Journal of Conflict Resolution* 36 (2): 284–308.
Mazarr, Michael J. 1999. "Virtual Nuclear Arsenals: A Second Look." In *Nuclear Weapons in a Transformed World*, edited by Michael J. Mazarr, 374–86. New York: St. Martin's Press.
McDermott, Rose, and Jonathan A. Cowden. 2001. "The Effects of Uncertainty and Sex in a Crisis Simulation Game." *International Interactions* 27 (4): 353–80.
McDermott, Rose, Jonathan A. Cowden, and C. Koopman. 2002. "Framing, Uncertainty and Hostile Communications in a Crisis Experiment." *Political Psychology* 23 (1): 133–49.

McDonald, H. Brooke, and Richard Rosecrance. 1985. "Alliance and Structural Balance in the International System: A Reinterpretation." *Journal of Conflict Resolution* 29 (1): 57–82.

McDonald, Patrick J. 2009. *The Invisible Hand of Peace: Capitalism, the War Machine, and International Relations Theory*. New York: Cambridge University Press.

——. 2015. "Great Powers, Hierarchy, and Endogenous Regimes: Rethinking the Domestic Causes of Peace." *International Organization* 69 (3): 557– 88.

McDowell, Steven. 2018. *"Law, Settlement Failure, and the Timing of Litigation in Interstate Territorial Disputes."* PhD dissertation, University of Notre Dame.

McManus, Roseanne W. 2018. "Making It Personal: The Role of Leader-Specific Signals in Extended Deterrence." *Journal of Politics* 80 (3): 982–95.

——. 2019. "Crazy Like a Fox? Are Leaders with Reputations for Madness More Successful at International Coercion?" *British Journal of Political Science*, forthcoming 51 (1): 275-93.

McManus, Roseanne W., and Mark David Nieman. 2019. "Identifying the Level of Major Power Support Signaled for Protégés: A Latent Measure Approach." *Journal of Peace Research* 56 (3): 364–78.

McManus, Roseanne W., and Keren Yarhi-Milo. 2017. "The Logic of 'Offstage' Signaling: Domestic Politics, Regime Type, and Major Power-Protégé Relations." *International Organization* 71 (4): 701–33.

McNamara, Robert S. 1983. "The Military Role of Nuclear Weapons: Perceptions and Misperceptions." *Foreign Affairs* 62 (1): 59–80.

Mearsheimer, John J. 1993. "The Case for a Ukrainian Nuclear Deterrent." *Foreign Affairs* 72 (3): 50–66.

——. 1994. "The False Promise of International Institutions." *International Security* 19 (3): 5–49.

——. 2001. *The Tragedy of Great Power Politics.* New York: W. W. Norton.

——. 2014. "Why the Ukraine Crisis Is the West's Fault: The Liberal Delusions That Provoked Putin." *Foreign Affairs* 93 (5): 77–89.

——. 2022. "The Causes and Consequences of the Ukraine War." *Horizons: Journal of International Relations and Sustainable Development* 21 (Summer): 12–27.

Mearsheimer, John J., and Stephen M. Walt. 2013. "Leaving Theory Behind: Why Simplistic Hypothesis Testing Is Bad for International Relations." *European Journal of International Relations* 19 (3): 427–57.

Mecklin, John. 2017. "Commentary: Can Congress Stop a President from Waging Nuclear War?" Reuters, November 30. https://www.reuters.com/article/us-mecklin-nuclear-commentary/commentary-can-congress-stop-a-president-waging-nuclear-war-idUSKBN1DU2HW.

Meernik, James, and Peter Waterman. 1996. "The Myth of Diversionary Use of Force by American Presidents." *Political Research Quarterly* 49 (3): 573–90.

Mehta, Rupal, and Rachel Whitlark. 2017. "The Benefits and Burdens of Nuclear Latency." *International Studies Quarterly* 61 (3): 517–28.

Meierrieks, Daniel. 2012. "Rooted in Urban Poverty? Failed Modernization and Terrorism." *Peace Economics, Peace Science, and Public Policy* 18 (3): 1–9.

Mercer, Jonathan. 1996. *Reputation and International Politics*. Ithaca, NY: Cornell University Press.

Merrills, John G. 2017. *International Dispute Settlement*, 6th edition. Cambridge: Cambridge University Press.

Merritt, Richard L., and Dina A. Zinnes. 1988. "Validity of Power Indices." *International Interactions* 14 (2): 141–51.

Merton, Robert K. 1968. "The Matthew Effect in Science: The Reward and Communication Systems of Science Are Considered." *Science* 159 (3810): 56–63.

Metzger, Shawna K., and Benjamin T. Jones. 2016. "Surviving Phases: Introducing Multistate Survival Models." *Political Analysis* 24 (4): 457–77.

Miguel, Edward, Shanker Satyanath, and Ernest Sergenti. 2004. "Economic Shocks and Civil Conflict: An Instrumental Variables Approach." *Journal of Political Economy* 112 (4): 725–53.

Milgrom, Paul R., Douglass C. North, and Barry R. Weingast. 1990. "The Role of Institutions in the Revival of Trade: The Law Merchant, Private Judges, and the Champagne Fairs." *Economics and Politics* 2 (1): 1–23.

Military Strength Ranking. 2020. "Global Firepower Index [Electronic resource]. 2019." URL: https://www.globalfirepower.com/countries-listing.asp.

Miller, Benjamin. 2018. "Theory of Regional War and Peace." In *The Oxford Encyclopedia of Empirical International Relations Theory*, edited by William R. Thompson, Vol. 4, 322–52. New York: Oxford University Press.

Miller, Steven E. 1984. *Strategy and Nuclear Deterrence*. Princeton, NJ: Princeton University Press.

Miller, Steven V. 2013. "Territorial Disputes and the Politics of Individual Well-Being." *Journal of Peace Research* 50 (6): 677–90.

———. 2017. "Individual-Level Expectations of Executive Authority under Territorial Threat." *Conflict Management and Peace Science* 34 (5): 526–45.

———. 2018. "External Territorial Threats and Tolerance of Corruption: A Private/Government Distinction." *Peace Economics, Peace Science and Public Policy* 24 (1): 1–13.

Miller, Steven V., and Douglas M. Gibler. 2011. "Democracies, Territory, and Negotiated Compromises." *Conflict Management and Peace Science* 28 (3): 261–79.

Miller, Steven V., Jaroslav Tir, and John A. Vasquez. 2020. "Geography, Territory, and Conflict." *Oxford Research Encyclopedia of International Studies*. doi: 10.1093/acrefore/9780190846626.013.320

Milner, Helen, and Dustin Tingley. 2010. "The Political Economy of U.S. Foreign Aid: American Legislators and the Domestic Politics of Aid." *Economics & Politics* 22 (2): 200–32.

Mingst, Karen A. 2004. *Essentials of International Relations*. New York: W. W. Norton.

Mintz, Alex, and Nehemia Geva. 1993. "Why Don't Democracies Fight Each Other? An Experimental Study." *Journal of Conflict Resolution* 37 (3): 484–503.

Mitani, John, David Watts, and Sylvia Amsler. 2010. "Lethal Intergroup Aggression Leads to Territorial Expansion in Chimpanzees." *Current Biology* 68 (2): 161–80.

Mitchell, Sara McLaughlin. 2002. "A Kantian System? Democracy and Third Party Conflict Resolution." *American Journal of Political Science* 46 (4): 749–59.

——. 2012. "Norms and the Democratic Peace." In *What Do We Know about War?* edited by John A. Vasquez, 167–89. Lanham, MD: Rowman and Littlefield.

——. 2020. "Clashes at Sea: Explaining the Onset, Militarization, and Resolution of Diplomatic Maritime Claims." *Security Studies* 29 (4): 637–70.

Mitchell, Sara McLaughlin, Scott Gates, and Håvard Hegre. 1999. "Evolution in Democracy-War Dynamics." *Journal of Conflict Resolution* 43 (6): 771–92.

Mitchell, Sara McLaughlin, and Paul R. Hensel. 2007. "International Institutions and Compliance with Agreements." *American Journal of Political Science* 51 (4): 721–37.

Mitchell, Sara McLaughlin, Kelly M. Kadera, and Mark J. C. Crescenzi. 2009. "Practicing Democratic Community Norms: Third Party Conflict Management and Successful Settlements." In *International Conflict Mediation: New Approaches and Findings*, edited by Jacob Bercovitch and Scott S. Gartner, 243–64. New York: Routledge.

Mitchell, Sara McLaughlin, and Emilia Justyna Powell. 2011. *Domestic Law Goes Global: Legal Traditions and International Courts*. New York: Cambridge University Press.

Mitchell, Sara McLaughlin, and Brandon C. Prins. 1999. "Beyond Territorial Contiguity: Issues at Stake in Democratic Militarized Interstate Disputes." *International Studies Quarterly* 43 (1): 169–83.

——. 2004. "Rivalry and Diversionary Uses of Force." *Journal of Conflict Resolution* 48 (6): 937–61.

Mitchell, Sara McLaughlin, and Cameron G. Thies. 2011. "Issue Rivalries." *Conflict Management and Peace Science* 28 (3): 230–60.

Mitchell, Sara McLaughlin, and Clayton L. Thyne. 2010. "Contentious Issues As Opportunities for Diversionary Behavior." *Conflict Management and Peace Science* 27 (5): 461–85.

Modelski, George. 1987. *Long Cycles in World Politics*. Seattle: University of Washington Press.

Modelski, George, and William R Thompson. 1988. *Seapower in Global Politics, 1494–1993.* New York: Springer.

Mohurle, Savita, and Manisha Patil. 2017. "A Brief Study of Wannacry Threat: Ransomware Attack 2017." *International Journal of Advanced Research in Computer Science* 8 (5): 1938–40.

Moon, Chungshik, and Mark Souva. 2016. "Audience Costs, Information, and Credible Commitment Problems." *Journal of Conflict Resolution* 60 (3): 434–58.

Moravcsik, Andrew. 1997. "Taking Preferences Seriously: A Liberal Theory of International Politics." *International Organization* 51 (4): 513–53.

Morey, Daniel. 2011. "When War Brings Peace: A Dynamic Model of the Rivalry Process." *American Journal of Political Science* 55 (2): 263–75.

——. 2016. "Military Coalitions and the Outcome of Interstate Wars." *Foreign Policy Analysis* 12 (4): 533–51.

——. 2020. "Centralized Command and Coalition Victory." *Conflict Management and Peace Science* 37(6): 716–34.

Morgan, E. Victor. 1952. *Studies in British Financial Policy, 1914–25*. New York: MacMillan.

Morgan, Patrick M. 1983. *Deterrence: A Conceptual Analysis*. Beverly Hills, CA: Sage.

——. 2018. "The Concept of Deterrence and Deterrence Theory." In *The Oxford Encyclopedia of Empirical International Relations Theory*, edited by William R. Thompson, Vol. 1, 293–317. New York: Oxford University Press.

Morgan, T. Clifton, and Kenneth N. Bickers. 1992. "Domestic Discontent and the External Use of Force." *Journal of Conflict Resolution* 36 (1): 25–52.

Morgan, T. Clifton, and Glenn Palmer. 2000. "A Model of Foreign Policy Substitutability: Selecting the Right Tools for the Job(s)." *Journal of Conflict Resolution* 44 (1): 11–32.

——. 2003. "To Protect and to Serve: Alliances and Foreign Policy Portfolios." *Journal of Conflict Resolution* 47 (2): 180–203.

Morgenthau, Hans J. 1948, 1960, 1978. 1st, 3rd, 5th editions. *Politics among Nations: The Struggle for Power and Peace*. New York: Knopf.

——. 1952. "Another Great Debate: The National Interest of the United States." *American Political Science Review* 46 (December): 961–88.

Morgenthau, Hans J., and Kenneth W. Thompson. 1985. *Politics among Nations: The Struggle for Power and Peace*, 6th edition. New York: Knopf.

Morrow, James D. 1985. "A Continuous-Outcome Expected Utility Theory of War." *Journal of Conflict Resolution* 29 (3): 473–502.

——. 1991. "Alliances and Asymmetry: An Alternative to the Capability Aggregation Model of Alliances." *American Journal of Political Science* 35 (4): 904–33.

——. 1993. "Arms versus Allies: Trade-Offs in the Search for Security." *International Organization* 47 (2): 207–33.

——. 1994. "Alliances, Credibility, and Peacetime Costs." *Journal of Conflict Resolution* 38 (2): 270–97.

——. 1999. "How Could Trade Affect Conflict?" *Journal of Peace Research* 36 (4): 481–89.

——. 2000. "Alliances: Why Write Them Down?" *Annual Review of Political Science* 3: 63–84.

——. 2017. "When Do Defensive Alliances Provoke Rather than Deter?" *Journal of Politics* 79 (1): 341–45.

Most, Benjamin A., and Harvey Starr. 1989, 2015. *Inquiry, Logic and International Politics*. 1st, 2nd editions. Columbia: University of South Carolina Press.

Mousseau, Michael. 1997. "Democracy and Militarized Interstate Collaboration." *Journal of Peace Research* 34 (1): 73–87.

——. 1998. "Democracy and Compromise in Militarized Interstate Conflicts, 1816–1992." *Journal of Conflict Resolution* 42 (2): 210–30.

——. 2000. "Market Prosperity, Democratic Consolidation, and Democratic Peace." *Journal of Conflict Resolution* 44 (4): 472–507.

——. 2005. "Comparing New Theory with Prior Beliefs: Market Civilization and the Democratic Peace." *Conflict Management and Peace Science* 22 (1): 63–77.

——. 2009. "The Social Market Roots of Democratic Peace." *International Security* 33 (4): 52–86.

——. 2012. "Capitalist Development and Civil War." *International Studies Quarterly* 56 (3): 470–83.

——. 2013. "The Democratic Peace Unraveled: It's the Economy." *International Studies Quarterly* 57 (1): 186–97.

——. 2019a. "Four Ways We Know the Democratic Peace Correlation Does Not Exist in the State of Knowledge." *Peace Economics, Peace Science and Public Policy* 45 (3): 1–8.

——. 2019b. "The End of War: How a Robust Marketplace and Liberal Hegemony Are Leading to Perpetual World Peace." *International Security* 44 (1): 160–96.

Mousseau, Michael, and Xionwei Cao. 2018."How the Contractualist Peace Overtook the Democratic Peace." In *The Oxford Encyclopedia of Empirical International Relations Theory*, edited by William R. Thompson, Vol. 2, 209–23. New York: Oxford University Press.

Mousseau, Michael, and Yuhang Shi. 1999. "A Test for Reverse Causality in the Democratic Peace Relationship." *Journal of Peace Research* 36 (6): 639–63.

Mueller, John E. 1985. *War, Presidents, and Public Opinion*. Lanham, MD: University Press of America.

——. 1989. *Retreat from Doomsday: The Obsolescence of Major War*. New York: Basic Books.

——. 1988. "The Essential Irrelevance of Nuclear Weapons: Stability in the Postwar World." *International Security* 13 (2): 55–79.

——. 2007. *The Remnants of War*. Ithaca, NY: Cornell University Press.

Murdie, Amanda. 2012. "The Bad, the Good, and the Ugly: The Curvilinear Effects of Civil-Military Conflict on International Crisis Outcome." *Armed Forces and Society* 39 (2): 233–54.

Murphy, Alexander B. 1990. "Historical Justifications for Territorial Claims." *Annals of the Association of American Geographers* 80 (4): 531–48.

Murphy, Martin N. 2010. *Small Boats, Weak States, Dirty Money: Piracy and Maritime Terrorism in the Modern World*. New York: Columbia University Press.

Murray, Shoon. 2018. "The 'Rally— 'Round the Flag' Phenomenon and the Diversionary Use of Force." In *The Oxford Encyclopedia of Empirical International Relations Theory*, edited by William R. Thompson, Vol. 3, 291–303. New York: Oxford University Press.

Nagl, John A. 2005. *Learning to Eat Soup with a Knife: Counterinsurgency Lessons from Malaya and Vietnam*. Chicago: University of Chicago Press.

Nail, Thomas. 2016. *Theory of the Border*. New York: Oxford University Press.

Nakashima, Ellen. 2019. "U.S. Cyber Command Operation Disrupted Internet Access of Russian Troll Factory on Day of 2018 Midterms," *Washington Post*, February 27. Available at https://www.washingtonpost.com/world/national-security/us-cyber-command-operation-disrupted-internet-access-of-russian-troll-factory-on-day-of-2018-midterms/2019/02/26/1827fc9e-36d6-11e9-af5b-b51b7ff322e9_story.html.

Narang, Neil, and Rupal N. Mehta. 2019. "The Unforeseen Consequences of Extended Deterrence: Moral Hazard in a Nuclear Client State." *Journal of Conflict Resolution* 63 (1): 218–50.

Narang, Vipin. 2009/2010. "Posturing for Peace? Pakistan's Nuclear Postures and South Asian Stability." *International Security* 34 (3): 38–78.

——. 2014. *Nuclear Strategy in the Modern Era: Regional Power Nuclear Postures and International Conflict*. Princeton, NJ: Princeton University Press.

Narang, Vipin, and Rebecca M. Nelson. 2009. "Who Are These Belligerent Democratizers? Reassessing the Impact of Democratization on War." *International Organization* 63 (2): 357–79.

Narang, Vipin, and Caitlin Talmadge. 2018. "Civil-Military Pathologies and Defeat in War: Tests Using New Data." *Journal of Conflict Resolution* 62 (7): 1379–405.

Narizny, Kevin. 2017. "On Systemic Paradigms and Domestic Politics: A Critique of the Newest Realism." *International Security* 42 (2): 155–90.

Nef, John U. 1950. *War and Human Progress*. Cambridge, MA: Harvard University Press.

Nemeth, Stephen C., Sara McLaughlin Mitchell, Elizabeth A. Nyman, and Paul R. Hensel. 2014. "Ruling the Sea: Managing Maritime Conflicts through UNCLOS and Exclusive Economic Zones." *International Interactions* 40 (5): 711–36.

Nexon, Daniel. 2009. *The Struggle for Power in Early Modern Europe: Religious Conflict, Dynastic Empires, and International Change*. Princeton, NJ: Princeton University Press.

Nitze, Paul H. 1956. "Atoms, Strategy, and Policy." *Foreign Affairs* 34 (2): 187–98.

Nolan, Janne E. 1989. *Guardians of the Arsenal: The Politics of Nuclear Strategy*. New York: Basic Books.

Nordås, Ragnhild, and Nils Petter Gleditsch. 2007. "Climate Change and Conflict." *Political Geography* 26 (6): 627–38.

Norrlof, Carla, and William C. Wohlforth. 2019. "Is US Grand Strategy Self-Defeating? Deep Engagement, Military Spending, and Sovereign Debt." *Conflict Management and Peace Science* 36 (3): 227–47.

North, Douglass C., and Barry R. Weingast. 1989. "Constitutions and Commitment: The Evolution of Institutions Governing Public Choice in 17th Century England." *Journal of Economic History* 49 (4): 803–32.

Northam, Jackie. 2015. "Iraqi Leader Visits Washington Looking for Help in Fight against Islamic State." *NPR* (April 14). Available at http://www.npr.org/sections/thetwo-way/2015/04/14/399655422/iraqi-leader-visits-washington-lookingfor-help-in-fight-against-islamic-state.

Nye, Joseph S., Jr. 2005. *Soft Power: The Means to Success in World Politics*. New York: PublicAffairs.

——. 2017. "Deterrence and Dissuasion in Cyberspace." *International Security* 41 (3): 44–71.

Nyman, Elizabeth A. 2015. "Offshore Oil Development and Maritime Conflict in the 20th Century: A Statistical Analysis of International Trends." *Energy Research and Social Science* 6: 1–7.

Obermeyer, Ziad, Christopher J. L. Murray, and Emmanuela Gakidou. 2008. "Fifty Years of Violent War Deaths from Vietnam to Bosnia: Analysis of Data from the World Health Survey Programme." *BMJ* 336 (7659): 1482–86.

Office of the Secretary of Defense. N.d. "Robert S. McNamara," https://history.defense.gov/Multimedia/Biographies/Article-View/Article/571271/robert-s-mcnamara/.

Ohanian, Lee E. 1997. "The Macroeconomic Effects of War Finance in the United States: World War II and the Korean War." *American Economic Review* 87 (1): 23–40.

——. 1998. *The Macroeconomic Effects of War Finance in the United States: Taxes, Inflation, and Deficit Finance*. New York: Garland.

O'Leary, Michael K. 1976. "The Role of Issues." In *Search of Global Patterns*, edited by James N. Rosenau, 318–25. New York: Free Press.

Olson, Mark W. 2004. "*Protecting the Financial Infrastructure*" Congressional Testimony, Washington DC, September 8. Committee on Financial Services,

U.S. House of Representatives. https://www.federalreserve.gov/boarddocs/testimony/2004/20040908/default.htm

Oneal, John R., and Bruce Russett. 1997. "The Classical Liberals Were Right: Democracy, Interdependence, and Conflict, 1950–1985." *International Studies Quarterly* 41 (2): 267–94.

Oren, Ido. 1995. "The Subjectivity of the 'Democratic Peace': Changing U.S. Perceptions of Imperial Germany." *International Security* 20 (2): 147–84.

Organski, A. F. K. 1958. *World Politics*. New York: Knopf.

Organski, A. F. K., and Jacek Kugler. 1977. "The Costs of Major Wars: The Phoenix Factor." *American Political Science Review* 71 (4): 1347–66.

——. 1978. "Davids and Goliaths: Predicting the Outcomes of International Wars." *Comparative Political Studies* 11 (2): 141–80.

——. 1980. *The War Ledger*. Chicago: University of Chicago Press.

Örsün, Ömer Faruk, Resat Bayer, and Michael Bernhard. 2018. "Democratization and Conflict." In *The Oxford Encyclopedia of Empirical International Relations Theory*, edited by William R. Thompson, Vol. 1, 450–68. New York: Oxford University Press.

Ostrom, Charles W., and Francis W. Hoole. 1978. "Alliances and Wars Revisited: A Research Note." *International Studies Quarterly* 22 (2): 215–36.

Ostrom, Charles W., and Brian Job. 1986. "The President and the Political Use of Force." *American Political Science Review* 80 (2): 541–66.

Ostrom, Elinor. 1990. *Governing the Commons: The Evolution of Institutions for Collective Action*. New York: Cambridge University Press.

——. 2000. "Collective Action and the Evolution of Social Norms." *Journal of Economic Perspectives* 14 (3): 137–58.

Overy, Richard. 1995. *Why the Allies Won*. New York: W. W. Norton.

Owsiak, Andrew P. 2012. "Signing Up for Peace: International Boundary Agreements, Democracy, and Militarized Interstate Conflict." *International Studies Quarterly* 56 (1): 51–66.

——. 2013. "Democratization and International Border Agreements." *Journal of Politics* 75 (3): 717–29.

——. 2018. "The Steps to War: Theory and Evidence." In *The Oxford Encyclopedia of Empirical International Relations Theory*, edited by William R. Thompson, Vol. 3, 484–508. New York: Oxford University Press.

——. 2019. "Foundations for Integrating the Democratic and Territorial Peace Arguments." *Conflict Management and Peace Science* 36 (1): 63–87.

Owsiak, Andrew P., Paul F. Diehl, and Gary Goertz. 2017. "Border Settlement and the Movement toward and from Negative Peace." *Conflict Management and Peace Science* 34 (2): 176–93.

Owsiak, Andrew P., and Derrick V. Frazier. 2014. "The Conflict Management Efforts of Allies in Interstate Disputes." *Foreign Policy Analysis* 10 (3): 243–64.

Owsiak, Andrew P., and Sara McLaughlin Mitchell. 2019. "Conflict Management in Land, River, and Maritime Claims." *Political Science Research and Methods* 7 (1): 43–61.

Owsiak, Andrew P., and Toby J. Rider. 2013. "Clearing the Hurdle: Border Settlement and Rivalry Termination." *Journal of Politics* 75 (3): 757–72.

Owsiak, Andrew P., and John A. Vasquez. 2019. "The Cart and the Horse Redux: The Timing of Border Settlement and Joint Democracy." *British Journal of Political Science* 49 (1): 339–54.

——. 2021. "Peaceful Dyads: A Territorial Perspective." Paper presented at the annual meeting of the Southern Political Science Association. January 7.

Packard, Gary C. 2009. "On the Use of Logarithmic Transformations in Allometric Analyses." *Journal of Theoretical Biology* 257 (3): 515–18.

Palmer, Glenn, Tamar R. London, and Patrick M. Regan. 2004. "What's Stopping You? The Sources of Political Constraints on International Conflict Behavior in Parliamentary Democracies." *International Interactions* 30 (1): 1–24.

Palmer, Glenn, Vito d'Orazio, Michael R. Kenwick, and Matthew Lane. 2015. "The MID4 Dataset, 2002–2010: Procedures, Coding Rules and Description." *Conflict Management and Peace Science* 32 (2): 222–42.

Palmer, Glenn, Vito D'Orazio, Michael R. Kenwick, and Roseanne W. McManus. 2020. "Updating the Militarized Interstate Dispute Data: A Response to Gibler, Miller, and Little." *International Studies Quarterly* 64 (2): 469–75.

Papayoanou, Paul A. 1997. "Economic Interdependence and The Balance of Power." *International Studies Quarterly* 41 (1): 113–40.

Pape, Robert A. 1996. *Bombing to Win: Air Power and Coercion in War*. Ithaca, NY: Cornell University Press.

Pardesi, Manjeet S. 2018. "Image Theory and Initiation of Strategic Rivalries." In *The Oxford Encyclopedia of Empirical International Relations Theory*, edited by William R. Thompson, Vol. 2, 225–44. New York: Oxford University Press.

Park, Johann, and Michael Colaresi. 2014. "Safe across the Border: The Continued Significance of the Democratic Peace When Controlling for Stable Borders." *International Studies Quarterly* 58 (1): 118–25.

Park, Johann, and Patrick James. 2015. "Democracy, Territory, and Armed Conflict, 1919–1995." *Foreign Policy Analysis* 11 (1): 85–107.

Partell, Peter. 1997. "Escalation at the Outset: An Analysis of Targets' Responses in Militarized Interstate Disputes." *International Interactions* 23 (1): 1–35.

Partell, Peter, and Glenn Palmer. 1999. "Audience Costs and Interstate Crisis: An Empirical Assessment of Fearon's Model of Dispute Outcome." *International Studies Quarterly* 43 (2): 389–405.

Paul, T. V. 2005a. "Causes of the India-Pakistan Enduring Rivalry." In *The India-Pakistan Conflict: An Enduring Rivalry*, edited by T. V. Paul, 3–26. Cambridge: Cambridge University Press.

——. 2005b. "Soft Balancing in the Age of US primacy." *International Security* 30 (1): 46–71.

——. 2009. *The Tradition of Non-use of Nuclear Weapons*. Stanford, CA: Stanford University Press.

——. 2018. *Restraining Great Powers: Soft Balancing from Empires to the Global Era*. New Haven, CT: Yale University Press.

Pauly, Reid. 2018. "Would U.S. Leaders Push the Button? Wargames and the Sources of Nuclear Restraint." *International Security* 43 (2): 151–92.

Peacock, Alan T., and Jack Wiseman. 1961. *The Growth of Public Expenditure in the United Kingdom*. Princeton, NJ: Princeton University Press.

Peceny, Mark. 1997. "A Constructivist Interpretation of the Liberal Peace: The Ambiguous Case of the Spanish-American War." *Journal of Peace Research* 34 (4): 415–30.

Peffley, Mark, and Robert Rohrschneider. 2003. "Democratization and Political Tolerance in Seventeen Countries: A Multilevel Model of Democratic Learning." *Political Research Quarterly* 56 (3): 243–57.

Peterson, Susan. 1996. *Crisis Bargaining and the State: The Domestic Politics of International Conflict*. Ann Arbor: University of Michigan Press.

Peterson, Tim, and Stephen L. Quackenbush. 2010. "Not All Peace Years Are Created Equal: Trade, Imposed Settlements, and Recurrent Conflict." *International Interactions* 36 (4): 363–83.

Pettersson, Therése, Stina Hogbladh, and Magnus Oberg. 2019. "Organized Violence, 1989–2018 and Peace Agreements." *Journal of Peace Research* 56 (4): 485–98.

Pickering, Jeffrey. 2002. "War-Weariness and Cumulative Effects: Victors, Vanquished, and Subsequent Interstate Intervention." *Journal of Peace Research* 39 (3): 313–37.

Pickering, Jeffrey, and David F. Mitchell. 2018. "Empirical Knowledge on Foreign Military Interventions." In *The Oxford Encyclopedia of Empirical International Relations Theory*, edited by William R. Thompson, Vol. 2, 714–36. New York: Oxford University Press.

Piketty, Thomas. 2017. *Capital in the Twenty-First Century*. Cambridge: Harvard University Press.

Pinker, Steven. 2011. *The Better Angels of Our Nature: Why Violence Has Declined*. New York: Viking.

——. 2012. "Fooled by Belligerence: Comments on Nassim Taleb's 'The Long Peace Is a Statistical Illusion.'" http://stevenpinker.com/files/comments_on_taleb_by_s_pinker.pdf

Poast, Paul. 2006. *The Economics of War*. New York: McGraw Hill-Irwin.

——. 2012."Does Issue Linkage Work? Evidence from European Alliance Negotiations, 1860 to 1945." *International Organization* 66 (2): 277–310.

——. 2013. "Can Issue Linkage Improve Treaty Credibility? Buffer State Alliances As a 'Hard Case.'" *Journal of Conflict Resolution* 57 (5): 739–64.

——. 2015. "Central Banks at War." *International Organization* 69 (1): 63–95.

——. 2019a. *Arguing about Alliances: The Art of Agreement in Military-Pact Negotiations*. Ithaca, NY: Cornell University Press.

——. 2019b. "Beyond the Sinew of War: The Political Economy of Security As a Subfield." *Annual Review of Political Science* 22 (1): 223–39.

Podvig, Pavel. 2008. "The Window of Vulnerability That Wasn't." *International Security* 33 (1): 118–38.

Polachek, Solomon. 1980. "Conflict and Trade." *Journal of Conflict Resolution* 24 (1): 55–78.

Polachek, Solomon, and Jun Xiang. 2010. "How Opportunity Costs Decrease the Probability of War in an Incomplete Information Game." *International Organization* 64 (Winter): 133–44.

Pomeroy, Robert, John Parks, Richard Pollnac, Tammy Campson, Emmanuel Genio, Cliff Marlessy, Elizabeth Holle, Michael Pido, Ayut Nissapa, Somsak

Boromthanarat, and Nguyen Thu Hue. 2007. "Fish Wars: Conflict and Collaboration in Fisheries Management in Southeast Asia." *Marine Policy* 31 (6): 645–56.
Porch, Douglas. 2013. *Counterinsurgency: Exposing the Myths of the New Way of War*. Cambridge: Cambridge University Press.
Posner, Eric, and John C. Yoo. 2005. "Judicial Independence in International Tribunals." *California Law Review* 93 (1): 1–74.
Potter, Philip B. K. 2007. "Does Experience Matter? American Presidential Experience, Age, and International Conflict." *Journal of Conflict Resolution* 51 (3): 351–78.
——. 2013. "Electoral Margins and American Foreign Policy." *International Studies Quarterly* 57 (3): 505–18.
Powell, Emilia Justyna. 2010. "Negotiating Military Alliances: Legal Systems and Alliance Formation." *International Interactions* 36 (1): 28–59.
——. 2015. "Islamic Law States and Peaceful Resolution of Territorial Disputes." *International Organization* 69 (4): 777–807.
——. 2020. *Islamic Law and International Law: Peaceful Resolution of Disputes*. New York: Oxford University Press.
Powell, Emilia Justyna, and Krista Wiegand. 2010. "Legal Systems and Peaceful Attempts to Resolve Territorial Disputes." *Conflict Management and Peace Science* 27 (2): 129–51.
Powell, Jonathan, and Mwita Chacha. 2016. "Investing in Stability: Economic Interdependence, Coups d'état, and the Capitalist Peace." *Journal of Peace Research* 53 (4): 525–38.
Powell, Robert. 1996. "Uncertainty, Shifting Power and Appeasement." *American Political Science Review* 90 (4): 749–64.
——. 1999. *In the Shadow of Power*. Princeton, NJ: Princeton University Press.
——. 2002. "Bargaining Theory and International Conflict." *Annual Review of Political Science* 5: 1–30.
——. 2004. "The Inefficient Use of Power: Costly Conflict with Complete Information." *American Political Science Review* 98 (2): 231–41.
——. 2006. "War As a Commitment Problem." *International Organization* 60 (1): 169–203.
Powers, Matthew Aderhold. 2018. "*The Impact of Yesterday's Territorial Shortages on Today's Adapted Mind.*" PhD dissertation, University of Illinois at Urbana-Champaign, http://hdl.handle.net/2142/102818.
Press, Daryl G. 2005. *Calculating Credibility: How Leaders Assess Military Threats*. Ithaca, NY: Cornell University Press.
Press, Daryl G., Scott D. Sagan, and Benjamin Valentino. 2013. "Atomic Aversion: Experimental Evidence on Taboos, Traditions, and the Non-Use of Nuclear Weapons." *American Political Science Review* 107 (1): 188–206.
Price, Richard. 2007. *The Chemical Weapons Taboo*. Ithaca, NY: Cornell University Press.
Prins, Brandon. 2005. "Interstate Rivalry and the Recurrence of Crises: A Comparison of Rival and Nonrival Crisis Behavior, 1918–1994." *Armed Forces and Society* 31 (3): 323–51.
Prorok, Alyssa, and Paul K. Huth. 2015. "International Law and the Consolidation of Peace Following Territorial Changes." *Journal of Politics* 77 (1): 161–74.

Quackenbush, Stephen L. 2010. "Territorial Issues and Recurrent Conflict." *Conflict Management and Peace Science* 27 (3): 239–52.

——. 2011. *Understanding General Deterrence: Theory and Application*. New York: Palgrave Macmillan.

——. 2015. *International Conflict: Logic and Evidence*. Washington, DC: CQ Press.

——. 2016. "Centers of Gravity and War Outcomes." *Conflict Management and Peace Science* 33 (4): 361–80.

——. 2018. "Empirical Analyses of Deterrence." In *The Oxford Encyclopedia of Empirical International Relations Theory*, edited by William R. Thompson, Vol. 1, 682–701. New York: Oxford University Press.

Quackenbush, Stephen L., and Amanda Murdie. 2015. "We Always Fight the Last War? Prior Experiences in Counterinsurgency and Conventional Warfare and War Outcomes." *International Interactions* 41(1): 183–200.

Quackenbush, Stephen L., and Michael Rudy. 2009. "Evaluating the Monadic Democratic Peace." *Conflict Management and Peace Science* 26 (3): 268–85.

Quackenbush, Stephen L., and Jerome F. Venteicher. 2008. "Settlements, Outcomes, and the Recurrence of Conflict." *Journal of Peace Research* 45 (6): 723–42.

Queralt, Didac. 2019. "War, International Finance, and Fiscal Capacity in the Long Run." *International Organization* 73 (4): 713–53.

Quester, George H. 2000. "Mismatched Deterrents: Preventing the Use of Nuclear, Biological, and Chemical Weapons." *International Studies Perspectives* 1 (2): 165–76.

Quinlan, Michael. 2007. "The Nuclear Proliferation Scene: Implications for NATO." In *NATO and the Future of the Nuclear Non-proliferation Treaty*, edited by Joseph F. Pilat and David S. Yost, NDC Occasional Papers no. 21, 99–105. Brussels: NATO Defence College.

Radchenko, Sergey. 2016. "The Soviet Union and the Cold War Arms Race." In *Arms Races in International Politics*, edited by Thomas Mahnken, Joseph Maiolo, and David Stevenson, 58–175. New York: Oxford University Press.

Raleigh, Clionadh, Lisa Jordan, and Idean Salehyan. 2008. "Assessing the Impact of Climate Change on Migration and Conflict." In *Papers Commissioned by the World Bank Group for the Social Dimensions of Climate Change Workshop*, 5–6. Washington, DC.

Raleigh, Clionadh, and Hendrik Urdal. 2007. "Climate Change, Environmental Degradation and Armed Conflict." *Political Geography* 26 (6): 674–94.

Ramsay, Christopher. 2004. "Politics at the Water's Edge." *Journal of Conflict Resolution* 48 (4): 459–86.

Randazzo, Kirk, Douglas M. Gibler, and Rebecca Reid. 2016. "Examining the Development of Judicial Independence." *Political Research Quarterly* 69 (3): 583–93.

Randle, Robert F. 1987. *Issues in the History of International Relations*. New York: Praeger.

Rasler, Karen A. 1986. "War, Accommodation, and Violence in the United States, 1870–1970." *American Political Science Review* 80 (3): 921–45.

——. 2000. "Shocks, Expectancy Revision, and the De-escalation of Protracted Conflicts: The Israeli-Palestinian Case." *Journal of Peace Research* 37 (6): 699–720.

Rasler, Karen A., and William R. Thompson. 1983. "Global Wars, Public Debts, and the Long Cycle." *World Politics* 35 (4): 489–516.
——. 1985. "War and the Economic Growth of Major Powers." *American Journal of Political Science* 29 (3): 513–38.
——. 1989. *War and State Making: The Shaping of the Global Powers*. Boston: Unwin Hyman.
——. 2005. *Puzzles of the Democratic Peace: Theory, Geopolitics, and the Transformation of World Politics*. London: Palgrave Macmillan.
——. 2006. "Contested Territory, Strategic Rivalries, and Conflict Escalation." *International Studies Quarterly* 50 (1): 145–67.
——. 2018. "War Making and the Building of State Capacity: Expanding the Bivariate Relationship." In *The Oxford Encyclopedia of Empirical International Relations Theory*, edited by William R. Thompson, Vol. 4, 495–519. New York: Oxford University Press.
Rasler, Karen A., William R. Thompson, and Sumit Ganguly. 2013. *How Rivalries End*. Philadelphia: University of Pennsylvania Press.
Rathbun, Brian C. 2012. "Politics and Paradigm Preferences: The Impact of Ideology of International Relations Scholars." *International Studies Quarterly* 56 (3): 607–22.
Rathbun, Brian C., and Rachel Stein. 2020. "Greater Goods: Morality and Attitudes toward the Use of Nuclear Weapons." *Journal of Conflict Resolution* 64 (5): 787–816.
Ratner, Steven R. 2006. "Land Feuds and Their Solutions: Finding International Law beyond the Tribunal Chamber." *American Journal of International Law* 100 (4): 808–29.
Rauchhaus, Robert. 2009. "Evaluating the Nuclear Peace Hypothesis: A Quantitative Approach." *Journal of Conflict Resolution* 53 (2): 258–77.
Rausch, Carsten. 2017. "Challenging the Power Consensus: GDP, CINC, and Power Transition Theory." *Security Studies* 26 (4): 642–64.
Ray, James Lee. 1995. *Democracy and International Conflict*. Columbia: University of South Carolina Press.
——. 2000. "Democracy: On the Level(s), Does Democracy Correlate with Peace?" In *What Do We Know about War?* edited by John A. Vasquez, 299–316. Lanham, MD: Rowman and Littlefield.
Ray, James Lee, and Allan Dafoe. 2018. "Democratic Peace versus Contractualism." *Conflict Management and Peace Science* 35 (2): 193–203.
Raymond, Gregory A. 1994. "Democracies, Disputes, and Third-Party Intermediaries." *Journal of Conflict Resolution* 38 (1): 24–42.
Reddie, Andrew W., Bethany L. Goldblum, Kiran Lakkaraju, Jason Reinhardt, Michael Nacht, and Laura Epifanovskaya. 2019. "Applying Wargames to Real-World Policies—Response." *Science* 363 (6434): 1406–407.
Reed, William. 2000. "A Unified Statistical Model of Conflict Onset and Escalation." *American Journal of Political Science* 44 (1): 84–93.
Reed, William, and David H. Clark. 2000. "War Initiators and War Winners: The Consequences of Linking Theories of Democratic War Success." *Journal of Conflict Resolution* 44 (3): 378–95.

Reeder, Bryce W. 2018. "The Political Geography of Rebellion: Using Event Data to Identify Insurgent Territory, Preferences, and Relocation Patterns." *International Studies Quarterly* 62 (3): 696–707.

Reiter, Dan. 2001. "Does Peace Nurture Democracy?" *Journal of Politics* 63 (3): 935–48.

——. 2003. "Exploring the Bargaining Model of War." *Perspectives on Politics* 1 (1): 27–43.

——. 2005. "Preventive Attacks against Nuclear Programs and the 'Success' at Osiraq." *Nonproliferation Review* 12 (2): 355–71.

——. 2009. *How Wars End*. Princeton, NJ: Princeton University Press.

——. 2018. "Is Democracy a Cause of Peace?" In *The Oxford Encyclopedia of Empirical International Relations Theory*, edited by William R. Thompson, Vol. 2, 352–69. New York: Oxford University Press.

Reiter, Dan, and Allan C. Stam III. 1998. "Democracy, War Initiation, and Victory." *American Political Science Review* 92 (2): 377–89.

——. 2002. *Democracies at War*. Princeton, NJ: Princeton University Press.

——. 2003. "Understanding Victory: Why Political Institutions Matter." *International Security* 28 (1): 168–79.

Reiter, Dan, Allan C. Stam III, and Michael C. Horowitz. 2016. "A Revised Look at Interstate Wars, 1816–2007." *Journal of Conflict Resolution* 60 (5): 956–76.

Reiter, Dan, and Scott Wolford. 2019. "*Leaders, Gender, and Crisis Bargaining*." Typescript, Emory University and the University of Texas.

Renshon, Jonathan, Allan Dafoe, and Paul K. Huth. 2018. "Influence and Reputation Formation in World Politics." *American Journal of Political Science* 62 (2): 325–39.

Reuveny, Rafael. 2001. "Bilateral Import, Export, and Conflict/Cooperation Simultaneity." *International Studies Quarterly* 45 (1): 131–58.

——. 2007. "Climate Change-Induced Migration and Violent Conflict." *Political Geography* 26 (6): 656–73.

Reuveny, Rafael, and Katherine Barbieri. 2014. "On the Effect of Natural Resources on Interstate War." *Progress in Physical Geography: Earth and Environment* 38 (6): 786–806.

Rhamey, J. Patrick, and Thomas J. Volgy. 2018. "Regions, Powers and Hierarchy from a Comparative Regional Analysis of Conflicts in International Politics." In *The Oxford Encyclopedia of Empirical International Relations Theory*, edited by William R. Thompson, Vol. 3, 335–48. New York: Oxford University Press.

Rhodes, Richard. 1986. *The Making of the Atomic Bomb*. New York: Simon and Schuster.

Richardson, Lewis F. 1944. "The Distribution of Wars in Time." *Journal of the Royal Statistical Society* 107 (3/4): 242–50.

——. 1948. "Variation of the Frequency of Fatal Quarrels with Magnitude." *Journal of the American Statistical Association* 43 (244): 523–46.

——. 1960a. *Arms and Insecurity*. Pacific Grove, CA: Boxwood Press.

——. 1960b. *Statistics of Deadly Quarrels*. Chicago: Quadrangle Books.

Rid, Thomas. 2012. "Cyber War Will Not Take Place." *Journal of Strategic Studies* 35 (1): 5–32.

——. 2020. *Active Measures: The Secret History of Disinformation and Political Warfare*. New York: Farrar, Straus and Giroux.

Rider, Toby J. 2013. "Uncertainty, Salient Stakes, and the Causes of Conventional Arms Races." *International Studies Quarterly* 57 (3): 580–91.

——. 2018. "Arms Races: An Assessment of Conceptual and Theoretical Challenges." In *The Oxford Encyclopedia of Empirical International Relations Theory*, edited by William R. Thompson, Vol. 1, 128–41. New York: Oxford University Press.

Rider, Toby J., Michael G. Findley, and Paul F. Diehl. 2011. "Understanding Arms Race Onset: Rivalry, Theory, and Territorial Competition." *Journal of Peace Research* 48 (1): 85–100.

Rider, Toby J., and Andrew P. Owsiak. 2015. "Border Settlement, Commitment Problems, and the Causes of Contiguous Rivalry." *Journal of Peace Research* 52 (4): 508–21.

——. Forthcoming. *On Dangerous Ground: A Theory of Bargaining, Border Settlement, and Rivalry*. NY: Cambridge University Press.

Ritter, Emily Hencken. 2014. "Policy Disputes, Political Survival, and the Onset and Severity of State Repression." *Journal of Conflict Resolution* 58 (2): 254–79.

Roberts, Anthea. 2017. *Is International Law International?* New York: Oxford University Press.

Roberts, Anthea, Paul B. Stephan, Pierre-Hugues Verdier, and Mila Versteeg (eds.). 2018. *Comparative International Law*. Oxford: Oxford University Press.

Roberts, David C., and Donald L Turcotte. 1998. "Fractality and Self-Organized Criticality of Wars." *Fractals* 6 (4): 351–57.

Rosato, Sebastian. 2003. "The Flawed Logic of Democratic Peace Theory." *American Political Science Review* 97 (4): 585–602.

——. 2005. "Explaining the Democratic Peace." *American Political Science Review* 99 (3): 467–72.

Rosecrance, Richard. 1985. *The Rise of the Trading State: Commerce and Conquest in the Modern World*. New York: Basic Books.

——. 1996. "The Rise of the Virtual State: Territory Becomes Passé." *Foreign Affairs* 85 (July/August): 45–61.

Rosen, Steven. 1972. "War Power and the Willingness to Suffer." In *Peace, War, and Numbers*, edited by Bruce Russett, 167–84. Beverly Hills, CA: Sage.

Rosenau, James N. 1966. "Pre-theories and Theories in Foreign Policy." In *Approaches to Comparative and International Politics*, edited by R. Barry Farrell, 27–93. Evanston, IL: Northwestern University Press.

——. 1967. "Foreign Policy As an Issue Area." In *Domestic Sources of Foreign Policy*, edited by James N. Rosenau, 11–50. New York: Free Press.

Rosenstone, Steven J., and John Mark Hansen. 1993. *Mobilization, Participation, and Democracy in America*. New York: Longman.

Rothgeb, John M., Jr. 1993. *Defining Power: Influence and Force in the Contemporary International System*. New York: St. Martin's.

Rousseau, David L. 2005. *Democracy and War: Institutions, Norms, and the Evolution of International Conflict*. Stanford, CA: Stanford University Press.

Rovner, Joshua. 2019. "Cyber War As an Intelligence Contest." *War on the Rocks*, September 16, https://warontherocks.com/2019/09/cyber-war-as-an-intelligence-contest/.

Rublee, Maria Rost. 2009. *Nonproliferation Norms: Why States Choose Nuclear Restraint.* Athens, GA: University of Georgia Press.

Rumer, Eugene. 2017. *Moldova Between Russia and the West: A Delicate Balance.* New York: Carnegie Endowment for Peace.

Rummel, Rudolph J. 1979. *War, Power, and Peace. Vol. 4 of Understanding Conflict and War*. Thousand Oaks, CA: Sage.

——. 1983. "Libertarianism and International Violence." *Journal of Conflict Resolution* 27 (March): 27–51.

Russett, Bruce M. 1963. "The Calculus of Deterrence." *Journal of Conflict Resolution* 7 (2): 97–109.

——. 1990. *Controlling the Sword: The Democratic Governance of National Security*. Cambridge, MA: Harvard University Press.

——. 1993. *Grasping the Democratic Peace: Principles for a Post–Cold War World.* Princeton, NJ: Princeton University Press.

Russett, Bruce, Christopher Layne, David E. Spiro, Michael W. Doyle. 1995. "The Democratic Peace." *International Security* 19 (4): 164–84.

Russett, Bruce, and John R. Oneal 2001. *Triangulating Peace: Democracy, Interdependence, and International Organizations*. New York: W. W. Norton.

Russett Bruce, John Oneal, and David Davis. 1998. "The Third Leg of the Kantian Tripod for Peace: International Organizations and Militarized Disputes, 1950–1985." *International Organizations* 52 (3): 441–48.

Sabrosky, Alan Ned. 1980. "Interstate Alliances: Their Reliability and the Expansion of War." In *The Correlates of War*, Vol. 2, edited by J. David Singer, 161–98. New York: Free Press.

Sack, Robert David. 1986. *Human Territoriality*. Cambridge: Cambridge University Press.

Sagan, Scott D. 1985. "Nuclear Alerts and Crisis Management." *International Security* 9 (4): 99–139.

——. 1990. *Moving Targets: Nuclear Strategy and National Security*. Princeton, NJ: Princeton University Press.

——. 1994. *The Limits of Safety: Organizations, Accidents, and Nuclear Weapons.* Princeton, NJ: Princeton University Press.

——. 2004. "Realist Perspectives on Ethical Norms and Weapons of Mass Destruction." In *Ethics and Weapons of Mass Destruction: Religious and Secular Perspectives*, edited by Sohail H. Hashmi, 73–95. New York: Cambridge University Press.

——. 2009. "The Case for No First Use." *Survival* 51 (3): 163–82.

——. 2010. "Nuclear Latency and Nuclear Proliferation," In *Forecasting Nuclear Proliferation in the 21st Century: The Role of Theory*, edited by William C. Potter and Gaukha Mukhatzhanova, 80–101. Stanford, CA: Stanford University Press.

Sagan, Scott D., and Jeremi Suri. 2003. "The Madman Nuclear Alert: Secrecy, Signaling, and Safety in October 1969." *International Security* 27 (4): 150–83.

Sagan, Scott D., and Benjamin Valentino. 2017. "Revisiting Hiroshima in Iran: What Americans Really Think about Using Nuclear Weapons and Killing Noncombatants." *International Security* 42 (1): 41–79.

Sagan, Scott D., and Kenneth N. Waltz. 2003. *The Spread of Nuclear Weapons: A Debate Renewed.* New York: W. W. Norton.

Sahlins, Peter. 1989. *Boundaries: The Making of France and Spain in the Pyrenees*. Berkeley: University of California Press.
Sakaguchi, Kendra, Anil Varughese, and Graeme Auld. 2017. "Climate Wars? A Systematic Review of Empirical Analyses on the Links between Climate Change and Violent Conflict." *International Studies Review* 19 (4): 622–45.
Sakuwa, Kentaro. 2018. "Approaches to Explaining Regional Conflict and Peace." In *The Oxford Encyclopedia of Empirical International Relations Theory*, edited by William R. Thompson, Vol. 1, 108–28. New York: Oxford University Press.
Salehyan, Idean. 2008. "From Climate Change to Conflict? No Consensus Yet." *Journal of Peace Research* 45 (3): 315–26.
Sample, Susan G. 1997. "Arms Races and Dispute Escalation: Resolving the Debate." *Journal of Peace Research* 34 (1): 7–22.
——. 1998. "Military Buildups, War, and Realpolitik: A Multivariate Model." *Journal of Conflict Resolution* 42 (2): 156–75.
——. 2000. "Military Buildups: Arming and War." In *What Do We Know about War?* edited by John A. Vasquez, 165–95. Lanham, MD: Rowman and Littlefield.
——. 2002. "The Outcomes of Military Buildups: Minor States vs. Major Powers." *Journal of Peace Research* 39 (6): 669–91.
——. 2012. "Arms Races: A Cause or a Symptom." In *What Do We Know about War?* 2nd edition, edited by John A. Vasquez, 111–38. Lanham, MD: Rowman and Littlefield.
——. 2014. "From Territorial Claim to War: Timing, Causation, and the Steps-to-War." *International Interactions* 40 (2): 270–85.
——. 2018a. "Anticipating War? War Preparations and the Steps-to-War Thesis." *British Journal of Political Science* 48 (2): 489–511.
——. 2018b. "Power, Wealth, and Satisfaction: When Do Power Transitions Lead to Conflict?" *Journal of Conflict Resolution* 62 (9): 1905–31.
——. 2021. "Arms Races." In *What Do We Know about War?* edited by John A. Vasquez. Lanham, MD: Rowman and Littlefield.
Sarkees, Meredith Reid. n.d. "The COW Typology of War: Defining and Categorizing Wars (Version 4 of the Data)." http://tinyurl.com/nzzu5dj (October 22, 2015).
Sarkees, Meredith Reid, and Frank Wayman. 2010. *Resort to War: 1816–2007*. Washington, DC: CQ Press.
Sarkees, Meredith Reid, Frank Whelon Wayman, and J. David Singer. 2003. "Inter-state, Intra-state, and Extra-state Wars: A Comprehensive Look at Their Distribution over Time, 1816–1997." *International Studies Quarterly* 47 (1): 49–70.
Saunders, Elizabeth N. 2011. *Leaders at War: How Presidents Shape Military Interventions*. Ithaca, NY: Cornell University Press.
Schell, Jonathan. 1984. *The Abolition*. New York: Knopf.
Schelling, Thomas C. 1966. *Arms and Influence*. New Haven, CT: Yale University Press.
——. 1978. *Micromotives and Macrobehavior*. New York: W. W. Norton.
——. 1980. *The Strategy of Conflict*. Cambridge, MA: Harvard University Press.
Schenoni, Luis L., Gary Goertz, Andrew P. Owsiak, and Paul F. Diehl. 2020. "Settling Resistant Territorial Disputes: The Territorial Boundary Peace in Latin America." *International Studies Quarterly* 64 (1): 57–70.

Scheve, Kenneth, and David Stasavage. 2010. "The Conscription of Wealth: Mass Warfare and the Demand for Progressive Taxation." *International Organization* 64 (4): 529–61.

——. 2012. "Democracy, War, and Wealth: Evidence from Two Centuries of Inheritance Taxation." *American Political Science Review* 106 (1): 81–102.

Schmidt, Cody J., Bomi K. Lee, and Sara McLaughlin Mitchell. "Climate Bones of Contention: How Climate Change Influences Territorial, Maritime, and River Interstate Conflicts." *Journal of Peace Research, forthcoming*. 2021 https://journals.sagepub.com/eprint/DU4JWEKYKQFC9BPVJBM5/full

Schneider, Gerald. 2018. "Capitalist Peace Theory: A Critical Approach." In *The Oxford Encyclopedia of Empirical International Relations Theory*, edited by William R. Thompson, Vol. 1, 181–94. New York: Oxford University Press.

Schneider, Jacquelyn. 2017. "Cyber and Crisis Escalation: Insights from Wargaming." In USASOC Futures Forum. https://paxsims.files.wordpress.com/2017/01/paper-cyber-and-crisis-escalation-insights-from-wargaming-schneider.pdf

Schrodt, Philip A. 2012. *Cameo: Conflict and Mediation Event Observations, Event and Actor Codebook*. University Park: Pennsylvania State University Press.

Schub, Robert. 2017. "Unfair Fights: Power Asymmetry, Nascent Nuclear Capability, and Preventive Conflict." *Conflict Management and Peace Science* 34 (4): 431–55.

——. 2020. "When Prospective Leader Turnover Promotes Peace." *International Studies Quarterly* 64 (3): 510–22.

Schultz, Kenneth A. 2001a. *Democracy and Coercive Diplomacy*. Cambridge: Cambridge University Press.

——. 2001b. "Looking for Audience Costs." *Journal of Conflict Resolution* 45 (1): 32–60.

——. 2005. "The Politics of Risking Peace: Do Hawks or Doves Deliver the Olive Branch?" *International Organization* 59 (1): 1–38.

——. 2014. "What's in a Claim? De jure vs. De facto Borders in Interstate Territorial Disputes." *Journal of Conflict Resolution* 58 (6): 1059–84.

——. 2015. "Borders, Conflict, and Trade." *Annual Review of Political Science* 18 (1): 125–45.

——. 2017. "Mapping Interstate Territorial Conflict: A New Data Set and Applications." *Journal of Conflict Resolution* 61 (7): 1565–90.

Schultz, Kenneth A., and Hein E. Goemans. 2019. "Aims, Claims and the Bargaining Model of War." *International Theory* 11 (3): 344–74.

Schultz, Kenneth A., and Justin S. Mankin. 2019. "Is Temperature Exogenous? The Impact of Civil Conflict on the Instrumental Climate Record in Sub-Saharan Africa." *American Journal of Political Science* 63 (4): 723–39.

Schultz, Kenneth A., and Barry R. Weingast. 2003. "The Democratic Advantage: Institutional Foundations of Financial Power in International Competition." *International Organization* 57 (1): 3–42.

Schweller, Randall L. 1998. *Deadly Imbalances: Tripolarity and Hitler's Strategy of World Conquest*. New York: Columbia University Press.

Scoblic, J. Peter. 2009. "Robert McNamara's Logical Legacy." Arms Control Today. https://www.armscontrol.org/act/2009_09/lookingback_McNamara.

Scott, James C. 2009. *The Art of Not Being Governed*. New Haven, CT: Yale University Press.

——.2017. *Nuclear Weapons and Coercive Diplomacy*. Cambridge: Cambridge University Press.

Seawright, Jason. 2016. *Multi-Method Social Science: Combining Qualitative and Quantitative Tools.* Cambridge: Cambridge University Press.

Sechser, Todd S. and Matthew Fuhrmann. 2017. *Nuclear Weapons and Coercive Diplomacy*. Cambridge: Cambridge University Press.

Sechser, Todd, Neil Narang, and Caitlin Talmadge. 2019. "Emerging Technologies and Strategic Stability in Peacetime, Crisis, and War." *Journal of Strategic Studies* 42 (6): 727–35.

Sen, Amartya K. 1990. "Individual Freedom As Social Commitment." *India International Centre Quarterly* 17 (1): 101–15.

Senese, Paul D. 1996. "Geographic Proximity and Issue Salience: The Effects on the Escalation of Militarized Conflict." *Conflict Management and Peace Science* 15 (2): 133–61.

——. 2005. "Territory, Contiguity and International Conflict: Assessing a New Joint Explanation." *American Journal of Political Science* 49 (4): 769–79.

Senese, Paul D., and Stephen L. Quackenbush. 2003. "Sowing the Seeds of Conflict: The Effect of Dispute Settlements on Durations of Peace." *Journal of Politics* 65 (3): 696–17.

Senese, Paul D., and John A. Vasquez. 2003. "A Unified Explanation of Territorial Conflict: Testing the Impact of Sampling Bias." *International Studies Quarterly* 47 (2): 275–98.

——. 2005. "Assessing the Steps to War." *British Journal of Political Science* 35 (4): 607–33.

——. 2008. *The Steps to War: An Empirical Study*. Princeton, NJ: Princeton University Press.

Shannon, Megan. 2009. "Preventing War and Providing the Peace? International Organizations and the Management of Territorial Disputes." *Conflict Management and Peace Science* 26 (2): 144–63.

Shaw, Malcolm. 2017. *International Law*, 8th edition. Cambridge: Cambridge University Press.

Shea, Patrick E. 2014. "Financing Victory: Sovereign Credit, Democracy, and War." *Journal of Conflict Resolution* 58 (5): 771–95.

Shea, Patrick E., and Paul Poast. 2018. "War and Default." *Journal of Conflict Resolution* 62 (9): 1876–904.

Shelef, Nadav. 2019. "How Homelands Change." *Journal of Conflict Resolution*. 64 (2–3): 490–517.

Shirkey, Zachary Z. 2018. "Military Intervention in Interstate and Civil Wars: A Unified Interpretation." In *The Oxford Encyclopedia of Empirical International Relations Theory*, edited by William R. Thompson, Vol. 2, 534–49. New York: Oxford University Press.

Siegel, Jennifer. 2014. *For Peace and Money: French and British Finance in the Service of Tsars and Commissars*. Oxford: Oxford University Press.

Sil, Rudra, and Peter J. Katzenstein. 2010. *Beyond Paradigms: Analytic Eclecticism in the Study of World Politics*. New York: Palgrave Macmillan.

Simmel, Georg. 1955. *Conflict*. Glencoe, IL: Free Press.

Simmons, Beth A. 1999. "See You in Court? The Appeal to Quasi-judicial Legal Processes in the Settlement of Territorial Disputes." In *A Roadmap To War: Territorial Dimensions of International Conflict,* edited by Paul F. Diehl, 205–37. Nashville, TN: Vanderbilt University Press.

——. 2002. "Capacity, Commitment and Compliance: International Law and the Settlement of Territorial Disputes." *Journal of Conflict Resolution* 46 (6): 829–56.

——. 2005. "Rules over Real Estate: Trade, Territorial Conflict, and International Border As Institution." *Journal of Conflict Resolution* 49 (6): 823–48.

Singer, J. David. 1972. *The Scientific Study of Politics: An Approach to Foreign Policy Analysis*. Morristown, NJ: General Learning Press.

——. 1979. *The Correlates of War, Vol. 1*. NY: Free Press.

——. 1987. "Reconstructing the Correlates of War Dataset on Material Capabilities of States, 1816–1985." *International Interactions* 14 (2): 115–32.

Singer, J. David, Stuart A. Bremer, and John Stuckey. 1972. "Capability Distribution, Uncertainty, and Major Power War, 1820–1965." In *Peace, War and Numbers*, edited by Bruce Russett, 19–48. Beverly Hills, CA: Sage.

Singer, J. David, and Melvin Small. 1966a. "Formal Alliances, 1815–1939: A Quantitative Description." *Journal of Peace Research* 3 (1): 1–31.

——. 1966b. "National Alliance Commitments and War Involvement, 1818–1945." *Papers, Peace Research Society* 5: 109–40.

——. 1972. *The Wages of War, 1816–1965: A Statistical Handbook*. New York: John Wiley.

Singer, Peter Warren, and August Cole. 2015. *Ghost Fleet: A Novel of the Next World War*. Boston: Houghton Mifflin Harcourt.

Sirin, Cigdem V., and Michael T. Koch. 2015. "Dictators and Death: Casualty Sensitivity of Autocracies in Militarized Interstate Disputes." *International Studies Quarterly* 59 (4): 802–14.

Siverson, Randolph M. 1980. "War and Change in the International System." In *Change in the International System*, edited by Ole R. Holsti, Randolph M. Siverson, and Alexander L. George, 211–29. Boulder, CO: Westview Press.

Siverson, Randolph M., and Juliann Emmons. 1991. "Birds of a Feather: Democratic Political Systems and Alliance Choices in the Twentieth Century." *Journal of Conflict Resolution* 35 (2): 285–306.

Siverson, Randolph M., and Joel King. 1980. "Attributes of National Alliance Membership and War Participation, 1815–1965." *American Journal of Political Science* 24 (1): 1–15.

Siverson, Randolph M., and Harvey Starr. 1989. "Alliance Behavior and Border Effects on the War Behavior of States: Refining the Interaction Opportunity Model." *Conflict Management and Peace Science* 10 (Spring): 21–46.

Siverson, Randolph M., and Michael R. Tennefoss. 1984. "Power, Alliance, and the Escalation of International Conflict, 1815–1965." *American Political Science Review* 78 (4): 1057–69.

Slantchev, Branislav L. 2006. "Politicians, the Media, and Domestic Audience Costs." *International Studies Quarterly* 50 (2): 445–77.

——. 2012. "Borrowed Power: Debt Finance and the Resort to Arms." *American Political Science Review* 106 (4): 787–809.

Slantchev, Branislav L., Anna Alexandrova, and Erik Gartzke. 2005. "Probabilistic Causality, Selection Bias, and the Logic of the Democratic Peace." *American Political Science Review* 99 (3): 459–62.

Slayton, Rebecca. 2017. "What Is the Cyber Offense-Defense Balance? Conceptions, Causes, and Assessment." *International Security* 41 (3): 72–109.

Small, Melvin, and J. David Singer. 1970. "Patterns in International Warfare, 1816–1965." *Annals of the American Academy of Political and Social Science* 391: 145–55.

——. 1976. "The War-Proneness of Democratic Regimes, 1816–1965." *Jerusalem Journal of International Relations* 1 (Summer): 50–68.

——. 1982. *Resort to Arms: International and Civil Wars, 1816–1980*, Beverly Hills, CA: Sage.

Smith, Alastair. 1995. "Alliance Formation and War." *International Studies Quarterly* 39 (4): 405–25.

——. 1996a. "Diversionary Foreign Policy in Democratic Systems." *International Studies Quarterly* 40 (1): 133–53.

——. 1996b. "To Intervene or Not to Intervene: A Biased Decision." *Journal of Conflict Resolution* 40 (1): 16–40.

——. 1998. "Extended Deterrence and Alliance Formation." *International Interactions* 24 (4): 315–43.

——. 1999. "Testing Theories of Strategic Choice: The Example of Crisis Escalation." *American Journal of Political Science* 43 (4): 1254–83.

——. 2003. "Election Timing in Majoritarian Parliaments." *British Journal of Political Science* 33 (3): 397–418.

Smith, Bradley C., and William Spaniel. 2019. "Militarized Disputes, Uncertainty, and Leader Tenure." *Journal of Conflict Resolution* 63 (5): 1222–52.

Smith, Justin H. 1919. *The War with Mexico, Vol. II*. New York: Macmillan.

Snyder, Glenn. 1960. "Balance of Power in the Missile Age." *Journal of International Affairs* 14 (1): 21–34.

——. 1965. *The Balance of Power and the Balance of Terror*. San Francisco: Chandler.

——. 1984. "The Security Dilemma in Alliance Politics." *World Politics* 36 (4): 461–95.

——. 2002. "Mearsheimer's World—Offensive Realism and the Struggle for Security: A Review Essay." *International Security* 27 (1): 149–73.

Snyder, Jack L. 1977. The Soviet Strategic Culture. Rand Report. R-2154-AF, Santa Monica.

——. 1984. "Civil-Military Relations and the Cult of the Offensive, 1914 and 1984." *International Security* 9 (1): 108–46.

Snyder, Jack, and Erica D. Borghard. 2011. "The Cost of Empty Threats: A Penny, Not a Pound." *American Political Science Review* 105 (3): 437–56.

Sobek, David, Dennis M. Foster, and Samuel B. Robison. 2012. "Conventional Wisdom? The Effect of Nuclear Proliferation on Armed Conflict, 1945–2001." *International Studies Quarterly* 56 (1): 149–62.

Song, Yann-Huei. 1997. "The Canada-European Union Turbot Dispute in the Northwest Atlantic: An Application of the Incident Approach." *Ocean Development and International Law* 28 (3): 269–311.

Souva, Mark. 2018. "Power Shifts and War." In *The Oxford Encyclopedia of Empirical International Relations Theory*, edited by William R. Thompson, Vol. 2, 138–53. New York: Oxford University Press.

Spagat, Michael, and Stijn van Weezel. 2020. "The Decline of War since 1950: New Evidence." In *Lewis Fry Richardson: His Intellectual Legacy and Influence in the Social Sciences*, edited by Nils Petter Gleditsch, 129–42. Cham: Springer International. http://link.springer.com/10.1007/978-3-030-31589-4_11

Spagat, Michael, Stijn van Weezel, Minzhang Zheng, and Neil F. Johnson. 2019. "Toward a Unified Understanding of Casualty Distributions in Human Conflict." arXiv:1911.01994 [physics]. http://arxiv.org/abs/1911.01994

Spaniel, William. 2019. *Bargaining over the Bomb: The Successes and Failures of Nuclear Negotiations*. Cambridge: Cambridge University Press.

Spaniel, William, and Bradley C. Smith. 2015. "Sanctions, Uncertainty, and Leader Tenure." *International Studies Quarterly* 59 (4): 735–49.

Sparrow, John C. 1952. *History of Personnel Demobilization in the United States Army*. Washington DC: Department of the Army.

Spiro, David E. 1994. "The Insignificance of the Liberal Peace." *International Security* 19 (2): 50–86.

Spruyt, Hendrik. 2002. "The Origins, Development, and Possible Decline of the Modern State." *Annual Review of Political Science* 5 (1): 127–49.

Stalley, Phillip. 2003. "Environmental Scarcity and International Conflict." *Conflict Management and Peace Science* 20 (1): 33–58.

Stam, Alan C., III. 1996. *Win, Lose or Draw: Domestic Politics and the Crucible of War*. Ann Arbor: University of Michigan Press.

Stanley, Elizabeth A., and John P. Sawyer. 2009. "The Equifinality of War Termination: Multiple Paths to Ending War." *Journal of Conflict Resolution* 53 (3): 651–76.

Starr, Joyce R. 1991. "Water Wars." *Foreign Policy* 82 (Spring): 17–36.

Stein, Arthur A. 1980. *The Nation at War*. Baltimore, MD: Johns Hopkins University Press.

——. 2016a. "Land Armaments in Europe, 1866–1914." In *Arms Races in International Politics*, edited by Thomas Mahnken, Joseph Maiolo, and David Stevenson, 41–58. New York: Oxford University Press.

Stevenson, David. 2016b. "Part I before 1914: Introduction." In *Arms Races in International Politics*, edited by Thomas Mahnken, Joseph Maiolo, and David Stevenson, 11–19. New York: Oxford University Press.

Stiglitz, Joseph E., and Linda J. Bilmes. 2008. *The Three Trillion Dollar War: The True Cost of the Iraq Conflict*. New York: W. W. Norton.

Strange, Susan. 1994. *States and Markets*, 2nd edition. London: Continuum.

Stoll, Richard J. 1982. "Let the Researcher Beware: The Use of the Richardsonian Equations to Estimate the Parameters of a Dyadic Arms Acquisition Process." *American Journal of Political Science* 26 (1): 77–89.

——. 1984a. "Bloc Concentration and Dispute Escalation among the Major Powers, 1830–1965." *Social Science Quarterly* 65 (1): 48–59.

——. 1984b. "The Guns of November: Presidential Reelections and the Use of Force, 1947–1982." *Journal of Conflict Resolution* 28 (2): 231–46.

——. 1992. "Steaming in the Dark? Rules, Rivals, and the British Navy, 1860–1913." *Journal of Conflict Resolution* 24 (3): 287–313.

——. 2018. "To Arms, to Arms: What Do We Know about Arms Races?" In *The Oxford Encyclopedia of Empirical International Relations Theory*, edited by William R. Thompson, Vol. 4, 352–71. New York: Oxford University Press.

Sullivan, Patricia L. 2007. "War Aims and War Outcomes: Why Powerful States Lose Limited Wars." *Journal of Conflict Resolution* 51 (3): 496–524.

——. 2012. *Who Wins? Predicting Strategic Success and Failure in Armed Conflict*. New York: Oxford University Press.

Summers, Harry. 1989. "A War Is a War Is a War Is a War." In *Low Intensity Conflict: The Pattern of Warfare in the Modern World,* edited by Loren B. Thompson, 27–30. Lexington, MA: Lexington.

Swain, Ashok. 1997. "Ethiopia, the Sudan, and Egypt: The Nile River Dispute." *Journal of Modern African Studies* 35 (4): 675–94.

Sweeney, Kevin J. 2003. "The Severity of Interstate Disputes: Are Dyadic Capability Preponderances Really More Pacific?" *Journal of Conflict Resolution* 47 (6): 728–50.

Talbot, Brent J. 2018. "Eliminating ICBMs As Part of a 21st-Century Deterrence Strategy." *Bulletin of the Atomic Scientists* 74: 1, 52–59.

Taleb, Nassim Nicholas. 2005. *Fooled by Randomness: The Hidden Role of Chance in Life and in the Markets*, 2nd edition, updated. New York: Random House.

——. 2007. *The Black Swan: The Impact of the Highly Improbable*, 1st edition. New York: Random House.

——. 2012. "The 'Long Peace' Is a Statistical Illusion." https://web.archive.org/web/20121117225617/http://www.fooledbyrandomness.com/longpeace.pdf.

Taliaferro, Jeffery W. 2001. "Security Seeking under Anarchy: Defensive Realism Revisited." *International Security* 25 (3): 128–61.

Talmadge, Caitlin. 2017. "Would China Go Nuclear? Assessing the Risk of Chinese Nuclear Escalation in a Conventional War with the United States." *International Security* 40 (4): 50–92.

——. 2019a. "Emerging Technology and Intra-war Escalation Risks: Evidence from the Cold War, Implications for Today." *Journal of Strategic Studies* 42 (6): 864–87.

——. 2019b. "The US-China Nuclear Relationship: Why Competition Is Likely to Intensify." Brookings Institution Report. Available at https://www.brookings.edu/wp-content/uploads/2019/09/FP_20190930_china_nuclear_weapons_talmadge-1.pdf

Tammen, Ronald L., Jacek Kugler, and Douglas Lemke. 2018. "Foundations of Power Transition Theory." In *The Oxford Encyclopedia of Empirical International Relations Theory*, edited by William R. Thompson, Vol. 2, 19–64. New York: Oxford University Press.

Tanaka, Seiki. 2016. "The Microfoundations of Territorial Disputes: Evidence from a Survey Experiment in Japan." *Conflict Management and Peace Science* 33 (5): 516–38.

Tannenwald, Nina. 1999. "The Nuclear Taboo: The United States and the Normative Basis of Nuclear Non-use." *International Organization* 53 (3): 433–68.

——. 2007. *The Nuclear Taboo: The United States and the Nonuse of Nuclear Weapons since 1945*. New York: Cambridge University Press.

Tarar, Ahmer. 2006. "Diversionary Incentives and the Bargaining Approach to War." *International Studies Quarterly* 50 (1): 169–88.

Terechshenko, Zhanna. 2020. "Hot under the Collar: A Latent Measure of Interstate Hostility." Working paper.

Theisen, Ole Magnus, Helge Holtermann, and Halvard Buhaug. 2012. "Climate Wars? Assessing the Claim That Drought Breeds Conflict." *International Security* 36 (3): 79–106.

Theisen, Ole Magnus, Nils Petter Gleditsch, and Halvard Buhaug. 2013. "Is Climate Change a Driver of Armed Conflict?" *Climate Change* 117: 613–25.

Thies, Cameron G. 2005. "War, Rivalry, and State Building in Latin America." *American Journal of Political Science* 49 (3): 451–65.

Thompson, Robert. 1966. *Defeating Communist Insurgency: The Lessons of Malaya and Vietnam*. New York: Praeger.

Thompson, William R. 1995. "The Consequences of War." In *The Process of War: Advancing the Scientific Study of War,* edited by Stuart A. Bremer and Thomas R. Cusack, 161–90. Amsterdam: Gordon and Breach.

——. 1996. "Democracy and Peace: Putting the Cart before the Horse?" *International Organization* 50 (1): 141–74.

——. 2001. "Identifying Rivals and Rivalries in World Politics." *International Studies Quarterly* 45 (4): 557–87.

——. 2007. "Powder Kegs, Sparks, and World War I." In *Explaining War and Peace: Case Studies and Necessary Condition Counterfactuals*, edited by Jack Levy and Gary Goertz, 113–46. New York: Routledge.

——, ed. 2018a. *The Oxford Encyclopedia of Empirical International Relations Theory*. New York: Oxford University Press.

——. 2018b. "Constructing a General Model Accounting for Interstate Rivalry Termination." In *The Oxford Encyclopedia of Empirical International Relations Theory*, edited by William R. Thompson, Vol. 1, 362–81. New York: Oxford University Press.

——. 2020. *Power Concentration in World Politics: The Political Economy of Systemic Leadership.* Cham, Switzerland: Springer.

——. 2021. *American Global Pre-eminence: The Development and Erosion of Systemic Leadership.* New York: Oxford University Press.

Thompson, William R. and Leila Zakhirova. 2018. "Systemic Leadership, Energy Considerations, and the Leadership Long-Cycle Perspective." In *The Oxford Encyclopedia of Empirical International Relations Theory*, edited by William R. Thompson, Vol. 3, 596–619. New York: Oxford University Press.

——. 2019. *Racing to the Top: How Energy Fuels Systemic Leadership in World Politics*. New York: Oxford University Press.

Thompson, William R., and Gary Zuk. 1982. "War, Inflation, and the Kondratieff Long Wave." *Journal of Conflict Resolution* 26 (4): 621–44.

Thucydides. 1972. *The History of the Peloponnesian War*. Translated by Rex Warner. Penguin.

——. 2006. *The History of the Peloponnesian War*. Translated by Richard Crawley and revised by Donald Lateiner. New York: Barnes and Noble Classics.

Tilly, Charles. 1975. *The Formation of National States in Western Europe*. Princeton, NJ: Princeton University Press.

——. 1978. *From Mobilization to Revolution*. Boston: Addison-Wesley.

——. 1985. "War Making and State Making As Organized Crime." In *Bringing the State Back In*, edited by Peter B. Evans, Dietrich Rueschemeyer, and Theda Skocpol, 169–91. New York: Cambridge University Press.

——. 1990. *Coercion, Capital, and European States, AD 990–1990*. Cambridge, MA: B. Blackwell.

Tir, Jaroslav. 2003. "Averting Armed International Conflicts through State-to-State Territorial Transfers." *Journal of Politics* 65 (4): 1235–57.

——. 2006a. "Domestic-Level Territorial Disputes: Conflict Management via Secessions." *Conflict Management and Peace Science* 23 (4): 309–28.

——. 2006b. *Redrawing the Map to Promote Peace*. New York: Lexington.

——. 2010. "Territorial Diversion: Diversionary Theory of War and Territorial Conflict." *Journal of Politics* 72 (2): 413–25.

Tir, Jaroslav, and Maureen Bailey. 2018. "Painting Too 'Rosie' a Picture: The Impact of External Threat on Women's Economic Welfare." *Conflict Management and Peace Science* 35 (3): 248–62.

Tir, Jaroslav, and Paul F. Diehl. 1998. "Demographic Pressure and Interstate Conflict: Linking Population Growth and Density to Militarized Disputes and Wars, 1930–89." *Journal of Peace Research* 35 (3): 319–39.

——. 2002. "Geographic Dimensions of Enduring Rivalries." *Political Geography* 21 (2): 263–86.

Tir, Jaroslav, and Douglas M. Stinnett. 2012. "Weathering Climate Change: Can Institutions Mitigate International Water Conflict?" *Journal of Peace Research* 49 (1): 211–25.

Tiron, Roxana, and Derek Wallbank. 2014. "House Passes $1 Billion Loan Guarantees to Aid Ukraine." *Bloomberg Business* (March 6). Available at http://www.bloomberg.com/news/articles/2014-03-06/house-passes-1-billion-in-loan-guarantees-to-aid-ukraine.

Toft, Monica Duffy. 2003. "Indivisible Territory, Geographic Concentration and Ethnic War." *Security Studies* 12 (2): 82–119.

______. 2007. "Population Shifts and Civil War: A Test of Power Transition Theory." *International Interactions* 33 (3): 243–69.

Tol, Richard S., and Sebastian Wagner. 2010. "Climate Change and Violent Conflict in Europe over the Last Millennium." *Climate Change* 99: 65–79.

Tomz, Michael. 2007a. "Domestic Audience Costs in International Relations: An Experimental Approach." *International Organization* 61 (4): 821–40.

_____. 2007b. *Reputation and International Cooperation: Sovereign Debt across Three Centuries*. Princeton, NJ: Princeton University Press.

Tomz, Michael, and Jessica L. P. Weeks. 2013. "Public Opinion and the Democratic Peace." *American Political Science Review* 107 (4): 849–65.

——. 2019. "Military Alliances and Public Support for War." Working paper.

Toset, Hans Petter Wollebaek, Nils Petter Gleditsch, and Håvard Hegre. 2000. "Shared Rivers and Interstate Conflict." *Political Geography* 19: 971–96.

Townsen, Ashly Adam, and Bryce W. Reeder. 2014. "Where Do Peacekeepers Go When They Go? Explaining the Spatial Heterogeneity of Peacekeeping Deployments." *Journal of International Peacekeeping* 18 (1–2): 69–91.

Trachtenberg, Marc. 2012. "Audience Costs: An Historical Analysis." *Security Studies* 21 (1): 3–42.

Tunsjø, Øystein. 2018. *The Return of Bipolarity in World Politics: China, the United States, and Geostructural Realism*. New York: Columbia University Press.

Twagiramungu, Noel, Allard Duursma, Mulugeta Gebrehiwot Berhe, and Alex de Waal. 2019. "Re-describing Transnational Conflict in Africa". *Journal of Modern African Studies* 57 (3): 377–91.

Ungerer, Jameson. 2012. "Assessing the Progress of the Democratic Peace Research Program." *International Studies Review* 14 (1): 1–31.

USAID Press Office. 2015. "U.S. Signs Loan Guarantee Agreement for Ukraine." USAID (May 18). Available at https://www.usaid.gov/news-information/press-releases/may-18-2015-us-signs-loan-guarantee-agreement-ukraine.

Uslaner, Eric M. 1976. "The Pitfalls of Per Capita." *American Journal of Political Science* 20 (1): 125–33.

Uzonyi, Gary, and Toby J. Rider. 2017. "Determinants of Foreign Aid: Rivalry and Domestic Instability." *International Interactions* 43 (2): 272–99.

Uzonyi, Gary, Mark Souva, and Sona N. Golder. 2012. "Domestic Institutions and Credible Signals." *International Studies Quarterly* 56 (4): 765–776.

Valentino, Benjamin A., Paul K. Huth, and Sarah E. Croco. 2010. "Bear Any Burden? How Democracies Minimize the Costs of War." *Journal of Politics* 72 (2): 528–44.

Valeriano, Brandon. 2013. *Becoming Rivals: The Process of Interstate Rivalry Development*. New York: Routledge.

Valeriano, Brandon, and Ryan C. Maness. 2012. "Persistent Enemies and Cybersecurity: The Future of Rivalry in an Age of Information Warfare." In *Cyberspace and National Security: Threats, Opportunities, and Power in a Virtual World*, edited by Derek S. Reveron, 139–58. Washington, DC: Georgetown University Press.

——. 2014. "The Dynamics of Cyber Conflict between Rival Antagonists, 2001–11." *Journal of Peace Research* 51 (3): 347–60.

Valeriano, Brandon, Benjamin Jensen, and Ryan C. Maness. 2018. *Cyber Strategy: The Evolving Character of Power and Coercion*. New York: Oxford University Press.

Valeriano, Brandon, Ryan C. Maness, and Benjamin Jensen. 2017. "Cyberwarfare Has Taken a New Turn: Yes, It's Time to Worry." *Washington Post: The Monkey Cage* 7, no. 13.

Valeriano, Brandon, and Allison Pytlak. 2016. "Closing the Internet Up: The Rise of Cyber Repression." *Council on Foreign Relations Net Politics*. https://www.cfr.org/blog/closing-internet-rise-cyber-repression

Valeriano, Brandon, and John A. Vasquez. 2010. "Identifying and Classifying Complex Wars." *International Studies Quarterly* 54 (June): 561–82.

Van Bavel, Bas J. P. et al. 2019. "Climate and Society in Long-Term Perspective: Opportunities and Pitfalls in the Use of Historical Datasets." *WIREs Climate Change* 10 (6): e611.

Van Valen, Leigh. 1973. "A New Evolutionary Law." *Evolutionary Theory* 1 (1): 1–30.

Vasquez, John A. 1979. "Coloring It Morgenthau: New Evidence for an Old Thesis on Quantitative International Politics." *British Journal of International Studies* 5 (October): 210–28.

——. 1983. "The Tangibility of Issues and Global Conflict: A Test of Rosenau's Issue Area Typology." *Journal of Peace Research* 20 (2): 179–92.

——. 1991. “The Deterrence Myth.” In *The Long Postwar Peace: Contending Explanations and Projections*, edited by Charles W. Kegley, Jr., 203–23. New York: HarperCollins.
——. 1993. *The War Puzzle*. New York: Cambridge University Press.
——. 1995. “Why Do Neighbors Fight? Proximity, Interaction, or Territoriality.” *Journal of Peace Research* 32 (3): 277–93.
——, ed. 2000, 2012. *What Do We Know about War? 1st and 2nd* editions. Lanham, MD: Rowman and Littlefield.
——. 2001. “Mapping the Probability of War and Analyzing the Possibility of Peace: The Role of Territorial Issues.” *Conflict Management and Peace Science* 18 (2): 145–74.
——. 2009. *The War Puzzle Revisited*. Cambridge: Cambridge University Press.
——. 2011b. “Territorial Paths to War: Their Probability of Escalation, 1816–2001.” In *Territory, War, and Peace,* edited by John A. Vasquez and Marie T. Henehan, 133–47. London: Routledge.
——. 2011a. “Paths to War: The Territorial Origins of War.” In *Territory, War, and Peace,* edited by John A. Vasquez and Marie T. Henehan, 148–76. London: Routledge.
——. 2018. *Contagion and War: Lessons from the First World War*. Cambridge: Cambridge University Press.
——. 2020. “*Anomalies of the Wilsonian (Monadic) Democratic Peace in the Nineteenth Century: What Can They Tell Us?*” Florence, Italy: European University Institute, EUI Working Papers [RSCAS 2020/31], hhttps://cadmus.eui.eu
Vasquez, John A., and Emily E. Barrett. 2015. “Peace As the Absence of Militarized Conflict: Comparing the Democratic and Territorial Peace.” *Journal of Territorial and Maritime Studies* 2 (1): 5–31.
Vasquez, John A., Paul F. Diehl, Colin Flint, Jürgen Scheffran, Sang-Hyun Chi, and Toby J. Rider. 2011. “The Conflict Space of Cataclysm: The International System and the Spread of War 1914–1917.” *Foreign Policy Analysis* 7 (2): 143–68.
Vasquez, John A., and Marie T. Henehan. 2001. “Territorial Disputes and the Probability of War.” *Journal of Peace Research* 38 (2): 123–38.
Vasquez, John A., and Choong-Nam Kang. 2013. “How and Why the Cold War Became a Long Peace: Some Statistical Insights.” *Cooperation and Conflict* 48 (1): 28–50.
Vasquez, John A., and Christopher S. Leskiw. 2001. “The Origins and War Proneness of Interstate Rivalries.” *Annual Review of Political Science* 4 (1): 295–316.
Vasquez, John A., and Ashlea Rundlett. 2016. “Alliances As a Necessary Condition of Multiparty Wars.” *Journal of Conflict Resolution* 60 (8): 1395–1418.
Vasquez, John A., and Brandon Valeriano. 2010. “Classification of Interstate Wars.” *Journal of Politics* 72 (April): 292–309.
Väyrynen, Raimo, ed. 2006. *The Waning of Major War: Theories and Debates*. London: Routledge.
Volgy, Thomas J., Kelly M. Gordell, Paul Bezerra, and Jon Patrick Rhamey Jr. 2018. “Conflict, Regions, and Regional Hierarchies.” In *The Oxford Encyclopedia of Empirical International Relations Theory*, edited by William R. Thompson, Vol. 1, 335–62. New York: Oxford University Press.
Volpe, Tristan. 2017. “Atomic Leverage: Compellence with Nuclear Latency.” *Security Studies* 26 (3): 517–44.

Von Bortkewitsch, Ladislas. 1898. *Das Gesetz Der Kleinen Zahlen*. Leipzig: B. G. Teubner.

Von Clausewitz, Carl. 1984. *On War*. Edited and translated by Michael Howard and Peter Paret. Princeton, NJ: Princeton University Press.

Vu Le, Quan, and Paul J. Zak. 2006. "Political Risk and Capital Flight." *Journal of International Money and Finance* 25 (2): 308–29.

Wagner, Harrison. 2000. "Bargaining and War." *American Journal of Political Science* 44 (3): 469–84.

——. 2001. "*Who's Afraid of Rational Choice Theory?*" Typescript, University of Texas.

Walker, Stephen G., Mark Schafer, and Michael D. Young. 1998. "Systematic Procedures for Operational Code Analysis: Measuring and Modeling Jimmy Carter's Operational Code." *International Studies Quarterly* 42 (1): 175–89.

Wallace, Michael D. 1979. "Arms Races and Escalation: Some New Evidence." *Journal of Conflict Resolution* 23 (1): 3–16.

——. 1980. "Some Persisting Findings." *Journal of Conflict Resolution* 24 (22): 289–92.

——. 1982. "Armaments and Escalation: Two Competing Hypotheses." *International Studies Quarterly* 26 (1): 37–51.

Wallensteen, Peter, and Margareta Sollenberg. 1995. "After the Cold War: Emerging Patterns of Armed Conflict 1989–94." *Journal of Peace Research* 32 (3): 345–60.

Wallerstein, Immanuel. 1974. "The Rise and Future Demise of the World Capitalist System." *Comparative Studies in Society and History* 16 (4): 387–415.

Walt, Stephen M. 1987. *The Origins of Alliances*. Ithaca, NY: Cornell University Press.

Walter, Barbara. 1997. "The Critical Barrier to Civil War Settlement." *International Organization* 51 (3): 335–64.

——. 2002. *Committing to Peace: The Successful Settlement of Civil Wars*. Princeton, NJ: Princeton University Press.

Waltz, Kenneth. 1959. *Man, the State, and War: A Theoretical Analysis*. New York: Columbia University Press.

——.1964. "The Stability of a Bipolar World." *Daedalus* 93 (3): 881–909.

——. 1979. *Theory of International Politics*. New York: Random House.

——. 1981. *The Spread of Nuclear Weapons: More May Be Better (Adelphi Paper No. 171)*. London: International Institute for Strategic Studies.

——. 1990. "Nuclear Myths and Political Realities." *American Political Science Review* 84 (3): 731–45.

——. 1997. "Thoughts about Virtual Nuclear Arsenals." *Washington Quarterly* 20 (3): 153–61.

Wang, Kevin, and James Lee Ray. 1994. "Beginners and Winners: The Fate of Initiators of Interstate Wars Involving Great Powers since 1495." *International Studies Quarterly* 38 (1): 139–54.

Ward, Michael. 1984. "Different Paths to Parity: A Study of the Contemporary Arms Race." *American Political Science Review* 78 (2): 297–317.

Wargames. 1983. Film. United States: MGM.

Wawro, Geoffrey. 2003. *The Franco-Prussian War: The German Conquest of France in 1870–1871*. New York: Cambridge University Press.

Wayman, Frank W. 1984. "Bipolarity and War: The Role of Capability Concentration and Alliance Patterns among Major Powers, 1816–1965." *Journal of Peace Research* 21 (1): 61–78.

——. 2000. "Rivalries: Recurrent Disputes and Explaining War." In *What Do We Know about War?* edited by John A. Vasquez, 219–34. Lanham, MD: Rowman and Littlefield.

Wayman, Frank Whelon, and Daniel M. Jones. 1991. "Evolution of Conflict in Enduring Rivalries." Paper presented at the annual meeting of the International Studies Association, Vancouver, British Columbia, March 20–23.

Wayman, Frank W., J. David Singer, and Gary Goertz. 1983. "Capabilities, Allocations, and Success in Militarized Disputes and Wars, 1816–1976." *International Studies Quarterly* 27 (4): 497–515.

Weart, Spencer R. 1998. *Never at War: Why Democracies Will Not Fight One Another*. New Haven, CT: Yale University Press.

Weber, Max. 2015. *Weber's Rationalism and Modern Society*. London: Palgrave MacMillan.

Weede, Erich. 1980. "Arms Races and Escalation: Some Persisting Doubts." *Journal of Conflict Resolution* 24 (June): 285–87.

——. 1984. "Democracy and War Involvement." *Journal of Conflict Resolution* 28 (4): 649–64.

——. 1996. *Economic Development, Social Order, and World Politics*. Boulder, CO: Lynne Rienner.

——. 2018. "The Expansion of Economic Freedom and the Capitalist Peace." In *The Oxford Encyclopedia of Empirical International Relations Theory*, edited by William R. Thompson, Vol. 1, 820–36. New York: Oxford University Press.

Weeks, Jessica L. P. 2008. "Autocratic Audience Costs: Regime Type and Signaling Resolve." *International Organization* 62 (1): 35–64.

——. 2012. "Strongmen and Straw Men: Authoritarian Regimes and the Initiation of International Conflict." *American Political Science Review* 106 (2): 326–47.

Weisiger, Alex. 2013. *Logics of War: Explanations for Limited and Unlimited Conflicts*. Ithaca, NY: Cornell University Press.

——. 2016. "Learning from the Battlefield: Information, Domestic Politics, and Interstate War Duration." *International Organization* 70 (2): 347–75.

Weiss, Jessica Chen. 2013. "Authoritarian Signaling, Mass Audiences and Nationalist Protest in China." *International Organization* 60 (2): 473–95.

Weiss, Jessica Chen and Allan Dafoe. 2019. "Authoritarian Audiences, Rhetoric, and Propaganda in International Crises: Evidence from China." *International Studies Quarterly* 63 (4): 963–973.

Wendt, Alexander. 1999. *Social Theory of International Politics*. Cambridge: Cambridge University Press.

Werner, Suzanne. 1998. "Negotiating the Terms of Settlement: War Aims and Bargaining Leverage." *Journal of Conflict Resolution* 42 (3): 321–43.

——. 1999. "The Precarious Nature of Peace: Resolving the Issues, Enforcing the Settlement, and Renegotiating the Terms." *American Journal of Political Science* 43 (3): 912–34.

Werner, Suzanne, and Amy Yuen. 2005. "Making and Keeping Peace." *International Organization* 59 (2): 262–93.

Wheeler, Hugh. 1975. "Effects of War on Industrial Growth." *Society* 12 (4): 48–52.
——. 1980. "Postwar Industrial Growth." In *The Correlates of War II: Testing Some Realpolitik Models*, edited by J. David Singer, 258–84. New York: Free Press.
Whiting, Allen S. 1960. *China Crosses the Yalu: The Decision to Enter the Korean War*. Stanford: Stanford University Press.
——. 1991. "The U.S.-China War in Korea." In *Avoiding War: Problems of Crisis Management*, edited by Alexander L. George, 103–25. Boulder: Westview Press.
Whitlark, Rachel Elizabeth. 2017. "Nuclear Beliefs: A Leader-Focused Theory of Counter-Proliferation." *Security Studies* 26 (4): 545–74.
Wiegand, Krista E. 2009. "China's Strategy in the Senkaku/Diaoyu Islands Dispute: Issue Linkage and Coercive Diplomacy." *Asian Security* 5 (2): 170–93.
——. 2011. "Militarized Territorial Disputes: States' Attempts to Transfer Reputation for Resolve." *Journal of Peace Research* 48 (1): 101–13.
____. 2014. "Mediation in Territorial, Maritime, and River Disputes," *International Negotiation* 19 (2): 343–70.
——. 2018. "Conflict Management of Territorial Disputes." In *The Oxford Encyclopedia of Empirical International Relations Theory*, edited by William R. Thompson, Vol. 1, 317–35. New York: Oxford University Press.
Wiegand, Krista E., and Emilia Justya Powell. 2011. "Past Experience, Quest for the Best Forum, and Peaceful Attempts to Resolve Territorial Disputes." *Journal of Conflict Resolution* 55 (1): 33–59.
Wiegand, Krista E., Emilia Justyna Powell, and Steven McDowell. 2020. "The Peaceful Resolution of Territorial Disputes Dataset, 1945–2015,"*Journal of Peace Research*, forthcoming. https://doi.org/10.1177/0022343319895560
Willard, Kristen L., Timothy W. Guinnane, and Harvey S. Rosen. 1996. "Turning Points in the Civil War: Views from the Greenback Market." *American Economic Review* 86 (4): 1001–18.
Williamson, Samuel R., Jr. 2014. "July 1914 Revisited and Revised: The Erosion of the German Paradigm." In *The Outbreak of the First World War: Structure, Politics, and Decision-Making*, edited by Jack S. Levy and John A. Vasquez, 30–64. Cambridge: Cambridge University Press.
Williamson, Samuel R., Jr., and Russel van Wyk. 2003. *July 1914: Soldiers, Statesmen and the Coming of the Great War: A Brief Documentary History*. Boston: Bedford/St. Martins.
Wilson, Thomas M, and Hastings Donnan. 2016. *A Companion to Border Studies*. New York: John Wiley
Wischnath, Gerdis, and Halvard Buhaug. 2014. "On Climate Variability and Civil War in Asia." *Climatic Change* 122: 709–21.
Wohlforth, William. 1999. "The Stability of a Unipolar World." *International Security* 24 (1): 5–41.
Wolf, Aaron. 1998. "Conflict and Cooperation along International Waterways." *Water Policy* 1 (2): 251–65.
Wolford, Scott. 2007. "The Turnover Trap: New Leaders, Reputation, and International Conflict." *American Journal of Political Science* 51 (4): 772–88.
——. 2012. "Incumbents, Successors, and Crisis Bargaining: Leadership Turnover As a Commitment Problem." *Journal of Peace Research* 49 (4): 517–30.

——. 2015. *The Politics of Military Coalitions*. New York: Cambridge University Press.

——. 2018. "Wars of Succession." *International Interactions* 44 (1): 173–87.

Wolford, Scott, and Emily Hencken Ritter. 2016. "National Leaders, Political Security, and the Formation of Military Coalitions." *International Studies Quarterly* 60 (3): 540–51.

World Bank Databank. Available at https://databank.worldbank.org/home.aspx.

Wright, Quincy. 1965 [1942]. *A Study of War*, 2nd edition. Chicago: University of Chicago Press.

Wright, Thorin M. 2014. "Territorial Revision and State Repression." *Journal of Peace Research* 51 (3): 375–87.

Wright, Thorin M., and Toby J. Rider. 2014. "Disputed Territory, Defensive Alliances and Conflict Initiation." *Conflict Management and Peace Science* 31 (2): 119–44.

Wright, Timothy, and Shweta Moorthy. 2018. "Refugees, Economic Capacity, and Host State Repression." *International Interactions* 44 (1): 132–55.

Wu, Cathy Xuanxuan, and Scott Wolford. 2018. "Leaders, States and Reputations." *Journal of Conflict Resolution* 62 (10): 2087–117.

Wucherpfennig, Julian, Nils B. Weidmann, Luc Girardin, Lars-Erik Cederman, and Andreas Wimmer. 2011. "Politically Relevant Ethnic Groups across Space and Time: Introducing the GeoEPR Dataset." *Conflict Management and Peace Science* 28 (5): 423–37.

Xiang, Jun, Xiaohong Xu, and George Keteku. 2007. "Power: The Missing Link in the Trade Conflict Relationship." *Journal of Conflict Resolution* 51 (4): 646–63.

Yarhi-Milo, Keren. 2018. *Who Fights for Reputation? The Psychology of Leaders in International Conflict*. Princeton, NJ: Princeton University Press.

Young, Arthur N. 1965. *China's Wartime Finance and Inflation, 1937–1945*. Cambridge, MA: Harvard University Press.

Yuen, Amy. 2009. "Target Concessions in the Shadow of Intervention." *Journal of Conflict Resolution* 53 (5): 745–73.

Zacher, Mark. 2001. "The Territorial Integrity Norm: International Boundaries and the Use of Force." *International Organization* 55 (2): 215–50.

Zagare, Frank C. 1987. *The Dynamics of Deterrence*. Chicago: University of Chicago Press.

——. 2018. "Perfect Deterrence Theory." In *The Oxford Encyclopedia of Empirical International Relations Theory*, edited by William R. Thompson, Vol. 3, 34–52. New York: Oxford University Press.

Zagare, Frank C., and D. Marc Kilgour. 2000. *Perfect Deterrence*. Cambridge: Cambridge University Press.

Zala, Benjamin. 2017. "Polarity Analysis and Collective Perceptions of Power: The Need for a New Approach." *Journal of Global Security Studies* 2 (1): 2–17.

Zarakol, Ayşe. 2017. *Hierarchies in World Politics*. Cambridge: Cambridge University Press.

Zawahri, Neda A. 2008. "Designing River Commissions to Implement Treaties and Manage International Rivers." *Water International* 33 (4): 464–74.

Zeitoun, Mark, and Jeroen Warner. 2006. "Hydro-hegemony: A Framework for Analysis of Trans-boundary Water Conflicts." *Water Policy* 8 (5): 435–60.

Zhang, David, Peter Brecke, Harry Lee, Yuan-Qing He, and Jane Zhang. 2007. "Global Climate Change, War, and Population Decline in Recent Human History." *Proceedings of the National Academy of Sciences in the United States of America* 104 (49): 19214–19.

Zimmerman, William. 1973. "Issue Area and Foreign-Policy Process: A Research Note in Search of a General Theory." *American Political Science Review* 67 (4): 1204–12.

Zinnes, Dina. 1976. *Contemporary Research in International Relations: A Perspective and a Critical Appraisal*. New York: Free Press.

Name Index

Subject Index

About the Contributors

Seden Akcinaroglu is associate professor at Binghamton University. She is the author of two books, *The Battle for Allegiance Governments, Terrorist Groups, and Constituencies in Conflict* and *Private Militaries and the Security Industry in Civil Wars*. Her research on civil wars, terrorism, and international conflict has been published in the *Journal of Conflict Resolution*, the *Journal of Peace Research*, and *International Interactions*.

Douglas B. Atkinson is a postdoctoral university assistant at the University of Salzburg (PhD University of Georgia). His areas of expertise include international conflict and foreign policy. His work has appeared in *Political Research Quarterly*, the *Journal of Global Security Studies*, and *Research and Politics*, among others.

Bear F. Braumoeller is professor of political science and the Baronov and Timashev Chair in Data Analytics at The Ohio State University. He conducts research in the areas of international relations, political methodology, and complexity and human behavior. His present work focuses on the decline-of-war thesis and the relationship between international order and international conflict. He currently directs the Modeling Emergent Social Order lab at Ohio State and codirects the Computational Social Science community of practice under the Translational Data Analytics Institute.

Rosella Cappella Zielinski is assistant professor of political science at Boston University. She is a specialist on the political economy of security. She is the author of *How States Pay for War*. She has published articles in *Conflict Management and Peace Science*, the *Journal of Peace Research*, *Security Studies*, the *European Journal of International Relations*, and *Foreign Affairs*, among others.

Paul F. Diehl is the Ashbel Smith Professor of Political Science at the University of Texas at Dallas, the Henning Larsen Professor Emeritus of Political Science at the University of Illinois at Urbana-Champaign, and a nonresident fellow at the Krulak Center for Innovation & Creativity at the Marine Corps University. He is a past president of the Peace Science Society (International) and of the International Studies Association. His areas of expertise include the causes of war, United Nations peacekeeping, and international law.

Matthew Fuhrmann is professor of political science and Presidential Impact Fellow at Texas A&M University. His research focuses on international security and conflict, nuclear proliferation, and military alliances. He is the author of *Atomic Assistance: How "Atoms for Peace" Programs Cause Nuclear Insecurity* and the coauthor of *Nuclear Weapons and Coercive Diplomacy*. His articles have been published in the *American Journal of Political Science*, the *British Journal of Political Science*, *International Organization*, *International Security*, the *Journal of Conflict Resolution*, and the *Journal of Politics*, among other journals.

Douglas M. Gibler is professor of political science in the Institute for Social Science Research at the University of Alabama. He is the author of *The Territorial Peace: Borders, State Development, and International Conflict*. His work has also appeared in the *American Political Science Review*, the *American Journal of Political Science*, *International Studies Quarterly*, the *Journal of Conflict Resolution*, and the *Journal of Politics*, and his work has been funded by the National Science Foundation and the Harry Frank Guggenheim Foundation.

Hein Goemans is professor of political science at the University of Rochester. He is the author of *War and Punishment* and of *Leaders and International Conflict*, coauthored with Giacomo Chiozza. His ongoing major research project focuses on the causes of territorial attachment and territorial conflict. He has previously examined the role of leaders in war termination and war initiation.

Gary Goertz is professor of political science and peace studies at the Kroc Institute for International Peace and the University of Notre Dame. He is a specialist on international conflict, rivalry, conflict management, peace, and qualitative methodology. He has published ten books, including *Contexts of International Politics*, *War and Peace in International Rivalry* (with Paul Diehl), *Multimethod Research, Causal Mechanisms, and Case Studies: An Integrated Approach*, and *The Puzzle of Peace: The Evolution of Peace in the International System* (with Paul Diehl and Alex Balas).

Paul R. Hensel is professor of political science at the University of North Texas and the founder of the Issue Correlates of War research project. He studies armed conflict and negotiations over territory, international rivers, and shared ethnic groups. His research has been published in the *American Journal of Political*

Science, *International Studies Quarterly*, the *Journal of Conflict Resolution*, the *Journal of Peace Research*, the *Journal of Politics*, and *Political Geography*, among others.

Benjamin Jensen holds a dual appointment as professor in the Marine Corps University School of Advanced Warfighting and scholar in residence at the American University School of International Service. He is a senior fellow at the Atlantic Council and served as the senior research director and lead author for the US Cyberspace Solarium Commission.

Kelly M. Kadera researches dynamic models of conflict, war contagion, civil war, power, democratic peace, gender and violence, and gender in the academy. She has won the American Political Science Association's Best Book in Conflict Processes Award, the International Studies Association's Susan Northcutt Award, and the University of Iowa's Outstanding Teacher Award. She is a former editor of the *International Studies Review* and an editor of the *American Political Science Review*. She has published articles in the *American Journal of Political Science*, *Conflict Management and Peace Science*, *International Studies Quarterly*, the *Journal of Conflict Resolution*, and *International Interactions*, among others. She is currently an associate professor at the University of Iowa.

Michael R. Kenwick is assistant professor of political science at Rutgers University and conducts research in international security focusing on alliances, civil-military relations, and border politics. His recent research has been published in the *Journal of Politics*, *International Organization*, *Political Analysis*, and *International Studies Quarterly*.

Ryan C. Maness is director of the Department of Defense's Information Operations for Research and an assistant professor in the Department of Defense Analysis at the Naval Postgraduate School. His research includes operations in the information environment, specifically in cyberspace, cyberstrategy, and power dynamics and interactions among states in cyberspace. His forthcoming book, *Cyber War versus Cyber Realities 2.0*, is a second edition of his 2015 book. He has also published *Cyber Strategy* (2018) and *Russia's Coercive Diplomacy* (2015), as well as articles in several top journals in security studies.

Roseanne W. McManus is associate professor of political science at the Pennsylvania State University. Her research focuses on signaling and coercion in an international security context. Before pursuing an academic career, she was a senior analyst at the Defense Intelligence Agency. She is the author of eleven articles in peer-reviewed journals and of *Statements of Resolve: Achieving Coercive Credibility in International Conflict*.

Steven V. Miller is associate professor of political science at Clemson University. His research explores the two distinct themes of how conflict and threat influence political attitudes and how scholars can better understand conflict processes between states. His publications have appeared in journals like *Comparative Political Studies*, the *Journal of Peace Research*, *Political Behavior*, and *Political Research Quarterly*, and his more recent research has also been highlighted by media outlets like the *New York Times* and the *Washington Post*.

Sara McLaughlin Mitchell is the Professor and Collegiate Scholar of Political Science at the University of Iowa. She is the author of five books and over fifty articles and book chapters. She has received over $1.1 million in external grants. Her areas of expertise include international conflict, political methodology, and gender issues in academia. She is a cofounder of the Journeys in World Politics workshop, a mentoring workshop for women involved in international relations. She received the International Studies Association's Quincy Wright Distinguished Scholar Award and a distinguished alumni award from Iowa State University, and she served as the president of the Peace Science Society.

Daniel S. Morey is associate professor of political science and a faculty member in the Lewis Honors College at the University of Kentucky. His research has appeared in the *American Journal of Political Science*, *International Studies Quarterly*, *Foreign Policy Analysis*, and *Conflict Management and Peace Science*. His primary research focus is the politics of war, exploring both the dynamics of war and the consequences of war for postwar relations. His current research projects focus on coalition warfare, exploring when coalitions are effective war fighters and how coalitions alter wartime bargaining.

Michael Mousseau is professor in the School of Politics, Security, and International Affairs at the University of Central Florida. His research shows how economic development can legitimate liberal democracy and cause peace within and among nations. He has published in *Conflict Management and Peace Science*, *Democratization*, the *European Journal of International Relations*, *International Interactions*, *International Security*, *International Studies Quarterly*, the *Journal of Conflict Resolution*, the *Journal of Peace Research*, and *Peace Economics, Peace Science and Public Policy*.

Andrew P. Owsiak is the Josiah Meigs Distinguished Teaching Professor and associate professor of international affairs at the University of Georgia. His areas of expertise include interstate conflict and conflict management processes. His recent books include *On Dangerous Ground* (coauthored with Toby J. Rider, 2021) and *International Conflict Management* (coauthored with J. Michael Greig and Paul F. Diehl, 2019). His published articles appear in the *Journal of Politics*, the *British Journal of Political Science*, and *International Studies Quarterly*, among others.

Paul Poast is associate professor of political science at the University of Chicago, where his research and writing focus on the political economy of security, alliance politics, and the use of data to study international politics. He is the author of three books, *The Economics of War* (2006), *Organizing Democracy* (2018), and *Arguing about Alliances* (2019).

Emilia Justyna Powell is associate professor of political science and a concurrent associate professor of law at the University of Notre Dame. She has written extensively on international law, international courts, international dispute resolution, the Islamic legal tradition, and Islamic constitutionalism. Her publications include the books *Islamic Law and International Law: Peaceful Resolution of Disputes* (2020) and *Domestic Law Goes Global: Legal Traditions and International Courts* (2011, with Sara McLaughlin Mitchell), and several articles in top political science academic journals as well as law journals.

Stephen L. Quackenbush is associate professor of political science at the University of Missouri. The focus of his research is international security, particularly in the areas of deterrence, recurrent conflict, and strategic studies. He is the author of *Understanding General Deterrence: Theory and Application* (2011) and *International Conflict: Logic and Evidence* (2015), and of articles that have appeared in journals such as the *Journal of Politics*, the *Journal of Conflict Resolution*, *Political Research Quarterly*, the *Journal of Peace Research*, *International Interactions*, *Conflict Management and Peace Science*, and *Review of International Studies*.

Elizabeth Radziszewski is a resident fellow at the Stockdale Center for Ethical Leadership at the US Naval Academy and assistant professor at Rider University. She is the author of *Private Militaries and the Security Industry in Civil Wars: Competition and Market Accountability* (2020) and *Social Networks and Public Support for the European Union* (2013). Her work on international conflict, civil wars, and innovation has been published in the *Wilson Quarterly*, the *Journal of Conflict Resolution*, and the *Journal of Global Security Studies*, among others.

Susan G. Sample is professor of political science at the University of the Pacific. She has published articles in the *British Journal of Political Science*, the *Journal of Conflict Resolution*, the *Journal of Peace Research*, and *International Interactions*, among others. Her research focuses on international conflict processes, with a primary emphasis on the impact of arms buildups.

William R. Thompson is Distinguished Professor and Rogers Professor of Political Science Emeritus at Indiana University. His recent books include *The Oxford Encyclopedia of Empirical International Relations Theory* (2018), *Racing to the Top: How Energy Fuels Systemic Leadership in World Politics*

(2019), *Shocks and Rivalries in the Middle East and North Africa* (2020), *Power Concentration in World Politics: The Political Economy of Systemic Leadership, Growth, and Conflict* (2020), *American Global Pre-eminence: The Development and Erosion of Systemic Leadership* (2021), *and Climate Change in the Middle East & North Africa from Pre-history to the Present: 15,000 Years of Crises, Setbacks, and Adaptation* (2021).

Brandon Valeriano is the Bren Chair of Military Innovation at the Marine Corps University. He is also a senior advisor for the Cyber Solarium Commission and a senior fellow at the Cato Institute. He has authored six books and has published articles in the *Journal of Politics*, *International Studies Quarterly*, and the *Journal of Peace Research*, among others. His two most recent books are *Cyber War versus Cyber Reality* and *Cyber Strategy*. He is the area editor in international relations for the *Journal of Cybersecurity* and the series editor of Disruptive Technology and International Security for Oxford University Press.

John A. Vasquez is Professor Emeritus of Political Science at the University of Illinois at Urbana-Champaign. He is the author of *The War Puzzle*, *The Power of Power Politics*, and *The Steps to War* (with Paul Senese) and of *Territory, War, and Peace* (with Marie Henehan). His most recent book is *Contagion and War: Lessons from the First World War*. He received the Lifetime Achievement Award from the APSA Conflict Processes Section, and he has served as the president of the Peace Science Society (International) and of the International Studies Association.

Krista E. Wiegand is associate professor of political science, the director of the Global Security Program at the Howard H. Baker Jr. Center for Public Policy, and the chair of Middle East studies at the University of Tennessee, Knoxville. She specializes in international conflict management and political violence, specifically conflict resolution, territorial and maritime disputes, mediation, rebel and terrorist group violence, and East Asian and Middle Eastern security. She has written two books and edited another book, and she has written more than forty articles and book chapters.

Scott Wolford is professor of government at the University of Texas at Austin. He has published two books, *The Politics of Military Coalitions* (2015) and *The Politics of the First World War: A Course in Game Theory and International Security* (2019). He is the editor of *Conflict Management and Peace Science*. His articles have appeared in the *American Journal of Political Science*, the *Journal of Politics*, *International Studies Quarterly*, the *Journal of Peace Research*, *International Organization*, and the *Journal of Conflict Resolution*, among others.

Yufan Yang is a PhD student in the political science department at the University of Iowa. Her research interests are international and civil conflict, social movements, and authoritarian regimes. She also takes a great interest in political methodology, including the usage of big data, text analysis, and machine learning in general in social sciences. Before coming to the United States, she earned a bachelor's degree from Fudan University in China and a master's degree from King's College London in the United Kingdom.